SYSTEM MODELING
and SIMULATION

SYSTEM MODELING and SIMULATION

An Introduction

Frank L. Severance, Ph.D.

Professor of Electrical and Computer Engineering
Western Michigan University

JOHN WILEY & SONS, LTD

Chichester • New York • Weinheim • Brisbane • Singapore • Toronto

Copyright © 2001 by John Wiley & Sons Ltd
            Baffins Lane, Chichester,
            West Sussex, PO19 1UD, England

        National   01243 779777
        International   (+44) 1243 779777

e-mail (for orders and customer service enquiries): cs-books@wiley.co.uk

Visit our Home Page on http://www.wiley.co.uk

Other Wiley Editorial Offices

John Wiley & Sons, Inc., 605 Third Avenue,
New York, NY 10158-0012, USA

Wiley-VCH Verlag GmbH, Pappelallee 3,
D-69469 Weinheim, Germany

John Wiley and Sons Australia, Ltd, 33 Park Road, Milton,
Queensland 4064, Australia

John Wiley & Sons (Asia) Pte Ltd, 2 Clementi Loop #02-01,
Jin Xing Distripark, Singapore 129809

John Wiley & Sons (Canada) Ltd, 22 Worcester Road,
Rexdale, Ontario, M9W 1L1, Canada

Library of Congress Cataloging-in-Publication Data
Severance, Frank L.
    System modeling and simulation: an introduction / Frank L. Severance.
        p. cm
    Includes bibliographical references and index.
    ISBN 0-471-49694-4
    1. System theory. I. Title.

Q295.S48 2001
003'.3–dc21

British Library of Cataloguing in Publication Data

A catalogue record for this book is available from the British Library

ISBN 0471-49694-4

Typeset in Times Roman by Techset Composition Limited, Salisbury, Wiltshire
Printed and bound in Great Britain by Biddles Ltd, Guildford and King's Lynn
This book is printed on acid-free paper responsibly manufactured from sustainable forestry in which at least two trees are planted for each one used for paper production.

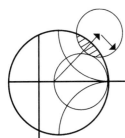

Contents

6 MODELING TIME-DRIVEN SYSTEMS 224

7 EXOGENOUS SIGNALS AND EVENTS 282

8 MARKOV PROCESSES 334

9 EVENT-DRIVEN MODELS 380

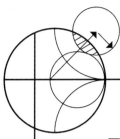

PREFACE

It is unlikely that this book would have been written 100 years ago. Even though there was a considerable amount of modeling going on at the time and the concept of signals was well understood, invariably the models that were used to describe systems tended to be simplified (usually assuming a linear response mechanism), with deterministic inputs. Systems were often considered in isolation, and the inter-relationships between them were ignored. Typically, the solutions to these idealized problems were highly mathematical and of limited value, but little else was possible at the time. However, since system linearity and deterministic signals are rather unrealistic restrictions, in this text we shall strive for more. The basic reason that we can accomplish more nowadays is that we have special help from the digital computer. This wonderful machine enables us to solve complicated problems quickly and accurately with a reasonable amount of precision.

For instance, consider a fairly elementary view of the so-called carbon cycle with the causal diagram shown on the next page. Every child in school knows the importance of this cycle of life and understands it at the conceptual level. However, "the devil is in the details", as they say. Without a rigorous understanding of the quantitative (mathematical) stimulus/response relationships, it will be impossible to actually use this system in any practical sense. For instance, is global warming a fact or a fiction? Only accurate modeling followed by realistic simulation will be able to answer that question.

It is evident from the diagram that animal respiration, plant respiration, and plant and animal decay all contribute to the carbon dioxide in the atmosphere. Photosynthesis affects the number of plants, which in turn affects the number of animals. Clearly, there are feedback loops, so that as one increases, the other decreases, which affects the first, and so on. Thus, even a qualitative model seems meaningful and might even lead to a degree of understanding at a superficial level of analysis. However, the apparent simplicity of the diagram is misleading. It is rare that the input–output relationship is simple, and usually each signal has a set of difference or differential equations that model the behavior. Also, the system input is usually non-deterministic, so it must be described by a random process. Even so, if these were linear relationships, there is a great body of theory by which closed-form mathematical solutions could, in principle, be derived.

We should be so lucky! Realistic systems are usually nonlinear, and realistic signals are noisy. Engineered systems especially are often discrete rather than continuous. They are often sampled so that time itself is discrete or mixed, leading to a system with multiple

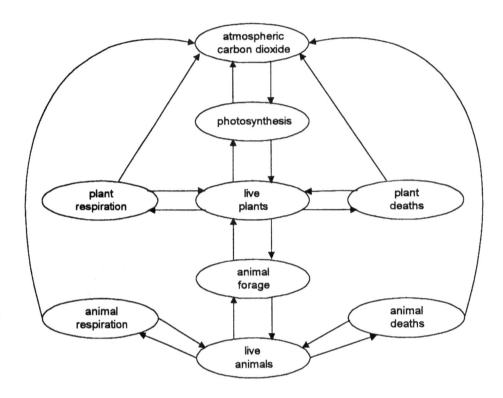

time bases. While this reality is nothing new and people have known this for some time, it is only recently that the computer could be employed to the degree necessary to perform the required simulations. This allows us to achieve realistic modeling so that predictable simulations can be performed to analyze existing systems and engineer new ones to a degree that classical theory was incapable of.

It is the philosophy of this text that no specific software package be espoused or used. The idea is that students should be developers new tools rather than simply users of existing ones. All efforts are aimed at understanding of first principles rather than simply finding an answer. The use of a Basic-like pseudocode affords straightforward implementation of the many procedures and algorithms given throughout the text using any standard procedural language such as C or Basic. Also, all algorithms are given in detail and operational programs are available on the book's Website in Visual Basic.

This book forms the basis of a first course in System Modeling and Simulation in which the principles of time-driven and event-driven models are both emphasized. It is suitable for the standard senior/first-year graduate course in simulation and modeling that is popular in so many modern university science and engineering programs. There is ample material for either a single-semester course of 4 credits emphasizing simulation and modeling techniques or two 3-credit courses where the text is supplemented with methodological material. If two semesters are available, a major project integrating the key course concepts is especially effective. If less time is available, it will likely be that a choice is necessary – either event-driven or time-driven models. An effective 3-credit course stressing event-driven models can be formed by using Chapter 1, the first half of Chapter 3, and Chapters 7–9, along with methodological issues and a project. If time-driven models are to be emphasized, Chapters 1–6 and 10 will handle both deterministic and non-deterministic input signals. If it is possible to ignore stochastic signals and Petri nets, a course in both time-driven and event-driven models is possible by using Chapters 1 and 2, the first half of Chapter 3, Chapter 4, and Chapters 8–10.

ACKNOWLEDGEMENTS

As with any project of this nature, many acknowledgments are in order. My students have been patient with a text in progress. Without their suggestions, corrections, and solutions to problems and examples, this book would have been impossible. Even more importantly, without their impatience, I would never have finished. I thank my students, one and all!

Frank L. Severance
Kalamazoo, Michigan

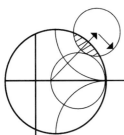

CHAPTER 1

Describing Systems

1.1 THE NATURE OF SYSTEMS

The word "system" is one that everyone claims to understand – whether it is a physiologist examining the human circulatory system, an engineer designing a transportation system, or a pundant playing the political system. All claim to know what systems are, how they work, and how to explain their corner of life. Unfortunately, the term system often means different things to different people, and this results in confusion and problems. Still there are commonalities. People who are "system thinkers" usually expect that systems are (1) based on a set of cause–effect relationships that can be (2) decomposed into subsystems and (3) applied over a restricted application domain. Each of these three expectations require some explanation.

Causes in systems nomenclature are usually referred to as inputs, and effects as outputs. The system approach assumes that all observed outputs are functions only of the system inputs. In practice, this is too strong a statement, since a ubiquitous background noise is often present as well. This, combined with the fact that we rarely, if ever, know everything about any system, means that the observed output is more often a function of the inputs and so-called white noise. From a scientific point of view, this means that there is always more to discover. From an engineering point of view, this means that proposed designs need to rely on models that are less than ideal. Whether the system model is adequate depends on its function. Regardless of this, any model is rarely perfect in the sense of exactness.

There are two basic means by which systems are designed: top-down and bottom-up. In top-down design, one begins with highly abstract modules and progressively decomposes these down to an atomic level. Just the opposite occurs in bottom-up design. Here the designer begins with indivisible atoms and builds ever more abstract structures until the entire system is defined. Regardless of the approach, the abstract structures encapsulate lower-level modules. Of course there is an underlying philosophical problem here. Do atomic elements really exist or are we doomed to forever incorporate white background noise into our models and call them good enough? At a practical level, this presents no problem, but in the quest for total understanding no atomic-level decomposition for any physically real system has ever been achieved!

The power of the systems approach and its wide acceptance are due primarily to the fact that it works. Engineering practice, combined with the large number of mathematically powerful tools, has made it a mainstay of science, commerce, and (many believe) western culture in general. Unfortunately, this need for practical results comes at a price. The price is that universal truth, just like atomic truth, is not achievable. There is always a restricted range or zone over which the system model is functional, while outside this application domain the model fails. For instance, even the most elegant model of a human being's circulatory system is doomed to failure after death. Similarly, a control system in an automobile going at 25 miles per hour is going to perform differently than one going at 100 miles per hour. This problem can be solved by treating each zone separately. Still there is a continuity problem at the zone interfaces, and, in principle, there needs to be an infinite number of zones. Again, good results make for acceptance, even though there is no universal theory.

Therefore, we shall start at the beginning, and at the fundamental question about just what constitutes a system. In forming a definition, it is first necessary to realize that systems are human creations. Nature is actually monolithic, and it is we, as human beings, who either view various natural components as systems or design our own mechanisms to be engineered systems. We usually view a system as a "black box", as illustrated in Figure 1.1. It is apparent from this diagram that a system is an entity *completely isolated from its environment* except for an entry point called the *input* and an exit point called the *output*. More specifically, we list the following system properties.

P1. All environmental influences on a system can be reduced to a vector of m real variables that vary with time, $x(t) = [x_1(t), \ldots, x_m(t)]$. In general, $x(t)$ is called the *input* and the components $x_i(t)$ are *input signals*.

P2. All system effects can be summarized by a vector of n real variables that vary with time, $z(t) = [z_1(t), \ldots, z_n(t)]$. In general, $z(t)$ is called the *output* and the components $z_i(t)$ are *output signals*.

P3. If the output signals are algebraic functions of only the current input, the system is said to be of *zeroth order*, since there can be no system dynamics. Accordingly, there is a state vector $y(t) = [y_1(t), \ldots, y_p(t)]$, and the system can be written as

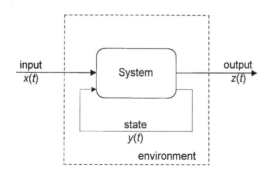

FIGURE 1.1   System block diagram.

two algebraic equations involving the input, state, and output:

$$y(t) = f_1(x(t)),$$
$$z(t) = f_2(x(t), y(t)),$$
(1.1)

for suitable functions f_1 and f_2. Since the state $y(t)$ is given explicitly, an equivalent algebraic input–output relationship can be found. That is, for a suitable function g,

$$z(t) = f_2(x(t), f_1(x(t))) \equiv g(x(t)).$$
(1.2)

P4. If the input signal depends dynamically on the output, there must also be system memory. For instance, suppose that the system samples a signal every $t = 0, 1, 2, \ldots$ seconds and that the output $z(t)$ depends on input $x(t-1)$. It follows that there must be two memory elements present in order to recall $x(t-1)$ and $x(t-2)$ as needed. Each such implied memory element increases the number of system state variables by one. Thus, the state and output equations comparable to Equations (1.1) and (1.2) are dynamic in that f_1 and f_2 now depend on time delays, advances, derivatives and integrals. This is illustrated diagrammatically in Figure 1.2.

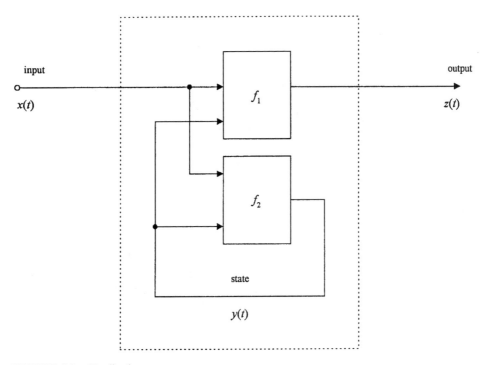

FIGURE 1.2   Feedback system.

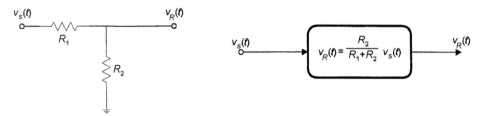

FIGURE 1.3   Electrical circuit as a SISO system, Example 1.1.

EXAMPLE 1.1

Consider the electrical resistive network shown in Figure 1.3, where the system is driven by an external voltage source $v_S(t)$. The output is taken as the voltage $v_R(t)$ across the second resistor R_2.

Since there is a Single Input and a Single Output, this system is called SISO. The input–output identification with physical variables gives

$$x(t) = v_S(t),$$
$$z(t) = v_R(t). \tag{1.3}$$

Since the network is a simple voltage divider circuit, the input–output relationship is clearly not dynamic, and is therefore of order zero:

$$z(t) = \frac{R_2}{R_1 + R_2} x(t). \tag{1.4}$$

In order to find the state and output equations, it is first necessary to define the state variable. For instance, one might simply choose the state to be the output, $y(t) = z(t)$. Or, choosing the current as the state variable, i.e., $y(t) = i(t)$, the state equation is $y(t) = x(t)/(R_1 + R_2)$ and the output equation is $z(t) = R_2 y(t)$.

Clearly, the state is not unique, and it is therefore usually chosen to be intuitively meaningful to the problem at hand.   ○

EXAMPLE 1.2

Consider the resistor–capacitor network shown in Figure 1.4. Since the capacitor is an energy storage element, the equations describing the system are dynamic. As in Example 1.1, let us take the input to be the source voltage $v_S(t)$ and the output as the voltage across the capacitor, $v_C(t)$. Thus, Equations (1.3) still hold. Also, elementary physics gives $RC dv_C/dt + v_C = v_S$. By defining the state to be the output, the state and output relationships corresponding to Equations (1.1) are

$$\dot{y}(t) = \frac{1}{RC}[x(t) - y(t)],$$
$$z(t) = y(t). \tag{1.5}$$

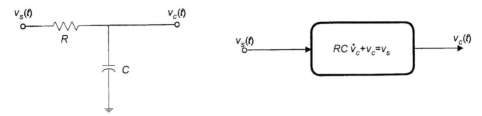

FIGURE 1.4   Electrical *RC* circuit as a first-order SISO system.

As will be customary throughout this text, dotted variables denote time derivatives. Thus,

$$\dot{y} = \frac{d}{dt}y(t), \qquad \ddot{y} = \frac{d^2}{dt^2}y(t), \qquad \ldots, \qquad y^{(n)} = \frac{d^n}{dt^n}y(t). \qquad \bigcirc$$

Electrical circuits form wonderful systems in the technical sense, since their voltage–current effects are confined to the wire and components carrying the charge. The effects of electrical and magnetic radiation on the environment can often be ignored, and all system properties are satisfied. However, we must be careful! Current traveling through a wire does affect the environment, especially at high frequencies. This is the basis of antenna operation. Accordingly, a new model would need to be made. Again, the input–output signals are based on abstractions over a certain range of operations.

One of the most popular system applications is that of control. Here we wish to cause a subsystem, which we call a *plant*, to behave in some prescribed manner. In order to do this, we design a *controller* subsystem to interpret desirable goals in the form of a *reference signal* into plant inputs. This construction, shown in Figure 1.5, is called an *open-loop* control system.

Of course, there is usually more input to the plant than just that provided by the controller. Environmental influences in the form of *noise* or a more overt signal usually cause the output to deviate from the desired response. In order to counteract this, an explicit *feedback* loop is often used so that the controller can make decisions on the basis of the reference input and the actual state of the plant. This situation, shown in Figure 1.6, is called a *feedback control* or *closed-loop control system*.

The design of feedback control systems is a major engineering activity and is a discipline in its own right. Therefore, we leave this to control engineers so that we can concentrate on the activity at hand: modeling and simulating system behavior. Actually, we

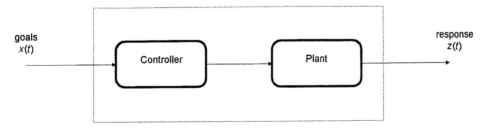

FIGURE 1.5   An open-loop control system.

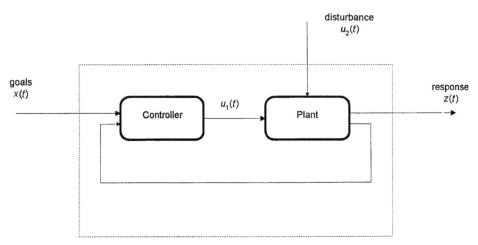

FIGURE 1.6   A closed-loop control system.

will still need to analyze control systems, but we will usually just assume that others have already designed the controllers.

The electrical circuit systems described in Examples 1.1 and 1.2 are cases of *continuous time-driven models*. Time-driven models are those in which the input is specified for all values of time. In this specific case, time t is continuous since the differential equation can be solved to give an explicit expression for the output:

$$v_C(t) = v_C(t_0) + \frac{1}{RC} \int_{t_0}^{t} v_S(\tau) e^{(\tau - t)/RC} \, d\tau, \qquad t \geq t_0 \tag{1.6}$$

Where $v_C(t_0)$ is the initial voltage accross the capacitor at time $t = t_0$. Thus, as time "marches on", successive output values can be found by simply applying Equation (1.6).

In many systems, time actually seems to march as if to a drum; system events occur only at regular time intervals. In these so-called discrete-time-based systems, the only times of interest are $t_k = t_0 + hk$ for $k = 0, 1, \ldots$. As k takes on successive non-negative integer values, t_k begins at initial time t_0 and the system signal remains unchanged until h units later, when the next drum beat occurs. The constant length of the time interval $t_{k+1} - t_k = h$ is the step size of the sampling interval.

The input signal at the critical event times is now $x(t_k) = x(t_0 + hk)$. However, for convenience, we write this as $x(t_k) = x(k)$, in which the functional form of the function x is not the same. Even so, we consider the variables t and k as meaning "continuous time" and "discrete time", respectively. The context should remove any ambiguity.

EXAMPLE 1.3

Consider a continuous signal $x(t) = \cos(\pi t)$, which is defined only at discrete times $t_k = 3 + \frac{1}{2}k$. Clearly the interval length is $h = \frac{1}{2}$ and the initial time is

$t_0 = 3$. Also,

$$\begin{aligned}
x(t_k) &= \cos[\pi(3 + \tfrac{1}{2}k)] \\
&= -\cos(\tfrac{1}{2}\pi k) \\
&= \begin{cases} 0, & k = \text{odd}, \\ (-1)^{(k+2)/2}, & k = \text{even}. \end{cases}
\end{aligned} \tag{1.7}$$

Thus, we write

$$x(k) = \begin{cases} 0, & k = \text{odd}, \\ (-1)^{(k+2)/2}, & k = \text{even}, \end{cases} \tag{1.8}$$

and observe the significant differences between the discrete form of $x(k)$ given in Equation (1.8) and the original continuous form $x(t) = \cos(\pi t)$. ○

EXAMPLE 1.4

Consider a factory conveyor system in which boxes arrive at the rate of one box each 10 seconds. Each box is one of the following weights: 5, 10, or 15 kg. However, there are twice as many 5 kg boxes and 15 kg boxes as 10 kg boxes. A graphic of this system is given in Figure 1.7. How do we model and simulate this?

Solution
From the description, the weight distribution of the boxes is

w	$\Pr[W = w]$
5	0.4
10	0.2
15	0.4
	1.0

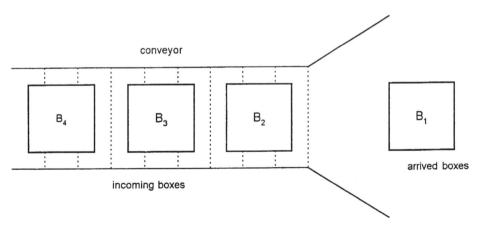

FIGURE 1.7 A deterministic conveyor system, Example 1.4.

where W is a "weight" random variable that can take on one of the three discrete values $W \in \{5, 10, 15\}$. The notation $\Pr[W = w]$ is read "the probability that the random variable W is w". The set $\{5, 10, 15\}$ is called the *sample space* of W, and is the set of all possible weights.

According to the description, these boxes arrive every 10 seconds, so $t = 10k$ gives the continuous time measured in successive k-values, assuming the initial time is zero. However, how do we describe the system output? The problem statement was rather vague on this point. Should it be the *number* of boxes that have arrived up to time t? Perhaps, but this is rather uninteresting. Figure 1.8 graphs $N(t) =$ number of boxes that have arrived up to and including time t as a function of time t.

A more interesting problem would be the weight of the boxes as they arrive. Unlike $N(t)$, the weight is a non-deterministic variable, and we can only hope to *simulate* the behavior of this variable $W(k) =$ weight of the kth event. This can be accomplished by using the RND function, which is a hypothetical random number generator that provides uniformly random distributed variates such that

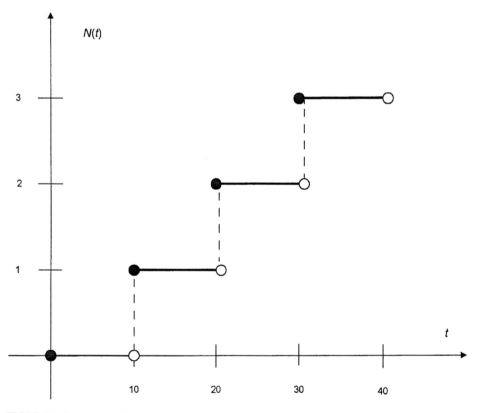

FIGURE 1.8   State $N(t)$ for constant inter-arrival times.

$0 \leqslant \mathrm{RND} < 1$. The following routine provides output $w(1), w(2), \ldots, w(n)$, which is the weight of the first n boxes:

```
for k=1 to n
  r=10*RND
  if r<4 then w(k)=5
  if 4 ≤ r<6 then w(k)=10
  if r ≥ 6 then w(k)=15
next k
```

While this routine simulates the action, useful results must be statistically analyzed since the input was non-deterministic in nature. After a sufficiently large simulation run, frequencies of each type of box arrival can be tallied and graphed. Figure 1.9 shows a run of 100 simulated arrivals compared against the ideal as defined by the problem statement.   ○

The system described in Example 1.4 has just a single state variable. By knowing $w(k)$, we know all there is to know about the system. For instance, we need not know any history in order to compute the output since only the kth signal is important. To see the significance of the state concept as exemplified by memory, consider one further example.

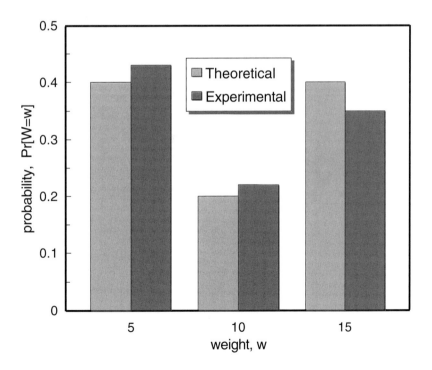

FIGURE 1.9   Histogram.

EXAMPLE 1.5

Consider a factory system with two conveyors: one brings in boxes as described in Example 1.4, but this time arriving boxes are placed on a short conveyor that holds exactly 3 boxes. Arriving boxes displace those on the conveyor, which presumably just fall off! In this case, we do not care about the individual box weight, rather we care about the total weight on the conveyor of interest. Thus, the input $x(k)$ in this example can be simulated by the output of Example 1.4. However, in order to determine the output $g(k)$, the system must remember the two previous inputs as well. Characterize this system.

Solution
Since the system input is a random variable, the output must be non-deterministic as well. At the same time, after the conveyor has been loaded, two of the three spots are known. Thus,

$$z(k) = \begin{cases} x(1), & k = 1, \\ x(1) + x(2), & k = 2, \\ x(k) + x(k-1) + x(k-2), & k > 2. \end{cases} \quad (1.9)$$

Mathematically Equation (1.9) is a second-order difference equation, $z(k) = x(k) + x(k-1) + x(k-2)$, subject to the two initial conditions $z(1) = x(1)$ and $z(2) = x(1) + x(2)$. This corresponds to the two memory elements required. A complete simulation of this model is shown in Listing 1.1. The first four statements inside the loop are identical to the single-conveyor system described in Example 1.4. The last three statements describe the second conveyor of this example. ○

```
for k=1 to n
        r=10*RND
        if r<4 then x(k)=5
        if 4 ≤ r<6 then x(k)=10
        if r ≥ 6 then x(k)=15

        if k=1 then z(k)=x(1)
        if k=2 then z(k)=x(1)+x(2)
        if k>2 then z(k)=x(k)+x(k-1)+x(k-2)
next k
```

LISTING 1.1   Simulation of the two-conveyor system, Example 1.5.

1.2   EVENT-DRIVEN MODELS

Examples 1.1, 1.2, and 1.5 of the previous section were examples of *time-driven systems*. This can be seen from the program sketch given in Listing 1.1, in which successive k-values in the for–next loop compute the response to equally spaced time events. In fact, the program structure is analogous to a microprocessor-based polling system where the computer endlessly loops, asking for a response in each cycle. This is in contrast to an interrupt approach, where the microprocessor goes about its business and only responds to events via interrupts. These interrupt-like programs create so-called *event-driven models*.

In an event-driven model, the system remains dormant except at non-regularly scheduled occurring events. For instance, in modeling the use of keyboard and mouse input devices on a computer, a user manipulates each device on an irregular basis. The time between successive events k and $k + 1$, $t_{k+1} - t_k$, is called the *inter-arrival time*. Unlike the case of time-driven models, where this difference is constant, here this difference generally varies and is non-deterministic. Thus, event-driven models and simulations are often based on stochastic methods.

In models of this type, there are two most interesting questions to be asked. First, we might ask how many n events occur in a fixed interval, say $[0, t]$. For a great number of problems, this number depends only on the *length* of the interval. That is, the expected number of events over the interval $[0, t]$ is the same as the number expected on $[\tau, \tau + t]$ for any $t \geqslant 0$. When this is true, the probability distribution thus defined is said to be stationary. Specifically, the answer is in the form of a probability statement: $P_n(t) =$ "probability there are n events during interval $[0, t]$", where $n = 0, 1, 2, \ldots$ is the sample space. Since the sample space is countably infinite, $P_n(t)$ is a discrete probability mass function.

A closely related question to the one above is the expected inter-event time. Even though the inter-event time is not constant, its statistical description is often known a priori. Denoting the inter-event time by the random variable $T = t_{k+1} - t_k$, it should be clear that T is continuous. Thus, we define the probability density function $f_T(t)$ rather than a probability mass function as we would for a discrete random variable. Recall that continuous random variables are more easily specified by their distribution function $F_T \equiv \Pr[T \leqslant t]$. The density function follows immediately since $f_T(t) = df_T(\tau)/dt$.

The probability distributions defined by $P_n(t)$ and $f_T(t)$ are most important, and will be considered in detail in the next section. But first let us see how straightforward the application of event-based models is to simulate. The problem is to calculate a sequence of event times $t_k =$ "time at which the kth event occurs". For simplicity, let us simply assume that the inter-arrival times are uniformly distributed over $[0, 1)$. In other words, we can use RND for our random intervals. Note that the statement $t_{k+1} - t_k = $ RND is equivalent to $t_k - t_{k-1} = $ RND. It follows that a code sequence to produce n random event times is

```
t_k=0
for k=1 to n
        t_k=t_{k-1}+RND
next k
```

From the sequence thus generated, answers to the $P_n(t)$ and $F_T(t)$ questions can be found.

EXAMPLE 1.6

Consider a conveyor system similar to the one described in Example 1.4. However, this time, all boxes are the same but they arrive with random (in the RND sense for now) inter-arrival times. This time, let us investigate the number of boxes $N(t)$ that have arrived up to and including time t. Our results should resemble those shown graphically in Figure 1.10.

Solution
The simulation input is the set of critical event times discussed earlier. The output is the function $N(t)$, which is the number of events up to and including time t. But, in order to compute the number of events at time t, it is necessary to check if t falls within the interval $[t_k, t_{k+1}]$. If it does, $N(t) = k$; otherwise, a next interval must be checked. Listing 1.2 shows code producing the required pairs t, N from which the graph in Figure 1.10 can be created. Notice that while the input times are continuous, the event times are discrete. Also, the output $N(t)$ is a monotonically increasing function taking on non-negative integral values.   ○

Modeling and simulating system behavior is fundamentally a statistical problem. Whether by Heisenberg's uncertainty principle on the nanoscale or simply by experimental error, the deterministic formulae given in traditional texts do not precisely predict practice.

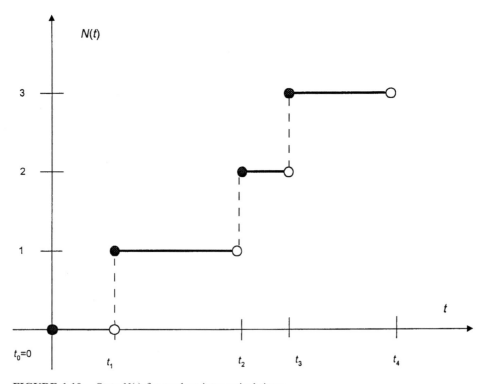

FIGURE 1.10   State $N(t)$ for random inter-arrival times.

```
t₀=0
for k=1 to n
        tₖ=tₖ₋₁+RND
next k
for t=0 to tₙ step h
        for k=1 to n
                if tₖ₋₁ ≤ t<tₖ then N=k
        next k
        print t, N
next t
```

LISTING 1.2   Simulation for Example 1.6.

Even more importantly, the inputs that drive these systems must often be characterized by their basically stochastic nature. However, even though this input may appear random, more often than not the input *distribution* is a random process with well-defined statistical parameters. Thus, we use a probabilistic approach.

For example, one of the central problems encountered in telephony is that of message traffic. Engineers need to know the state of the telephone communications network (that is, the number of callers being serviced) at all times. Over a period of time, callers initiate service (enter the system) and hang up (depart the system). The number of callers in the system (arrivals minus departures) describes the *state*. Thus, if the state is known, questions such as "over a period of time, just how many phone calls can be expected?" and "how many messages will be terminated in so many minutes?" can be addressed. The answer to these questions is not precisely predictable, but, at the same time, average or *expected values* can be determined. Thus, while not deterministic, the state is statistically remarkably regular and is true regardless of message rate. That is, even though an operator in New York City will experience far more traffic than one in Kalamazoo, Michigan, relatively speaking their probability distributions are nearly identical.

1.3   CHARACTERIZING SYSTEMS

Models are characterized by their system behavior and the type of input accepted by the system. Once both the input and system behavior is known, the output can be found. This is known as the analysis problem. For instance, a model might be a simple function machine that doubles the input and adds one as follows:

Such a model, which simply algebraically transforms the input, is a zeroth-order, time-independent system, since there are no states and the formula relating input and output is independent of time. Regardless of what the input type is and regardless of the historical record, the transformation f is always the same.

Since the above system is a zeroth-order model, a natural question is just what constitutes a first- or second-order model. Higher-order models all implicitly involve time. For instance, the input–output relationship defined by $z(k) + 4z(k-1) = x(k)$ characterizes a first-order discrete system, because the output at any discrete time k depends not only on the input but on the output at the previous time as well. Thus, there is an implied memory or system state. Similarly, the input–output relationship defined by $z(k) + 4z(k-1) + 3z(k-2) = x(k)$ is a second-order system, since the history required is two epochs.

There are two subtle problems raised in defining higher-order systems like this. First, if it is necessary to always know the history, how does one start? That is, the output $z(n)$ for an nth-order system is easy to find if one already knows $z(0), z(1), \ldots, z(n-1)$. Therefore, these initial states need to be given a priori. This is a good news–bad news sort of question. It turns out that if the system is linear and stable, the initial conditions become irrelevant in the long run or steady state. Of course, the bad news is that not all systems are linear or stable; indeed, sensitivity to initial conditions tends to be a hallmark of nonlinear systems, which may even be chaotic.

The other subtlety is the nature of time. The discrete-time structure implied by the variable k above is akin to that of a drum beating out a regular rhythm. In this case, time "starts" at $k = 0$ and just keeps going and going until it "ends" at time $k = n$. This is a useful concept for models such as a game of chess where $z(k)$ represents the state of a match on the kth move. Time k has little or no relationship to chronological time t. In contrast, real or chronological time tends to be continuous, and is most familiar to us as humans in that we age over time and time seems to be infinitely divisible. The variable t is commonly used to denote chronological time, and dynamical systems describing continuous-time phenomena are represented by differential equations rather than the difference equations of the discrete-time case. In this case, the differential equation $\ddot{z} + 4\dot{z} + 3z = x(t)$ is a second-order system, because we recognize that it requires two initial conditions to define a unique solution. In general, an nth-order linear differential equation will define an nth-order system, and there will be n states, each of which will require an initial condition $z(0), \dot{z}(0), \ldots, z^{(n-1)}(0)$.

Whether by algebraic, difference, or differential equations, continuous and discrete models as described above are called *regular*, since time marches on as to a drum beat. Assuming that the inter-event time is a constant δ time unit for each cycle, the frequency of the beat is $f = 1/\delta$ cycles per time unit. In the limiting case of continuous time, this frequency is infinity, and the inter-event time interval is zero. However, not all models are regular.

As we have seen earlier, some models are defined by their inter-event (often called inter-arrival) times. Such systems lie dormant between beats, and only change state on receipt of a new event; thus they are *event-driven* models. Event-driven models are characterized by difference equations involving time rather than output variables. By denoting the time at which the kth event occurs by t_k, the system defined by $t_{k+1} - t_k = k^2$ has ever-increasing inter-event intervals. Similarly, the system defined by $t_{k+1} - t_k = 2$ is regular, since the inter-event time is constant and the system defined by $t_{k+1} - t_k = \text{RND}$, where RND is a random number uniformly distributed on the interval $[0, 1]$, is stochastic. Stochastic systems are especially rich, and will be considered in detail later.

1.4 SIMULATION DIAGRAMS

As in mathematics generally, equations give precise meaning to a model's definition, but a conceptual drawing is often useful to convey the underlying intent and motivation. Since most people find graphical descriptions intuitively pleasing, it is often helpful to describe systems graphically rather than by equations. The biggest problem is that the terms *system* and *model* are so very broad that no single diagram can describe them all. Even so, models in this text for the most part can be systematically defined using intuitively useful structures, thus making simulation diagrams a most attractive approach.

System diagrams have two basic entities: signals – represented by directed line segments – and transformations – represented as boxes, circles, or other geometric shapes. In general, signals connect the boxes, which in turn produce new signals in a meaningful way. By defining each signal and each transform carefully, one hopes to uniquely define the system model. By *unique* it is meant that the state variables of the model match the state variables of the underlying system, since the drawing per se will never be unique. However, if the states, signals, and transforms coincide, the schematic is said to be *well posed* and the system and model are *isomorphic*. Of course obtaining the schematic is largely an art form, and one should never underestimate the difficulty of this problem. Even so, once accomplished, the rest of the process is largely mechanical and simulations can be straightforward.

There are several different types of signals. It is always important to keep precise track of which signal type is under discussion, since many system models are heterogeneous. Perhaps the primary distinction is whether a signal is an *across* variable or a *through* variable. An across signal has the same value at all points not separated by a transformer. In contrast, a through signal has the same value at all points not separated by a node. Thus, an across variable is just a value of the signal produced by an input or transformer. It can be thought of as akin to voltage in an electrical circuit where every point on the same or connected arcs (physically, wires) has the same signal value.

A *through* signal takes an entirely different view. Rather than thinking of a function at a given time, we envision various events occurring within the system. For instance, the conveyor in Example 1.3 would be such a system, and the times at which the boxes enter constitute the signal. Here the signal description is given by a formula for t_k, which is the time at which event k occurs. That is, given an event number, the time at which that particular event occurs is returned. A through signal can be visualized as the messages in a communications link where each message flows through various conduits. As messages come to a junction, they go to only one of several direction choices. So, rather than the same signal throughout the wire as in an across signal, a through signal has a conservation principle: *the number of messages leaving a junction equals the number of messages entering the junction.*

An across signal represents a common value (of voltage), while a through signal represents an amount (of messages). Notice how the through description is the inversion of the across description:

across signal: given time, return level
through signal: given a level, return time

All this is illustrated in Figure 1.11. Within connected arcs, an across signal is always the same. For instance in Figure 1.11(a), even though the paths split, the values along each path are the same and $x(t) = y(t) = z(t)$. In contrast, in Figure 1.11(b), as the through signal t_k comes to the junction, some of the messages (or people, boxes, events, or whatever) go up and the rest go down. This requires some sort of distributor D to select which object goes where. There are a number of choices here, but in the end the number that enter a node is the sum of those going up and those going down. However, our description gives times, not objects. Thus we consider the each explicit listing of each time sequence, $[t_k] = [t_1, t_2, \ldots, t_{m+n}]$, $[r_k] = [r_1, r_2, \ldots, r_m]$, and $[s_k] = [s_1, s_2, \ldots, s_n]$, where each vectored sequence is listed in ascending order by convention. It follows from this that $[t_k] = [r_k] \cup [s_k]$, where the operation $\cup$ is called the *merge union*.

EXAMPLE 1.7

Consider the system below, in which two through signals are combined at a collector junction C. Find and graph each signal as a function of chronological time.

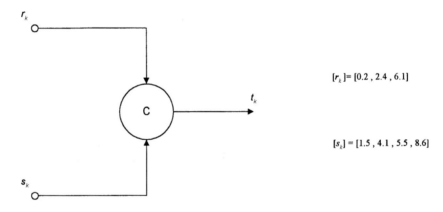

$[r_k] = [0.2, 2.4, 6.1]$

$[s_k] = [1.5, 4.1, 5.5, 8.6]$

Solution
The merge union of sequences r_k and s_k is $[t_k] = [0.2, 1.5, 2.4, 4.1, 5.5, 6.1, 8.6]$. Each graph is shown in Figures 1.12–1.14. Notice that each graph is actually an integral function of time, $k = k(t)$, and that there is no direct relationship between t_k, which is simply the kth component of the vector $[t_k]$, and t, which is continuous time.   ○

The names *across* and *through* signals are motivated by voltage and current as encountered in electrical circuits. Voltage is a potential difference between two points. As such, any two points in the same wire not separated by a component (such as a resistor or capacitor) are electrically equivalent. Thus, within the same or connected arcs, a voltage signal is the same. From an electrical perspective, the way by which voltage is measured is *across* an element *between* two points. This contrasts with current, which is a through signal. Current is an absolute (charge per unit time) rather than a relative difference, and thus conservation principles apply. It is much like water in a plumbing system: what goes

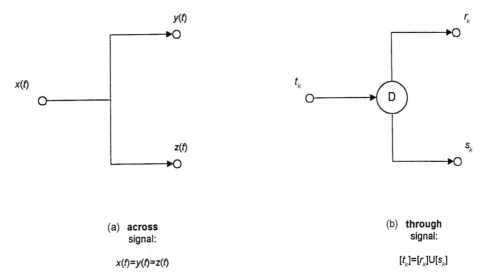

(a) **across**
     signal:

$x(t)=y(t)=z(t)$

(b) **through**
     signal:

$[t_k]=[r_k]\cup[s_k]$

FIGURE 1.11   Across (a) versus through (b) signals.

in must come out. In measuring current, it is necessary not to measure *between* two points, but *at* a single point. Thus we speak of current through a point or, in the more general sense, a through signal.

The nature of a dynamic system is that it will evolve over time, and, therefore, interesting results tend to be state-versus-time graphs. In cases where the signal is discrete, some correlation is also made with continuous time as well. For instance, in studying queuing systems in which customers line up at a service counter, the input signals are often characterized by inter-arrival times. As an example, the mathematical statement $t_k - t_{k-1} = \text{RND}$ can be interpreted as "the inter-arrival time between two successive events is a random number uniformly distributed on the interval $[0, 1]$". A useful simulation to this model will be able to find $N(t)$, the number of events that have occurred up to time t. How to go from a statement regarding signals in one form to another is important.

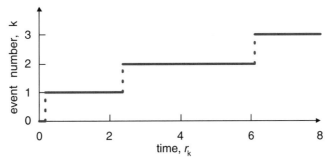

FIGURE 1.12   Graph of $[r_k]$, Example 1.7.

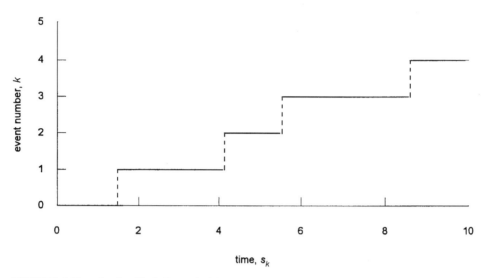

FIGURE 1.13   Graph of $[s_k]$, Example 1.7.

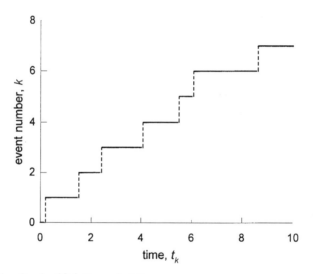

FIGURE 1.14   Graph of $[t_k]$, Example 1.7.

Two other important classes of contrasting signals are those that use continuous time versus those that use discrete time. Continuous time is the time that we as human beings are used to. Given any two time instants $[t_1, t_2]$, we can always conceptualize another time instant t in between: $t_1 < t < t_2$. In this so-called chronological time, even time concepts such as $t = \sqrt{2}$ seconds are not unreasonable. However, not all systems are dynamic with respect to chronological time. Biological systems are dynamic in that they change over time, but different organisms see their physiological times quite differently. For that matter, not all systems are even dynamic. In modeling a chess game, the only time that makes sense is the number of moves into the game. How long it takes to decide on the move is rather irrelevant. This, of course, is an example of discrete time. There are consecutive times instants, and in between any two consecutive moments there is nothing.

Just as there is a relationship between chronological time and physiological time, there are often relationships between continuous time and discrete time. In an engineered system involving microprocessors that acquire, process, and control signals, it is common to sample the continuous signal presented by the world. This sampled signal is stored and processed by a computer using the sampled value. Presumably, the cost of losing information is compensated by the power and versatility of the computer in processing. However, if control is required, the discrete signal must be converted back into a continuous signal and put back into the continuous world from which it came. This process is called *desampling*. Sampled signals are called *regular discrete signals*, since their instances are chronologically similar to a drum beat – they are periodic and predictable.

In contrast to regularly sampled signals, there are event-driven discrete signals where the signal itself is predictable, but the time at which it occurs is not. For example, the signal $N(t)$ that is "the number of customers waiting in line for tickets to a football game" has a predictable state (consecutive integers as arrivals come), but the times at which arrivals occur is random. Assuming that the times for consecutive arrival are t_k, $N(t_{k+1}) = N(t_k) + 1$. However, $t_{k+1} - t_k = \text{RND}$ is not a regular sampled signal, since the next continuous time of occurrence is not predictable. Even so, from the system's point of view, nothing happens between event occurrences, so it is both discrete and regular. Here the event instances occur at discretized time k, but the exact time of the next occurrence is not predictable, even though we know that the next event will bump up the event counter by one. In short, sampling only makes sense from the external continuous time's view. From an internal view, it is just discrete time, and there is no worry about any reference to an external view.

There are other contrasting signal types as well, the most notable being deterministic versus random signals. Random signals are extremely important, since it is impossible to model all aspects of a system. Unless we know all there is to know about every process that impinges on a system, there will be some cause–effect relationships that are unaccounted for. The logical means of dealing with this uncertainty is by incorporating some random error into the system. Of course this isn't as trivial as it might seem at first. Suppose we incorporate a set of effects into our model, but we know there are still others unaccounted for. Rather than search endlessly for the remaining (as if there are only a finite set of effects anyway!), we statistically analyze the residuals and simply deal with them as averages using a set of simulation runs. In this way, the effects that we do know will be validated and our model is useful, even if it is only for a limited reality. Of course, all models are only of limited value, since our knowledge is of limited extent. In any case, random signals and statistical methods are essential to good system modeling.

As signals meander throughout a modeling schematic, they encounter different components, which act as transformers. There are a number of specific transformers, but in general there are only four basic types, each of which will be developed further in this text. They are *algebraic*, *memory*, *type converter*, and *data flow*. If we were to be extremely pedantic, all transformers could be developed axiomatically. That is, beginning

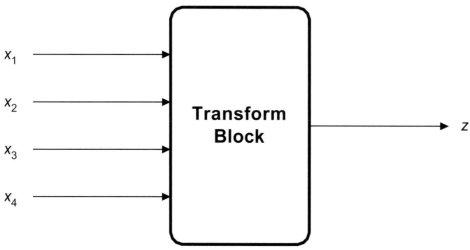

FIGURE 1.15   Generic data flow transform block.

with only an adder and subtracter, all algebraic transformers could be produced. By including a memory unit in our list of primitives, the memory types could all be created. This could continue up the ladder until all transformer types are characterized. This will not be done here; rather, we will build them throughout the text as necessary.

Consider the generic transformer shown in Figure 1.15. Note that there can be several inputs, but there is only a single output. If a multiple-output unit is required, it can always be created by combining several single-output transformers within a single module.

The simplest of these is the data flow transformer, which essentially acts like a function machine. The data flow transformer changes the input into an output according to some well-defined formula that depends on only the input and perhaps time. For example, $z = x_1 + 3x_2 - x_3^2 + \sin x_4$ is a time-independent, nonlinear algebraic transform, since the formula is obviously nonlinear and there is no explicit mention of time in the equation. Usually the formulas aren't so involved. More typical algebraic transforms include summation and multiplication blocks.

One step up from the algebraic block are blocks that require memory. These are called memory transformers, since they apply the historical record and must therefore also require state descriptions. This permits difference and differential equations along with integrators and integral equations. For instance, the transform $z(k) = z(k - 1) + 2x_1(k) + 2x_3(k - 2) + \sin k$ is a linear, time-dependent, second-order memory transform, since the output depends on only previous values of both the input and output along with a nonlinear reference to discrete time k.

There are a number of very useful memory transformers, including delays, accumulators and filters. In their purest form, there are only a memory unit and algebraic transformers from which to build the larger memory transformer family. These, plus ubiquitous time, can make the set much richer.

EXAMPLE 1.8

Consider the transformer defined by the following input–output difference equation: $z(k) = z(k-1) + 2x_1(k) + 2x_3(k-2) + \sin k$. Create, using only memory units, algebraic transformers, and a ubiquitous time reference, a system representation.

Solution

In the spirit of academic exercises, it is always possible to create a larger module from an encapsulation of primitive elements such as +, *, and M (memory). Of course if there is memory, there also must be initial conditions so as to begin the simulation. Such a diagram is given in Figure 1.16. ○

There is no primitive to explicitly handle derivatives and integrals. Even so, since they arise so often, it is customary to handle them by encapsulated modules. Later, we will show difference equation techniques with which to handle these as well.

Type converter transformers change continuous signals to discrete and vice versa. This is usually done by means of sampling and desampling, but there are also other means. Regardless of technique, the important thing is to know what type of signal is being used at each stage of the model. Purely continuous and purely discrete systems are the exception – most systems are hybrid.

The sampled signal is relatively straightforward. The value of a continuous signal is sampled every $t_k = hk + t_0$ seconds and retained as a function of discrete time k. These sampled values can then be manipulated using a computer for integral k. On the other hand, in desampling there are many options. The simplest is the so-called zero-order hold (ZOH), where the sampled value is simply held constant until the next sample is taken. For speedy real-time computing systems where sampling frequencies are on the order of

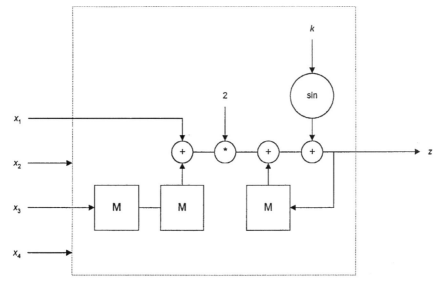

FIGURE 1.16   Encapsulation of primitive elements.

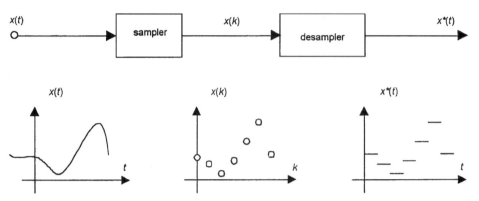

FIGURE 1.17   Sampled and desampled signals.

milliseconds, this is rarely a problem. But, for very long sampling intervals (equivalently, very small sampling frequencies), some means of anticipating the continuous signal value is useful. For instance, the Dow Jones Industrial Averages are posted on a daily basis, but it would be of obvious benefit to estimate the trading behavior throughout the day as well. Thinking of the opening Dow posting as the sampled signal, the desampler would be a formula for estimating the actual average for any time of day. This is illustrated in Figure 1.17, where we see a continuous signal $x(t)$ entering a sampler. The sampler transforms $x(t)$ into the sampled signal $x(k)$. The sampled signal enters a desampler, creating another continuous signal $x^*(t)$, which is only approximately $x(t)$. Even though the signals are given the same name, the continuous, sampled, and desampled signals are not defined by functions of the same form. For example, recall Example 1.3.

The final transformation type is a data flow. Data flow transformations are usually used in event-driven rather than time-driven systems. Thinking of signals as discrete messages rather than continuous functional values, signal arcs are more like pipelines carrying each message to one destination or another. Data flow transformations usually execute flow control. In this model, messages might queue up in a transformation box and form a first-in–first-out (FIFO) queue or maybe a last-in–first-out (LIFO) stack. Typically, a data flow box only allows a message to exit after it has received an authorization signal from each of its inputs. In this way, models can be created to demonstrate the reasoned ebb and flow of messages throughout the system.

The simplest data flow transforms occur in state machines. These systems are akin to having a single message in an event-driven system; where the message resides corresponds to the system state. This is best illustrated by an example.

EXAMPLE 1.9

Model a 2-bit binary counter using a finite state machine. If the input is $x = 0$ then the counter stops counting, and if $x = 1$ then the counter continues on from where it last left off. The output of the counter should produce the sequence 3, 1, 5, 2, 3, 1, 5, 2, ... Since there are four different output values, there are four different state values too. Clearly there are only two input values. Model this state machine using bit vectors.

Solution

Since there are at most $2^2 = 4$ different states, this is a 2-bit binary counter. A suitable block diagram is shown in Figure 1.18. The state is shown in binary form as the vector $y = (y_1, y_2)$. The input is x and the output is variable z. Each of the variables x, y_1, and z have two values, 0 and 1. Output z is not shown as a bit vector, but takes on one of the values $\{1, 2, 3, 5\}$.

The actual working of the system is best understood using a *transition diagram* as shown in Figure 1.19. In this diagram, states are shown as vertical lines with a different output associated with each one. Transitions from one state to another are represented by horizontal arrows with the input required to achieve each. In this case, the counter progresses through the states in a binary form: 00, 01, 10, 11, 00, etc. However, the output that is actually observable is 3, 1, 5, 2, 3, etc., as required. This is a state machine, since for the same input (1) there are different outputs depending on which state the system is in.

In general, there is not simply one input, but a sequence of inputs $x(k)$. For instance, if the input string is $x = [1, 1, 0, 0, 1, 0, 1, 0, 1, 1, 1]$ and the initial state is $y = 10$, the output string is $z = [5, 2, 2, 2, 3, 3, 1, 1, 5, 2, 3]$. If the output z is also required to be a bit vector, $z = [z_1, z_2, z_3]$ can be used. $\bigcirc$

State machines are important, since their response to an input can vary. Unlike function machines, which use the same formula under all circumstances, state machine responses depend on both the input and the current state. The archetypical finite state machine (FSM) is the digital computer. It is a state machine, since its output depends on the memory configuration as well as user initiative. The FSM is finite since there are a finite number of memory cells in any computer. Engineers design and program using state machine methods, and they will be explored later in this text.

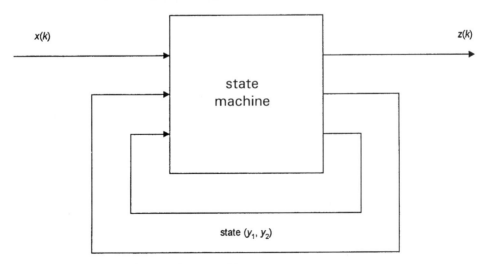

FIGURE 1.18   Block diagram of state machine.

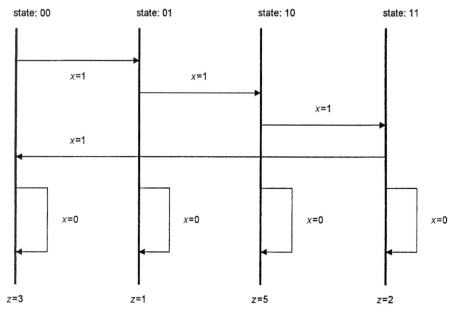

FIGURE 1.19   Transition diagram for state machine.

However, state machines are restrictive in the sense that there can be only one "token" in the system at a time. By *token* we mean a generic entity such as a process in a multitasking operating system or a message in communication network. Describing systems with multiple, often autonomous, tokens, requires a more general structure than an FSM diagram. This is especially true when the tokens have a need for inter-process as well as intra-process communication. Modeling and simulation of such systems with parallel processes requires devices called Petri nets. These will also be discussed later in the text.

1.5   THE SYSTEMS APPROACH

Every system has three basic components: input, output, and the system description. If any two of these are specified, the remaining one follows. Each possibility results in a different problem, analysis, design, and management, with a different view. We list these possibilities below.

Specified entity		Unspecified entity	Name
input	system	output	analysis
input	output	system	design
system	output	input	control

A scientist will tend to use analysis. From this perspective, the system is part of nature and is given a priori. It only remains to discover just what the system actually is. By bombarding (sometimes literally in particle physics) with a number of different inputs and analyzing the results, it is hoped that the system description will reveal itself. Supposedly after much study and many trials, the scientist gets an idea that can be described as a mathematical formulation. Using the scientific method, a hypothesis is conjectured for a specific input. If the output matches that predicted by the system model, our scientist gets to publish a paper and perhaps win a grant; otherwise he is relegated to more testing, a revision of the mathematical model, and another hypothesis.

An engineer takes a different view. For him, inputs and outputs are basically known from engineering judgement, past practice, and specifications. It remains for him to design a system that produces the desired output when a given input is presented. Of course, designing such a thing mathematically is one thing, but creating it physically using non-ideal concepts adds a number of constraints as well. For instance, an electrical rather than mechanical system might be required. If a system can be designed so that the input–output pairs are produced and the constraints are met, he gets a raise. Otherwise, another line of work might be in order.

A manager takes another view. Whether a manager of people, a computer network, or a natural resource system, the system in already in place. Also, the output is either specified directly (a number of units will be manufactured in a certain period of time) or indirectly (maximize profit and minimize costs). It is the manager's duty to take these edicts and provide inputs in such a way as to achieve the required ends. If he is unable to satisfy his goals, the system might need adjusting, but this is a design function. He is only in charge of marshaling resources to achieve the requirements.

Each of these views is correct, and in fact there are books written on each of them. However, most are discipline-specific and lack generality. Therefore, the system scientist will address specific problems with each view in mind. After mastering the basic mathematics and system tools described in this text, it is only natural to look to literature addressing each problem. For instance, the *system identification* problem studies how best to "discover" a correct system description. This includes both the mathematical form and parameter values of the description.

System optimization, on the other hand, assumes that the mathematical form is known, but strives to find the parameter values so that a given objective function is optimized. In practice, system designers have to know not only how to design systems but also how to identify them and do so in an optimal manner. Thus, even though design is but one facet, designers are usually well versed in all systems aspects, including analysis and management as well as design. Each of these views will be investigated throughout this book.

BIBLIOGRAPHY

Andrews, J. G. and R. R. McLone, *Mathematical Modeling*. Butterworth, 1971.
Aris, R., *Mathematical Modeling*, Vol. VI. Academic Press, 1999.
Aris, R., *Mathematical Modeling Techniques*. Dover, 1994.

Close, C. M. and Frederick, D. K., *Modeling and Analysis of Dynamic Systems*, 2nd edn. Wiley, 1994.

Cundy, H. H. and A. P. Rollett, *Mathematical Models*. Oxford University Press, 1952.

Director, S. W. and R. A. Rohrer, *Introduction to Systems Theory*. McGraw-Hill, 1988.

Gernshenfeld, *The Nature of Mathematical Modeling*. Cambridge University Press, 1999.

Law, A. and D. Kelton, *Simulation, Modeling and Analysis*. McGraw-Hill, 1991.

Ljung, L., *System Identification – Theory for the User*, 2nd edn. Prentice-Hall, 1999.

Profozich, P. M., *Managing Change with Business Process Simulation*. Prentice-Hall, 1997.

Roberts, N., D. Andersen, R. Deal, M. Garet, and W. Shaffer, *Introduction to Computer Simulation*. Addison-Wesley, 1983.

Sage, A. P., *Systems Engineering*. Wiley, 1995.

Sage, A. P., *Decision Support Systems Engineering*. McGraw-Hill, 1991.

Sage, A. P. and Armstrong, *An Introduction to Systems Engineering*. Wiley, 2000.

Sandquist, G. M., *Introduction to System Science*. Prentice-Hall, 1985.

Thompson, *Simulation: a Modeler's Approach*. Wiley Interscience, 2000.

Vemuri, V., *Modeling of Complex Systems*. Academic Press, 1978.

Watson, H. J., *Computer Simulation*, 2nd edn. Wiley, 1989.

White, H. J., *Systems Analysis*. W.B. Saunders, 1969.

Zeigler, B. P., *Theory of Modeling and Simulation*, 2nd edn. Academic Press, 2000.

EXERCISES

1.1 Consider the *RL* circuit shown, in which the input is a source voltage $v_S(t)$ and the output is the voltage across the inductor $v_L(t)$. Assuming that the state variable is the current $i(t)$, find state and output equations analogous to Equation (1.5).

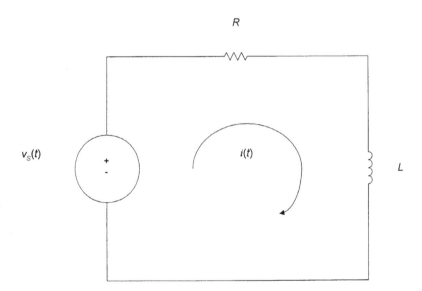

1.2 Consider a linear, time-independent, first-order SISO system described by the following input–output relationship:

$$\frac{dz}{dt} + az = bx(t).$$

(a) Derive a general explicit solution for $z(t)$ in terms of the initial output $z(0)$.

(b) Apply the results of part (a) to the following differential equation:

$$4\frac{dz}{dt} = 3z = 5u(t),$$

$$z(0) = 2,$$

where $u(t)$ is the unit step function: $u(t) = 1$, $t \geqslant 0$; 0, $t < 0$.

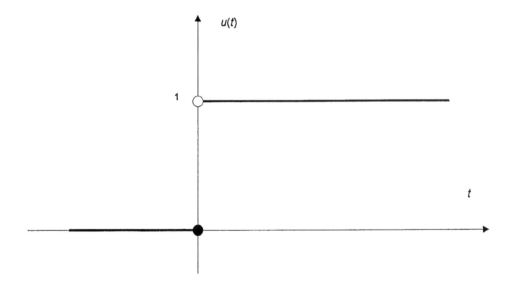

(c) Apply the results of part (a) to the following differential equation:

$$\frac{dz}{dt} + 4z = r(t),$$

$$z(0) = 0,$$

where $r(t)$ is a rectangular wave: $r(t) = 1$, $t \in [2k, 2k + 1]$; -1, $t \in [2k + 1, 2k + 2]$, $k = 0, 1, 2, \ldots$ (see the graph on the next page).

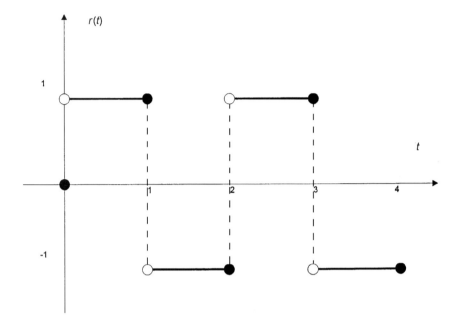

1.3 Consider the *RC* circuit shown, in which there are two source voltage inputs: $v_1(t)$ and $v_2(t)$. Use the voltage across the capacitor $v_C(t)$ as the state variable and find the state and output equations analogous to Equation (1.5).

(a) Assume the output is the current going through the capacitor, $i_C(t)$.

(b) Assume there are three outputs: the voltage across each resistor and the capacitor.

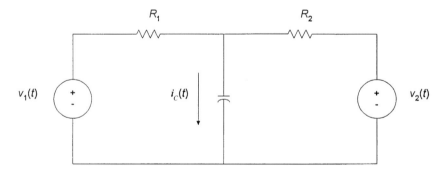

1.4 Consider a system of two inputs x_1, x_2 and two outputs z_1, z_2 described by the following input–output relationships:

$$\ddot{z}_1 + 3\dot{z}_1 + 2z_1 = x_1 + 3x_2,$$
$$\ddot{z}_2 + 4\dot{z}_2 + 3z_2 = -x_1 + x_2.$$

Define column vectors x, y, and z for the input, state, and output so that this system is in *standard linear form*. That is for matrices A, B, C, and D, find

$$\dot{y} = Ay + Bx,$$
$$z = Cy + Dx.$$

1.5 Consider a system with input $x(k)$ and outputs $z(k)$ described by the following difference equation relating the input to the output:

$$z(k) + 3z(k-1) + 2z(k-2) = x(k) + 3x(k-1).$$

Define the column vector y for the state variable so that this system is in *standard linear form*. That is, for matrices A, B, C, and D,

$$y(k) = Ay(k-1) + Bx(k),$$
$$z(k) = Cy(k) + Dx(k).$$

1.6 Consider a linear, time-independent, first-order SISO discrete system described by the difference equation $z(k) = az(k-1) + x(k)$.

(a) Show that the explicit solution in terms of the initial output is given by

$$z(k) = a^k z(a) + \sum_{j=1}^{k} a^{k-j} x(j).$$

(b) Apply the results of part (a) to the system description $z(k) = z(k-1) + 2k$; $z(0) = 1$.

(c) Apply the results of part (a) to the system description $z(k) + z(k-1) = 2k$; $z(0) = 1$.

1.7 Consider the continuous signal $x(t) = 5\sin(2\pi t) \ln t$, which is sampled every $t_k = \frac{1}{8}t$ seconds. Find an explicit equation for the sampled signal $x(k)$.

1.8 Recall the model and simulation outlined in Example 1.4. Implement this simulation and perform it for $n = 10$, $n = 100$, $n = 1000$, and $n = 10\,000$ iterations. Compare your results against the theoretical expectations. Present your conclusions in the form of a graph of error versus number of iterations. Notice that the exponential form of n implies a logarithmic scale for the iteration axis.

1.9 Consider a factory system similar to that of example 1.4 in which there are boxes of three different weights: 5, 10, and 15 pounds. The probability an incoming box has a given weight is as follows:

w	$\Pr[W = w]$
5	0.5
10	0.2
15	0.3
	1.0

(a) Create a simulation of 200 boxes being placed on the conveyor and the total weight recorded.

(b) Summarize the total weight distribution so that the relative number of times each wieght (15, 20, 25, 30, 35, 40 or 45 pounds) occurs.

(c) Calculate the theoretical distribution corresponding to the simulation of part (b). Compare the two distributions by forming a distribution and a histogram.

(d) Using a Chi-square test at the 98% confidence level, determine whether or not the simulation is valid.

1.10 Implement and perform the simulation outlined in Example 1.5 for various values of n.

(a) Make a study (note Exercise 1.8) by which a reasonable result can be guaranteed.

(b) Using the suitable n found in part (a), compute the mean and standard deviation of this random process after stationarity is achieved.

1.11 Consider the sequence of inter-event times generated by the formula $t_k = t_{k-1} - \ln(\text{RND})$.

(a) Using this formula, create a simulation similar to that in Example 1.6 where the number of events $N(t)$ is graphed as a function of time.

(b) Using the simulation, compute an average inter-event time.

(c) Using the simulation, compute the standard deviation of the inter-event times.

(d) Repeat parts (a)–(c) for the inter-event sequence defined by $t_k = t_{k-1} - 2\ln(\text{RND}) - 3\ln(\text{RND})$.

1.12 Write a program that will find the merge union of two event-time sequences $[t_k] = [r_k] \cup [s_k]$, where $[r_k] = [r_1, r_2, \ldots, r_m]$ and $[s_k] = [s_1, s_2, \ldots, s_n]$ are the inputs and $[t_k] = [t_1, t_2, \ldots, t_{m+n}]$ is the output.

(a) Using your merge union program, create a simulation that generates two event sequences $r_k = r_{k-1} + 2\ln(\text{RND})$ and $s_k = s_{k-1} + 3\ln(\text{RND})$, and generates the results as a sequence $[t_k]$.

(b) Create graphs of your results similar to Figures 1.12–1.14 of Example 1.7.

1.13 Consider transformers defined by the following input–output relations. Implement each at the atomic level using only + (addition), * (multiplication), and M (memory) units. Try to use as few memory units as possible.

(a) $z(k) + 2z(k-1) + 4z(k-2) = x(k)$;

(b) $z(k) + 4z(k-2) = x(k) + 3x(k-1)$;

(c) $z(k) = x(k) + x(k-1) + x(k-2)$.

1.14 Consider the most general input–output relationship for a linear, discrete-time, time-invariant SISO system:

$$z(k) + \sum_{j=1}^{n} a_j z(k-j) = \sum_{i=0}^{m} b_i x(k-j).$$

(a) Show that it is possible to create a simulation diagram for this system using only $\max(m, n)$ memory units.

(b) Apply the technique of part (a) to the two-input, one-output system defined in Example 1.8: $z(k) = z(k-1) + 2x_1(k) + 2x_3(k-2) + \sin k$, thereby making a simulation diagram with $\max(2, 1) = 2$ memory units.

1.15 Consider the finite state machine defined in Example 1.9, but this time with three possible inputs:

if $x = 00$, the machine stays in place;
if $x = 01$, the machine goes sequences forward;
if $x = 10$, the machine goes backward;
the input $x = 11$ is disallowed.

Create the transition diagram of the new state machine.

1.16 Consider a finite state machine that works like the one in Example 1.9, except that it has two outputs instead of one. The first output z_1 behaves exactly like z and generates the sequence 3, 1, 5, 2, 3, 1, 5, 2, ... The second output z_2 also behaves like z, but generates the sequence 3, 1, 5, 2, 7, 8, 3, 1, 5, 2, 7, 8, ...

(a) Create the transition diagram of this state machine.

(b) Generalize this result.

1.17 Two stations, P_1 and P_2, located on the x axis at points $P_1(d_1, 0)$ and $P_2(d_2, 0)$, sight a target whose *actual* location is at point $P(x, y)$. However, as a result of an angular quantization error that is uniformly distributed over $[-\delta, \delta]$, this pair of stations calculate an *apparent* position $P'(x', y')$. Specifically, the observed angles are given by $\theta_1 + \mu\delta$ and $\theta_2 + \mu\delta$, where θ_1 and θ_2 are the actual angles, respectively. μ is a random variate that is uniformly distributed on $[-1, 1]$. Simulate this system mathematically and analyze the results.

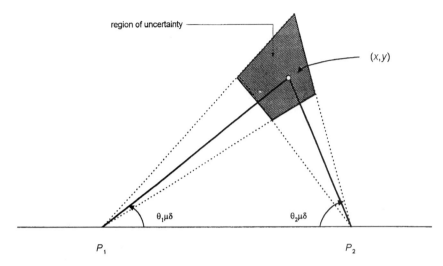

Write a program, with inputs δ and n (the number of points to be sampled), that reads the n actual points (x, y); these are tabulated below. The program should calculate the apparent points (x', y') as seen by each of the two stations P_1 and P_2. Using the points defining the actual trajectory defined in the table below, compute and tabulate the apparent coordinates. Graph the actual and apparent trajectories for several different quantization error sizes δ.

x	y	x	y	x	y
20.6	13.2	19.3	10.3	8.3	7.8
17.2	7.6	14.5	5.5	11.4	11.6
10.7	4.9	8.6	6.7	9.5	5.7
8.1	10.6	10.8	12.3	3.5	1.6
11.5	9.3	9.9	6.3	18.4	8.9
7.6	3.9	4.3	1.9	12.5	4.8
20.3	12.4	18.9	9.6	8.1	8.7
16.6	6.9	13.5	5.2	11.6	10.8
10.1	5.2	10.8	7.6	8.9	5.1
8.4	11.4	6.0	2.8	2.7	1.5
11.2	8.4	19.7	11.0	17.8	8.2
6.8	3.4	15.2	5.9	11.6	4.8
20.0	11.7	9.0	6.2	8.0	9.6
15.9	6.4	9.9	12.5	11.6	10.0
9.4	5.7	10.4	6.9	8.2	4.5
9.1	12.2	5.2	2.4		

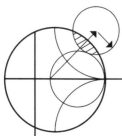

CHAPTER 2

Dynamical Systems

Mathematical models of continuous systems are often defined in terms of differential equations. Differential equations are particularly elegant, since they are able to describe continuous dynamic environments with precision. In an ideal world, it would be possible to solve these equations explicitly, but unfortunately this is rarely the case. Even so, reasonable approximations using numerical difference methods are usually sufficient in practice. This chapter presents a series of straightforward numerical techniques by which a great many models can be approximated using a computer.

2.1 INITIAL-VALUE PROBLEMS

One general class of models is that of dynamical systems. Dynamical systems are characterized by their system state and are often described by a set of differential equations. If the differential equations, combined with their initial conditions, uniquely specify the system, the variables specified by the initial conditions constitute the system state variables. In general, suppose there are m differential equations, each of order n_i. There are $n = \sum_{i=1}^{m} n_i$ initial conditions. Equivalently, there are n first-order differential equations.

Equation	Order
1	n_1
2	n_2
$\vdots$	$\vdots$
m	n_m

The output variables of each of these first-order differential equations, each along with their single initial condition, comprise a set of system state variables. However, since the equations are not unique, neither are the state variables themselves. Nonetheless, there are exactly n of them. Therefore, we begin by considering the first-order *initial-value problem*.

$$\frac{d\mathbf{x}}{dt} = f(t, \mathbf{x}),$$
$$\mathbf{x}(t_0) = \mathbf{x}_0,$$

$$(2.1)$$

where $\mathbf{x}(t) = [x_1(t), x_2(t), \ldots, x_n(t)]$ is the system state vector and $\mathbf{x}(0) = [x_1(0), x_2(0), \ldots, x_n(0)]$ are the corresponding initial conditions.

EXAMPLE 2.1

Consider a system defined by the following differential equations:

$$\ddot{\alpha} + 2\beta\dot{\alpha} + \beta^2\alpha = \cos t,$$
$$\dot{\beta} + \alpha\beta = 4, \qquad\qquad\qquad (2.2)$$

subject to the initial conditions

$$\alpha(0) = 2,$$
$$\dot{\alpha}(0) = -1, \qquad\qquad (2.3)$$
$$\beta(0) = 1.$$

Since there are two dynamic state variables (those involving derivatives) and first- and second-order differential equations, this is a third-order system. Therefore, it is possible to re-define this as a system of three first-order differential equations. Letting $x_1 = \alpha(t)$, $x_2 = \dot{\alpha}(t)$, $x_3 = \beta(t)$, Equations (2.2) can be re-written as

$$\dot{x}_2 + 2x_2x_3 + x_3^2 x_1 = \cos t,$$
$$\dot{x}_3 + x_1x_3 = 4.$$

Noting that x_2 is the derivative of x_1, Equations (2.2) and (2.3) may be redefined as the following system of first-order differential equations:

$$\dot{x}_1 = x_2,$$
$$\dot{x}_2 = -2x_2x_3 - x_1x_3^2 + \cos t, \qquad (2.4a)$$
$$\dot{x}_3 = -x_1x_3 + 4;$$

$$x_1(0) = 2,$$
$$x_2(0) = -1, \qquad\qquad (2.4b)$$
$$x_3(0) = 1.$$

Defining the three-component state vector as $\mathbf{x} = [x_1, x_2, x_3]$, this system may be written in the form of Equation (2.1), where

$$\mathbf{f} = [x_2, -2x_2x_3 - x_1x_3^2 + \cos t, -x_1x_2 + 4],$$
$$\mathbf{x}_0 = [2, -1, 1],$$
$$t_0 = 0.$$

It is straightforward to generalize this technique to a great number of systems.

○

Euler's Method

Since the technique of Example 2.1 is so general, we now consider only first-order initial-value problems. Further, without loss of generality, we consider the scalar version of Equation (2.1): $\dot{x} = f(t, x)$. Using the definition of derivative,

$$\lim_{h \to 0} \frac{x(t+h) - x(t)}{h} = f(t, x).$$

Therefore, for small h, $x(t+h) \approx x(t) + hf(t, x)$. It is convenient to define a new discrete variable k in place of the continuous time t as $t \equiv t_k = hk + t_0$, where $t_0 \leqslant t \leqslant t_n$ and $k = 0, 1, 2, \ldots, n$.

This linear transformation of time may be thought of as sampling the independent time variable t at $n+1$ sampling points, as illustrated in Figure 2.1. Formally,

$$x(h(k+1) + t_0) \approx x(hk + t_0) + hf[hk + t_0, x(hk + t_0)]. \tag{2.5}$$

We also introduce a new discrete dependent variable $x(k)$ as

$$x(k+1) = x(k) + hf[t(k), x(k)], \tag{2.6}$$

for $k = 0, 1, 2, \ldots, n$. Whether we are discussing continuous or discrete time, the sense of the variable x should be clear from the context. If the time variable is t, the signal is taken as continuous or analog and the state is $x(t)$. Or, if the time is discrete, the state variable is $x(k)$. The process of replacing continuous time by discrete time is called *discretization*. Accordingly, $x(k) \approx x(tk) \equiv x(t)$ for *small enough* step size h.

Solving Equation (2.5) iteratively using Equation (2.6) will approximate the solution of Equation (2.1). Notice that all variables on the right-hand side of Equation (2.6) are at time k, whereas those on the left-hand side are at time $k+1$. Therefore, we refer to this expression as an *update* of variable x and often do not even retain the index k. For instance,

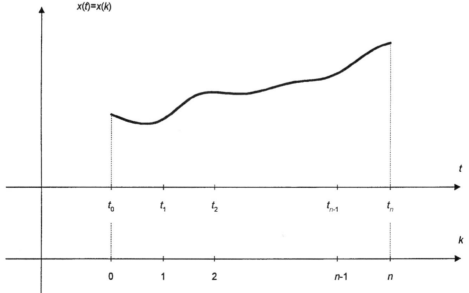

FIGURE 2.1   Relationship between continuous time t and discrete sampled time k.

in writing a computer program, a typical assignment statement can express this recursively as an update of x, $x = x + hf(t, x)$, followed by an update of t, $t = t + h$. The variables x and t on the right-hand sides of the assignments are called "old" while those on the left-hand sides are called "new". This technique, called Euler's method, owes its popularity to this simplicity.

EXAMPLE 2.2

Consider the system described by

$$\dot{x} = x^2 t,$$
$$x(1) = 3. \tag{2.7}$$

Using elementary techniques, it is easy to show that the exact solution of this system is

$$x(t) = \frac{6}{5 - 3t^2}. \tag{2.8}$$

However, assuming that the explicit solution is unknown, Euler's method proceeds as follows. Arbitrarily letting the integration step size be $h = 0.05$, the equivalent discrete system is characterized by the initial conditions

$$t_0 = 1,$$
$$x(0) = 3$$

and the difference equations

$$t_{k+1} = t_k + \tfrac{1}{20},$$
$$x(k + 1) = x(k) + \tfrac{1}{20} x^2(k) t_k,$$

for $k = 1, 2, \ldots, n$. Notice that the old value of t is required to update x. However, x is not required in the update of t. Therefore, by updating x before t, there is no need to use subscripts to maintain the bookkeeping details. This example is solved algorithmically as in Listing 2.1.

```
t=1
x=3
print t, x
for k=1 to n
        x=x+hx²t
        t=t+h
        print t, x
next k
```

LISTING 2.1   Euler's method applied to the system (2.7).

The exact solution given by Equation (2.8) and the approximate solution generated by Listing 2.1 for $n = 6$ are tabulated in Table 2.1.

TABLE 2.1 Exact Solution $x(t)$ and Euler Approximation $x(k)$ to System (2.7)

k	0	1	2	3	4	5	6
t_k	1.00	1.05	1.10	1.15	1.20	1.25	1.30
$x(t)$	3.00	3.55	4.38	5.81	8.82	19.20	-85.71
$x(k)$	3.00	3.45	4.07	4.99	6.42	8.89	13.83

These data are also reproduced graphically in Figure 2.2, where it should be noticed that there is a compounding effect on the error. Although the approximate solution starts correctly at $t_0 = 1$, each successive step deviates further from the exact. Therefore, in applying the Euler method, it is important to not stray too far from the initial time. It is also necessary to choose the integration step size h wisely.

The solution given by Equation (2.8) reveals a singularity at $t_{\text{crit}} = \sqrt{\frac{5}{3}} \approx 1.29$. Therefore, as t approaches the neighborhood of t_{crit}, the numerical results become increasingly precarious. This leads to increasingly large deviations between $x(t_k)$ and $x(k)$ as t approaches t_{crit} from the left. It is clear from Figure 2.2 that the Euler approach leads to poor results in this neighborhood. Continuing past t_{crit} (see $k = 6$ in Table 2.1), the error is even more obvious and the values obtained meaningless. ○

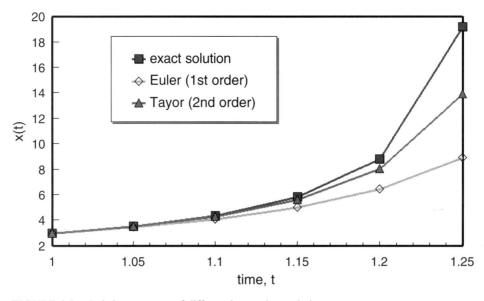

FIGURE 2.2 Relative accuracy of different integration techniques.

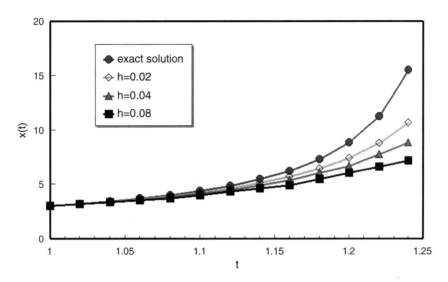

FIGURE 2.3   Effect of varying step size h on the system (2.7).

One way to improve the accuracy is to decrease the step size h. The effect of this is illustrated in Figure 2.3. Even so, reducing h has two major drawbacks: first, there will necessarily be many more computations to estimate the solution at a given point. Secondly, due to inherent machine limitations in data representation, h can also be too small. Thus, if h is large, inherent difference approximations lead to problems, and if h is too small, *truncation and rounding errors* occur. The trick is to get it just right by making h small enough.

EXAMPLE 2.3

Rather than printing the results of a procedure at the end of every computation, it is often useful to print the results of a computation periodically. This is accomplished by using a device called a *control break*. A control break works by means of nested loops. After an initial print, the outside loop prints n times and, for each print, the inside loop produces m computations. Thus, there are mn computations and $n + 1$ prints in total.

For example, suppose it is necessary to print the $n + 1$ computations for Example 2.2, m iterations for each print until the entire simulation is complete. The solution is to control the iteration with two loops rather than one. The outer or print loop is controlled using index i ($i = 1, 2, \ldots, n$) and the inner or compute loop uses index j ($j = 1, 2, \ldots, m$). Figure 2.4 shows this structure. The implementation of the control break is given in Listing 2.2.   ○

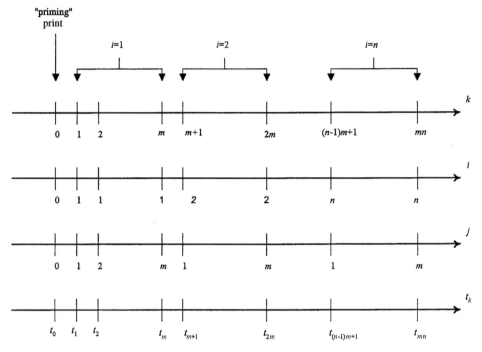

FIGURE 2.4   Structure of a control break.

```
t=1
x=3
print t, x
for i=1 to n
        for j=1 to m
                x=x+hx²t
                t=t+h
        next j
        print t, x
next i
```

LISTING 2.2   The control break using Euler's method on the system (2.7).

Taylor's Method

Euler's method can be thought of as a special case of Taylor's expansion theorem, which relates a function's neighboring value $x(t + h)$ to its value and higher-order attributes at time t. Specifically,

$$x(t + h) = \sum_{i=0}^{\infty} \frac{x^{(i)}(t)}{i!} h^i. \tag{2.9}$$

Taking the first-order case by summing for $i = 0, 1$ and noting that $\dot{x}(t) = f(t, x)$. Equation (2.9) reduces to Euler's update formula, $x(t + h) = x(t) + hf(t, x)$.

This may be extended using a higher-order approximation. For example, the *second-order* approximation gives

$$x(t + h) \approx x(t) + h\dot{x}(t) + \tfrac{1}{2}h^2\ddot{x}(t)$$

Using Equation (2.1),

$$\ddot{x} = \frac{d}{dt}f(t, x)$$

$$= \frac{\partial f}{\partial t} + \frac{\partial f}{\partial x}\frac{dx}{dt}$$

$$\equiv \frac{\partial f}{\partial t} + \frac{\partial f}{\partial x}f(t, x).$$

Accordingly,

$$x(t + h) \approx x(t) + hf(t, x) + \frac{1}{2}h^2\left[\frac{\partial f}{\partial t} + \frac{\partial f}{\partial x}f(t, x)\right].$$

This can be discretized as with Euler as follows:

$$t_{k+1} = t_k + h$$

$$x(k + 1) = x(k) + hf + \frac{1}{2}h^2\left(\frac{\partial f}{\partial t} + \frac{\partial f}{\partial x}f\right). \qquad (2.10)$$

EXAMPLE 2.4

Apply the Taylor technique to the system (2.7). Since $\partial f/\partial t = x^2$ and $\partial f/\partial x = 2tx$, Equation (2.10) reduces to

$$x(k + 1) = x(k) + x^2(k)t_k h + x^2(k)\left[x(k)t_k^2 + \tfrac{1}{2}\right]h^2.$$

Using $h = 0.05$, the Taylor procedure gives the data shown in Table 2.2. The graph is shown in Figure 2.2, where it will be noticed that Taylor is more accurate than Euler. Even so, the singularity at $t_{\text{crit}} = 1.29$ leads to intolerable inaccuracies.   ○

TABLE 2.2   Solution to the System (2.7) Using Taylor's Method

k	0	1	2	3	4	5	6
t_k	1.00	1.05	1.10	1.15	1.20	1.25	1.30
$x(t)$	3.00	3.55	4.38	5.81	8.82	19.20	−85.71
$x(k)$	3.00	3.53	4.32	5.61	8.05	13.89	36.66

Runge–Kutta Methods

The second-order (Taylor) results of Example 2.4 compare favorably against those of Example 2.3, where the first-order (Euler) technique was applied. If even more terms of the Taylor series are used, the results are even more dramatic. In fact, often a fourth-order approximation is used:

$$x(t+h) = x(t) + h\dot{x}(t) + \tfrac{1}{2}h^2\ddot{x}(t) + \tfrac{1}{6}h^3 x^{(3)}(t) + \tfrac{1}{24}h^4 x^{(4)}(t).$$

However, in order to apply this formula, it is first necessary to differentiate f several times, as was done for Taylor. This can be tedious at best, and is often impossible since an analytical formula for f may be unavailable. This is often the case in realistic systems, since we can directly measure only signal values, not their derivatives, which must be inferred.

An alternative to explicit differentiation is that of expressing f at intermediate arguments; that is, a *functional* approach. It can be shown that the following discretized algorithm approximation is equivalent to the fourth-order Taylor method, and just as accurate:

$$\begin{aligned}
\kappa_1 &= f(t_k, x(k)), \\
\kappa_2 &= f(t_k + \tfrac{1}{2}h, x(k) + \tfrac{1}{2}h\kappa_1), \\
\kappa_3 &= f(t_k + \tfrac{1}{2}h, x(k) + \tfrac{1}{2}h\kappa_2), \\
\kappa_4 &= f(t_k + h, x(k) + h\kappa_3);
\end{aligned} \tag{2.11}$$

$$\begin{aligned}
t_{k+1} &= t_k + h, \\
x(k+1) &= x(k) + \tfrac{1}{6}h(\kappa_1 + 2\kappa_2 + 2\kappa_3 + \kappa_4).
\end{aligned} \tag{2.12}$$

This so-called *Runge–Kutta* algorithm is classic. It offers two advantages: it is accurate (equivalent to the fourth-order Taylor), and it is easy to use since no derivatives need be evaluated. It serves as the de facto technique of choice in much numerical work.

EXAMPLE 2.5

Let us apply Runge–Kutta to the system (2.7). Listing 2.3 is a program to implement Runge–Kutta. Notice that the order of execution is critical, since κ_1 must be calculated before κ_2, κ_2 before κ_3, and κ_3 before κ_4, before updating x. The alternative to this strict sequential requirement is to do the bookkeeping by means of an array for x, but that is wasteful of computer memory and should be avoided. As a comparison, Table 2.3 shows the various approaches described thus far using $h = 0.05$. These are also compared in Figure 2.5, where the superior features of Runge–Kutta are evident.

It is evident from Table 2.3 and Figure 2.5 that higher-order integration techniques give superior accuracy without the disadvantages encountered by merely reducing the step size h. However, the singularity at $t_{\text{crit}} = 1.29$ still causes significant problems.   ○

```
input t, x
print t, x
for k=1 to n
        κ₁=tx²
        κ₂=(t+½h)(x+½hκ₁)²
        κ₃=(t+½h)(x+½hκ₂)²
        κ₄=(t+h)(x+hκ₃)²
        x=x+⅙h(κ₁+2κ₂+2κ₃+κ₄)
        t=t+h
        print t, x
next k
```

LISTING 2.3   The Runge–Kutta algorithm for numerical integration.

TABLE 2.3   Results of Euler, Taylor, and Runge–Kutta Methods for Approximating System (2.7)

k	0	1	2	3	4	5	6
t_k	1.00	1.05	1.10	1.15	1.20	1.25	1.30
Euler (order 1)	3.00	3.45	4.07	4.99	6.42	8.89	13.83
Taylor (order 2)	3.00	3.53	4.32	5.61	8.05	13.89	36.66
Runge–Kutta (order 4)	3.00	3.55	4.38	5.81	8.82	18.98	392.55
Exact solution	3.00	3.55	4.38	5.81	8.82	19.20	−85.71

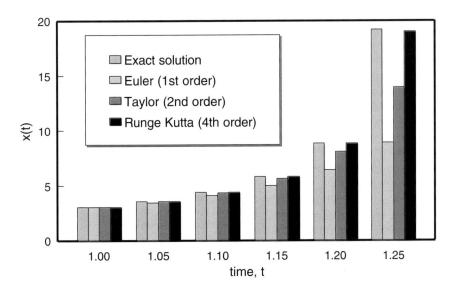

FIGURE 2.5   Comparison of integration techniques.

The reason that singularities cause unrecoverable problems is because the function's slope tends to infinity as the singularity is reached. Accordingly, it is necessary to use all terms (if that were possible) in the Taylor series (2.9) to follow the function. Without assuming a specific form, such as that of a *rational function*, it is not possible to achieve accuracy across a singularity (e.g., at $t = 1.3$).

Consider the function illustrated in Figure 2.6. It is evident that different regions will require different integration procedures. On the interval $[t_0, t_1]$ a zeroth-order approach would probably be sufficient, since the function is essentially constant. Certainly a first-order approach would lead to an accurate solution. However, on $[t_1, t_2]$, a second-order method would be better since the slope (first derivative) changes and the concavity (second derivative) changes as well. On $[t_2, t_3]$, there are more frequent changes, so a smaller step size h is necessary, and probably a higher-order, say Runge–Kutta, approach is a good idea. In any case, values of $x(t)$ on $[t_3, t_4]$ would be totally erroneous because of the singularity at t_3. This is true regardless of the order of the technique.

This interpretation is easy if the solution $x(t)$ is known. Of course, this cannot really happen, since if $x(t)$ were known then the problem would be solved a priori. It would be most useful if those intuitions were available for a given function f. Unfortunately, few elementary or general results of this type exist, and practitioners must be satisfied using heuristics. For instance, by using first-, second-, and fourth-order approaches to the same problem, one would expect three different answers. However, by continually decreasing the step size h, eventually the results would approximately match. This "solution" would then be considered satisfactory.

Adaptive Runge–Kutta Methods

One of the more popular heuristics is to compute a solution using the different orders, say using a fourth-order and a fifth-order Runge–Kutta. If the results are similar, the integration proceeds using a relaxed step size (tripling h is typical). On the other hand, if the results are significantly different, the step size is reduced (perhaps by a factor of 10).

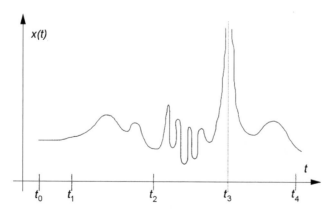

FIGURE 2.6   A generic solution $x(t)$.

Thus, if the function f is sufficiently smooth, large steps are taken. However, as the terrain gets rougher, smaller steps ensure that accuracy is not lost. The fourth-order Runge–Kutta formulae are given as Equations (2.11) and (2.12). The fifth-order Runge–Kutta–Fehlberg formulae are as follows:

$$
\begin{aligned}
x(k+1) &= x(k) + h\left(\tfrac{25}{216}\kappa_1 + \tfrac{1408}{2565}\kappa_3 + \tfrac{2197}{4104}\kappa_4 - \tfrac{1}{5}\kappa_5\right), \\
y(k+1) &= x(k) + h\left(\tfrac{16}{135}\kappa_1 + \tfrac{6656}{12825}\kappa_3 + \tfrac{28561}{56430}\kappa_4 - \tfrac{9}{50}\kappa_5 + \tfrac{2}{55}\kappa_6\right);
\end{aligned}
\tag{2.13}
$$

$$
\begin{aligned}
\kappa_1 &= f(t_k, x(k)), \\
\kappa_2 &= f(t_k + \tfrac{1}{4}h, x(k) + \tfrac{1}{4}h\kappa_1), \\
\kappa_3 &= f(t_k + \tfrac{3}{8}h, x(k) + h[\tfrac{3}{32}\kappa_1 + \tfrac{9}{32}\kappa_2]), \\
\kappa_4 &= f(t_k + \tfrac{12}{13}h, x(k) + h[\tfrac{1932}{2197}\kappa_1 - \tfrac{7200}{2197}\kappa_2 + \tfrac{7296}{2197}\kappa_3]), \\
\kappa_5 &= f(t_k + h, x(k) + h[\tfrac{439}{216}\kappa_1 - 8\kappa_2 + \tfrac{3680}{513}\kappa_3 - \tfrac{845}{4104}\kappa_4]), \\
\kappa_6 &= f(t_k + \tfrac{1}{2}h, x(k) + h[-\tfrac{8}{27}\kappa_1 + 2\kappa_2 - \tfrac{3544}{2565}\kappa_3 + \tfrac{1859}{4104}\kappa_4 - \tfrac{11}{40}\kappa_5]).
\end{aligned}
\tag{2.14}
$$

Using fourth- and fifth-order Runge–Kutta, the so-called *Runge–Kutta–Fehlberg* algorithm described above is shown in Listing 2.4. In order to apply this program, it is first necessary to prescribe a tolerance ϵ. As is clear from the listing, a small ϵ will ensure greater accuracy by reducing h when necessary. Similarly, if the two computations are close enough, the size of h is relaxed and accuracy is maintained with fewer computations required.

```
t=t₀
x=x₀
print t, x
for k=1 to n
[1]         compute 4th order Runge-Kutta
                        using Equations (2.11), (2.12); x
            compute 5th order Runge-Kutta-Fehlberg
                        using Equations (2.13), (2.14); y
            if |x-y|<ε then
                x=y
                t=t+h
                h=3h
            else
                h=h/10
                goto [1]
            end if
            print t, x
    next k
```

LISTING 2.4   Adaptive Runge–Kutta–Fehlberg algorithm.

The adaptive approach is powerful and accurate. However, it should be remembered that accuracy is relative and it may be over-kill for the problem at hand. Specific problems have specific requirements and *Runge–Kutta–Fehlberg* is still not useful when crossing a singularity.

2.2   HIGHER-ORDER SYSTEMS

As illustrated in Example 2.1, it is often possible to reduce any collection of differential equations to an equivalent set of first order equations. Once this is done, any of the techniques described above may be applied in order to attain a numerical approximation to the solution. For instance, a system of two first-order equations is described as follows:

$$\dot{x} = f(t, x, y),$$
$$\dot{y} = g(t, x, y),$$
$$x(t_0) = x_0,$$
$$y(t_0) = y_0.$$

By associating κ_1, κ_2, κ_3, and κ_4 with the variable x as before and introducing λ_1, λ_2, λ_3, and λ_4 to associate with y, the extension of Runge–Kutta to this system is straightforward. The discrete solution is extended from Equations (2.11) and (2.12) as follows:

$$
\begin{aligned}
\kappa_1 &= f(t_k, x(k), y(k)), \\
\lambda_1 &= g(t_k, x(k), y(k)), \\
\kappa_2 &= f(t_k + \tfrac{1}{2}h, x(k) + \tfrac{1}{2}h\kappa_1, y(k) + \tfrac{1}{2}h\lambda_1), \\
\lambda_2 &= g(t_k + \tfrac{1}{2}h, x(k) + \tfrac{1}{2}h\kappa_1, y(k) + \tfrac{1}{2}h\lambda_1), \\
\kappa_3 &= f(t_k + \tfrac{1}{2}h, x(k) + \tfrac{1}{2}h\kappa_2, y(k) + \tfrac{1}{2}h\lambda_2), \\
\lambda_3 &= g(t_k + \tfrac{1}{2}h, x(k) + \tfrac{1}{2}h\kappa_2, y(k)k + \tfrac{1}{2}h\lambda_2), \\
\kappa_4 &= g(t_k + h, x(k) + h\kappa_3, y(k) + h\lambda_3), \\
\lambda_4 &= f(t_k + h, x(k) + h\kappa_3, y(k) + h\lambda_3);
\end{aligned}
\tag{2.15}
$$

$$
\begin{aligned}
x(k+1) &= x(k) + \tfrac{1}{6}h(\kappa_1 + 2\kappa_2 + 2\kappa_3 + \kappa_4), \\
y(k+1) &= y(k) + \tfrac{1}{6}h(\lambda_1 + 2\lambda_2 + 2\lambda_3 + \lambda_4).
\end{aligned}
\tag{2.16}
$$

These concepts are implemented as shown in the following examples.

EXAMPLE 2.6

Consider the system described by the block diagram shown in the figure below, where the system is described by

$$\ddot{x} + 3x\dot{x} = u(t),$$
$$u(t) = t, \qquad t \geqslant 0,$$
$$x(0) = 2,$$
$$\dot{x}(0) = 1.$$

(2.17)

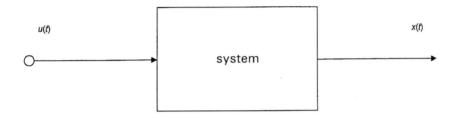

u(t)   system   x(t)

Solution
Since the differential equation is of second order, there must be two state variables. We can take one of these to be the output $x(t)$ and define the other as $y(t) = \dot{x}(t)$. Thus, we create an equivalent definition of the system as

$$\dot{x}(t) = y,$$
$$\dot{y} = t - 3xy,$$
$$x(0) = 2,$$
$$y(0) = 1.$$

(2.18)

Using Euler, the discretized system is

$$t_{k+1} = t_k + h,$$
$$x(k+1) = x(k) + hy(k),$$
$$y(k+1) = y(k) + h[t_k - 3x(k)y(k)].$$

It will be noted that these equations are coupled in that x apparently cannot be updated before y, and vice versa, without retaining the subscripts. Thus, in implementing this system, care must be taken if the use of subscripts is to be avoided. This can be done by introducing a temporary variables x_1 and y_1 which are computed prior to the actual update of x and y. This technique is demostrated in Listing 2.5.

Using the updated value $x(k+1)$, rather than the "old" x-value $x(k)$, usually causes minimum difficulty, since they are approximately equal. Even so, as a rule, it is simply best to avoid this inaccuracy.

Runge–Kutta can also be used to solve the system (2.17). This is done by applying Equations (2.15) and (2.16) to the equivalent first-order system (2.18).

```
t=0
x=2
y=1
print t, x, y
for k=1 to n
          x₁=x+hy
          y₁=y+h(t-3xy)
          x=x₁
          y=y₁
          t=t+h
          print t, x, y
next k
```

LISTING 2.5   Euler implementation of the system (2.17).

Therefore,

$$
\begin{aligned}
\kappa_1 &= y(k), \\
\lambda_1 &= t_k - 3x(k)y(k), \\
\kappa_2 &= y(k) + \tfrac{1}{2}h\lambda_1, \\
\lambda_2 &= t_k + \tfrac{1}{2}h - 3[x(k) + \tfrac{1}{2}h\kappa_1][y(k) + \tfrac{1}{2}h\lambda_1], \\
\kappa_3 &= y(k) + \tfrac{1}{2}h\lambda_2, \\
\lambda_3 &= t_k + \tfrac{1}{2}h - 3[x(k) + \tfrac{1}{2}\kappa_2 h][y(k) + \tfrac{1}{2}h\lambda_2], \\
\kappa_4 &= y(k) + h\lambda_3, \\
\lambda_4 &= t_k + h - 3[x(k) + h\kappa_3][y(k) + h\lambda_3];
\end{aligned}
$$

$$
\begin{aligned}
x(k+1) &= x(k) + \tfrac{1}{6}h(\kappa_1 + 2\kappa_2 + 2\kappa_3 + \kappa_4), \\
y(k+1) &= y(k) + \tfrac{1}{6}h(\lambda_1 + 2\lambda_2 + 2\lambda_3 + \lambda_4).
\end{aligned}
$$

These are implemented without the use of temporary variables in Listing 2.6.

○

This technique for integrating a set of differential equations is straightforward, and can be adapted to a wide variety of systems. However, special circumstances abound! Therefore, simulation problems involving dynamical systems often require considerable work in prescribing precision parameters such as integration step size h and, if the adaptive methods are applied, tolerance ϵ.

2.3 AUTONOMOUS DYNAMIC SYSTEMS

From a simulation perspective, the most interesting systems are dynamic. Rather than simply static structures, they vary with time. To model these systems, we use differential equations in time, since derivatives describe the rates at which variables change. For example, Newton's laws describe motion in terms of momentum: the time rate of change in

```
t=0
x=2
y=1
print t, x, y
for k=1 to n
    κ₁=y
    λ₁=t-3xy
    κ₂=y+½hλ₁
    λ₂=t+½h-3(x+½hκ₁)(y+½hλ₁)
    κ₃=y+½hλ₂
    λ₃=t+½h-3(x+½hκ₂)(y+½hλ₂)
    κ₄=y+hλ₃
    λ₄=t+h-3(x+hκ₃)(y+hλ₃)
    x=x+⅙h(κ₁+2κ₂+2κ₃+κ₄)
    y=y+⅙h(λ₁+2λ₂+2λ₃+λ₄)
    t=t+h
    print t, x, y
next k
```

LISTING 2.6   Runge–Kutta solution of the system (2.17).

the product of mass and velocity, $p = d(mv)/dt$. Also, in biological systems, population dynamics descriptions are typically stated as "the rate the population changes is proportional to the population size". This translates to the mathematical statement $dN/dt = \lambda N$, where N is the population size at time t and λ is a proportionality constant.

Systems can be driven by either endogenous (internal) or exogenous (external) inputs. These inputs are either synchronous if the system is time-driven, or asynchronous if it is event-based. For instance, the classical second-order linear system $\ddot{x} + 2\zeta\omega\dot{x} + \omega^2 x = r(t)$ has input $r(t)$ and output $x(t)$. ζ and ω are problem-specific constants. If $r(t)$ is a deterministic function of time such as $r(t) = 4 - t$, the system is said to be time-driven, since the system dynamics (described by the left-hand side) vary with time as specified by the right-hand side. On the other hand, if $r(t)$ is defined stochastically as a random process such as

$$r(t) = \begin{cases} 1, & \text{event occurs,} \\ 0, & \text{event does not occur,} \end{cases}$$

with the mean event inter-occurrence time given, the model is event-driven. In this case, the system is not deterministic, yet it is statistically meaningful.

It is also possible that a system has no input at all. We call such systems *autonomous*, since they are totally self-contained and their behavior is independent of external influence. Accordingly, the solution to an autonomous system is called the system's *natural response*. If the system is linear, it can be shown that the natural response is one of three types: (1) stable – meaning that after a short transient phase the output approaches zero; (2) unstable – meaning that the natural response increases without bound; or (3) marginal – meaning that the response is periodic and bounded. For instance, consider the classical second-order system with constant coefficients in which $r(t)$ is the input:

$$\ddot{x} + 2\zeta\omega\dot{x} + \omega^2 x = r(t).$$

In the natural response case where $r(t) = 0$, it can be shown that the solution is one of three types, depending on parameters ζ and ω. Specifically,

$$x(t) = \begin{cases} Ae^{-\zeta\omega t}\cos(\omega\sqrt{1 - \zeta^2}\,t) + Be^{-\zeta\omega t}\sin(\omega\sqrt{1 - \zeta^2}\,t), & |\zeta| < 1, \\ A\cos(\omega t) + B\sin(\omega t), & |\zeta| = 1, \\ Ae^{\zeta\omega t} + Be^{-\zeta\omega t}, & |\zeta| > 1, \end{cases} \qquad (2.19)$$

where the constants A and B are determined from the initial conditions. In the case of non-autonomous systems, the superposition principle applies. This states that the total solution for a linear system is given by the sum of the natural response and the forced response, as shown in the following example.

EXAMPLE 2.7

Consider the following second-order system:

$$\ddot{x} + 4\dot{x} + 5x = 15t + 22,$$
$$x(0) = 0, \qquad\qquad\qquad (2.20)$$
$$\dot{x}(0) = 1.$$

Solve the system using superposition. In doing so, show that the total solution is the sum of two terms: (1) a term involving the initial conditions but independent of the input; and (2) a term involving the input but independent of the initial conditions.

Solution
The natural response is the solution to the equation $\ddot{x} + 4\dot{x} + 5x = 0$, so that $\omega = \sqrt{5}$ and $\zeta = 2/\sqrt{5}$. Using Equation (2.19), $x_{\text{natural}}(t) = Ae^{-2t}\cos t + Be^{-2t}\sin t$.

The forced response can be found by noting that the output is driven by the input, and will therefore be of the same form as the input and its derivatives. In this case, $x_{\text{forced}} = C + Dt$, since the input is linear and its derivative is constant, implying that the sum will also be linear. Substituting into Equation (2.20) and equating coefficients, $x_{\text{forced}} = 3t + 2$.

Using superposition, the total solution is $x(t) = Ae^{-2t}\cos t + Be^{-2t}\sin t + 3t + 2$. The constants A and B can be found by substituting the initial conditions. Thus, the final solution is $x(t) = 2e^{-2t}\cos t - e^{-2t}\sin t + 3t + 2$. Notice that the two components to the solution are such that the natural response $x_{\text{natural}}(t)$ is dependent on the initial conditions and simply goes to zero as time goes to infinity, while the forced response $x_{\text{forced}}(t)$ is independent of the initial conditions but takes over completely in the steady state. This result holds in general in stable linear systems: $x(t) = x_{\text{natural}}(t) + x_{\text{forced}}(t)$, where $x_{\text{natural}}(t)$

depends on only the initial conditions and $x_{forced}(t)$ depends only on the input and takes over in the steady state.   ○

The case where the steady-state solution is independent of the initial conditions can only be guaranteed for linear systems. In fact, in many nonlinear systems, the *final solution* (another name for steady state) is so sensitive to where the system begins that we call the system *chaotic*. This, combined with the fact that there are no known general analytical approaches to nonlinear differential equations, means that numerical techniques with extremely tight control on the integration step size are required.

Consider the problem of modeling the population dynamics of a certain species. Although there are a number of factors controlling this population, as a first try we note that the rate at which a population's size increases is roughly proportional to the population size at any time. Letting $x(t)$ represent the population size at time t and λ be a proportionality constant, $\dot{x} = \lambda x$. Assuming an initial size x_0 at time $t = 0$, the solution is $x(t) = x_0 e^{\lambda t}$ for $t \geqslant 0$. Clearly, this population increases exponentially over time. This model is sometimes called the Malthusian model after the 18th century English economist Thomas Malthus, who used the above argument to predict a worldwide population explosion.

Many examples exist confirming Malthus' model. Even so, it is not hard to find reasons why and cases where this model doesn't work especially well. Some significant factors, in addition to population size, that affect growth rates include the capacity of the environment to support the population, interactions with competing and supporting populations, and the energy supplied by heat over time. For example, a population might be too large for its food supply or there might be predators in the area where it lives.

As a first modification, suppose the system can support a population maximum of size x_m, called the carrying capacity. The ratio $x(t)/x_m$ is the proportion of the system that is full and $1 - x(t)/x_m$ is the proportion of the system that remains available for growth. It follows that a more realistic model is that the growth rate is proportional to both the population size and the proportion available for expansion. Specifically,

$$\dot{x} = \lambda x \left(1 - \frac{x}{x_m} \right). \tag{2.21}$$

This equation, which is called the *logistic equation*, is clearly nonlinear. It will be noted that if $x(t)$ is small, the equation is almost Malthusian, since the factor $1 - x(t)/x_m \approx 1$ is approximately unity. Similarly, as the population size becomes close to the capacity x_m, $1 - x(t)/x_m \approx 0$ becomes almost zero and the growth rate drops off. In other words, this model fits our intuition.

Even though Equation (2.21) is nonlinear, its solution is straightforward. Elementary calculus leads to the following explicit formula for $x(t)$:

$$x(t) = \frac{x_m}{1 - (1 - x_m/x_0)e^{-\lambda t}}.$$

Figure 2.7 shows $x(t)$ for x_m fixed at $x_m = 25$ and several different values of x_0. Notice that if x_0 is less than x_m, $x(t)$ grows to approach its carrying capacity x_m asymptotically, while if

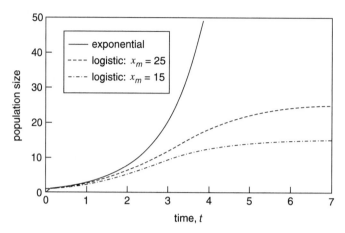

FIGURE 2.7   Logistic equation for differing initial populations.

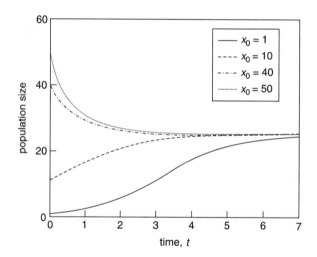

FIGURE 2.8   Logistic equation for differing carrying capacities.

x_0 is larger than x_m, the curve approaches x_m from above. Figure 2.8 shows $x(t)$ for x_0 fixed at $x_0 = 1$ for several different values of x_m. In each case, $x(t)$ approaches the carrying capacity x_m, regardless of the initial value. Thus, in each case, the initial and final values are as expected.

The logistic model performs well in environments where there is one population and a finite food source. However, a more typical situation is one where there is a predator and a prey competing to stay alive by eating and not being eaten, depending on the point of view! A predator will grow proportionally to its own population size and its food supply – the prey.

Let $x(t)$ and $y(t)$ be the sizes of the prey and predator populations, respectively. Then the number of predator–prey interactions is proportional to the product $x(t)y(t)$. It follows then that the rates of growth of the predator–prey populations are given by

$$\dot{x} = \alpha_1 x \left(1 - \frac{y}{\beta_1}\right),$$
$$\dot{y} = \alpha_2 y \left(-1 + \frac{x}{\beta_2}\right), \qquad (2.22)$$

where α_1, α_2 and β_1, β_2 are positive so-called proportionality capacity constants for each population. It should be noted that these names are misnomers, since α_1 is only an approximation to a proportionality constant when y is small relative to x. Similarly, α_2 is almost inversely proportional when x is small. The β-constants are mathematically analogous to the system capacity in the logistic model, but they have no direct counterparts in the predator–prey model.

The predator–prey model described in Equations (2.22) is credited to A. J. Lotka and Vito Volterra, whose primary work was in integral equations, and these equations are therefore referred to as the Lotka–Volterra equations. They are clearly nonlinear and coupled. Unfortunately, there is no known analytical solution, and so numerical techniques as described earlier must be employed.

Any of the techniques described earlier can be used, so let us begin applying the easiest: Euler. Since the two equations are coupled, this requires the use of temporary variables so that the updates are applied after they are computed. Assuming a suitable integration step size h, this algorithm is shown in Listing 2.7. Since this is Euler, the h-value will have to be quite small – perhaps even too small. Assuming that results are required over the time interval $[t_0, t_n]$ and the interval is h, $n = (t_n - t_0)/h$ is a rather large number. For instance if $t_0 = 0$, $t_n = 10$ and $h = 0.001$, then the $n = 10\,000$ points calculated could hardly all be printed or plotted. Probably only 50 would be sufficient.

This problem can be taken care of by the use of a control break. Listing 2.8 produces, in addition to the priming print in the initialization stage, n additional prints, each with m calculations in between. This will lead to mn calculations covering the interval $[t_0, t_n]$. Assuming a step size h, $m = (t_n - t_0)/nh$. Programmatically, this is accomplished by the code, where the outer loop (i) is performed n times, each of which prints the results obtained by the inner loop (j), which is performed m times for each of the n i-values. Thus,

```
read t, x, y
print t, x, y
for k=1 to n
          x₁=x (1+α₁h-α₁hy/β₁)
          y₁=y (1-α₂h+α₂hx/β₂)
          x=x₁
          y=y₁
          t=t+h
          print t, x, y
     next k
```

LISTING 2.7   Simulation of the Lotka–Volterra model.

```
read t, x, y
print t, x, y
for i=1 to n
        for j=1 to m
                x₁=x(1+α₁h-α₁hy/β₁)
                y₁=y(1-α₂h+α₂hx/β₂)
                x=x₁
                y=y₁
                t=t+h
        next j
        print t, x, y
next i
```

LISTING 2.8   Simulation of the Lotka–Volterra model utilizing a control break.

there are a total of *mn* updates. Notice that within the computation loop, each update is evaluated before it is actually performed. This is because that the update computation requires the "old" variables rather than the new ones, and therefore no subscripted variables are required.

EXAMPLE 2.8

Solve the following Lotka–Volterra system numerically over the time interval $[0, 5]$:

$$\dot{x} = 3x(1 - \tfrac{1}{10}y),$$
$$\dot{y} = 1.2y(-1 + \tfrac{1}{25}x),$$
$$x(0) = 10,$$
$$y(0) = 5.$$

Graph the results in both the time and state-space domains.

Solution
Even though we do not yet know how to choose a step size, let us assume that $h = 0.001$ is acceptable. Since the extent of the graph is 5 time units and 10 points per interval seems reasonable, let's also plan on $5*10 = 50$ points to be plotted in addition to the initial points at $t = 0$. Thus, we choose $n = 50$ prints and $m = (5 - 0)/50*0.001 = 100$ updates per print, giving a total of 5000 updates. The first few calculations are as follows, and the graph is sketched as a function of time in Figure 2.9.   ○

t	x	y
0.0	10.00	5.00
0.1	11.67	4.67
0.2	13.76	4.40
0.3	16.32	4.19

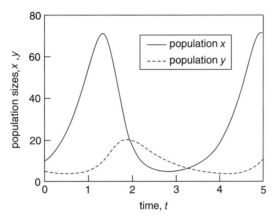

FIGURE 2.9   Time domain description of the Lotka–Volterra model of Example 2.8.

There is another view of the predator–prey model that is especially illustrative. This is the so-called *phase plot* of y versus x given in Figure 2.10. As one would expect, as the predator's food source (the prey) increases, the predator population increases, but at the expense of the prey, which in turn causes the predator population to drop off. Therefore, as one population achieves a bountiful food supply, it becomes a victim of its own success. This is evident from the closed loop of the phase plot. Such a loop implies an endless periodic cycle. In this nonlinear case, the term *stability* has to mean that neither population becomes overly dominant, nor does either population become extinct; hopefully, this would lead to a periodic ebb and flow. Notice that here the (x, y) points connect in a closed path called an *orbit*, which indicates periodic behavior.

Orbits are especially useful devices to describe system stability. For instance, if (x, y) spirals inward to a single point (x_{ss}, y_{ss}), the values x_{ss} and y_{ss} are the steady states of the system. This is how a linear system driven by constant inputs functions. Similarly, if the

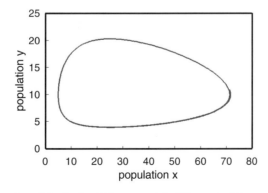

FIGURE 2.10   Phase plot of the Lotka–Volterra model of Example 2.8.

orbit spirals outward, the system becomes unbounded and thus unstable. This is a serious topic that is beyond the scope of this text, but let us say that the interpretation of orbits is very helpful in characterizing nonlinear systems in general. In particular, except for linear systems, different initial conditions can lead to very different behaviors. Only in linear systems is the stability independent of the initial conditions.

Of course there is no best value of integration step size h. In principle, h should be zero – so the closer the better – but if h is too small (approximately machine zero) then numerical stability (not to be confused with system stability) makes the results extremely inaccurate. Also, as h becomes smaller, more time is expended in execution, which is undesirable. Thus, the question becomes one of "what h is accurate enough"? To observe the effect of differing h-values, consider the previous example for a set of step sizes that are each $\frac{1}{5}$ of the previous h each iteration:

$$h_0 = 0.1,$$
$$h_1 = \tfrac{1}{5}h_0 = 0.02,$$
$$h_2 = \tfrac{1}{5}h_1 = 0.004,$$
$$h_3 = \tfrac{1}{5}h_2 = 0.0008,$$
$$h_4 = \tfrac{1}{5}h_3 = 0.00016.$$

Considering the same Lotka–Volterra system as in Example 2.8, we plot $x(t)$ over time, using these different h-values, in Figure 2.11.

Figure 2.11 shows clearly that the results differ greatly for different step sizes. This is a critical decision, to say the least. However, it will be noticed that as h becomes smaller, the graphs begin to converge, just as one should expect. In fact, there is virtually no difference between $h_3 = 0.0008$ and $h_4 = 0.00016$. If this is done for $y(t)$, a similar result holds, so a reasonable conclusion would be to use $h_3 = 0.0008$.

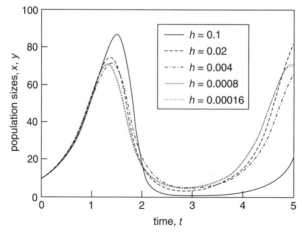

FIGURE 2.11   Partial solution $x(t)$ plotted for different step sizes h, Example 2.8.

A better and more systematic procedure goes as follows. Again form a geometric sequence of h-values, each with common ratio r:

$$h_0,$$
$$h_1 = h_0 r,$$
$$h_2 = h_1 r = h_0 r^2,$$
$$h_3 = h_2 r = h_0 r^3,$$
$$\vdots$$
$$h_l = h_{l-1} r = h_0 r^l.$$

Compute n values using $m = (t_n - t_0)/nh$ updates for each point of the system state variables over the range $[t_0, t_n]$. Choose one state variable, say $x(t)$, and tabulate it for each of the $i = 0, 1, 2, \ldots, l$ different h-values. This table is formed below for the system given in Example 2.8 using $l = 4$. Notice that the solution for $t = 0.1$ is close for all h-values, but, by the time $t = 5.0$, the solutions differ significantly.

t	$h_0 = 0.1$	$h_1 = 0.05$	$h_2 = 0.0125$	$h_3 = 0.00625$	$h_4 = 0.003125$
0.0	10.0	10.0	10.0	10.0	10.0
0.1	11.5	11.58527	11.6305	11.65383	11.662568
0.2	13.3495	13.54552	13.65014	13.70421	13.7317
⋮	⋮	⋮	⋮	⋮	⋮
5.0	20.08104	72.0827	81.77159	77.91122	74.51112

It is our goal to compare the *relative root mean square error* (RMS) for each method. This is done by finding the difference between adjacent columns, squaring and totaling them over the time interval $[0, 5]$. This result is then divided as follows so as to find the relative RMS for $i = 0, 1, \ldots, l$ as specified here:

$$\text{relative RMS error} = \sqrt{\frac{\sum_{k=0}^{n} [x(h_i) - x(h_{i+1})]^2}{n(h_i - h_{i+1})^2}}, \quad i = 0, 1, \ldots, l.$$

Since h_i was chosen to form a geometric sequence, there are other equivalent formulas, so do not be surprised to see this result stated in other forms in other texts.

For the system specified in Example 2.8, the calculations for the $l = 8$ differences proceed as in the following table.

i	h_i to h_{i+1}	Relative RMS
0	0.1 to 0.05	275.45
1	0.50 to 0.025	215.46
2	0.025 to 0.0125	157.42
3	0.0125 to 0.00625	152.99
4	0.00625 to 0.003125	153.85
5	0.003125 to 0.0015625	154.34
6	0.0015625 to 0.00078125	154.43
7	0.00078125 to 0.000390625	154.66
8	0.000390625 to 0.0001953125	155.10

There are two important things to notice regarding the relative RMS table. First, the relative RMS values tend to stabilize at about $l = 5$ ($h = 0.003125$). Second, they actually seem worse after that point. It is easy to see why the seemingly arbitrary value of $h = 0.001$ was chosen as numerically sufficient for Example 2.8. Such studies are always appropriate in simulation research.

There is no need to limit population models to just one predator, one prey. Populations can feed on and are fed on by more than one species, and the whole issue becomes quite complex. Of course, this is good, since that is exactly what we as modelers are trying to describe! At this level, the mathematics is still nonlinear and the differential equations are coupled. Without going into detailed explanations as to the physical interpretation of the model, another example is in order.

EXAMPLE 2.9

Consider the Lotka–Volterra system described by the following equations:

$$\dot{x} = 3x(1 - \tfrac{1}{10}y + \tfrac{1}{15}z),$$
$$\dot{y} = 1.2y(1 + \tfrac{1}{20}x - \tfrac{1}{25}z),$$
$$\dot{z} = 2z(1 - \tfrac{1}{30}x + \tfrac{1}{35}y),$$

$$(2.23)$$

$$x(0) = 5,$$
$$y(0) = 10,$$
$$z(0) = 8.$$

Create a suitable simulation and interpret the graphical results for the time interval $[0, 25]$.

Solution
Especially since this is a nonlinear system whose results are unknown, a Runge–Kutta technique is employed, with $h = 0.0001$. The code is given in Listing 2.9. The procedure outlined above, where we found a leveling-off point for the relative RMS error curve, indicates this is a good step size. Even so, we expect some deterioration in the long run due to numerical estimation concerns. The resulting population-versus-time graph is shown in Figure 2.12.

As would be expected, there is a transient phase, in this case lasting for about 18 time units, followed by a cyclical steady-state phase where the populations ebb and flow over time.

The three phase plots are shown in Figures 2.13–2.15. Notice that the initial transient phases build into ever-widening oscillations, which are followed by periodic limit cycles in the steady state. However, the cycles do not seem to close as they did for the two-dimensional example considered earlier. This is probably because of numerical problems rather than system instability. In order to do the phase plots, it is necessary to plot each state (x, y, z) against its derivative $(\dot{x}, \dot{y}, \dot{z})$. Since these derivatives are given explicitly by Equations (2.23), the phase plots follow immediately from the integrated differential equations. Suitable code is given in Listing 2.9.   ○

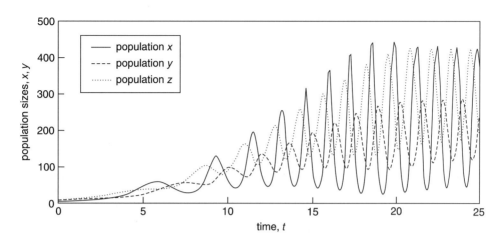

FIGURE 2.12   Time plot of a Lotka–Volterra system of three populations, Example 2.9.

2.4   MULTIPLE-TIME-BASED SYSTEMS

The critical variable in dynamic systems is time. All rates, and thus all derivatives, are changes with respect to time. Normally, we think of a system's time somewhat like a

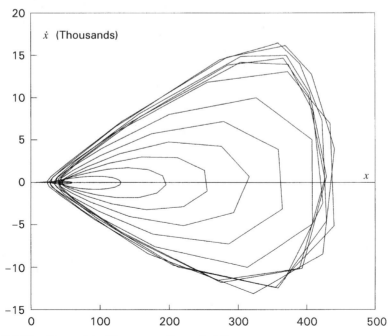

FIGURE 2.13   Phase plot of $\dot{x}$ versus x, Example 2.9.

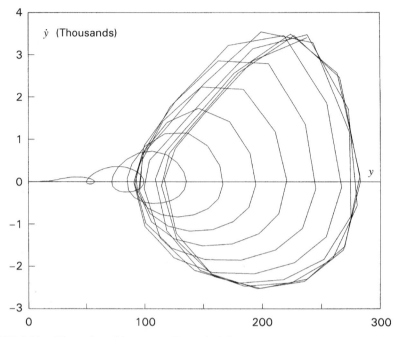

FIGURE 2.14   Phase plot of $\dot{y}$ versus y, Example 2.9.

master drumbeat in which all subsystems measure all events with respect to the same base. This is not always an accurate model. For instance, plants grow and mature with respect to how much heat energy (temperature) is in the system rather than the "real" chronological time. This heat energy is usually thought of as physiological time, and the dynamic equations measure rates with respect to physiological rather than chronological time. It follows that in some systems, each subsystem measures time differently.

Consider a biological growth model in which certain bacteria exist in an environment where the temperature $T(t)$ varies over time t. It has been found that below a threshold temperature T_0 there is no growth, but for temperatures greater than T_0 the *relative* growth rate $\dot{x}/x$ varies directly with $T(t) - T_0$. Stated mathematically,

$$\frac{\dot{x}}{x} = \begin{cases} 0, & T(t) < T_0, \\ r[T(t) - T_0], & T(t) \geq T_0, \end{cases} \tag{2.24}$$

for a threshold temperature T_0 and proportionality constant r.

It is useful to define a new time, called the *physiological time* τ, as follows:

$$\tau(t) = \int_{\Gamma} [T(t) - T_0]\, dt, \tag{2.25}$$

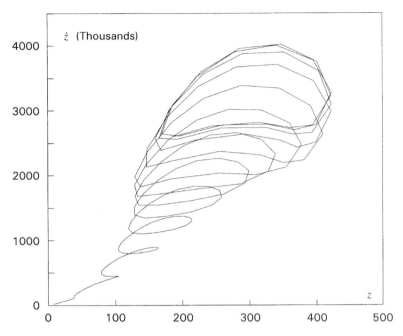

FIGURE 2.15　Phase plot of $\dot{z}$ versus z, Example 2.9.

where $\Gamma(t) = \{t | T(t) \geq T_0\}$ is the set where the temperature exceeds the threshold. Clearly, from this definition, the physiological time is the cumulative heat units summed over time. Thus, the typical unit for $\tau(t)$ is the degree-day. Differentiating Equation (2.24),

$$\frac{d\tau}{dt} = \begin{cases} 0, & T(t) < T_0, \\ T(t) - T_0, & T(t) \geq T_0, \end{cases}$$

from which $\dot{x}/x = r\, d\tau/dt$, leading immediately to

$$x(\tau) = x_0 e^{r\tau}. \tag{2.26}$$

The importance of Equation (2.26) is that the exponential growth model described earlier for chronological time also holds true for physiological time. In fact, it has been shown that all growth models discussed earlier, including the logistic and Lotka–Volterra models, behave in the same way. By defining the physiological time τ by Equation (2.25), the subsystem "sees" time τ as its natural time base. Thus, we have derived a sort of "relativity" principle where the external observer referenced in the master system sees non-standard behavior, but the subsystem sees internal events in the predicted manner. This is summarized in Table 2.4, where suitable activity functions are shown for each growth model considered earlier.

```
t=t₀
x=x₀
y=y₀
z=z₀
m=(tₙ-t₀)/(hn)
print t, x, y, z
for i=1 to n
        for j=1 to m
```
$$K_{x1}=3x(1-\tfrac{1}{10}y+\tfrac{1}{15}z)$$
$$K_{y1}=1.2y(1+\tfrac{1}{20}x-\tfrac{1}{25}z)$$
$$K_{z1}=2z(1-\tfrac{1}{30}x+\tfrac{1}{35}y)$$

$$K_{x2}=3(\tfrac{1}{2}x+hK_{x1})(1-(y+hK_{y1}/2)/10+(\tfrac{1}{2}z+hK_{z1})/15)$$
$$K_{y2}=1.2(\tfrac{1}{2}y+hK_{y1})(1+(x+hK_{x1}/2)/20-(\tfrac{1}{2}z+hK_{z1})/25)$$
$$K_{z2}=2(\tfrac{1}{2}z+hK_{z1})(1-(x+hK_{x1}/2)/30+(\tfrac{1}{2}y+hK_{y1})/35)$$

$$K_{x3}=3(\tfrac{1}{2}x+hK_{x2})(1-(\tfrac{1}{2}y+hK_{y2})/10+(\tfrac{1}{2}z+hK_{z2})/15)$$
$$K_{y3}=1.2(\tfrac{1}{2}y+hK_{y2})(1+(\tfrac{1}{2}x+hK_{x2})/20-(\tfrac{1}{2}z+hK_{z2})/25)$$
$$K_{z3}=2(\tfrac{1}{2}z+hK_{z2})(1-(\tfrac{1}{2}x+hK_{x2})/30+(\tfrac{1}{2}y+hK_{y2})/35)$$

$$K_{x4}=3(x+hK_{x3})(1-(y+hK_{y3})/10+(z+hK_{z3})/15)$$
$$K_{y4}=1.2(y+hK_{y3})(1+(x+hK_{x3})/20-(z+hK_{z3})/25)$$
$$K_{z4}=2(z+hK_{z3})(1-(x+hK_{x3})/30+(y+hK_{y3})/35)$$

$$x=x+h\tfrac{1}{6}(K_{x1}+2K_{x2}+2K_{x3}+K_{x4})$$
$$y=y+h\tfrac{1}{6}(K_{y1}+2K_{y2}+2K_{y3}+K_{y4})$$
$$z=z+h\tfrac{1}{6}(K_{z1}+2K_{z2}+2K_{z3}+K_{z4})$$
```
                t=t+h

        next j
```
$$x_1=3x(1-y\tfrac{1}{10}+z/15)$$
$$y_1=1.2y(1+x\tfrac{1}{20}-z/25)$$
$$z_1=2z(1-x\tfrac{1}{30}+y/35)$$
```
        print t, x, x₁, y, y₁, z, z₁
next i
```

LISTING 2.9   Simulation for the Lotka–Volterra system of Example 2.9.

It should be noted that the chronological description of each model is the same as in the previous section except for the factor of $d\tau/dt$, and that these reduce to the traditional model if $\tau = t$. The physiological descriptions are the same as the classical ones, except that the systems "see" the physiological time τ rather than the chronological time t.

TABLE 2.4 Chronological-Time-Based Heat-Unit Models for Various Population Systems

Model name	Chronological-time description	Physiological-time description
Malthusian	$\dfrac{\dot{x}}{x} = r\dfrac{d\tau}{dt}$	$\dfrac{dx}{d\tau} = rx(\tau)$
Logistic	$\dfrac{\dot{x}}{x(1 - x/x_m)} = r\dfrac{d\tau}{dt}$	$\dfrac{dx}{d\tau} = rx(\tau)\left[1 - \dfrac{x(\tau)}{x_m}\right]$
Lotka–Volterra	$\dfrac{\dot{x}}{x(1 - y/x_m)} = r_1\dfrac{d\tau}{dt}$	$\dfrac{dx}{d\tau} = r_1 x(\tau)\left[1 - \dfrac{y(\tau)}{x_m}\right]$
	$\dfrac{\dot{y}}{y(-1 + x/y_m)} = r_2\dfrac{d\tau}{dt}$	$\dfrac{dy}{d\tau} = r_2 y(\tau)\left[-1 + \dfrac{x(\tau)}{y_m}\right]$

EXAMPLE 2.10

Consider an idealized model in which the temperature profile is given by the following formula:

$$T(t) = A \sin\left(\pi\frac{t - t_1}{t_4 - t_1}\right),$$

where A is the amplitude, and t_1 and t_4 are zero-crossing points. Find a formula for the physiological time, assuming a generic threshold T_0. From this, graph a series of physiological profiles for different thresholds.

Solution
It is convenient to define the constant $\rho = T_0/A$, which we assume here is such that $|\rho| < 1$. The cases where $|\rho| \geqslant 1$ are not usually encountered, even though they are not difficult to handle. A graph showing this profile and threshold is given in Figure 2.16.

Assuming that $-A < T_0 < A$, which is equivalent to saying $|\rho| < 1$, observe that the temperature profile $T(t)$ exceeds the threshold T_0 only between points t_2 and t_3. These two critical points can be calculated by solving

$$A \sin\left(\pi\frac{t - t_1}{t_4 - t_1}\right) = T_0$$

for t. This gives the two values:

$$t_2 = t_1 + \frac{t_4 - t_1}{\pi}\sin^{-1}\rho, \qquad t_3 = t_4 - \frac{t_4 - t_1}{\pi}\sin^{-1}\rho.$$

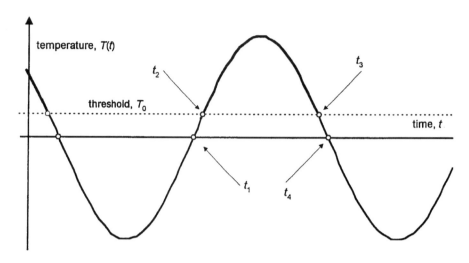

FIGURE 2.16   Temperature profile and threshold for Example 2.10.

Defining the argument as $\psi(t) = \pi(t - t_1)/(t_4 - t_1)$, it can be shown that

$$\sin \psi(t_2) = \rho,$$
$$\sin \psi(t_3) = \rho,$$
$$\cos \psi(t_2) = \sqrt{1 - \rho^2},$$
$$\cos \psi(t_3) = -\sqrt{1 - \rho^2}.$$

From the definition of physiological time, if $t < t_2$ then $\tau(t) = 0$. For $t_2 \leqslant t \leqslant t_3$, the physiological time is the cumulative temperature only over those times between t_2 and t_3. Thus,

$$\tau(t) = \int_{t_2}^{t} \left[A \sin \left(\pi \frac{t - t_1}{t_4 - t_1} \right) - T_0 \right] dt$$

$$= \frac{A(t_4 - t_1)}{\pi} [\cos \psi(t_2) - \cos \psi(t)] - T_0(t - t_2)$$

$$= \frac{A(t_4 - t_1)}{\pi} [\cos \psi(t_2) - \cos \psi(t)] - T_0(t - t_1) + \frac{T_0(t_4 - t_1)}{\pi} \sin^{-1} \rho$$

$$= \frac{A(t_4 - t_1)}{\pi} [\sqrt{1 - \rho^2} + \rho \sin^{-1} \rho - \cos \psi(t) - \rho \psi(t)].$$

For $t > t_3$, physiological time stands still since the temperature remains below the threshold. Thus,

$$\tau(t) = \tau(t_3) = \frac{A(t_4 - t_1)}{\pi} (2\sqrt{1 - \rho^2} + 2\rho \sin^{-1} \rho - \pi \rho).$$

Thus, in general,

$$\tau(t) = \begin{cases} 0, & t < t_2, \\ \dfrac{A(t_4 - t_1)}{\pi}(\sqrt{1 - \rho^2} + \rho \sin^{-1} \rho - \rho\psi - \cos \psi), & t_2 \leqslant t \leqslant t_3, \\ \dfrac{A(t_4 - t_1)}{\pi}(2\sqrt{1 - \rho^2} + 2\rho \sin^{-1} \rho - \pi\rho), & t > t_3, \end{cases} \quad (2.27)$$

where

$$\rho = \frac{T_0}{A}, \qquad \psi(t) = \pi \frac{t - t_1}{t_4 - t_1}.$$

Using a nominal mid-latitude annual cycle as a specific example, graphs of the physiological time $\tau(t)$ versus chronological time t are plotted in Figure 2.17. The values $t_1 = 90$ days, $t_4 = 270$ days, and $A = 100°F$ are fixed, and ρ varies from 0.0 to 0.8 in steps of 0.2. This establishes temperature thresholds of $T_0 = 0, 20,$ 40, 60, and 80°F. This model clearly shows how, for higher thresholds, both more heat units are generated and the initial growth time begins earlier.   ○

According to Equation (2.26), the formulas in Table 2.4 can be simulated using the new degree-time units. The only change is that now the integration step size changes over time. Thus, if the physiological time is available as an explicit function, say *physio(t)*, and the chronological step size is δ, the update of a logistic model is

```
h=δ[physio(t+δ)-physio(t)]
x=x[1+ah(1-x/xm)]
t=t+δ
print t, x
```

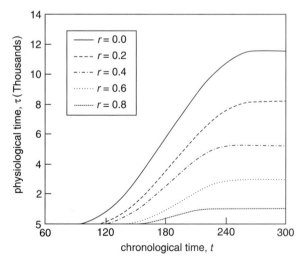

FIGURE 2.17   Physiological time τ versus chronological time t for the temperature profile in Figure 2.16.

It is also possible to implement a simulation using a subroutine. Assuming that *physio*(t, τ) is a subroutine that accepts chronological time t and computes the physiological time τ, the logistic update is now

```
call physio(t,τ₁)
call physio(t+δ,τ₂)
h=δ(τ₁-τ₂)
x=x[1+ah(1-x/xₘ)]
t=t+δ
print t, x
```

where $h = \delta(\tau_1 - \tau_2)$ is the number of degree-days between successive chronological times t and $t + \delta$.

As would be expected, the threshold temperature plays a significant role in heat-unit models. The higher T_0 is, the later growth begins and the less growth there is overall.

EXAMPLE 2.11

Consider the logistic model defined by

$$\dot{x} = 5x(1 - \tfrac{1}{25}x),$$
$$x(0) = 5.$$

Write a simulation program and sketch the population-versus-time graph over one year's time, assuming the heat-unit model with temperature profile

$$T(t) = 100 \sin\left(\pi \frac{t - 90}{180}\right).$$

Solution
This temperature profile is the one used in Example 2.10 with $t_1 = 90$, $t_4 = 270$, and $A = 100$. Applying Equation (2.27), a suitable function is given in Listing 2.10. Of course, the variables t_2, t_3, and ρ, which are computed in the main program, must be defined as global variables for the subroutine to know their values without parameters being passed. For a given threshold T_0, the main program proceeds algorithmically much as previously. Listing 2.11 creates a one-year simulation using function physio(t) as required.

```
function physio(t)
      if t<t₂ then physio=0
      if t₂⩽t⩽t₃ then physio=5730[√(1-ρ²)+ρsin⁻¹ρ-(ρ(t-90)/57 - cos((t-90)/57))]
      if t>t₃ then physio=5730(2√(1-ρ²)+2ρsin⁻¹ρ-πρ)
return
```

LISTING 2.10   The function *physio* for the logistic heat-unit hypothesis model, Example 2.11.

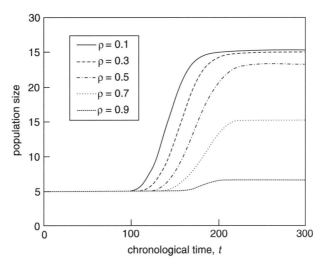

FIGURE 2.18   Time plot of results for the logistic heat-unit hypothesis model, Example 2.11.

Figure 2.18 is a graph of this system over [60,270], using $\rho = 0.1$, 0.3, 0.5, 0.7, and 0.9. It should be noted that in order to produce this family of trajectories as shown, a structure inversion would need to be introduced. Specifically, Listing 2.11 would need to be embedded in a loop: for $\rho = 0.1$ to 0.9 step 0.2; and the population stored as a subscripted variable. The two loops would then be inverted for printing.   ○

In predator–prey models, each species can have a different threshold. If this is the case, the function *physio* now has two variables: *physio*(t, ρ). As in physio(t), t is the chronological time and $\rho = T_0/A$ is the relative threshold characteristic of a specific species. The simulation algorithm is best shown by example.

```
t=0
x=5
ρ=T₀/100
t₂=90+116 sin⁻¹(ρ)
t₃=270-116 sin⁻¹(ρ)
print t, x
for k=1 to 365
        h=δ[physio(t+δ) - physio(t)]
        t=t+δ
        x=x[1+5h(1-x/25)]
        print t, x
next k
```

LISTING 2.11 Main program for the logistic heat-unit hypothesis model, Example 2.11.

EXAMPLE 2.12

Consider the following Lotka–Volterra system:

$$\dot{x} = 0.005x(1 - \tfrac{1}{10}y),$$
$$\dot{y} = 0.15y(-1 + \tfrac{1}{15}x),$$
$$x(0) = 10,$$
$$y(0) = 8.$$

Solve this system numerically using the heat-unit hypothesis, with the temperature profile as defined in Examples 2.10 and 2.11. Assume that population x uses temperature threshold 50°F and population y uses 20°F. Obviously, y gets quite a head start!

Solution
In this case, each population reacts differently, depending on the threshold temperature. Accordingly, the new function is *physio(t, ρ)*, where ρ is calculated in the main program and passed to the *physio* function. In this way, the physiological time for each species will be computed locally inside the function rather than externally in the main program. A partial listing is shown in Listing 2.12.

The results are shown in Figures 2.19 and 2.20. Figure 2.19 gives the population-versus-time graph where both cyclic behavior and seasonal temperature effects are evident. Figure 2.20 is a phase plot demonstrating the periodic nature of the system.

```
t=0
x=10
y=8
print t, x, y
for k=1 to 365
        h=δ[physio(t+δ,0.5) - physio(t,0.5)]
        x=x[1+0.005h(1-y/10)]
        h=δ[physio(t+δ,0.2) - physio(t,0.2)]
        y=y[1+0.15h(-1+x/15)]
        t=t+δ
        print t, x, y
    next k
```

LISTING 2.12 Main program for the Lotka–Volterra heat-unit hypothesis model, Example 2.12.

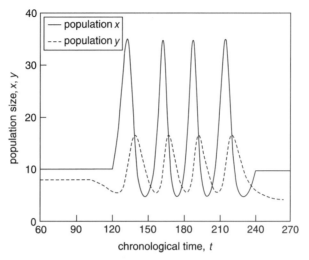

FIGURE 2.19   Time plot of results for the Lotka–Volterra heat-unit hypothesis model, Example 2.12.

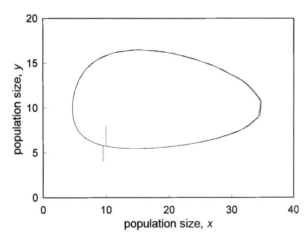

FIGURE 2.20   Phase plot of results for the Lotka–Volterra heat-unit hypothesis model, Example 2.12.

2.5   HANDLING EMPIRICAL DATA

Even though models are idealizations of phenomena, it is often desirable to apply real data to the model to produce a realistic simulation. In the case of a logistic model, we assume that populations grow in accordance with the equation

$$\frac{d\tau}{dt} = ax\left(1 - \frac{x}{x_m}\right).$$

However, in actually using this model, a real temperature profile can be obtained by taking physical measurements. These measurements can lead to much more realistic results. Thus, we choose not to go with a formula such as the one developed in Example 2.10, but rather to use an empirically gathered set of $n+1$ points such as those tabulated below. While the values of $T(t)$ are not known between adjacent data points, it is possible to perform a linear interpolation between the $n+1$ points with n line segments.

t	$\hat{T}(t)$
t_0	$\hat{T}_0$
t_1	$\hat{T}_1$
t_2	$\hat{T}_2$
$\vdots$	$\vdots$
t_n	$\hat{T}_n$

Consider the function segment in Figure 2.21, showing the two adjacent data points $(t_k, \hat{T}_k)$ and $(t_{k+1}, \hat{T}_{k+1})$. A generic point (t, T) on the adjoining line segment defines similar triangles, the slopes of which are equal:

$$\frac{T - \hat{T}_k}{t - t_k} = \frac{\hat{T}_{k+1} - \hat{T}_k}{t_{k+1} - t_k}.$$

Solving for $T(t)$,

$$T(t) = \frac{\hat{T}_{k+1} - \hat{T}_k}{t_{k+1} - t_k}(t - t_k) + \hat{T}_k \tag{2.28}$$

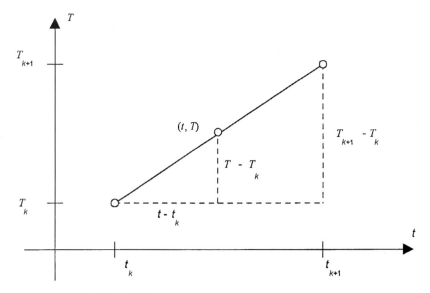

FIGURE 2.21   Linear interpolation.

Equation (2.28) defines estimated signal values for all times t, even those that were not defined in the original data set. However, at the data points themselves, the function matches the empirical value $T(t_k) = \hat{T}_k$.

EXAMPLE 2.13

Consider the following data set, which is taken to be an approximation to a temperature profile. Apply linear interpolation to the data set. From this, define the piecewise-linear function that connects the points. Also write a subroutine *interp*(t, T) that accepts the variable t and computes the interpolated variable T.

k	t_k	$\hat{T}_k$
0	0	0
1	1	5
2	2	8
3	4	-6
4	5	-6
5	7	6
6	8	3
7	9	-4

Solution

The temperature profile approximated by the data is graphed in Figure 2.22. Notice that, without further information, there is no basis for extrapolation (estimation of the profile beyond the points given), and therefore the approximation is only defined on the interval $[0, 9]$.

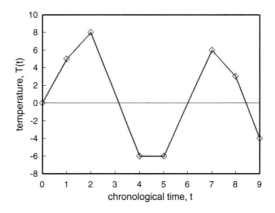

FIGURE 2.22   Empirical temperature profile for Example 2.13.

```
function interp(t)
          if t<0 then "undefined"
          if 0 ≤ t<1 then interp=5t
          if 1 ≤ t<2 then interp=3t+2
          if 2 ≤ t<4 then interp=-7t+22
          if 4 ≤ t<5 then interp=-6
          if 5 ≤ t<7 then interp=6t-36
          if 7 ≤ t<8 then interp=-3t+27
          if 8 ≤ t<9 then interp=-7t+59
          if t>9 then "undefined"
     return
```

LISTING 2.13   The function *interp* for Example 2.13.

A straightforward application of Equation (2.28) provides the following temperature profile estimator:

$$
T(t) = \begin{cases}
\text{undefined}, & t < 0, \\
5t, & 0 \leqslant t < 1, \\
3t + 2, & 1 \leqslant t < 2, \\
-7t + 22, & 2 \leqslant t < 4 \\
-6, & 4 \leqslant t < 7, \\
6t - 36, & 5 \leqslant t < 7, \\
-3t + 27, & 7 \leqslant t < 8, \\
-7t + 59, & 8 \leqslant t \leqslant 9, \\
\text{undefined}, & t > 9.
\end{cases}
$$

A function $interp(t)$ that computes the interpolated estimate of T for any chronological time t is given in Listing 2.13.   ○

Example 2.13 can be misleading, since often empirical data comprise much larger data sets than just eight points. In these cases, it is customary to store the $(t, \hat{T})$ ordered pairs as two vectors of length $n + 1$: $[t(0), t(1), \ldots, t(n)]$ and $[\hat{T}(0), \hat{T}(1), \ldots, \hat{T}(n)]$, where the t-vector is stored in ascending order and the $\hat{T}$-vector is sorted based on the t-values. If the data sets are not too long, a simple linear search can be performed to recover the adjacent points surrounding time t and a linear interpolation performed. Such a function is given in Listing 2.14. If the data set is too large or algorithm speed is important, a binary search is preferable.

Of course, the goal is to obtain results for the population model, not just to find the temperature profile. This requires one more function, named $physio(t)$, which calculates the physiological time τ corresponding to chronological time t. The $physio$ subroutine is called by the main program, which runs the actual simulation. This is best illustrated by a final example.

```
function interp(t)
      input t
      if t<t(0) or t>t(n) then
            "error"
      else
            k=1
            while t>t(k) and t⩽t(n)
                  k=k+1
            end while
            interp=(T̂_{k+1} − T̂_k)/(t_{k+1} − t_k) (t-t_k) +T̂_k
      end if
return
```

LISTING 2.14 A generalized linear interpolation function, *interp*.

EXAMPLE 2.14

Consider the following logistic model of a population system driven by physiological time τ:

$$\dot{x} = 0.8x(1 - \tfrac{1}{25}x),$$
$$x(0) = 1.$$

Assume that the temperature profile is given empirically by the data set of Example 2.13 and use a temperature threshold of $T_0 = 1$. Write appropriate code solving this system and graph the results compared with the same system driven by chronological time t.

Solution
This solution will require a program structure similar to that shown in Figure 2.23. The main program obtains the chronological time τ by passing the chronological time t to the function *physio(t)*. The function *physio(t)* obtains the interpolated temperature T by passing time t on to the *interp(t)* subroutine described in Example 2.13.

The function *physio* uses Eulerian integration, as shown in Listing 2.15. Notice that it requires chronological step size δ, which is a global variable common to all modules. Since there is only one species in this example, the threshold T_0 is also assumed common. On the other hand, if this were a Lotka–Volterra mode, T_0 would need to be passed into each subroutine so that the chronological time could be computed locally.

The main program is straightforward, and is given in Listing 2.16. The results are shown in Figure 2.24. Notice the distinctly different results between the classical and heat-unit models. It is clear that the heat-unit hypothesis is significant. Also, the bimodal nature of the temperature profile provides an interesting effect on the population growth.

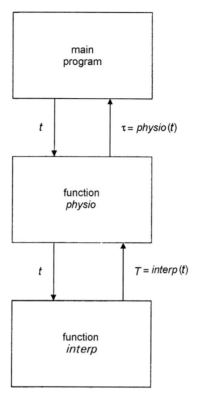

FIGURE 2.23   Communication structure, Example 2.14.

```
function physio(t)
        if t<0 then "error"
        τ=0
        for u=0 TO t step δ
                v=interp(u)
                if v>T₀ then τ=τ+δ(v-T₀)
        next u
    return
```

LISTING 2.15 The function *physio*, Example 2.14.

Creating and simulating models is a multistep process. Except for only the most ideal systems, this requires many iterations before achieving useful results. This is because real phenomena that we can observe exist in the messy environment we call reality. For instance, taking a cue from the Lotka–Volterra predator–prey models, organisms can exist either in isolation or in cooperative–competitive situations. However, the models in each of these cases can be considerably different. For the scientist, this is important – knowledge of the system behavior over a large parametric range for all interactions is paramount to

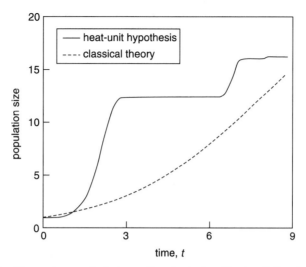

FIGURE 2.24   Comparing results between the classical and heat-unit theories, Example 2.14.

```
t=t₀
x=x₀
print t, x
for k=1 to n
        h=δ[physio(t+δ) - physio(t)]
        x=x[(1+ah(1-x/xₘ)]
        t=t+δ
        print t, x
next k
```

LISTING 2.16   The main program for Example 2.14.

understanding. However, for the engineer, a limited dynamic range with an isolated entity is often acceptable, since he can simply agree at the outset not to exploit extremes.

Regardless of the point of view – scientific or engineering – modeling and simulation studies tend to include the following steps and features:

1. *Define the model form.* This includes isolating bounds between what will and will not be included in the model. Equations describing modeled entities are developed here. This is the inspirational and insightful part of the process.
2. *Model the historical inputs and outputs.* This step will assist in:
   (a) *System identification.* This is where the model's constants and parameters are evaluated.
   (b) *Model evaluation.* Before putting the model to use, testing and validation are essential. This will often require statistical studies so as to infer reasonable empirical signals.

    3. *Model future inputs*. In order to use the model, realistic data are needed.

    4. *Create a set of questions to be posed*. This list should include stability studies, transient and steady-state behavior, as well as time series prediction and other questions.

Notice that there are really two models: the system model as discussed in this chapter and the input–output signal models, which often have strong stochastic characteristics. These will be discussed in future sections of the text. Closely related to modeling are the simulations. Simulations are essential for the system identification, model evaluation, and questions to be posed.

BIBLIOGRAPHY

Bronson, R., *Theory and Problems of Modern Introductory Differential Equations* [Schaum's Outline Series]. McGraw-Hill, 1973.

Burden, R. L., J. Douglas Faires and S. Reynolds, *Numerical Analysis*, 2nd edn. Prindle, Weber & Schmidt, 1981.

Close, C. M. and Frederick, D. K., *Modeling and Analysis of Dynamic Systems*, 2nd edn. Wiley, 1994.

Conte, C. M., S. K. Dean and C. de Boor, *Elementary Numerical Analysis*, 3rd edn. McGraw-Hill, 1980.

Drazin, P. G., *Nonlinear Systems*. Cambridge University Press, 1992.

Fausett, L. V., *Applied Numerical Analysis Using MATLAB*. Prentice-Hall, 1999.

Johnson, D. E., J. R. Johnson and J. L. Hilburn, *Electric Circuit Analysis*, 3rd edn. Prentice-Hall, 1992.

Juang, J.-N., *Applied System Identification*. Prentice-Hall, 1994.

Kalmus, H., *Regulation and Control in Living Systems*. Wiley, 1966.

Khalil, H. K., *Nonlinear Systems*. MacMillan, 1992.

Mesarovic, M. D., *Systems Theory and Biology*. Springer-Verlag, 1968.

Neff, H. P., *Continuous and Discrete Linear Systems*. Harper and Row, 1984.

Press, W. H., B. P. Flannery, S. A. Teukolsky, and W. T. Vettering, *Numerical Recipes: The Art of Scientific Computing*. Cambridge University Press, 1986.

Rosen, R., *Dynamical System Theory in Biology*. Wiley Interscience, 1970.

Scheid, F., *Theory and Problems of Numerical Analysis* [Schaum's Outline Series]. McGraw-Hill, 1968.

Thompson, J. R., *Simulation: A Modeler's Approach*. Wiley Interscience, 2000.

Yakowitx, S. and F. Szidarovszky, *An Introduction to Numerical Computations*. MacMillan, 1986.

EXERCISES

2.1 Consider the following system of linear differential equations with initial conditions $\alpha(1) = 0$, $\dot{\alpha}(1) = 2$, $\beta(1) = -1$, $\dot{\beta}(1) = 0$, $\ddot{\beta}(1) = 4$.

$$\ddot{\alpha} + 2\alpha - 4\alpha = 5, \ \dddot{\beta} + 3\ddot{\beta} + \beta = 2, \ \alpha(1) = 2,$$

(a) Write this system as five first-order differential equations along with their initial conditions.
(b) Express this system as a single linear equation of the form $\dot{\mathbf{x}} = A\mathbf{x} + \mathbf{B}$, $\mathbf{x}(0) = \mathbf{x}_0$ for a suitably defined state vector $\mathbf{x}(t)$.

2.2 A vibrating string can be described using the following second-order differential equation:

$$\frac{W}{g}\ddot{\alpha}(t) + b\dot{\alpha}(t) + k\alpha(t) = f(t),$$

where $\alpha(t)$ is the displacement as a function of time, and W, g, b, and k are physical constants depending on the medium and other factors. The function $f(t)$ is the input that drives the system.
(a) Express this as a system of two first-order differential equations.
(b) Write the equations of part (a) as a single linear equation of the form $\dot{\mathbf{x}} = \mathbf{A}\mathbf{x} + \mathbf{B}f$ for a suitably defined state vector $\mathbf{x}(t)$.

2.3 Consider the following first-order initial-value problem:

$$\frac{dx}{dt} = \cos 2t + \sin 2t, \qquad x(0) = 2.$$

(a) Derive an explicit analytical solution for $x(t)$.
(b) Write a program by which to solve this system using the Euler integration method. Specifically, print 31 points on the interval $[0,6]$, where each point has m calculation iterations.
(c) Using the program created in part (b), run simulations with $m = 1, 2, 5, 10$.
(d) Compare your results by
(i) Superimposing the time plots for parts (a) and (c).
(ii) Computing the RMS error between the actual and approximated solution for each case.

2.4 Consider the differential equation $\ddot{x} = 2\dot{x} - 8x = e^t$, subject to the initial conditions $x(0) = 1$ and $\dot{x}(0) = -4$.
(a) Rewrite this as a system of two first-order initial-value problems.
(b) Show that the explicit solution is given by

$$x(t) = \tfrac{1}{30}(3e^{2t} - 6e^t + 31e^{-4t}).$$

(c) Write a program to approximate a solution on the interval $[0,2]$ using Euler's method. Execute using the following integration step sizes: $h = 0.1$, 0.5 and 0.01.
(d) Compare your results by overlaying the graphs of parts (b) and (c) on the same plot.

2.5 The voltage across a capacitor in a simple series RC circuit with voltage source $v_s(t)$ follows the relationship

$$v_s(t) = v(t) + RC\frac{dv}{dt}$$

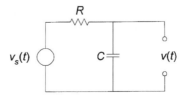

(a) Show that the explicit solution for $v(t)$ is given by

$$v(t) = v(0)e^{-t/RC} + \frac{1}{RC}e^{-t/RC}\int_0^t v_s(t)\, dt$$

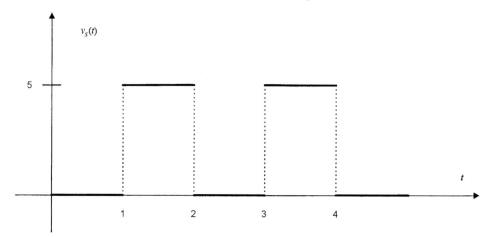

(b) If $v(0) = 5$ volts, $RC = 2$, and $v_s(t)$ is a pulse function as shown in input waveform as sketched above, show that the analytical solution is given by

$$v(t) = \begin{cases} 5e^{-t}, & 0 \leqslant t < 1, \\ 5te^{-t}, & 1 \leqslant t < 2, \\ 10e^{-t}, & 2 \leqslant t < 3, \\ 5(t-1)e^{-t/2}, & 3 \leqslant t < 4, \\ 15e^{-t/2}, & 4 \leqslant t \leqslant 5. \end{cases}$$

(c) Using a value of $h = 0.1$, numerically solve for $v(t)$ using Euler.

(d) Compare the results of parts (b) and (c).

2.6 Consider a system of two first-order differential equations in the following form:

$$\dot{x} = f(t, x, y),$$
$$\dot{y} = g(t, x, y).$$

(a) Show that the second-order Taylor approximation is given by

$$x(k+1) = x(k) + hf + \tfrac{1}{2}h^2\left(\frac{\partial f}{\partial t} + f\frac{\partial f}{\partial x} + g\frac{\partial f}{\partial y}\right),$$

$$y(k+1) = y(k) + hg + \tfrac{1}{2}h^2\left(\frac{\partial g}{\partial t} + f\frac{\partial g}{\partial x} + g\frac{\partial g}{\partial y}\right).$$

(b) Apply the results of part (a) using an integration step size of $h = 0.1$ over the interval $[0, 5]$ to the initial-value problem

$$\dot{x} = xy + 2t, \quad \dot{y} = x + 2y + \sin t, \quad x(0) = 1, \quad y(0) = 0.$$

(c) Apply Euler's formula using the same h-value. Graph and compare.

2.7 Show that if $\dot{x}(t) = f(t, x)$, it follows that

$$\ddot{x}(t) = \frac{\partial^2 f}{\partial t^2} + 2f\frac{\partial^2 f}{\partial x \partial t} + \frac{\partial f}{\partial t}\frac{\partial f}{\partial x} + f\left(\frac{\partial f}{\partial x}\right)^2 + f^2\frac{\partial^2 f}{\partial x^2}.$$

Thus, derive Taylor's third-order formula, which is analogous to Equation (2.10). Apply this to the system

$$\dot{x}(t) = x^2 t,$$
$$x(1) = 3.$$

(a) Using $h = 0.05$ over the interval $1 \leqslant t \leqslant 1.3$, compare your results using the second-order Taylor and the fourth-order Runge–Kutta in Table 2.3.

(b) Make a study of the results of the third-order Taylor for different h-values. You might try a geometric sequence of h-values as illustrated in the text. Justify your results.

2.8 Consider the following initial-value problem defined on the interval $[0, 10]$:

$$\dot{x} = 2x^2 + \cos t, \quad x(0) = 4.$$

Using an integration step size of $h = 0.1$:

(a) Find an approximate solution using the Euler integration method.

(b) Find an approximate solution using the second-order Taylor integration method.

(c) Find an approximate solution using the Runge–Kutta integration method.

(d) Compare your results by superimposing the time plots of each.

2.9 Write a program to implement the adaptive Runge–Kutta (Runge–Kutta–Fehlberg) integration technique outlined in Section 2.1 that prints the values of t, h, and x at the end of each iteration. Using an initial value of $h = 0.5$ and a tolerance $\varepsilon = 0.01$:

(a) Apply your program to the following nonlinear initial-value problem:

$$\dot{x} = 2x^2 + \cos t, \quad x(0) = 4.$$

(b) Graph the results $x(t)$ and h from part (a) versus t. Compare, and explain the correlation between features of the two graphs.

(c) Repeat the simulation for various different tolerances, $\varepsilon = 0.5, 0.1, 0.05, 0.01, 0.005, 0.001$. For each tolerance, graph the time plot of $x(t)$ versus t. Superimpose the results.

(d) For each simulation experiment of part (c), graph the step size versus t. Superimpose the results.

(e) Explain the features of the graphs produced in parts (d) and (e).

2.10 Consider the initial-value problem

$$\ddot{x} + 2\dot{x} + 5y = 0,$$
$$\dot{x} + 2y = \dot{y},$$
$$x(0) = 0,$$
$$\dot{x}(0) = 0,$$
$$y(0) = 1.$$

(a) Rewrite this as a system of first order differential equations.

(b) Show that the exact solution is given by

$$x(t) = 2\cos t + 6\sin t - 2 - 6t,$$
$$y(t) = -2\cos t + 2\sin t + 3.$$

(c) Approximate the solution on $[0, 10]$ using $h = 0.1$ in Euler's procedure.

(d) Approximate the solution on $[0, 10]$ using $h = 0.1$ in Taylor's second-order procedure.

(e) Approximate the solution on $[0, 10]$ using $h = 0.1$ in the Runge–Kutta algorithm.

(f) Compare the results of parts (c), (d) and (e) against the exact solution.

2.11 Generalize the result of Problem 2.7 to the case of three first-order differential equations,

$$\dot{x} = \alpha(t, x, y, z),$$
$$\dot{y} = \beta(t, x, y, z),$$
$$\dot{z} = \gamma(t, x, y, z),$$

using:
(a) second-order Taylor, and
(b) fourth-order Runge–Kutta.

2.12 Consider the initial-value problem

$$\dddot{x} - 2t\ddot{x}^2 + 4\dot{x} = 1 + 2\cos t, \quad x(0) = 1, \quad \dot{x}(0) = 2, \quad \ddot{x}(0) = 3.$$

(a) Rewrite this as a system of first-order differential equations.
(b) Approximate the solution on $[0, 5]$ using Euler's procedure. Empirically determine an appropriate step size h; justify your choice.
(c) Using the results of Problem 2.11, approximate the solution on $[0, 5]$ using Taylor's second-order procedure. Empirically determine an appropriate step size h; justify your choice.
(d) Using the results of Problem 2.11, approximate the solution on $[0, 5]$ using the Runge–Kutta algorithm. Empirically determine an appropriate step size h; justify your choice.

2.13 Implement the adaptive Runge–Kutta (Runge–Kutta–Fehlberg) algorithm described in Listing 2.4.
(a) Apply this to the system discussed throughout this chapter:

$$\dot{x}(t) = x^2 t,$$
$$x(1) = 3.$$

(b) Using an initial $h = 1$ over the interval $1 \leqslant t \leqslant 1.28$, compare your results against the actual solution.
(c) Make a time–complexity study of the adaptive technique versus the Euler, Taylor, and non-adaptive Runge–Kutta procedures. Is there a relationship between each of these approaches and the time it takes for a given accuracy?

2.14 Consider the following generic second-order linear system with input $f(t)$ and output $x(t)$:

$$\ddot{x} + 2\zeta\omega\dot{x} + \omega^2 x = r(t),$$

where ζ and ω are constants.
(a) In the case where $r(t) = 0$, derive Equation (2.19).
(b) Find the general solution when $r(t) = 1$, $x(0) = 1$, and $\dot{x}(0) = 0$.
(c) Assuming that $x(0) = 2$ and $\dot{x}(0) = 0$, find $x(t)$ for each of the following special cases:
   (i) $\ddot{x} + 4\dot{x} + 4x = 1$,
   (ii) $\ddot{x} + 2\dot{x} + 2x = 2 + t^2$,
   (iii) $\ddot{x} + 3\dot{x} + 2x = 4\cos 2t$.

2.15 Show that the solution to the basic logistic equation

$$\frac{dx}{dt} = \lambda x \left(1 - \frac{x}{x_m}\right),$$

where $x(0) = x_0$, is given by

$$x(t) = \frac{x_m}{1 - (1 - x_m/x_0)e^{-\lambda t}}$$

2.16 Implement the generalized linear interpolation function *interp(t)* described in Listing 2.14. Verify the working results by writing a program to read the following tabulated data and graph the interpolated points on $[1, 11]$:

k	t_k	$\hat{T}_k$
0	1	3
1	3	4
2	5	7
3	7	5
4	9	3
5	11	0

2.17 Solve and graph the solution (both time plots and phase plots) of the following Lotka–Volterra system over $[0, 10]$:

$$\frac{dx}{dt} = 5x(1 - \tfrac{1}{6}y),$$
$$\frac{dy}{dt} = 2y(-1 + \tfrac{1}{7}x),$$
$$x(0) = 20,$$
$$y(0) = 20.$$

Use $h = 0.1$, with:
(a) second-order Taylor;
(b) fourth-order Runge–Kutta.

2.18 Consider the following Lotka–Volterra system:

$$\frac{dx}{dt} = 5x(1 - \tfrac{1}{6}y),$$
$$\frac{dy}{dt} = 2y(-1 + \tfrac{1}{7}x),$$
$$x(0) = 20,$$
$$y(0) = 20.$$

It is required to determine a reasonable h-value so as to make an accurate simulation of this model. Using the method of creating and analyzing a geometric sequence of h_i outlined in this chapter:
(a) Write a generic Runge–Kutta program to create the geometric sequence of h_i, run the simulation over $[0, 20]$ and compare the associated relative RMS error of adjacent simulations.
(b) Create tables similar to those in the text, and graph the relative RMS error as a function of h_i. From these, justify a choice of h.
(c) Plot the simulation using both time and phase plots for the chosen h.

2.19 Consider an ideal sinusoidal temperature profile of the form

$$T(t) = A \sin\left(\pi \frac{t - t_1}{t_4 - t_1}\right)$$

as described in Example 2.10.
(a) Write a program to plot the physiological time $\tau(t)$ given crossing times t_1 and t_4, amplitude A, and temperature threshold T_0.
(b) To validate the results, reproduce Figure 2.17.
(c) Convert the program of part (a) into a function *physio(t)*, assuming the threshold variable T_0 to be common to both the main program and *physio*.
(d) Convert the program of part (a) into a function *physio(t, ρ)* that passes the relative threshold ρ from *main* to *physio*.
(e) Verify proper working of the two subroutines. This will require careful scoping of variables t_1, t_4, A, and T_0.

2.20 Recall Example 2.9 in which there are three interacting populations x, y and z that behave according to the following Lotka–Volterra equations:

$$\dot{x} = 3x\left(1 - \frac{y}{10} + \frac{z}{15}\right)$$

$$\dot{y} = 1.2y\left(1 + \frac{x}{20} - \frac{z}{25}\right)$$

$$\dot{z} = 2z\left(1 - \frac{x}{30} + \frac{y}{35}\right)$$

$$x(0) = 5$$
$$y(0) = 10$$
$$z(0) = 8$$

However, now assume a heat unit hypothesis in which x, y and z have temperature thresholds of 45°, 40° and 50° respectively. Also assume a temperature profile of

$$T(t) = 100 \sin\left(\pi \frac{t - 90}{180}\right)$$

as described in Example 2.10. Using the function $physio(t, \rho)$ defined in problem 2.19 (c).
(a) Create a simulation of populations x, y and z over a year's time.
(b) Graph the time-doman results as an overlay showing all populations over time. This graph should be comparable to that shown in Figure 2.12.
(c) Graph the phase-space results for each population: x versus $\dot{x}$, y versus $\dot{y}$, and z versus $\dot{z}$. These should be comparable to those shown in Figures 2.13, 2.14 and 2.15.

2.21 Consider the following data:

k	t_k	$\hat{T}_k$
0	1	3
1	3	4
2	5	7
3	7	5
4	9	3
5	11	0

(a) From the table, derive an explicit formula $T(t)$ for the piecewise-linear function passing through each point.
(b) Write a linear interpolation function using the results of part (a).
(c) Using (b), graph $T(t)$ as a function of time t.

2.22 Consider the following temperature profile with 36 data points taken 10 days apart for one year:

	01	11	21
January	5	18	40
February	35	22	25
March	35	37	45
April	50	53	40
May	58	54	65
June	53	69	73

July	90	84	75
August	83	86	71
September	67	51	45
October	60	55	38
November	37	43	23
December	38	24	12

(a) Using the *interp* subroutine of Problem 2.16, graph the empirical temperature profile.

(b) Using the *physio* and *interp* functions of Problems 2.16, 2.19(c) and 2.21 along with the tabulated temperature data, simulate the following logistic model using the heat-unit hypothesis. Graph the population-versus-time results.

$$\dot{x} = 0.8x(1 - \tfrac{1}{25}x), \qquad\qquad\qquad x(0) = 1.$$

(c) Using the *physio* and *interp* functions of Problems 2.16, 2.19(d) and 2.21 along with the tabulated temperature data, simulate the following logistic model using the heat-unit hypothesis. Graph the population-versus-time results.

$$\dot{x} = 0.005x(1 - \tfrac{1}{10}y), \qquad \dot{y} = 0.15y(-1 + \tfrac{1}{15}x), \qquad x(0) = 10, \qquad y(0) = 8.$$

2.23 Investigate the orbital dynamics of a simple planetary system by simulating the Earth and Sun as point masses in a *two-body problem*. For convenience, assume that the orientation of the coordinate system is such that the initial velocity is only in the y direction.

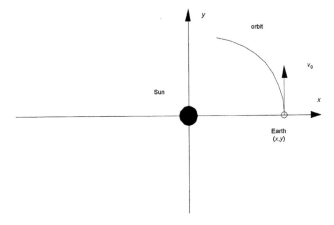

Recall that Newton's Laws of gravitation reduce to

$$\frac{d^2x}{dt^2} = \frac{-Gmx}{(x^2+y^2)^{3/2}},$$

$$\frac{d^2y}{dt^2} = \frac{-Gmy}{(x^2+y^2)^{3/2}},$$

(2.29)

where the position of the Earth is expressed in Cartesian coordinates (x, y), G is the universal gravitational constant, and m is the Sun's mass. Since this system is of order two and has two degrees of freedom, it is necessary to have four initial conditions. These are $x(0) = r$, $y(0) = 0$, $\dot{(0)} = 0$, and $\dot{y}(0) = v$, where r is the initial distance of the Earth from the Sun and v is the initial velocity of the Earth.

(a) Look up and compute (using units of kilometers and days) the quantities G, m, and r. Further, if the orbit is circular, $v^2 = Gm/r$. Compute v.

(b) Determine the step size by assuming an h that works as expected for a circular orbit is good enough throughout the entire problem. In particular, if the position at the end of 365 days is within 0.1% of the point at which we started, we'll be satisfied! Write a program to numerically integrate Equation (2.29) using the initial conditions described above. For various values of h, find one that is suitable.

(c) By increasing or decreasing the initial velocity, you will obtain the more general case of an elliptical orbit. The shape (as defined by the *eccentricity*) and *period* will change.

    (i) Increase the initial velocity by 10%, 20%, and 30%. Plot one orbit of the resulting trajectories. Repeat this by decreasing the initial velocity by the same amounts.

    (ii) For each of the resulting orbits in part (i), graphically determine the period P (the time that it takes to complete the orbit) and eccentricity ε, where

$$\varepsilon = \sqrt{1 - \left(\frac{\text{length of the minor axis}}{\text{length of the major axis}}\right)^2}.$$

(d) Using the results of part (c), plot a graph of P versus ε.

2.24 Investigate the dynamics of a simple planar pendulum by simulating the pendulum bob as a point mass in a uniform gravitational field. However, rather than make the traditional linearizing assumption, study the nonlinear case.

Recall that Newton's laws applied to a simple pendulum reduce to

$$\ddot{\theta} + \frac{g}{l}\sin\theta = 0,$$
$$\theta(0) = \theta_0, \tag{2.30}$$
$$\dot{\theta}(0) = 0,$$

where the angle θ of the pendulum is measured from the vertical. The mass of the bob is m and the length of the suspension is l. The constant g is the acceleration due to gravity.

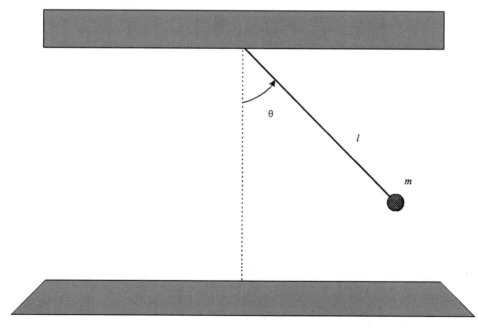

(a) Look up and compute (use units of kilometers and seconds) the value of g. Also obtain results pertaining to simple harmonic motion.

(b) Making the traditional linearizing assumption,

$$\frac{d^2\theta}{dt^2} + \frac{g}{l}\theta = 0, \tag{2.31}$$

derive the theoretical solution

$$\theta(t) = \theta_0 \cos\left(\sqrt{\frac{g}{l}}t\right).$$

(c) Write a program to integrate Equation (2.31). Determine an appropriate step size h by assuming that an h that works as expected for simple harmonic motion is good enough for the given initial angle. In particular, if the position at the end of one period is within 0.1% of the original point, we'll be satisfied!

(d) Write a program to numerically integrate the nonlinear case: Equation (2.30). Unlike the linear case, the period will now change as a function of the initial angle. Repeatedly change the initial angle θ_0 from $0°$ to $90°$ in steps of $5°$. Show the resulting $\theta(t)$ (y-axis) versus time t (x-axis) graphs. Plot three or four distinct trajectories.

(e) For each of the resulting curves of part (d), determine the period P graphically. Graph P as a function of initial angle θ_0. Repeat this for various ratios of g/l. Compare your results.

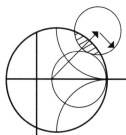

CHAPTER 3

Stochastic Generators

It is traditional to describe nature in deterministic terms. However, neither nature nor engineered systems behave in a precisely predictable fashion. Systems are almost always innately "noisy". Therefore, in order to model a system realistically, a degree of randomness must be incorporated into the model. Even so, contrary to popular opinion, there is definite structure to randomness. Even though one cannot precisely predict a next event, one can predict how next events will be distributed. That is, even though you do not know when you will die, your life insurer does – at least statistically!

In this chapter, we consider a sort of inverse problem. Unlike traditional data analysis where statistics (mean, standard deviation, and the like) are computed from given data, here we wish to generate a set of data having pre-specified statistics. The reason that we do this is to create a realistic input signal for our models. From historical studies, we can statistically analyze input data. Using these same statistics, we can then generate realistic, although not identical, scenarios. This is important, since the actual input signal will likely never be replicated, but its statistics will most likely still be stationary. For instance, in climatological models, the actual temperature profile will vary from year to year, but on average it will probably remain relatively unchanged. So, in modeling the effects of global warming, we change not only the data, but the statistical character of the data as well.

3.1   UNIFORMLY DISTRIBUTED RANDOM NUMBERS

Most language compilers have a facility to generate random numbers that are uniformly distributed on the interval $[0, 1]$. Such routines are called $U[0, 1]$ generators. Specifically, upon an RND call using the BASIC programming language, a fraction x will be returned, where $0 \leqslant x < 1$. Strictly speaking, this is a discrete random variable (by virtue of the digital nature of computers), but for practical purposes it can be assumed to be continuous. This means that in 100 RND function calls, roughly 10% will fall between 0.0 and 0.1, another 10% between 0.1 and 0.2, etc.

The majority of uniform random number generators are based on the *linear congruential generators* (LCG). These are actually deterministic, as is any algorithmic procedure, but for the most part they behave in a pseudorandom fashion. The LCG creates a sequence of numbers that appear unpredictable. An LCG requires an initial "seed" Z_0 to begin. This seed, and successive terms of the sequence Z_k, are recursively applied to an LCG formula. The Z_k are then normalized to produce output U_k, which are statistically uniform on the interval $0 \le U_k < 1$. Formally, the sequence is generated as follows:

$$Z_0 = \text{"seed"},$$
$$Z_{k+1} = (aZ_k + c)\bmod(m), \qquad (3.1)$$
$$U_k = \frac{Z_k}{m},$$

where a, c, and m are referred to as the *multiplier*, *increment*, and *modulus* respectively. The infix operator mod is the *modulo* function, and is defined to be the remainder formed by the quotient $(aZ_k + c)/m$. For example, $32\bmod(5)$ is 2, since on dividing 32 by 5, the remainder is 2.

Although this procedure seems rather cut-and-dried, this is not the case. For instance, much research has been done to determine "good" values for a, c, and m. Some produce better pseudorandom sequences than others.

EXAMPLE 3.1

Determine the sequence of numbers generated by the LCG with $a = 5$, $c = 3$, $m = 16$, and $Z_0 = 7$.

Solution
Using Equation (3.1) and rounding to three decimal places, the first random number is $U_0 = \frac{7}{16} \approx 0.437$. Also, Equation (3.1) recursively defines the sequence by $Z_{k+1} = (5Z_k + 3)\bmod(16)$. Since the priming seed is $Z_0 = 7$, $Z_1 = (5 \cdot 7 + 3)\bmod(16) = (38)\bmod(16) = 6$. It follows from this that $U_1 = \frac{6}{16} = 0.375$.
Similarly, letting $k = 1$ results in $Z_2 = 1$ and $U_2 = 0.062$. Continuing in this manner, Table 3.1 lists the first 16 numbers in sequence. After this point, the sequence repeats again and again. ○

Since Z_k is obtained as the remainder of a division by m, there are only m remainders possible. Thus, in Example 3.1, there are a maximum of 16 possible random numbers. Clearly, a large value of m is necessary for a useful sequence. In the event that m different random numbers occur for m repetitions, the LCG chosen (i.e., the one determined by a, b, and m) is said to have *full period*. This is because once a Z_k repeats, the entire cycle follows. The LCG in Example 3.1 has a full period of 16.

While the LCG algorithm is stated recursively by Equation (3.1), it can also be formulated explicitly. It is left to the student to verify that the explicit formula

$$Z_k = \left[a^k Z_0 + \frac{c(a^k - 1)}{a - 1} \right] \bmod(m) \qquad (3.2)$$

results in the same sequence. This further demonstrates the deterministic nature of pseudorandom numbers generated by a LCG.

TABLE 3.1   Pseudorandom
Sequence Generated by the LCG
of Example 3.1

k	Z_k	U_k
0	7	0.437
1	6	0.375
2	1	0.062
3	8	0.500
4	11	0.688
5	10	0.625
6	5	0.313
7	12	0.750
8	15	0.938
9	14	0.875
10	9	0.563
11	0	0.000
12	3	0.188
13	2	0.125
14	13	0.813
15	4	0.250

An especially useful result regarding the choice of parameters is summarized by the Hull–Dobell theorem. This theorem is powerful, since it provides both necessary and sufficient conditions in order that full period be attained.

Hull–Dobell Theorem   *The LCG has full period if and only if the following three conditions hold:*

  (i) a and c are relatively prime.
  (ii) All prime numbers q that divide m also divide a − 1.
  (iii) If 4 divides m, then 4 also divides a − 1.

The proof of this theorem is beyond the scope of this text. However, one can quickly apply the results. Example 3.1 provides one such illustration. Condition (i) holds, since 5 and 3 (a and c) are relatively prime. Since $m = 16 = 2^4$, the only prime number dividing m is 2. Two also divides $a − 1 = 4$; so much for condition (ii). Four also divides 16, but then it also divides $a − 1 = 4$. All conditions are met, thereby implying this LCD is full period, as was demonstrated in the listing of Table 3.1.

Serious simulations require a large number of non-repeating random numbers. Again, we cannot tolerate a repetition, since once a pseudorandom number repeats, the whole sequence also repeats from that point on. A computer implementation should ideally handle this algorithm at the hardware level, since the resulting speedups are needed in the computationally intensive business of computer simulation. This can be done if m is a power of 2, preferably the machine word size. Using an elementary shift register, a simple but effective technique for doing this is demonstrated in the following example.

EXAMPLE 3.2

Again consider the linear congruential generator of Example 3.1 with $a = 5$, $c = 3$, and $m = 16$. Since $2^4 = 16$, it is possible to represent the integer LCG numbers using a 4-bit shift register. This is because integer representation in computers is traditionally binary. So, the contents of register R can be represented as a 4-bit vector R: $[r_{-1}r_{-2}r_{-3}r_{-4}]$, where $r_{-j} = 0$ or 1 is the coefficient of 2^{4-j} in the binary representation, as illustrated in Figure 3.1.

For instance, since $Z_6 = 5$, the register R initially contains R: $[0101]$. To obtain Z_7, R must be replaced by $5Z_6 + 3$. This operation is accomplished in digital computers by a sequence of shifts (to multiply) and additions; in this case resulting in R: $[1 \quad 1100]$, which is 28 in decimal notation. However, the leading "1" is lost, since it overflows the 4-bit register, and only the R: $[1100]$, which is a decimal 12, remains. This has the effect of dividing 28 by 16 and retaining the remainder 12, which is by definition mod(16). Repeating, $R \leftarrow 5R + 3$: $[1111]$ produces $Z_8 = 15$, etc. A shift register can perform the required multiplication and addition operations with no problem.

However, since, in general, computer divisions require many clock cycles, one might expect that computing U_k would create a bottleneck. This is not the case. By assuming an implied binary point to the left of the leading zero, the divisions, which are right shifts using a shift register, are done with no cycles whatsoeever! In this example, note that $(0.1100)_2$ is 0.75 in base 10, as expected from Example 3.1.

In fact, this technique works for integers other than 16. In general a b-bit shift register performs the modulo operation for $m = 2^b$. By assuming the implied binary point in the leftmost bit position, the modulo and divide operations are performed in a single clock cycle! Thus, there is a real motivation to use LCGs with modulus chosen to accommodate the word size of the host machine.   ○

If the increment is set to $c = 0$, the LCG is called a *multiplicative generator*. In this special case, the Hull–Dobell theorem cannot be satisfied, since condition (i) fails because any factor of m automatically divides $c = 0$. However, it has been shown that by making an astute choice of multiplier a, it is possible to find multiplicative LCGs that have period $m - 1$. This, plus the fact that a significant amount of research has been done for the zero-increment case, makes many traditional generators multiplicative.

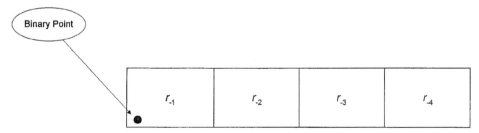

FIGURE 3.1   Machine representation of a 4-bit binary random number using an LCG.

There are a number of different generators used in real computers. The case of $a = 5^{15}$, $b = 36$, and $c = 0$ (Coveyoo and MacPherson) is used on the UNIVAC 1100 series machine, and has desirable statistical properties. The traditional RANDU generator used in IBM's old Scientific Subroutine Package uses $a = 2^{16} + 3$, $c = 0$, and $m = 2^{31}$. Unfortunately, this does not have full period and has poor statistical properties. IBM's newer entry, the Mathematics Subroutine Library (MSL), is far superior, with $a = 7^5$, $c = 0$, and $m = 2^{31} - 1$.

3.2   STATISTICAL PROPERTIES OF $U[0, 1]$ GENERATORS

In addition to having a large period and a modulus chosen for hardware computability, a $U[0, 1]$ generator must behave well in a statistical sense. Behaviors can be defined rather rigorously, but suffice it to say that two properties are most important:

(i) The generator must be uniform. That is, the proportion of numbers generated on any length interval L must be close to the proportion generated on all other intervals of length L.

(ii) The sequence must be independent. Specifically, any number should not appear to influence the next. In cases where this is not true, the sequence demonstrates clustering and gap tendencies.

There are both theoretical and empirical means by which to evaluate generators. Even so, since to prove a property is to convince, ultimately empirical demonstrations must be performed. We consider only empirical tests here.

In order to test for uniformity (property (i)), one can apply the chi-square test discussed in Appendix A; this is easily understood from the *frequency distribution table* (FDT) shown in Table 3.2. The entries of an FDT are as shown, but it should be noted that, ultimately, there is an association between each interval and the theoretical and empirical frequency of occurrences f_k and e_k. We begin by partitioning the interval [0, 1] into m subintervals: $[0, 1/m], [1/m, 2/m], \ldots, [(m-1)/m, 1]$, each of length $1/m$. By generating n random numbers and assigning each to one of the m classes, frequencies $f_1, f_2, \ldots, f_m$ can be tabulated. These must be compared against the expected frequencies $e_k = n/m$ for each class.

TABLE 3.2   Frequency Distribution Table

Interval number k	Interval	Empirical frequency f_k	Expected frequency e_k
1	$[0, 1/m]$	f_1	e_1
2	$[1/m, 2/m]$	f_2	e_2
3	$[2/m, 3/m]$	f_3	e_3
$\vdots$	$\vdots$ $\vdots$	$\vdots$	$\vdots$
m	$[(m-1)/m, 1]$	f_m	e_m

The chi-square statistic follows:

$$\chi^2 = \sum_{k=1}^{m} \frac{(f_k - e_k)^2}{e_k}$$

$$= \frac{m}{n} \sum_{k=1}^{m} \left(f_k - \frac{n}{m} \right)^2, \tag{3.3}$$

with $v = m - 1$ degrees of freedom. It will be recalled from Appendix A that in general the number of degrees of freedom, v, is given by $v = m - l - 1$, where m is the number of cells and l is the number of distribution parameters to be estimated. In this case, no parameters need be calculated, so $l = 0$ and $v = m - 1$.

EXAMPLE 3.3

A proposed $U[0, 1]$ generator called SNAFU is tested by generating 100 numbers and counting the frequencies in each of the following ranges:

$$0.00 \leqslant x < 0.25,$$
$$0.25 \leqslant x < 0.50,$$
$$0.50 \leqslant x < 0.75,$$
$$0.75 \leqslant x < 1.00.$$

The results are $f_1 = 21, f_2 = 31, f_3 = 26, f_4 = 22$. Is this "close enough" to be uniform?

Solution
Here, there are $n = 100$ trials and $m = 4$ classes. Obviously we would have preferred $n/m = 25$ numbers in each class, but this was not the case. Even so, using Equation (3.3) gives

$$\chi^2 = \tfrac{4}{100}[(21 - 25)^2 + (31 - 25)^2 + (26 - 25)^2 + (22 - 25)^2] = 2.48,$$

with $v = 4 - 1 = 3$. From Appendix F, the critical χ^2 value for an $\alpha = 95\%$ confidence level is $\chi_c^2 = 7.81$. Since $\chi^2 < \chi_c^2$, we accept the hypothesis that this is indeed uniform. ○

In practice, one should strive for far more precision. At least $m = 100$ classes should be used, and a sequence of at least $n = 1000$ is not unreasonable. Since this leads to significantly large values for the number of degrees of freedom, v, Appendix F can be ignored, and the following asymptotic approximation for χ_c^2 used:

$$\chi_c^2 = v \left(1 - \frac{2}{9v} + \frac{z_c}{3} \sqrt{\frac{2}{v}} \right)^3,$$

where z_c is the upper $1 - \alpha$ critical point of the normalized Gaussian distribution tabulated in Appendix E.

Even when the set of numbers generated is uniform, the generator is useless if the sequence produces clusters or has gaps. As well as being uniform in the large, it must also be uniform locally. This property of independence is often checked using the *runs test*.

The runs test begins by generating a sequence of n proposed random numbers, then checking the length of ascending runs throughout. For instance, the sequence 5, 3, 6, 9, 8, 7 has three runs of length 1 (the 5, the 8, and the 7) and one run of length 3 (the 3, 6, 9). In general, each of the n numbers in the list belongs to exactly one "run". However, what is important is the number of runs of a given length. The frequencies r_k are defined as

$$r_1 = \text{number of runs of length 1,}$$
$$r_2 = \text{number of runs of length 2,}$$
$$r_3 = \text{number of runs of length 3,}$$
$$r_4 = \text{number of runs of length 4,}$$
$$r_5 = \text{number of runs of length 5,}$$
$$r_6 = \text{number of runs of length greater than 5,}$$

and the vector $\mathbf{r} = [r_1, r_2, r_3, r_4, r_5, r_6]$ is formed. The test statistic R is then calculated using either of the following formulas:

$$R = \frac{1}{n} \sum_{i=1}^{6} \sum_{j=1}^{6} a_{ij}(r_i - nb_i)(r_j - nb_j)$$

$$= \frac{1}{n}(\mathbf{r} - n\mathbf{B})\mathbf{A}(\mathbf{r} - n\mathbf{B})^T \tag{3.4}$$

where the matrix $\mathbf{A} = [a_{ij}]$ is given by

$$\mathbf{A} = \begin{bmatrix} 4529 & 9045 & 13568 & 18091 & 22615 & 27892 \\ 9045 & 18097 & 27139 & 36187 & 45234 & 55789 \\ 13568 & 27139 & 40721 & 54281 & 67852 & 83685 \\ 18091 & 36187 & 54281 & 72414 & 90470 & 111580 \\ 22615 & 45234 & 67852 & 90470 & 113262 & 139476 \\ 27892 & 55789 & 83685 & 111580 & 139476 & 172860 \end{bmatrix}$$

and the vector $\mathbf{B} = [b_i]$ by

$$\mathbf{B} = \left[\frac{1}{6}, \frac{5}{24}, \frac{11}{120}, \frac{19}{720}, \frac{29}{5040}, \frac{1}{840} \right]$$

It can be shown that the vector $\mathbf{B} = [b_i]$ is the theoretical probability of achieving a run of length k. Clearly, R compares the closeness of frequency r_k with the expected frequency nb_k. A large difference implies a large R, and a small difference gives a small R value. Not surprisingly, the R statistic is approximated by a chi-square distribution with $v = 6$ degrees of freedom.

It is recommended that the runs test be applied to sequences of at least $n = 4000$ samples. But, visually checking for runs is at best tedious! An algorithm to count frequencies $r(k)$ is given in Listing 3.1, where the RND statements (lines 2 and 6) are uniform $[0, 1]$ random number generator calls. The runs test is applicable to a variety of generators, and the distribution need not even be uniform. For this reason, one should consider doing the runs test before checking the distribution itself.

```
                 for i=1 to 6
                         r(i)=0
                 next i
                 a=RND
                 i=1
                 j=1
         [1]     i=i+1
                 if i>n then goto [3]
                 b=RND
                 if a<b then goto [2]
                 if j>6 then j=6
                 r(j)=r(j)+1
                 j=1
                 a=b
                 goto [1]
         [2]     j=j+1
                 a=b
                 goto [1]
         [3]     if j>6 then j=6
                 r(j)=r(j)+1
```

LISTING 3.1   Frequency-counting algorithm for the runs test.

EXAMPLE 3.4

The random number generator on a typical PC is invoked by the RND call of Microsoft's Visual Basic[TM]. By using the *randomize* function, the sequence of random numbers was initiated using an initial seed of $Z_0 = 0$. A list of $n = 10\,000$ supposedly random numbers was generated and the run frequencies r_k were calculated using the algorithm in Listing 3.1. These are given below:

$$r_1 = 1670,$$
$$r_2 = 2147,$$
$$r_3 = 891,$$
$$r_4 = 262,$$
$$r_5 = 52,$$
$$r_6 = 9.$$

Although it looks formidable, Equation (3.4) is straightforward, especially using MatLab, Minitab, MathCad, or similar software. Using Equation (3.4), $R = 5.60$. The question now is whether or not the proportions r_i/n are close enough to the theoretical b_i. Equivalently, is R small enough in the χ^2 sense? To test this, let us choose an $\alpha = 90\%$ confidence level. This, along with $v = 5$, gives $\chi_c^2 = 10.645$ from Appendix F. Since $R = 5.60 < 10.645 = \chi_c^2$, we accept that the sequence of numbers passes the runs test at the 90% confidence level. $\bigcirc$

3.3   GENERATION OF NON-UNIFORM RANDOM VARIATES

In practice, it is reasonable to assume that a good $U[0, 1]$ generator is available. What is important is to be able to generate numbers in an arbitrary statistical distribution. To this end, there are several popular means for establishing effective algorithms.

Formula Method

For instance, certain random variates can be expressed simply as a formula involving $U[0, 1]$ generators. Letting RND be such a generator, it is possible to establish a sequence of random numbers X that are uniform on the arbitrary interval $[a, b]$. In this case, the algorithm is simply a formula

$$X = (b - a)\text{RND} + a. \tag{3.5}$$

It is easy to see that this is the case by simply considering the extremes of U. Since the minimum of U is 0, the minimum of X is $X = (b - a)(0) + a = a$. Similarly, the maximum of U is 1, and it follows that the maximum of X is $X = (b - a)(1) + a = b$. Therefore, the endpoints are mapped appropriately. Combined with the fact that this is a linear transformation, X is uniform on $[a, b]$, as required.

Formulas exist for non-uniform distributions too. It can be shown that a normalized Gaussian random variate is generated by the following formula:

$$Z = \sqrt{-2\ln(\text{RND})}\cos(2\pi\text{RND}). \tag{3.6}$$

Z is a random variate with mean zero and standard deviation one. This is easily demonstrated by generating a large number of Zs, then computing the mean and standard deviation of the sample. By creating a histogram, one can "eyeball" the result to see that it is Gaussian. Alternatively, this can be done formally by using a chi-square test.

In general, a Gaussian with mean μ and standard deviation σ is required. A standardized variable X can be created from normalized Z by applying the linear transformation $X = \sigma Z + \mu$. This can be verified by using the expectation operator E as follows. First for the mean:

$$\mu_X = E[X] = E[\sigma Z + \mu]$$
$$= \sigma E[Z] + \mu$$
$$= \sigma(0) + \mu = \mu,$$

and next for the variance:

$$\sigma_X^2 = E[(X - \mu_X)^2]$$
$$= E[\{(\sigma Z + \mu) - \mu_X\}^2]$$
$$= \sigma^2 E[X^2]$$
$$= \sigma^2(1) = \sigma^2.$$

Therefore, we are justified in simply standardizing the normalized random variate Z to get the required random variate X.

One of the ways in which a formula can be derived for a random variate is by the so-called *inversion method*. This method requires computation and inversion of the distribution function F. In cases where it is possible to find F^{-1} explicitly, the random variate X found by

$$X = F^{-1}(\text{RND}) \tag{3.7}$$

will have the required distribution. This is verified as follows: let Y be the random variable generated by $F_U^{-1}(X)$. Then, by definition,

$$
\begin{aligned}
F_Y(x) &= \Pr[Y < x] \\
&= \Pr[F_X^{-1}(U) < x] \\
&= \Pr[U < F_x(x)] \\
&= F_U(F_X(x)) \\
&= F_X(x),
\end{aligned}
$$

since $f_U(x) = 1$ and $F_U(x) = x$ is monotonic increasing on $0 \leqslant x \leqslant 1$. Therefore, the random variate Y is the same as X. The application of Equation (3.7) is straightforward, as is illustrated in the following example.

EXAMPLE 3.5

Derive a formula by which to generate exponentially distributed random variates with mean μ.

Solution
Recall that the density function $f(x)$ for an exponential random variable with mean $E[X] = 1/\lambda = \mu$ is

$$
f(x) = \begin{cases} \dfrac{1}{\mu} e^{-x/\mu}, & x \geqslant 0, \\ 0, & x < 0. \end{cases}
$$

The distribution function is

$$
\begin{aligned}
F(x) &= \int_{-\infty}^{x} f(t)\, dt \\
&= \int_{0}^{x} \frac{1}{\mu} e^{-t/\mu}\, dt \\
&= 1 - e^{-x/\mu}, \qquad x \geqslant 0.
\end{aligned}
$$

The inverse of this function is found by solving for x, then interchanging the role of F and x. Solving for x, $x = -\mu \ln(1 - F)$. In order to find the required inverse, the role of X is replaced by $F^{-1}(x)$ and that of F by X. Further, note that if X is uniform on $[0, 1]$, then so is $1 - X$. Thus, a suitable algorithm is given by a single line of code:

$$X = -\mu \ln(\text{RND}). \tag{3.8}$$

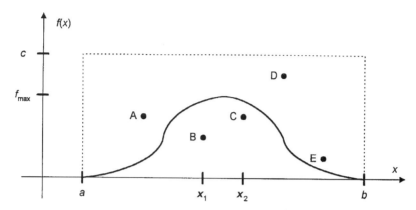

FIGURE 3.2   Density function and target area for the rejection algorithm.

Rejection Method

Unfortunately, finding an explicit formula for an arbitrary inverse of the distribution function is not always possible. Some random variables, such as the Gaussian and Beta distributions, have no explicit analytical formulas for their distribution functions. Even so, methods exist by which random variates can still be generated. One of the more interesting alternatives is based on the idea of throwing darts. To illustrate this, consider the density function contained within the box bounded by $a \leqslant x \leqslant b, 0 \leqslant y \leqslant c$ shown in Figure 3.2. The rectangular box is called the *target* and must enclose the density function in question.

The algorithm proceeds by first selecting a sequence of points that fall randomly within the target area. Specifically, the coordinates (x, y) are chosen by $x = a + (b - a)\text{RND}$ and $y = c\text{RND}$, so that they are uniformly distributed within the target. It follows that points falling under the curve possess an x-value that is a "legitimate" random variate. That is, x is the random variate whose density function is given by $f(x)$. Although this seems like an arbitrary method, it produces excellent results. Specifically, the algorithm proceeds as shown in Listing 3.2.

```
[1]    x=a+(b-a)*RND
       y=c*RND
       if y>f(x) then goto [1]
       print x
```

LISTING 3.2   The rejection algorithm.

The random point is (x, y). Only those points falling under the curve are selected, and if the point is selected, then $x = x_1$ is the random variate. For instance, of the points A, B, C, D, and E shown in Figure 3.2, points A, D, and E are ignored since they are each above the curve $f(x)$, even though they are within the rectangular target. However, the x_1 and x_2 coordinates of B and C form two random variates with density function $f(x)$.

EXAMPLE 3.6

Apply the rejection method to generate 100 random variates for the Beta distribution with $a = 3$ and $b = 2$.

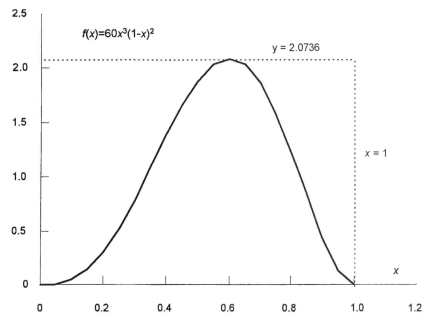

f(x)=60x³(1-x)²

y = 2.0736

x = 1

x

FIGURE 3.3   The Beta($a = 3, b = 2$) density function.

Solution
Referring to Appendix C, the required Beta density function is

$$f(x) = \frac{\Gamma(7)}{\Gamma(4)\Gamma(3)} x^3 (1 - x)^2$$

$$= 60x^3(1 - x)^2, \quad 0 \leqslant x \leqslant 1.$$

While any-sized rectangle that surrounds the entirety of $f(x)$ will work, fewer points will be rejected if the rectangle is as small as possible. Since the domain of $f(x)$ is [0, 1], the best rectangle has vertical sides of $x = 0$ and $x = 1$. Also, since all density functions are non-negative, the best rectangle has a base of $y = 0$, and a height of what ever the maximum value of $f(x)$ happens to be. Both the density function and the best rectangle are shown in Figure 3.3.

In order to find the optimum y-interval, the maximum of $f(x)$ needs to be found. This is a straightforward calculus procedure in that the extremes (either maximum or minimum points) will occur where the derivative is zero (or at the endpoints [0, 1]). In this case, the derivative of the density function vanishes when

$$f'(x) = -120x^3(1 - x) + 180x^2(1 - x)^2$$

$$= 60x^2(1 - x)(3 - 5x) = 0.$$

Solving for x, we have $x = 0$, $x = 1$, or $x = 0.6$. Since we know that the extremes $x = 0$ and $x = 1$ are minimum points, $x = 0.6$ must be the maximum, as is clear from the graph of Figure 3.3. From this, the height is $f(0.6) \approx 2.0736$, and a reasonable target height is given by $y = 2.1$. A program to print 1000 suitable random numbers is as follows:

```
          for i=1 to 1000
[1]               x=RND
                  y=2.1*RND
                  if y>60x³(1-x)² then goto [1]
                  print x
          next i
```

In order to check the validity of this algorithm, a frequency distribution table similar to Table 3.2 should be compiled. In forming this table, one first defines a reasonable number of equal-sized *class intervals* or *bins*. Next, an experiment producing a large number of random variates is performed and the frequency of variates in each bin is tallied. This forms the basis of the *empirical frequency*. This needs to be compared against the theoretical ideal, which can be computed using the density function $f(x)$.

A suitable frequency distribution table is given in Table 3.3, where an arbitrary bin size of $\Delta x = 0.05$ over the interval $[0, 1]$ is chosen and an experiment of $n = 1000$ random variates is performed. The expected (theoretical) frequency assuming an arbitrary interval $[a, b]$ and n random variates is found by applying the following formula:

$$
\begin{aligned}
E_{[a,b]} &= n \int_a^b f(x)\, dx \\
&= n \int_a^b 60x^3(1-x)^2\, dx \\
&= n[15(b^4 - a^4) - 24(b^5 - a^5) + 10(b^6 - a^6)].
\end{aligned}
$$

The results are displayed graphically in Figure 3.4, where each of the frequencies is plotted as a function of x over the interval $[0, 1]$. It appears that the results confirm that the rejection method gives a good set of random variates. If 10 000 variates were generated instead of the 1000, we would expect to see even better results.   ○

There is actually no real need for the target region to be rectangular. Any target will do, as long as it completely surrounds the density function. Of course, inefficiencies are inevitable if there is excess rejection area, since a larger proportion of random points will be rejected. This is because there will be more iterations and overall execution will be slower.

One further complication arises if the density function has "tails". For example, neither the Gamma (one tail) nor the Gaussian (two tails) can be totally surrounded with any rectangle. Even so, by choosing a sufficiently broad target, the distribution can be approximated by ignoring the very unlikely occurrences at the extremes.

TABLE 3.3 Frequency Distribution
Table for Example 3.6

Interval		Empirical frequency	Expected frequency
[0.00,	0.05]	4	0.09
[0.05,	0.10]	5	1.15
[0.10,	0.15]	2	4.62
[0.15,	0.20]	5	11.07
[0.20,	0.25]	23	20.64
[0.25,	0.30]	38	32.87
[0.30,	0.35]	46	46.95
[0.35,	0.40]	64	61.78
[0.40,	0.45]	69	76.06
[0.45,	0.50]	86	88.49
[0.50,	0.55]	110	97.77
[0.55,	0.60]	111	102.81
[0.60,	0.65]	94	102.77
[0.65,	0.70]	81	97.22
[0.70,	0.75]	78	86.26
[0.75,	0.80]	85	70.55
[0.80,	0.85]	44	51.54
[0.85,	0.90]	52	31.49
[0.90,	0.95]	5	13.62
[0.95,	1.00]	6	2.23

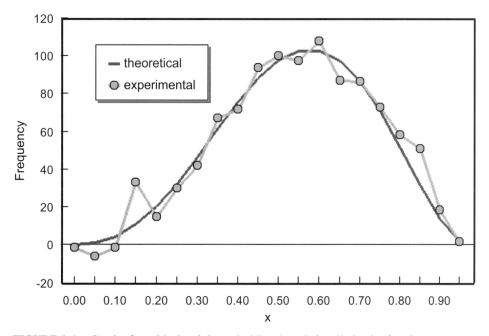

FIGURE 3.4   Graph of empirical and theoretical Beta($a = 3, b = 2$) density functions.

Convolution Method

Now consider the case of a random variable X defined as the sum of n other independent and identically distributed (IID) random variables $X_1, X_2, \ldots, X_n$. Specifically, if X_i has the same density function $f_i(x)$ for $i = 1, 2, \ldots, n$, the density function $f(x)$ of X is the convolution of each of all the n basis density functions. Formally, this means that

$$\text{if} \quad X = \sum_{k=1}^{n} X_k, \quad \text{then} \quad f(x) = f_1(x) \otimes f_2(x) \otimes \cdots \otimes f_n(x),$$

where $f_i(x)$ is the density function of X_i and "$\otimes$" is the convolution operator defined by

$$f_1(x) \otimes f_2(x) = \int_{-\infty}^{\infty} f_1(\lambda) f_2(x - \lambda)\, d\lambda.$$

Thus, in practice, the random variate itself can be found by adding the n IID variates $X = \sum_{k=1}^{n} X_k$, each of which is presumably found by other means. One specific case is the m-Erlang distribution (see Appendix C), which, by definition, is the sum of m IID exponential random variates. The mean of an m-Erlang distribution is

$$\mu = E\left[\sum_{k=1}^{m} X_k\right]$$
$$= \sum_{k=1}^{m} E[X_k]$$
$$= \frac{m}{\lambda},$$

where λ is the reciprocal of the exponential distribution's mean. An algorithm to generate an m-Erlang random variate with mean μ by adding m exponential random variates is shown in Listing 3.3.

```
x=0
for k=1 to m
        x = x - μ ln(RND)/m
next k
print x
```

LISTING 3.3   Convolution algorithm to generate an m-Erlang random variate.

EXAMPLE 3.7

Generate a sequence of 1000 2-Erlang random variates with mean 5, and compare against the expected results. Quantify this experiment by using the chi-square test for goodness of fit.

Solution

The 2-Erlang distribution (see Appendix C) is a special case of the Gamma distribution with $\alpha = 2$. The mean is $2/\lambda = 5$, which implies that $\lambda = 0.4$. Thus, the density function for the 2-Erlang distribution is

$$f(x) = \tfrac{4}{25} x e^{-2x/5}, \qquad x \geqslant 0. \tag{3.9}$$

The convolution method states that an Erlang random variate is the sum of $-2.5\ln(\text{RND})$ and $-2.5\ln(\text{RND})$, which is algebraically equivalent to $-2.5\ln(\text{RND} \times \text{RND})$. Again, recall that the two RND function calls are not the same as squaring one function call, since each will generate a different uniform random variate. Thus, an appropriate program segment for generating 1000 2-Erlang random variates X is given as follows:

```
for k=1 to 1000
     x=-2.5*ln(RND*RND)
     print x
next k
```

In order to verify the results of this experiment, the required $n = 1000$ random variates are summarized in the frequency distribution table given is Table 3.4, where the interval size is $\Delta x = 0.5$ units from $x = 0$ to $x = 1$. The expected frequency over each interval can be computed by

$$E_{[a,b]} = n\int_a^b f(x)\,dx = \frac{4n}{25}\int_a^b xe^{-2x/5}\,dx$$
$$= n[e^{-2x/5}(1 - \tfrac{2}{5}x)]_a^b = n[(e^{-2b/5} - e^{-2a/5}) + \tfrac{2}{5}(ae^{-2a/5} - be^{-2b/5})].$$

TABLE 3.4   Frequency Distribution Table for Example 3.7

Interval	Empirical frequency	Expected frequency
[0.0, 0.5]	8	17.52
[0.5, 1.0]	37	44.03
[1.0, 1.5]	56	60.35
[1.5, 2.0]	64	69.31
[2.0, 2.5]	76	73.03
[2.5, 3.0]	64	73.13
[3.0, 3.5]	77	70.79
[3.5, 4.0]	78	66.90
[4.0, 4.5]	64	62.09
[4.5, 5.0]	49	56.83
[5.0, 5.5]	53	51.44
[5.5, 6.0]	46	46.13
[6.0, 6.5]	50	41.06
[6.5, 7.0]	35	36.31
[7.0, 7.5]	27	31.93
[7.5, 8.0]	29	27.95
[8.0, 8.5]	21	24.36
[8.5, 9.0]	22	21.15
[9.0, 9.5]	9	18.31
[9.5, 10.0]	25	15.80
[10.0, 10.5]	21	13.60
[10.5, 11.0]	9	11.68
[11.0, 11.5]	4	10.01
[11.5, 12.0]	6	8.56
[12.0, 12.5]	3	7.30
[12.5, 13.0]	5	6.22
[13.0, 13.5]	10	5.30
[13.5, 14.0]	10	4.50
[14.0, 14.5]	5	3.82
[14.5, 15.0]	5	3.24

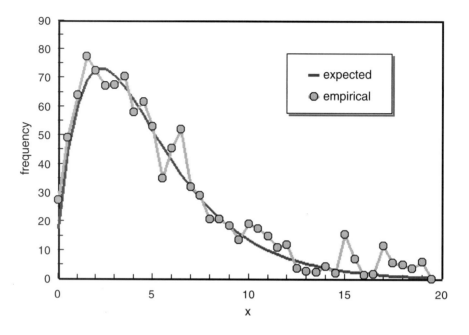

FIGURE 3.5   Theoretical and empirical density functions for Example 3.7.

Substituting the respective a and b values for each interval $[a, b]$ and $n = 1000$ gives the theoretical expected frequency for each interval. These can then be compared against the experimentally acquired frequencies listed. Both are graphed in Figure 3.5, where the goodness of fit appears obvious.

However, the "eye-ball" technique is never really good enough. The χ^2 test is best for measuring goodness of fit. Especially once in a spreadsheet format, this statistic is easily computed as

$$\chi^2 = \sum_{i=1}^{20} \frac{(E_i - F_i)^2}{E_i} = 29.29.$$

Since there are $m = 20$ classes, the number of degrees of freedom is $v = 20 - 1 = 19$. A check of Appendix F gives the following critical values:

$$\chi^2_{0.90} = 1 - F(0.10) = 27.204,$$
$$\chi^2_{0.95} = 1 - F(0.05) = 30.144,$$
$$\chi^2_{0.98} = 1 - F(0.02) = 35.020,$$
$$\chi^2_{0.99} = 1 - F(0.01) = 37.566.$$

Clearly, the value $\chi^2 = 29.29$, while technically "good enough" at the 90% level, is a little too close to borderline for comfort. However, rather than be marginally satisfied, it is probably a good idea to simply increase the sample size to, perhaps, 10 000 random variates. In doing this, the new χ^2 is considerably lower, and the test hypothesis that the two distributions are essentially the same is acceptable.

○

3.4   GENERATION OF ARBITRARY RANDOM VARIATES

In a number of practical situations, there is no explicit formula for the density function. Instead, only a set of empirical variates are known. For instance, there is no recognized theoretical model of the temperature in Chicago over a year's time. However, the historical record for Chicago is well established, and we should be able to create a series of random variates that have the same statistics as those of the historical record. While such models may not provide insight, they do have one very important application. In order to validate a model, one must verify system performance from the historical record. Given that the inputs will likely be random processes, it is important that the process statistics match the historical validation set. By generating empirical random variates, validation is easily accomplished.

It is possible to generate random variates based solely on an empirical record using a two-step process. We consider a set of data $\{x_1, x_2, x_3, \ldots, x_n\}$ as given and sorted in ascending order. From this data, we can derive an explicit, albeit empirical, distribution function $F(x)$ that is piecewise-linear and continuous and fits the data perfectly. This is given by the formula

$$F(x) = \begin{cases} 0, & x < x_1, \\ \dfrac{i-1}{n-1} + \dfrac{x - x_i}{(n-1)(x_{i+1} - x_i)}, & x_i \leqslant x < x_{i+1} \quad \text{for } i = 1, \ldots, n-1, \\ 1, & x \geqslant x_n. \end{cases} \quad (3.10)$$

The second step is to find $F^{-1}(x)$. This can be done in a straightforward manner, since $F(x)$ is piecewise-linear and easily inverted. From $F^{-1}(x)$, random variates can be generated using Equations (3.5): $X = F^{-1}(\text{RND})$. This process is best illustrated with an example.

EXAMPLE 3.8

Consider the following (sorted) set of random observations from an unknown process: $\{1, 2, 4, 5, 7, 7, 9\}$. Determine the distribution function and its inverse.

Solution
This problem has $n = 7$ data points, so, using Equation (3.10) for $x < 1$, $F(x) = 0$. Letting $i = 1$ for $1 \leqslant x < 2$,

$$F(x) = \frac{1-1}{7-1} + \frac{x-1}{(7-1)(2-1)}$$

$$= \tfrac{1}{6}(x - 1).$$

Similarly, when $i = 2$, $2 \leqslant x < 4$ and

$$F(x) = \frac{2-1}{7-1} + \frac{x-2}{(7-1)(4-2)}$$

$$= \tfrac{1}{12}x.$$

Continuing in this manner and summarizing,

$$F(x) = \begin{cases} 0, & x < 1, \\ \tfrac{1}{6}(x-1), & 1 \leqslant x < 2, \\ \tfrac{1}{12}x, & 2 \leqslant x < 4, \\ \tfrac{1}{6}(x-2), & 4 \leqslant x < 5, \\ \tfrac{1}{12}(x+1), & 5 \leqslant x < 7, \\ \tfrac{1}{12}(x+3), & 7 \leqslant x < 9, \\ 1, & x \geqslant 9. \end{cases} \qquad (3.11)$$

As a check, notice that this function is piecewise-continuous at each endpoint and monotonic, just as the distribution function must always be. Its graph is shown in Figure 3.6.

It is straightforward to invert Equation (3.11). Recall that to invert an equation $y = F(x)$, one simply solves for x (this is especially easy in our example, where the function is piecewise-linear), then interchanges x and y: $y = F^{-1}(x)$. This needs to be done for each case of the example at hand. For instance, in case 2, $y = F(x) = \tfrac{1}{6}(x-1)$ for $1 \leqslant x < 2$. This leads to $x = 6y + 1$; thus, $F^{-1}(x) = 6x + 1$. It is also necessary to re-establish the appropriate interval. For example, in case 2, $x = 1 \Rightarrow y = 0$ and $x = 2 \Rightarrow y = \tfrac{1}{6}$. Thus, $0 \leqslant y < \tfrac{1}{6}$ for y or $0 \leqslant x < \tfrac{1}{6}$ for $F^{-1}(x)$. Completing the remaining six cases results in the

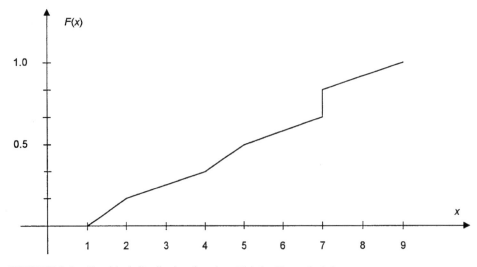

FIGURE 3.6   Empirical distribution function $F(x)$ for Example 3.8.

following explicit formula for $F^{-1}(x)$, which is graphed in Figure 3.7:

$$F^{-1}(x) = \begin{cases} \text{undefined}, & x < 0, \\ 6x + 1, & 0 \leqslant x < \frac{1}{6}, \\ 12x, & \frac{1}{6} \leqslant x < \frac{1}{3}, \\ 6x + 2, & \frac{1}{3} \leqslant x < \frac{1}{2}, \\ 12x + 1, & \frac{1}{2} \leqslant x < \frac{2}{3}, \\ 7, & \frac{2}{3} \leqslant x < \frac{5}{6}, \\ 12x - 3 & \frac{5}{6} \leqslant x < 1, \\ \text{undefined}, & x > 1. \end{cases} \tag{3.12}$$

Using the inverse function F^{-1} given by Equation (3.12), an appropriate algorithm is given in Listing 3.4.

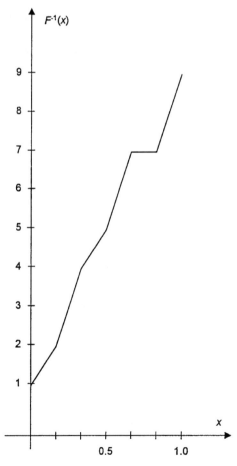

FIGURE 3.7   Empirical inverse of the distribution in Example 3.8.

```
U=RND
if  0 ⩽ U<⅙  then X=6U+1
if  ⅙ ⩽ U<⅓  then X=12U
if  ⅓ ⩽ U<½  then X=6U+2
if  ½ ⩽ U>⅔  then X=12U-1
if  ⅔ ⩽ U<⅚  then X=7
if  ⅚ ⩽ U<1  then X=12U-3
```

LISTING 3.4   Random variates from an empirical distribution, Example 3.8.

    The results of 1000 random variates generated by this method are shown graphically in Figures 3.8 and 3.9. Figure 3.8 reproduces the density function and Figure 3.9 gives the distribution function using interval size $\Delta x = 0.2$. The similarity between Figures 3.6 and 3.9 is clear, and should serve as a convincing demonstration that this method produces excellent and reliable results.   ○

    Similar results can be obtained for discrete distributions. Suppose a mass function has values $p(0), p(1), p(2), \ldots$ that sum to unity. The algorithm given in Listing 3.5 produces n random variates $X(i)$, which are distributed in the same way as the function $p(k)$. This algorithm simply loops until the right mass point is found; the index provides the variate. This algorithm works for either empirical or theoretical distributions. In the case of empirical distributions, it is necessary to first compute each probability mass point, then create an array $p = [p(x)]$. If the probability mass function is known a priori, $p(x)$ is simply that function.

EXAMPLE 3.9

Consider the probability distribution defined below. Generate 2000 random variates using the algorithm given in Listing 3.5, and compare against the theoretical results.

x	$p(x)$
0	0.25
1	0.10
2	0.20
3	0.40
4	0.05

Solution
Since this example has a small finite number of mass points, implementation algorithm proceeds by prefacing the loop with the five lines of code: $p(0) = 0.25$, $p(1) = 0.1$, $p(2) = 0.2$, $p(3) = 0.4$, $p(4) = 0.05$. Results of the random variate distribution are summarized in Figure 3.10 for $n = 100$, 500, and 1000, and compared against the ideal, thus validating the algorithm. If the list of mass points were long or possibly infinite, an explicit formula for $p(x)$ could be imbedded within the loop instead.   ○

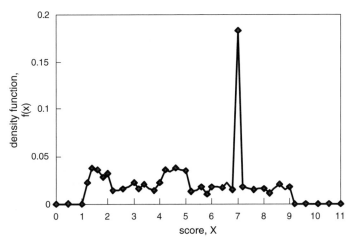

FIGURE 3.8   Approximation of the density function using empirical data $\{1, 2, 4, 5, 7, 7, 9\}$.

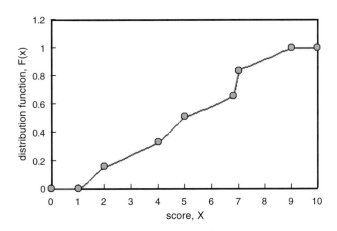

FIGURE 3.9   Approximation of the distribution function using empirical data $\{1, 2, 4, 5, 7, 7, 9\}$.

```
         p(k) given for k=0,1,2,...

         for i = 1 TO n
                 u = RND
                 k = 0
                 t = 0
[1]              t = t + p(k)
                 if t < u then k = k + 1: goto [1]
                 print k
         next i
```

LISTING 3.5   Program segment to generate a discrete random variate, each with probability $p(0), p(1), p(2), \ldots$.

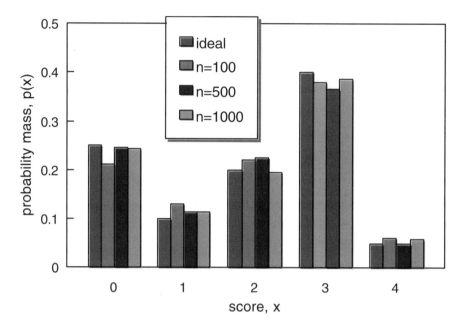

FIGURE 3.10   Generating discrete probability mass functions.

When modeling, one of the first questions to answer is whether simply mimicking history is sufficient. Certainly, in modeling a system, one must understand the mechanism of the system – not simply duplicate experience of it. Failure to do this will create an excellent model over a mere subset of the complete spectrum of working environments. This is like training a doctor in nothing but surgery, then discovering too late that he recommends amputation for symptoms of a cold!

On the other hand, systems are driven by inputs, and, as we have seen, these inputs are often stochastic. In these situations, it is often sufficient to drive a system using known experience, but not understand the *input* mechanism. For instance, the Earth's weather that we experience every day is a notorious example of a complex, nonlinear system. Indeed, no model yet usefully explains climate at the micro-level. On the other hand, a number of practical stochastic models are able to mimic weather behavior. By simply observing weather statistics and stochastically reproducing them as input over many runs, one can validate an agricultural model and estimate appropriate parameters.

In cases such as these, the generation of arbitrary random variates make sense. First one produces an historical record such as daily temperature highs. After constructing the empirical distribution function of daily high temperatures and inverting as in Example 3.7, one can reproduce statistically reliable stochastic data by which to drive a system.

3.5   RANDOM PROCESSES

In simulating dynamic systems, it is important to model not only the system itself but also the input that drives the system. The input signals are rarely deterministic. In such cases, we call them *random processes* because input signals have an inherent fluctuation built around an underlying unpredictable process. Up to now, we have considered these processes to be a sequence of uncorrelated random values. That is, if $x(k)$ and $x(k+1)$ are two successive input signals, $E[x(k)x(k+1)] = E[x(k)]E[x(k+1)] = \mu^2$. This is a very nice property, but it is simply too much to ask for. In a good many models, the current value of the signal greatly influences the value that the signal takes on in the future. Take stock market values for instance, where today's value directly influences tomorrow's.

The fact that the signal is stochastic does not mean that it is totally random. It is actually a collection of signals called the process *ensemble*. Each particular representative of the ensemble is simply one *instance* of the signal in question, and the whole thing is called a *random process*. By choosing the statistics of the input signal ensemble carefully, one can discover the corresponding statistics of the output ensemble. A continuous-time random process is denoted by $X(t)$. $X(t)$ represents the whole ensemble, so we denote by $x(t)$ any particular instance of the process ensemble in question. Similarly, in discrete time, the process and instance are denoted by $X(k)$ and $x(k)$, respectively. One view is to simply think of $X(t)$ as a random variable that is a function of time. Thus, all the normal statistics such as moments, means, and standard deviations are available. Of these, two are most important. They are the *mean* and the *autocorrelation*, and are defined as follows:

$$\mu_x(t) = E[X(t)],$$
$$R_{xx}(t, \tau) = E[X(t)X(t + \tau)]. \tag{3.13}$$

It can be seen that the autocorrelation is the expected value of the product of the signal with itself, just evaluated at τ time units later – hence the name "autocorrelation". Clearly the maximum value that the autocorrelation can have is when $\tau = 0$, since at this time, the correlation of a signal with itself is the best that can be done. Only when the signal is periodic can the autocorrelation match the value at $\tau = 0$.

In general, $R_{xx}(t, \tau)$ is a function of both time t and lag time τ. However, in the event that R is a function only of τ, we call the random process *autocorrelated*. If the mean $\mu_{X(t)} = E[X(t)]$ is also a constant, the process is called *wide-sense stationary*. Intuitively, the stationarity property is most important, since this means that the system has achieved steady-state behavior.

EXAMPLE 3.10

Consider an ensemble consisting of four equally likely signals defined over the time interval [0 ,1] as follows:

$$X(t) = \{2t + 1, t + 2, 3t + 2, 4t + 1\}.$$

Find the mean, second moment, variance, and autocorrelation of $X(t)$.

Solution

Since each member of the ensemble is equally likely, the expected value is simply the average. Therefore, computation of the required statistics proceeds as follows:

$$\mu = \tfrac{1}{4}[(4t+1)+(2t+1)+(t+2)+(3t+2)]$$

$$= \tfrac{1}{2}(5t+3), \tag{3.14}$$

$$E[X^2(t)] = \tfrac{1}{4}[(4t+1)^2+(2t+1)^2+(t+2)^2+(3t+2)^2]$$

$$= \tfrac{1}{2}(15t^2+14t+5), \tag{3.15}$$

$$\sigma^2 = E[X^2(t)] - \mu^2$$

$$= \tfrac{1}{2}(15t^2+14t+5) - \tfrac{1}{4}(5t+3)^2$$

$$= \tfrac{1}{4}(5t^2-2t+1), \tag{3.16}$$

$$R_{xx}(t,\tau) = E[X(t)X(t+\tau)]$$

$$= \tfrac{1}{4}[(4t+1)(4t+4\tau+1)+(2t+1)(2t+2\tau+1)$$

$$+ (t+2)(t+\tau+2)+(3t+2)(3t+3\tau+2)]$$

$$= \tfrac{1}{2}(15t^2+14t+5+15t\tau+7\tau) \tag{3.17}$$

It should be noticed that the autocorrelation evaluated at $\tau = 0$ is the second moment: $R_{xx}(t,0) = E[X^2(t)]$. Thus, Equation (3.15) is simply a special case of (3.17). Also, since $R_{xx}(t,\tau)$ is a function of t, it follows that the process in question is non-stationary. The random process $X(t)$ along with mean $\mu(t)$ and standard error bars $\mu(t) \pm \sigma(t)$ are graphed in Figure 3.11.   ○

In more realistic situations, one never has a simple finite set of instances of an ensemble. Instead, one usually has a description of the ensemble that just *characterizes* $X(t)$. From this characterization, a formula or probabilistic statement follows in which random *variables* inevitably play a prominent role. This characterization usually amounts to a theoretical or empirical derivation of the mean $\mu(t)$ and autocorrelation $R_{xx}(t,\tau)$. From

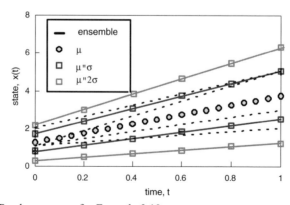

FIGURE 3.11   Random process for Example 3.10.

these, the variance follows if it is desired. In the best of worlds – the one in which we choose to live in this text – the mean μ is a constant and the autocorrelation $R_{xx}(\tau)$ is independent of t, so the random process $X(t)$ is wide-sense stationary.

EXAMPLE 3.11

Consider a sinusoidal random process of the form $X(t) = 5\sin(2t + \theta)$, where θ is a random variable uniformly distributed on the interval $[0, 2\pi]$. In this case, we intuitively see an infinite number of sine waves with amplitude 5 and frequency 2 radians per second or $1/\pi$ hertz. So, the ensemble comprises all such waves, each with a phase shift somewhere between 0 and 2π. See Figure 3.12, where $X(t)$ is graphed on the interval $[0, 10]$. Find the mean and autocorrelation of this process.

Solution
It is evident from the graph of this process that the mean is a constant, since the cross-section of any time slice will produce the same distribution with a mean of zero and a finite standard deviation. Clearly, the autocorrelation will depend on only the lag time, since all time slices are the same and therefore the process is wide-sense stationary, as hoped. However, let us compute μ and $R_{xx}(\tau)$ directly from Equation (3.13).

Since the random variable θ is uniform on $[0, 2\pi]$, the probability density function must be $1/2\pi$, and the mean is

$$\mu(t) = \int_0^{2\pi} \frac{1}{2\pi} 5\sin(2t + \theta)\, d\theta$$

$$= \left[-\frac{5}{2\pi}\cos(2t + \theta) \right]_0^{2\pi}$$

$$= -\frac{5}{2\pi}[\cos(2t + 2\pi) - \cos 2t]$$

$$= 0, \tag{3.18}$$

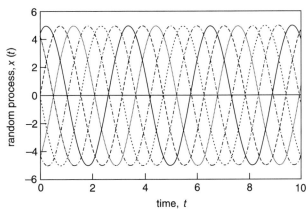

FIGURE 3.12   Representative instances of the random process in Example 3.11, θ uniform on $[0, 2\pi]$.

$$R_{xx}(\tau) = E[X(t)X(t+\tau)]$$

$$= \int_0^{2\pi} \frac{1}{2\pi} 5\sin(2t+\theta)\,5\sin(2t+2\tau+\theta)\,d\theta$$

$$= \frac{25}{4\pi}\int_0^{2\pi} [\cos 2\tau - \cos(4t+2\tau+2\theta)]\,d\theta$$

$$= \frac{25}{4\pi}\left[\theta\cos 2\tau - \tfrac{1}{2}\sin(4t+2\theta+2)\right]_0^{2\pi}$$

$$= \frac{25}{4\pi}[2\pi\cos(2t+2\pi) - \sin(4t+4\pi+2) + \sin(4t+2)]$$

$$= \tfrac{25}{2}\cos(2\tau). \tag{3.19}$$

As expected, the mean is a constant, and the autocorrelation is dependent on lag time τ and independent of time t. Therefore, the process is wide-sense stationary. We also note that $R_{xx}(\tau)$ is periodic, with the same frequency $\omega = 2$ radians per second as $X(t)$. Therefore, not only does the maximum value, which occurs at $\tau = 0$, give the variance (since $\mu = 0$) as $R_{xx}(\tau) = 12.5$, but this will repeat every time the unlagged and lagged signals coincide.   ○

The autocorrelation is a statistic that has a number of special properties. In the case where it is wide-sense stationary, the following properties hold (they are given without proof):

P_1: $R_{xx}(0) = E[X^2(t)]$.
P_2: $\lim_{\tau\to\infty} R_{xx}(\tau) = \mu^2 + $ sinusoidal terms.
P_3: $R_{xx}(\tau)$ is an even function.
P_4: $\max_\tau [R_{xx}(\tau)] = R_{xx}(0)$.
P_5: If $X(t)$ is periodic, then $R_{xx}(\tau)$ is periodic, with the same period as X.

From the above properties, it is possible to deduce an intuitive notion of the nature of the autocorrelation. First, knowledge of the autocorrelation leads immediately to the mean and variance of $X(t)$. For instance, if $R_{xx}(\tau) = 9 + \cos 7\tau + 5e^{-2|\tau|}$, the second moment is $R_{xx}(0) = 15$ by Property P_1 and the mean is $\mu = \pm 3$ by Property P_2. Also notice that $R_{xx}(\tau)$ is an even function, and therefore its graph will be symmetric with respect to the vertical axis. Intuitively, this particular autocorrelation function will be maximum at $\tau = 0$, then decay to the square of the mean. The fact that it is oscillatory with frequency 7 radians per second implies that the original process X also has frequency 7 radians per second. The vertical distance between $\tau = 0$ and the limiting value of τ, in this case $15 - 9 = 6$, is the variance. Note the graph of this typical autocorrelation function given in Figure 3.13.

3.6   CHARACTERIZING RANDOM PROCESSES

It will be recalled that random processes are slightly different from random variables. Random variables are a theoretical collection of *numbers* $\{x_i\}$ from which we randomly

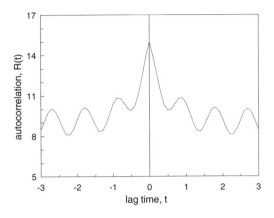

FIGURE 3.13   Graph of the autocorrelation function.

select one instance X and ask questions about probabilities such as $\Pr[X = x_i]$ and $\Pr[X < x_i]$. In the case of random processes, we have a theoretical collection of *signals* or *functions* $\{x_i(t)\}$. Here we tend to ask different sorts of questions, namely questions about the signal itself. We usually do not have an explicit formula for the signal, but rather a characterization of the signal. Therefore, in modeling and simulation studies, it is less important to reproduce accurate signals than to reproduce the signal characteristics. For instance, in generating random signals, it might be most important to create a sequence having a given autocorrelation. The signals, per se, are less important. Thus, simulation studies are based on these two analogous principles:

- *Random variates*:   Create a sequence of numbers (called random variates) with a predefined density function.
- *Random processes*:   Create a sequence of signals (called random processes) with a predefined autocorrelation.

There are an infinite number of statistics by which we can characterize the distribution of a random variable, the most popular being the mean and the standard deviation, or, equivalently, the variance. In the case of a random process, things are slightly different. Again the mean is most important, but the autocorrelation replaces the variance as the statistic of choice. The autocorrelation measures not only the variance (if the process is wide-sense stationary and the mean is zero, the autocorrelation is a function only of the lag time and $R_{xx}(0) = \sigma^2$), but also the signal's relation to itself. It is also possible to characterize a process by the autocovariance, which is defined as $C_{xx}(t, \tau) = E[(x(t) - \mu_x)(x(t + \tau) - \mu_x)]$. It follows that C_{xx} measures the autocorrelation of the process *relative to the mean*. Thus, using C_{xx} in place of R_{xx} adds little information, especially since a great many processes have mean $\mu_x = 0$, in which case $C_{xx} = R_{xx}$. Thus, the autocorrelation is the traditional measure used in practice.

The autocorrelation of a wide-sense stationary process is a function of lag time. As such, it gives important information regarding the relationship of a signal with its historical values. For instance, if $R_{xx}(3)$ and $R_{xx}(5)$ are large but the other autocorrelated values are small, a reasonable model for this signal might be of the form $x(t) = \alpha x(t - 3) + \beta x(t - 5) + w(t)$, where $w(t)$ is a residual signal and α and β are constants to be determined experimentally. At the same time, it should not be forgotten that $R_{xx}(0)$ is always a maximum and that $R_{xx}(-3)$ and $R_{xx}(-5)$ are large too, since the autocorrelation is an even function. Models of stochastic signals such as this are called autoregressive moving-average (ARMA) models, and will be studied further later in this text.

It follows that the computation of the autocorrelation of a random process is important, just as is the case for the mean and standard deviation for a random variable. However, even in the case of simulations, where one has the luxury of exhaustively repeating the simulation time and time again, $R_{xx}(\tau)$ takes inordinate amounts of time to compute. This is because the autocorrelation is defined as a statistical average, and computing it using n trials,

$$R_{xx}(\tau) = E[(X(t)X(t+\tau)]$$

$$\approx \frac{1}{n} \sum_{i=1}^{n} x_i(t)x_i(t+\tau), \tag{3.20}$$

can give a reasonable estimate only when n is large, say 100 runs or so.

However, there sometimes exists another alternative. If the random process is *ergodic*, all statistical averages can be computed as time averages. Thus, for wide-sense stationary ergodic processes, both the mean and autocorrelation can be calculated using the time average, and, for continuous time t,

$$\mu_x = \langle x(t) \rangle$$

$$= \lim_{T \to \infty} \frac{1}{T} \int_0^T x(t)\, dt, \tag{3.21}$$

$$R_{xx}(\tau) = \langle x(t)x(t+\tau) \rangle$$

$$= \lim_{T \to \infty} \frac{1}{T} \int_0^T x(t)x(t+\tau)\, dt. \tag{3.22}$$

For discrete time k, the analogous equations are

$$\mu_x = \langle x(k) \rangle$$

$$= \lim_{n \to \infty} \frac{1}{n} \sum_{k=0}^{n} x(k), \tag{3.23}$$

$$R_{xx}(\tau) = \langle x(k)x(k+\tau) \rangle$$

$$= \lim_{n \to \infty} \frac{1}{n+1} \sum_{k=0}^{n} x(k)x(k+\tau) \tag{3.24}$$

In applying these equations, note that instead of taking the average over many samples, the time average requires only one instance of the process. Since wide-sense stationarity is assumed, the average is taken over each of the lag products, and a result is obtained. In general, there is no easy way by which to prove that this sort of ergodic behavior is valid. However, it is generally simply assumed that if a process is wide-sense stationary, it is also ergodic and the time average can be applied. Experience validates this practice.

In order to see the validity of this approach, consider a random process that is not explicitly known, but whose empirical instances are tabulated. Specifically, suppose there

TABLE 3.5 Four Instances of an Empirical Random Process for $t = 0.1, 2, \ldots, 10$

t	1	2	3	4	Statistical average
0	0.00	0.00	0.00	0.00	0.00
1	0.00	0.00	0.00	0.00	0.00
2	0.87	0.36	0.39	0.08	0.43
3	−0.00	0.04	0.69	0.45	0.29
4	0.39	0.81	0.09	−0.18	0.28
5	0.55	−0.21	0.62	0.72	0.42
6	−0.03	0.55	0.65	−0.01	0.29
7	−0.06	0.27	0.15	0.67	0.26
8	0.55	0.44	−0.05	0.57	0.38
9	0.31	0.36	0.97	0.49	0.53
10	0.50	0.37	−0.37	0.56	0.27
	0.30	0.17	0.30	0.18	Time average, $t = 0, \ldots, 10$
	0.28	0.33	0.34	0.41	Time average, $t = 3, \ldots, 10$

are m instances $i = 1, \ldots, m$ of the process with $n + 1$ time steps $k = 0, 1, \ldots, n$ organized in tabular form as a table with $n + 1$ rows and m columns. A specific example is shown in Table 3.5, where four empirical instances (columns) of this process are given by tabulating the signal values for 11 ($t = 0$ to 10) time slices.

For the process described in Table 3.5, row-wise averages are placed in the right margin and column-wise averages are in the bottom margin. It should come as no surprise that these averages are different, but, except for the first two rows, they are close. Each row average corresponds to the statistical average for each time slice. Therefore, the fact that each row average is comparable for $t = 3, 4, \ldots, 10$ implies that the process is stationary in the mean after a short transient phase from $t = 0$ to $t = 2$ time units. Similarly, each column average corresponds to the time average for each instance of the process. Since each column average is comparable, any time slice will produce the same average estimate. If the time averages are computed without using the transient phase times, both time and statistical averages are roughly the same. Thus, this process is probably ergodic.

These results are shown graphically in Figures 3.14 and 3.15. Figure 3.14 shows that after the transient phase, the process seems to settle into an oscillatory steady state. Since the transient phase ($t = 0, 1, 2$) is significant when compared with the total time horizon ($t = 0, 1, 2, \ldots, 10$), its time-averaging effect is pronounced. Figure 3.15 demonstrates that transients should be deleted from consideration when approximating the time average empirically.

In principle, a similar table is computed for each lag time of an autocorrelation calculation. The following example shows how to create such a table and compute appropriate averages.

EXAMPLE 3.12

Consider the random process defined by the tollowing recursive formula:

$$x(k) = w(k) - \tfrac{1}{2}x(k - 1) - \tfrac{1}{4}x(k - 2),$$

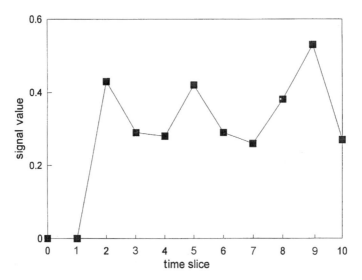

FIGURE 3.14   Statistical averages, Table 3.5.

where $x(0) = 0$, $x(1) = 0$, and $w(k)$ is "noise" (to be discussed later) defined by $w(k) = \text{RND}$, which is uniformly distributed on the interval $[0, 1)$. Generate m instances of $n + 1$ times for the given process. Compute:

(a) the statistical average for each time slice;

(b) the time average for each instance;

(c) the time average for the autocorrelation for each instance. In so doing, create a table similar to Table 3.5 in which instances are represented as columns and time

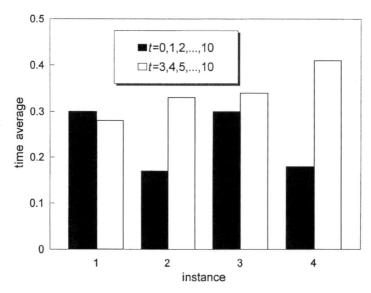

FIGURE 3.15   Time averages for total time horizon and post transient phase times, Table 3.5.

slices by rows. Accordingly, the time averages are vertical (column) averages of each instance and statistical averages are horigontal (row) averages of each time slice.

Solution
First it is necessary to produce the random process itself. Since the process is defined for $i = 1, \ldots, m$ and $k = 0, 1, \ldots, n$, the process will be a function of two variables, $x(i, k)$. This is accomplished by the following algorithm:

```
for i=1 to m
        x(i,0)=0
        x(i,1)=0
        for k=2 to n
                w=RND
                x(i,k) = w-½x(i,k-1)-¼x(i,k-2)
        next k
next i
```

In order to print a table body analogous to that of Table 3.5, a structure inversion in which the loop roles are reversed. Now for each time slice, all instances must be printed. Notice the post-fixed comma [print x(i, k),] used to suppress a line feed as well as the blank print [print] to force one

```
for k=0 to n
        for i=1 to m
                print x(i,k),
        next i
        print
next k
```

Part (a) requires the statistical average $\mu(k)$ for each of the $k = 0, 1, \ldots, n$ time slices. These are found by introducing an accumulator *tot* and averaging:

```
for k=0 to n
        tot=0
        for i=1 to m
                tot=tot+x(i,k)
        next i
        μ(k)=tot/m
next k
```

Time averages need to be computed after the apparent initial transient phase of $k = 0, 1, \ldots, k_0 - 1$, and start accumulation with $k = k_0$. In this example, k_0 is at least 2, since the defining difference equation is of second order. Thus, time averages are computed as follows:

```
for i=1 to m
        tot=0
        for k=k₀ to n
                tot=tot+x(i,k)
        next k
        μ(i)=tot/(n-k₀+1)
next i
```

The step to the time-average autocorrelation for lag times $\tau = 0, \ldots, p$ is also straightforward. However, note that the lag time must be short compared to the reference time since the number of products averaged decreases with p.

```
for i=1 to m
        for τ=0 to p
                tot=0
                for k=k₀ to n-τ
                        tot=tot+x(i,k)*x(i,k+τ)
                next k
                R(i,τ)=tot/(n-τ-k₀+1)
        next τ
next i
```

Since the process was assumed to be ergodic, the autocorrelation is not a function of the reference time – only of the lag time τ.   ○

In principle, a given process can be found to be ergodic or not. However, in applications, this is either impossible or impractical to show. This is illustrated in the following example, which is a continuation of Example 3.11.

EXAMPLE 3.13

Consider the random process $X(t) = 5\sin(2t + \theta)$, where θ is uniformly distributed on $[0, 2\pi]$. By Example 3.11, we know that $X(t)$ is wide-sense stationary with mean $E[X(t)] = 0$ and autocorrelation $R_{xx}(\tau) = \frac{25}{2}\cos(2\tau)$. We now find $R_{xx}(\tau)$ using the time average rather than the statistical average used previously.
(a) Calculate the mean using the time average.
(b) Calculate the autocorrelation using the time average.

Solution
Whereas the statistical average must be independent of the random variable θ, the time average must be independent of the time t. In the case where neither depends on θ or t, the process has a good chance of being ergodic and therefore wide-sense stationary as well. To calculate the mean, it is useful to note that owing to the periodic nature of the process, the limit in the time formula can be replaced with an integral over a single cycle. In this example the period is π, so

$$\mu_x(\text{time average}) = \lim_{T \to \infty} \frac{1}{T} \int_0^T 5\sin(2t + \theta) \, dt$$

$$= \frac{1}{\pi} \int_0^\pi 5\sin(2t + \theta) \, dt$$

$$= -\frac{5}{2\pi} [\cos(2t + \theta)]_0^\pi$$

$$= \frac{5}{2\pi} [\cos\theta - \cos(2\pi + \theta)]$$

$$= 0.$$

Similarly, for the autocorrelation,

$$R_{xx}(\text{time average}) = \lim_{T \to \infty} \int_0^T [5\sin(2t+\theta)][5\sin(2t+2\tau+\theta)]\, dt$$

$$= \frac{1}{\pi} \int_0^\pi [5\sin(2t+\theta)][5\sin(2t+2\tau+\theta)]\, dt$$

$$= \frac{25}{\pi} \int_0^\pi \sin(2t+\theta)\sin(2t+2\tau+\theta)\, dt$$

$$= \frac{25}{2\pi} \int_0^\pi [\cos 2\tau - \cos(4t+2\tau+2\theta)]\, dt$$

$$= \frac{25}{2\pi} [t\cos 2\tau - \tfrac{1}{4}\sin(4t+2\tau+2\theta)]_0^\pi$$

$$= \frac{25}{2\pi} [\pi\cos 2\tau - \tfrac{1}{4}\sin(4\pi+2\tau+2\theta) + \tfrac{1}{4}\sin(2\tau+2\theta)]$$

$$= \tfrac{25}{2}\cos 2\tau.$$

Even though the previous examples show processes that are stationary in the mean (the mean is constant), wide-sense stationary (the mean is constant and the autocorrelation depends only on the lag time), and ergodic (time and statistical averages give identical results), it should be noted that is often not the case. Here we consider an example where the process is not stationary.

EXAMPLE 3.14

Previously we have shown that the process $X(t) = 5\sin(2t+\theta)$, where θ is a random variable, is stationary in the mean, wide-sense stationary (Example 3.11), and ergodic (Example 3.13), as long as θ is uniformly distributed on $[0, 2\pi]$. Find what kinds of stationarity apply to the process in which θ is uniformly distributed on $[0, \pi]$.

Solution
It is easy to see that this process is not stationary in the mean, since the sine function has period 2π and neither of the θ-intervals is only half this large. Formally, this can be demonstrated as follows:

$$\mu_x(\text{statistical average}) = \int_0^\pi \frac{1}{\pi} 5\sin(2t+\theta)\, d\theta$$

$$= -\frac{5}{\pi}[\cos(2t+\theta)|_0^\pi$$

$$= \frac{-5}{\pi}[\cos(2t+\pi) - \cos 2t]$$

$$= -\frac{10}{\pi}\cos 2t.$$

Since the mean is not a constant, this process is not stationary in the mean, and therefore it follows that this is not wide-sense stationary either. On the other hand, computing the time average of this process results in calculations identical to those in Example 3.13, and therefore μ_x(time average) $= 0$. It follows that this process cannot be ergodic either. In fact, in order for any process to be ergodic, it must first be wide-sense stationary, since the time average can never depend on time.

Calculation of the autocorrelation is similar:

$$R_{xx}(\text{statistical average}) = \int_0^\pi \frac{1}{\pi} [5\sin(2t + \theta)][5\sin(2t + 2\tau + \theta)] \, d\theta$$

$$= \frac{25}{2\pi} \int_0^\pi [\cos 2t - \cos(4t + 2\tau + 2\theta)] \, d\theta$$

$$= \frac{25}{2\pi} \left[\theta \cos 2\tau - \tfrac{1}{2}\sin(4t + 2\tau + 2\theta)\right]_0^\pi$$

$$= \frac{25}{2\pi} \left[\pi \cos 2\tau - \tfrac{1}{2}\sin(4t + 2\tau + 2\pi) + \tfrac{1}{2}\sin(4t + 2\tau)\right]$$

$$= \tfrac{25}{2} \cos 2\tau$$

As on the interval $[0, 2\pi]$, the autocorrelation matches that of the time average. Even so, the process is not wide-sense stationary or ergodic, since it is not stationary in the mean.   ○

3.7   GENERATING RANDOM PROCESSES

From a simulation point of view, random processes are random signals. There are two basic types of random signals: regular and episodic. In the regular case, time is considered to be a synchronous clock-like device that produces a regular beat. At each beat, a random signal is generated, resulting in a sequence of values at equally spaced time increments. This procedure, called a *regular random process*, can be applied in either discrete time or continuous time. The second type of signal considers the case where events do not occur regularly, but only at irregular asynchronous times. When the events occur, a change (be it deterministic or random) occurs in the signal value. In between these events, the system value remains unchanged. Processes generated in this manner are called *episodic random processes*.

Real systems in which control is either monolithic or autonomous can usually be represented by a deterministic or regular random process, since all driving decisions are local at a single unit. Therefore, we refer to these systems as *time-driven*, since decision making can only depend on time. This is in contrast to networks of peer-type units, where events external to a specific subunit can affect decision making at random times. Such systems are called *event-driven*, since decisions must be based on external events rather than an internal agenda. Clearly there is one other possibility as well: a system may be a hybrid in which time-driven methods and event-driven methods both apply. This is a topic beyond the scope of this text, and will only be considered on an ad hoc basis when needed.

From an engineering point of view, control systems come in two varieties. So-called *open-loop* control is a system design that does not exploit feedback. All decisions are made autonomously or change only with time. This is in contrast to *closed-loop* control, in which feedback from other system components or sensors to the external environment can influence a control decision. The execution of an open-loop system often takes the form of sequential computer code, while closed-loop, event-driven systems are written using interrupts. Thus, open-loop systems tend to be time-driven models and closed-loop systems tend to be event-driven.

Episodic Random Processes

Episodic random processes are characterized by non-deterministic inter-event times. If these times are exponentially distributed, the process is called Markovian. Further, if the signal value $x(t)$ is the number of events up to that time, the process is called Poisson. Simulating such a system requires that both the event occurrence times and signal values at these time be generated. In particular, we say that it is necessary to *schedule* event times $\{t_k\}$ and *generate* the event sequence $\{x_k\}$. For a Poisson process, this proceeds as follows:

- Schedule: Since each event occurs as a succession of exponentially distributed random event times, $t_k = t_{k-1} - \mu \ln(\text{RND})$ by Equation (3.8) for events $k > 0$. Beginning at time $t_0 = 0$ results in the required Markovian sequence.
- Generate: The signal value only changes at event times. Further, a Poisson process requires that the signal value be the previous state incremented by one. Combined with the fact that initially there are no events, $x_0 = 0$ and $x_k = x_{k-1} + 1$, $k > 0$.

Listing 3.6 produces a Poisson random process with mean inter-event time of $\mu = 2$ seconds for $m = 10$ events. The results are tabulated below. However, if it is necessary to generate the signal values $x(t)$ for regular time intervals so that the ordered pairs (t, x) can be plotted over a continuous-time interval $[0, t(m)]$. This is also done using the algorithm given in Listing 3.6 and graphed in Figure 3.16.

Time, $t(k)$	Signal, $x(k)$
0.00	0
1.44	1
3.28	2
5.89	3
9.95	4
10.91	5
16.26	6
18.27	7
19.78	8
21.17	9
22.64	10

Telegraph Processes

There are a number of random processes that follow more or less the same recipe as the Poisson process. While they cannot be expressed by a simple mathematical equation,

```
t(0)=0
x(0)=0
for k = 1 to m
        t(k)=t(k-1) - μ*ln(RND)
        x(k)=x(k) + 1
next k
for k = 0 to m-1
        for t = t(k) to t(k+1) step h
                print t, x(k)
        next t
next k
```

LISTING 3.6   Generating a single instance of a Poisson process with mean μ on the time horizon $[0, t(m)]$.

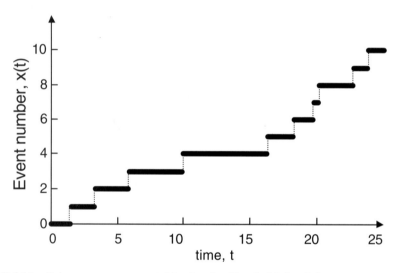

FIGURE 3.16   Poisson process generated by the algorithm in Listing 3.6.

they are still easily modeled and simulated. An important one is the telegraph process, which is a digital counterpart to the continuous-time sinusoidal process described in Examples 3.11 and 3.13. The telegraph process models a digital pulse train of bipolar bits as they proceed *synchronously* (at regular, predetermined time intervals) over a communications link. As with the sinusoidal process, the amplitude and frequency are fixed, known constants. However, the phase and the sign of the amplitude are each random.

Let $X(t)$ be either $+1$ or -1 with equal probability. Every T seconds, a new x-value arrives, but we do not know just when the data stream starts. Let it start somewhere between 0 and T with equal probability. Thus the randomness of this process is twofold: the initial time t_0 and the x-value. Proceeding as with Poisson, a reasonable algorithm by which to schedule times $\{t_k\}$ and *generate* events $\{x_k\}$ is the following:

- Schedule: The phrase "equal probability" for t_0 must be translated into a technical counterpart. More precisely, we should say that t_0 is uniformly distributed on the interval $[0, T]$, where the density function is $1/T$ for a fixed inter-event time T. Since successive wave pulses arrive each T time units,

$$t_0 = \frac{1}{T}\text{RND}, \qquad t_k = t_{k-1} + T.$$

- Generate: The amplitude of the signal is known to be unity, but the sign is random "with equal probability". In this case, the amplitude random variable is discrete, so two equal values means that $\Pr[X(t) = 1] = \frac{1}{2}$ and $\Pr[X(t) = -1] = \frac{1}{2}$. It follows that if $b = \text{RND}$, then $x_k = -1$, $b < 0.5$, and $x_k = 1$, $b \geqslant 0.5$, for all k.

From these descriptions, the telegraph signal can be simulated by changing the algorithm of Listing 3.6 in the obvious manner.

In executing a simulation involving random processes, signal statistics are usually computed in order to study various modeling effects. Typically these statistics are the process mean and autocorrelation. For the case of the synchronous telegraph process defined above, the mean is $\mu_X = 0$ and the autocorrelation is

$$R_{XX}(\tau) = \begin{cases} 1 - \dfrac{|\tau|}{T}, & |\tau| < T, \\ 0, & |\tau| \geqslant T. \end{cases} \tag{3.25}$$

It will be noted that, from the autocorrelation function, the mean is $\mu_X = \sqrt{\lim_{\tau \to \infty} R_{xx}(\tau)} = 0$ and the variance $\sigma_X^2 = R_{xx}(0) - \mu_X^2 = 1$. The graph of the theoretical autocorrelation is shown in Figure 3.17.

It should come as no surprise that the synchronous telegraph process is wide-sense stationary, since, from the defining description, there is no mention of reference time t. Thus, the formulation will only involve the lag time τ. Combined with the fact that the time average is obviously zero (note that every instance goes through a sequence of ± 1 signals), ergodicity is a reasonable assumption. These things are borne out theoretically as well. On the other hand, if the theoretical autocorrelation were not available, one would have to use the experimental methods outlined earlier to calculate both the mean and auto-

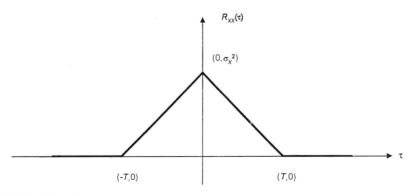

FIGURE 3.17　Autocorrelation function for the synchronous telegraph process.

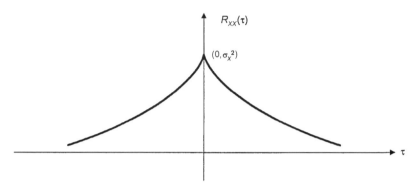

FIGURE 3.18   Autocorrelation function for the asynchronous telegraph process.

correlation. Fortunately, since ergodic behavior is reasonable from the process description, time averages are easiest, and a similar graph comes from such an empirical evaluation.

Another variation on the episodic random process theme is the *asynchronous telegraph process*. This process is identical to the synchronous one defined above except that an asynchronous process models an asynchronous signal in that each pulse is of varying (positive) length. As might be expected, the length varies according to an appropriate probabilistic model, which typically is Poisson. That is, the inter-event times are exponentially distributed. Using the rational for the Poisson and synchronous telegraph hypotheses, this is modeled mathematically as follows:

- Schedule: $t_0 = \dfrac{1}{T}$RND and $t_k = t_{k-1} - \mu \ln(\text{RND})$ for all k.
- Generate: if $b = \text{RND}$, then $x_k = -1$ for $b < 0.5$, and $x_k = 1$ for $b \geqslant 0.5$, for all k.

This model actually matches practice remarkably well. Even so, the major reason that it is employed is that the Poisson process is one of the few models that leads to an explicit formula for the autocorrelation. Again, given without proof are the facts that $\mu_X = 0$ and $R_{xx}(\tau) = e^{-2\lambda|\tau|}$. The graph of the autocorrelation is similar in shape to Figure 3.17, except that the "teepee" shape is now a sloped curve that approaches the τ axis asymptotically as shown in Figure 3.18.

In creating episodic models, there are two scheduling parameters that can be randomized: the initial time t_0 and the inter-arrival time $t_k - t_{k-1}$. If only t_0 is random, the resulting process is Poisson. On the other hand, if only the inter-event time $t_k - t_{k-1}$ is random, it is a synchronous telegraph. Finally, if both t_0 and $t_k - t_{k-1}$ are random, it is asynchronous. Of course, there is still one more possibility, where both t_0 and $t_k - t_{k-1}$ are deterministic. When this happens, we have a *regular* random process. This is summarized in Table 3.6.

TABLE 3.6   Classification of Episodic Signals

	$t_k - t_{k-1}$ deterministic	$t_k - t_{k-1}$ random
t_0 deterministic	Regular	Poisson
t_0 random	Synchronous telegraph	Synchronous telegraph

Regular Random Processes

Scheduling in regular random process model is automatic. Both the initial time and the inter-arrival times are given, and are usually fixed throughout the simulation. In this situation, there are three important cases of note:

1. If the signal is discrete, time is defined on the basis of a non-negative integer k. Between successive times k and $k + 1$, it is assumed that the system is dormant and there are no signal changes.
2. It is also possible to have a system in which system dynamics can occur between discrete times k. In this case, the discrete time is related to the continuous time t by $t = kT$, where T is the sampling interval and $1/T$ is the sampling frequency. By defining time in this way, it is possible to handle several concurrent time bases at once.
3. The last option is continuous time t. With continuous time, models are often described by differential equations. Since, in general, it is impossible to solve such mathematical systems analytically, numerical methods are often employed. For this reason, we define the integration step size h to be a small time increment over which no system dynamics can take place and by which appropriate numerical methods can be employed where $t = hk$.

3.8 ____ RANDOM WALKS

A straightforward and rather interesting discrete random process is the *random walk*. In its most basic form, the random walk process is one in which the signal value is either incremented or decremented by 1 at each time step. In this way, it imitates a walker who takes one step in a random direction each second of time. A series of such signals are shown in Figure 3.19 for the case where each direction is equally probable and the initial state is $x(0) = 0$.

Figure 3.19 illustrates two features of random walks. First, the mean is zero, since advance and retreat are equally likely. In spite of this, as time increases, the sequences spread out – that is, the standard deviation increases directly with time k.

In a more general context, if the probability that the walker's position advances is p, the probability that retreats is $1 - p$. Regardless of the value of p, if the walker is currently in position $x = 0$, the only allowed next positions are $x = 1$ and $x = -1$. Similarly, after the position $x = 2$, the next position must be either $x = 1$ or $x = 3$. Still, there is considerable symmetry, and it is straightforward to derive an expression for the probability of being in position n at time k, $\Pr[X(k) = n]$. Assuming there are no barriers in either the positive or negative directions, the probability is

$$\Pr[X(k) = n] = \binom{k}{\frac{1}{2}(k + n)} p^{(k+n)/2}(1 - p)^{(k-n)/2} \quad \text{for} \quad n = -k, -k + 2, \ldots, k - 2, k.$$

$$(3.26)$$

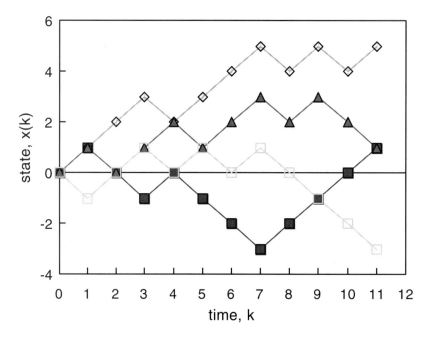

FIGURE 3.19   Four random walks: $p = \frac{1}{2}$.

The striking similarity of Equation (3.26) to the binomial distribution (see Appendix B) is no coincidence, and substituting the variable $i = \frac{1}{2}(k + n)$ results in the formula for the binomial mass function. Notice that, for any time, the allowed positions are either all even or all odd. For instance, the probability of a random walk reaching position x at time $k = 3$ is $\Pr[X(3) = 1] = 3p^2(1 - p)$. However, we take $\Pr[X(3) = 2]$ to be 0, since the combinatoric formula $\binom{3}{2.5}$ is undefined.

The mean of the random walk is $\mu_X(k) = k(2p - 1)$. This can be shown by taking advantage of our knowledge of the binomial distribution as follows:

$$E[X(k)] = \sum_{n=-k(\text{step } 2)}^{k} n \binom{k}{\frac{1}{2}(k + n)} p^{(k+n)/2}(1 - p)^{(k-n)/2}$$

$$= \sum_{i=0}^{k} (2i - k) \binom{k}{i} p^i(1 - p)^{k-i}$$

$$= 2 \sum_{i=0}^{k} i \binom{k}{i} p^i(1 - p)^{k-i} - k \sum_{i=0}^{k} \binom{k}{i} p^i(1 - p)^{k-i}$$

$$= 2(kp) - k(1)$$

$$= k(2p - 1). \tag{3.27}$$

Proceeding in the same manner, it can be shown that the variance for a random walk is $\sigma^2 = 4kp(1-p)$, where it will be noted that the variance, like the mean, is time-dependent. It follows that the random walk process is not stationary in any sense.

In the special case where $p = \frac{1}{2}$, $\mu_x(k) = 0$ and $\sigma_x(k) = \sqrt{k}$. As anticipated earlier, the standard deviation is a still a function of time, even though the mean is not. However, note that the standard deviation continues to increase with time, implying that the asymptotic case of the random walk is concave-right, even though it never levels off and converges. Therefore, in this simplest case, the random walk process is not wide-sense stationary.

EXAMPLE 3.15

Consider the case of a random walk in which $p = 0.4$. Ten instances are graphed in Figure 3.20. It will be noted that since the probability of advancement is less than the probability of retreat, the general tenancy is down. This can be anticipated by plotting *standard error* curves of $\mu_x \pm \sigma_x$ and $\mu_x \pm 2\sigma_x$ along with the walks themselves to demonstrate how the process is contained within the variance. ○

There are a number of creative ways by which to generalize random walks. One of these is to allow for two dimensions, in which the walker steps either left or right and either up or down at each time k. For that matter, there is nothing sacred about having an integral

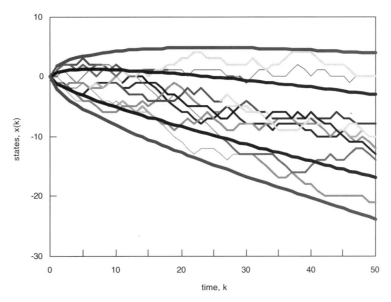

FIGURE 3.20   Ten instances of the random walk process, $p = 0.4$.

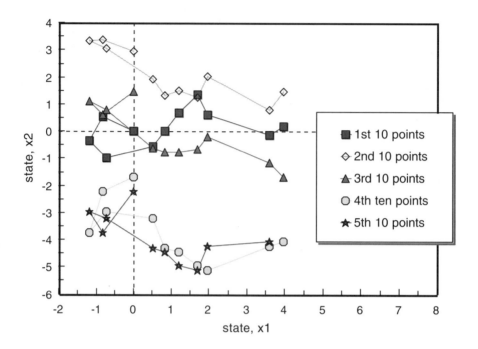

FIGURE 3.21   A pseudo random walk in two dimensions and standardized Gaussian increments.

position either. It is entirely possible that the walker journeys in random step sizes too. Figure 3.21 shows state space for a simulation in which the up/down and left/right protocol is taken and each increment is a normalized Gaussian random variable of mean 0 and standard deviation 1. This algorithm is implemented in Listing 3.7.

```
x₁=0
x₂=0
for k=1 to n
        x₁ = x₁ + √(−2*ln(RND))*cos(2π*RND)
        x₂ = x₂ + √(−2*ln(RND))*cos(2π*RND)
        print x₁, x₂
next k
```

LISTING 3.7   Algorithm for a two-dimensional random walk with Gaussian-sized steps at each time step.

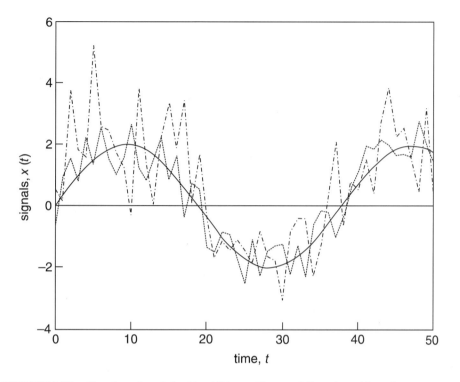

FIGURE 3.22   Signal confounded with additive uniform and Gaussian white noise.

tendencies, standard random walk error curves $\pm\sqrt{k}$ and $\pm 2\sqrt{k}$ are shown too. The reason that random walk standard errors are used is because in the limit as $k \to \infty$, the Gaussian distribution is asymptotically approximated by the binomial distribution.

White noise is important for several reasons. First, as we have seen above, it presents a realistic noise source, and realistic simulations will require such a facility. Since discrete white noise is the result of a sequence of independent and identically distributed random variates with mean zero, it reasonably models an ideal signal contaminated with random noise that cannot be anticipated. Another reason that white noise is so important is that it can be used to generate a large set of random processes besides white noise. Even though white noise presents a time series of independent events (property $\mathbf{P_3}$), these other random processes are not. Therefore, they are collectively referred to as *pink* or *colored noise*.

A colored noise signal $x(t)$ is often characterized by its spectral density $S_{xx}(\omega)$, which is the Fourier transform of the autocorrelation $R_x(\tau)$:

$$S_x(\omega) = \int_{-\infty}^{\infty} R_{xx}(\tau) \cos \omega\tau \; d\tau. \qquad (3.29)$$

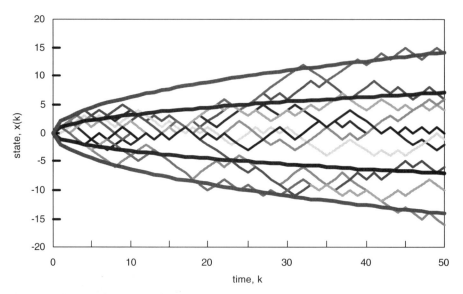

FIGURE 3.23 Ten instances of Brownian motion using the standardized Gaussian, $\sigma = 1$. Standardized random walk error curves are included for comparison.

This converts the time-domain autocorrelation into an equivalent frequency-domain description. The frequency ω has units of radians per second. Alternatively, the frequency can be expressed as $f = \omega/2\pi$, which is in cycles per second or hertz. Because of this equivalence, we can also extract $R_{xx}(\tau)$ from $S_{xx}(\omega)$ using the inverse Fourier transform. In other words, knowing one is equivalent to knowing the other:

$$R_{xx}(\tau) = \frac{1}{2\pi} \int_{-\infty}^{\infty} S_{xx}(\omega) \cos \omega \tau \, d\omega \qquad (3.30)$$

EXAMPLE 3.17

Consider random processes $X(t)$ and $Y(t)$ characterized by each of the following:
(a) autocorrelation $R_{xx}(\tau) = 4\delta(t)$;
(b) spectral density $S_{yy}(\omega) = \delta(\omega + 3) + \delta(\omega - 3)$,
where $\delta(\cdot)$ is the Dirac delta or impulse function. Find the spectral density $S_{xx}(\omega)$ and autocorrelation $R_{yy}(\tau)$.

Solution
The autocorrelation and spectral density can each be found using Equations (3.29) and (3.30). Since each problem is defined in terms of impulse functions, the so-called *sifting property* for impulse functions is especially useful here. This property states that for any continuous function $f(t)$, $\int_{-\infty}^{\infty} \delta(t - a)f(t) \, dt = f(a)$. In other words, integrals with delta-function integrands are especially easy to integrate.

(a) The random process X is evidently white noise with variance $\sigma_x^2 = 4$. The spectral density follows immediately from Equation (3.29):

$$S_{xx}(\omega) = \int_{-\infty}^{\infty} R_{xx}(\tau) \cos \omega \tau \, d\tau$$

$$= 4 \cos 0 = 4.$$

Since $S_{xx}(\omega)$ is constant, it is independent of the frequency. This illustrates a rather general relationship between the autocorrelation and spectral density functions. $R_{xx}(\tau)$ being non-zero at a single point (the origin) is analogous to $S_{xx}(\omega)$ being uniformly defined over the entire frequency spectrum. Accordingly, if $R_{xx}(\tau)$ were non-zero over a finite interval, $S_{xx}(\omega)$ would be band-limited in that there would be a set of frequencies over which it would be much more significant than others. As R expands its range, the range of S contracts. This example also illustrates why we refer to white noise as "white". Since the spectral density is uniform over the infinite range of frequencies, the signal is like that of white light in that it includes all wavelengths (or frequencies) at a constant intensity.

(b) The sifting property is also useful in going from the frequency domain to the time domain. $R_{yy}(\tau)$ follows directly from $S_{yy}(\omega)$:

$$R_{yy}(\tau) = \frac{1}{2\pi} \int_{-\infty}^{\infty} [\delta(\omega + 3) + \delta(\omega - 3)] \cos \omega \tau \, d\omega$$

$$= \frac{1}{2\pi} [\cos(-3\tau) + \cos 3\tau]$$

$$= \frac{1}{\pi} \cos 3\tau.$$

Again we note the reciprocal relationship between R and S. In this example, $S_{yy}(\omega)$ is narrowly non-zero at two points, while $R_{yy}(\tau)$ is persistent for all time τ. Also, as required, both the autocorrelation and spectral density are even functions.

○

Knowledge of the spectral density is important, since it gives the behavior of a process at different frequencies. For instance, a random process with $S_1(\omega) = 4/(\omega^2 + 1)$ acts as a low-pass filter (LPF), since for small ω, $S_1(\omega)$ is large while for large ω, $S_1(\omega)$ is, to all intents, zero. Thus, low-frequency signals are amplified and high-frequency signals are attenuated. Similarly, $S_2(\omega) = 4\omega^2/(\omega^2 + 1)$ is a high-pass filter. In fact, $S_1(\omega)$ and $S_2(\omega)$ are typical spectral density functions, since they exhibit the following properties:

P₁: $S(\omega) \geqslant 0$.
P₂: $S(\omega) = S(-\omega)$.
P₃: The signal power between frequencies a and b is $\int_a^b S(\omega) \, d\omega$.
P₄: $S(\omega) = 1$ for a white noise signal with unit variance.

EXAMPLE 3.18

Consider the family of low-pass filters defined by

$$S(\omega) = \frac{1}{1 + \tau^2\omega^2},$$

for a constant unknown parameter τ. $S(\omega)$ is graphed in Figure 3.24 for $\tau = 0.1$, 0.2, 0.5, and 1.0. Notice that the width of the *pass band*, that is, the set of frequencies for which $S(\omega)$ is comparatively large, varies inversely with τ. It follows that the *stop band*, that is, the set of frequencies for which $S(\omega)$ is very small, becomes more confining as τ becomes large, and vice versa.   ○

$S_{xx}(\omega)$ describes the theoretical frequency characteristics of signal $x(t)$. However, in practice, only the sampled signal $X(k)$ is available in the form of a time series. From $x(k)$, calculation of $R_{xx}(\tau)$ follows using Equation (3.24). A quick application of the fast Fourier transform (FFT) produces an estimate for $S_{xx}(\omega)$. This process is illustrated in Figure 3.25.

However, the inverse problem is not so easy. In simulation practice, it is often necessary to produce a time series $x(k)$ with a given spectral density. That is, we are given the frequency characteristics, and from these we must create an algorithm by which to produce the required time series. The questions whether it is possible to produce such a series and how this can be done are answered by the so-called *Shaping Theorem*:

Shaping Theorem *If $S_{xx}(\omega)$ is factorizable such that $S_{xx}(\omega) = H(s)H(-s)$, where $s = j\omega$, then $X(s) = H(s)W(s)$, where X, H, and W are the Laplace transforms of the required signal $x(t)$, impulse response $h(t)$, and unit white noise $w(t)$.*

In other words, it is sometimes possible to find a filter (defined by $H(s)$) such that the required signal $x(t)$ can be found. If such a filter exists, simply driving the filter with unity white noise input will produce the desired signal. We say that the *filter $H(s)$ shapes the*

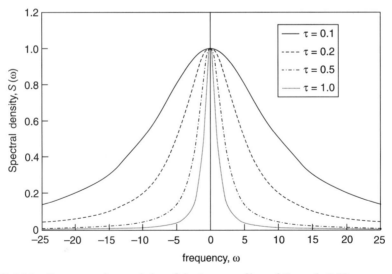

FIGURE 3.24   Frequency characteristics of the low-pass filter of Example 3.18.

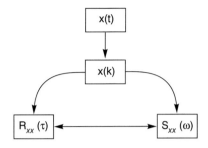

FIGURE 3.25   Finding the spectral density of a continuous-time signal $x(t)$.

white noise into $x(t)$. Sampling the result produces the required series $x(k)$. The application of this theorem is best illustrated with an example.

EXAMPLE 3.19

Create an algorithm that generates a time series $x(k)$ with spectral density:

$$S_{xx}(\omega) = \frac{9}{\omega^2 + 4}.$$

Solution

The first step is to factor the spectral density:

$$S_{xx}(\omega) = \frac{9}{\omega^2 + 4}$$

$$= \frac{9}{4 - s^2}$$

$$= \left(\frac{3}{2 + s}\right)\left(\frac{3}{2 - s}\right).$$

It follows that $H(s) = 3/(2 + s)$, from which

$$X(s) = \frac{3}{s + 2} W(s) \quad \text{and} \quad sX(s) + 2X(s) = 3W(s).$$

This can be returned to the time domain by recalling that the inverse Laplace transform of $sX(s)$ is the derivative $\dot{x}(t)$. Thus, $\dot{x}(t) + 2x(t) = 3w(t)$ is the differential equation that generates the signal $x(t)$ producing the required spectral density.

In order to generate the actual time series $x(k)$, a numerical integration technique, such as Euler, produces

$$x(k + 1) = x(k) + h[3w(k) - 2x(k)].$$

```
k=0
x=0
print k, x
for k=1 to n
        w=√3(2*RND-1)
        x=x+h(3w-2x)
        print k, x
next k
```

LISTING 3.8   Implementation of a shaping filter, Example 3.19.

Since $w(k)$ is white noise with unit variance, the algorithm given in Listing 3.8 generates the required $x(k)$, where we have arbitrarily assumed a uniform distribution for the white noise $w(k)$.

In actuality, there are an infinite number of possible time series signals $x(k)$. This is because there is no requirement on the probability distribution of the white noise random variate $w(k)$. Any probability distribution will suffice, as long as its mean $\mu_w = 0$ and its variance $\sigma_w^2 = 1$. For instance, $w(k) = \sqrt{-2\ln(\text{RND})}\cos(2\pi\text{RND})$ generates white noise based on a Gaussian distribution rather than the uniform distribution specified in Listing 3.8. Both produce time series with the required spectral density.   ○

Techniques such as that shown in Example 3.19 make white noise important not only in its own right, but also because it can create many different noise models by the simple application of a shaping filter. This filter can be used to mimic a noisy signal by matching either its time-domain (autocorrelation) or frequency-domain (spectral density) characteristics. This stands in contrast to the probabilistic techniques used to handle inter-arrival times of event-driven models. Combined, these two approaches form a very well-stocked tool box for the system modeler in need of simulating and driving systems with realistic signals.

BIBLIOGRAPHY

Childers, D. G., *Probability and Random Processes*. McGraw-Hill–Irwin, 1997.

Cinlar, E., *Introduction to Stochastic Processes*. Prentice-Hall, 1975.

Cooper, G. R. and C. D. McGillem, *Probabilistic Methods of Signal and System Analysis*, 3rd edn. Oxford University Press, 1998.

Feller, W., *An Introduction to Probability Theory and Its Applications*. Wiley, 1971.

Gardener, W., *Introduction to Random Processes*, 2nd edn. McGraw-Hill, 1990.

Goodman, R., *Introduction to Stochastic Models*. Benjamin-Cummings, 1988.

Helstrom, C. W., *Probability and Stochastic Processes for Engineers*, 2nd edn. MacMillan, 1991.

Hoel, P. G., *Introduction to Mathematical Statistics*, 4th edn. Wiley, 1971.

Knuth, D. E., *The Art of Computer Programming*, Vol. 2: *Seminumerical Algorithms*. Addison-Wesley, 1969.

Law, A. and D. Kelton, *Simulation, Modeling and Analysis*. McGraw-Hill, 1991.

Leon-Garcia, A., *Probability and Random Processes for Electrical Engineering*. Addison-Wesley, 1989.

Nelson, B., *Stochastic Modeling, Analysis and Simulation*. McGraw-Hill, 1995.

Press, W. H., B. P. Flannery, S. A. Teukolsky, and W. T. Vettering, *Numerical Recipes: The Art of Scientific Computing*. Cambridge University Press, 1986.

Proakis, J. G. and D. G. Manolakis, *Digital Signal Processing, Principles, Algorithms and Applications*, 3rd edn. Prentice-Hall, 1996.

Shoup, T. E., *Applied Numerical Methods for the Micro-Computer*. Prentice-Hall, 1984.

Spiegel, M. R., *Probability and Statistics* [Schaum's Outline Series]. McGraw-Hill, 1975.

EXERCISES

3.1 Formulate a linear congruential generator (LCG) with $a = 13$, $c = 7$, and $m = 32$.

  (a) Using $Z_0 = 9$, write a computer program to generate a sequence of $U(0, 1)$ random numbers. List the first 32 pseudorandom numbers in a table similar to Table 3.1.

  (b) Generate 100 of the pseudorandom numbers defined by this LCG. Complete a frequency distribution table similar to that of Table 3.2 using a bin size of $\Delta x = 0.2$ over the range $[0, 1]$. In other words, use the following classes and class marks:

Interval number	Class interval	Class mark
1	[0.0, 0.2]	0.1
2	[0.2, 0.4]	0.3
3	[0.4, 0.6]	0.5
4	[0.6, 0.8]	0.7
5	[0.8, 1.0]	0.9

  (c) Test the random numbers tabulated in part (b) for uniformity at the 5% and 10% levels of significance, using the chi-square test.

  (d) Implement a *runs test* on the data generated in part (b). Find the frequency of each sequence length, the statistic R, and your conclusion.

3.2 Repeat Exercise 3.1 using an LCG with $a = 19$, $c = 11$, and $m = 256$, and an initial seed of $Z_0 = 119$. Test for uniformity and runs using 10000 random numbers. This time use bins of size $\Delta x = 0.05$ over $[0, 1]$.

3.3 Generate 10000 pseudorandom numbers using the random number generators resident in your compiler.

  (a) Create a frequency distribution table of size $\Delta x = 0.01$ over $[0, 1]$ and sketch the histogram of your results.

  (b) Test the uniformity of your random number generator at the 5% level.

  (c) Perform the runs test on the data generated. Tabulate the frequency of each sequence length and find the statistic R. What is your conclusion?

3.4 Another means by which to test the sequential independence of an LCG is to compute the correlation between successive pseudorandom numbers. Recall that the correlation coefficient of two random variables is

$$\rho = E\left[\frac{X - \mu_x}{\sigma_x} \frac{Y - \mu_y}{\sigma_y}\right].$$

(a) Show that if $\mu_x = \mu_y \equiv \mu$ and $\sigma_x = \sigma_y \equiv \sigma$, then $\rho = (E[XY] - \mu^2)/\sigma^2$.

(b) Note that if $X = X(k)$ and $Y = X(k+1)$, then the results of part (a) follow immediately. Assuming a discrete data set $\{x(k)\}$ of n points, show that the correlation coefficient can be computed by

$$\rho = \frac{n \sum_{k=1}^{n} x(k)x(k+1) - \left(\sum_{k=1}^{n} x(k)\right)^2}{n \sum_{k=1}^{n} x^2(k) - \left(\sum_{k=1}^{n} x(k)\right)^2}.$$

(c) A *scattergram* of two random variables X and Y is a plot of (x_i, y_i) for $i = 1, 2, \ldots, n$. Using the data generated from the pseudorandom numbers in Exercise 3.3, create the scattergram of each random variate $x(k)$ with its successor $x(k+1)$.

(d) Assuming the result of part (b), compute the correlation coefficient for the pseudorandom numbers generated in Exercise 3.3. Interpret your result.

3.5 Using the Gaussian formula $Z = \sqrt{-2\ln(\text{RND})}\cos(2\pi\text{RND})$,

(a) Generate 1000 Gaussian-distributed random numbers with mean 3 and standard deviation 2.

(b) Plot the histogram of the data found in part (a) using 60 bins of size $\Delta x = 0.2$ on the interval $[-3, 9]$.

(c) Compute the mean and standard deviation of the generated data.

3.6 Assuming the definition of an LCG, Equation (3.1), show that

$$z_k = \left[a^k z_0 + \frac{c(a^k - 1)}{a - 1} \right] \text{mod}(m).$$

3.7 Implement the rejection method to generate 1000 random variates for the Beta distribution with $a = 2.7$ and $b = 4.3$.

(a) Use the target region bounded vertically by $x = 0$ and $x = 1$, and horizontally by the x-axis along with each of the following heights: (i) $y = 10$, (ii) $y = 9$, (iii) $y = 8$, (iv) $y = 7$, (v) $y = 6$.

(b) Find the optimal height for the target rectangle and generate the 1000 variates.

(c) Define the efficiency of a rejection algorithm as the ratio of accepted random variates to total candidate random variates. Compare the efficiencies of each region described in parts (a) and (b) by graphing the efficiency as a function of height.

3.8 Use the rejection method to generate 1000 random variates for the 3-Erlang distribution with mean 4. Use the region in the first quadrant bounded by $x = 3$ and $y = 2$ as your target region.

(a) Tabulate your results in a frequency distribution table similar to that of Table 3.3.

(b) Derive a formula for the expected frequency in an arbitrary bin $[a, b]$.

(c) Using the results of parts (a) and (b), complete the frequency distribution table and graph the corresponding histogram.

(d) Use chi-square tests at the 5% and 1% levels to compare the empirical results against those predicted theoretically.

3.9 Use the formula method to generate 1000 random variates for the normalized Gaussian distribution of mean 0 and standard deviation 1.

(a) Create a frequency distribution table with $\Delta x = 0.5$ on the interval $[-3, 3]$ similar to that of Table 3.3 to tabulate your results.

(b) Use the table given in Appendix E to compute the expected frequencies for each bin for the table in part (a). Create a comparative histogram similar to that of Figure 3.4.

(c) Use chi-square tests at the 5% and 1% levels to compare the empirical results against those predicted theoretically.

3.10 Consider the empirically generated random variates $\{1, 3, 3, 4, 6, 7, 10, 15\}$.

(a) Assuming the above data to characteristic of the actual distribution, find the distribution function $F(x)$.

(b) Find the inverse distribution function $F^{-1}(x)$.

(c) Generate 1000 random variates with the distribution characterized by the given variates.

(d) Sketch the resulting distribution and density functions based on the random variates of part (c).

3.11 Consider a discrete distribution with the following probability mass points:

x	$p(x)$
0	$\frac{1}{243}$
1	$\frac{18}{243}$
2	$\frac{40}{243}$
3	$\frac{80}{243}$
4	$\frac{80}{243}$
5	$\frac{24}{243}$

(a) Implement an algorithm to generate 10 000 random variates with the given distribution.

(b) Using a frequency distribution table, tabulate the results and graph the resulting histograms.

(c) Test the goodness of fit of the results using a chi-square test at the 1% confidence level.

3.12 Recall that a geometric distribution gives the probability that it will take exactly x trials to succeed in an experiment where the probability of success in a single trial is p. The probability mass function (see Appendix B) is $p(x) = p(1-p)^{x-1}$, $x = 0, 1, \ldots$. Since p is a probability, $0 \leqslant p \leqslant 1$.

(a) Let $p = \frac{1}{3}$ and create a sequence of 100 random variates.

(b) Create a histogram comparing the expected and experimental frequencies for the data formed in part (a).

(c) Apply the chi-square test at the 2% level to test the goodness of fit of the empirically generated distribution.

3.13 Find the mean, variance, and autocorrelation of a random process consisting of the following five equally likely signals: $X(t) = \{t^2, 4 - 4t^2, t(1-t), t+3, 2t^2 - 5\}$.

3.14 Consider a random process of the form $X(t) = 3\sin(5t + \theta) + 4\cos(5t + 2\theta)$, where θ is a random variable with mass function defined as follows:

θ	$\Pr(\Theta = \theta)$
0	$\frac{2}{21}$
$\frac{1}{6}\pi$	$\frac{1}{21}$
$\frac{1}{4}\pi$	$\frac{1}{7}$
$\frac{1}{3}\pi$	$\frac{4}{21}$
$\frac{1}{2}\pi$	$\frac{2}{7}$
π	$\frac{5}{21}$

(a) Compute the mean, variance and autocorrelation of this random process.

(b) Find the time average of this process.

(c) Is the process ergodic? Is the process wide-sense stationary?

3.15 Write a program based on the algorithm given in Listing 3.5 to generate a sequence of Poisson-distributed random variates.

(a) Generate a sequence of 100 such variates with mean 2 evens per unit time.

(b) Create a frequency distribution table of the empirically generated variates of part (a) along with the theoretically expected number.

(c) Calculate χ^2 for the data tabulated in part (b). Test for goodness of fit and draw the appropriate conclusion.

3.16 Consider an arbitrary string $S = [s_1, s_2, \ldots, s_n]$ with a finite, possibly empty, set of elements. Write a routine to randomly select an element from S. For instance, if $S = [1, 2, 4]$, the number "2" is returned with probability $\frac{1}{3}$. If $S = [\ ]$, the note "empty string" is always returned.

(a) Using the above algorithm with $S = [1, 2, 2, 4, 4, 4, 6, 6, 7, 8]$, generate a sequence of 100 random variates.

(b) Create a frequency distribution table of the random variates along with their theoretical distribution.

(c) Calculate χ^2 for the data tabulated in part (b). Test for goodness of fit and draw the appropriate conclusion.

3.17 A reasonable approximation to a random variable having a Gaussian distribution can be obtained by averaging together a number of independent random variables having a uniform probability density function. It follows that one means by which to generate a Gaussian random variate is to use numerical convolution of several independent uniformly distributed probability density functions.

(a) Using a bin size of $\delta = 0.1$, generate 600 random variates found by averaging six uniform random variates uniformly distributed over $[0, 1]$. Plot the resulting density function along with the theoretical Gaussian density function having the same mean and variance.

(b) Using the Chi square goodness of fit test, test the results of part (a).

(c) Repeat steps (a) and (b) using twelve uniformly distributed random variates and comment on your results.

3.18 Consider the random process $X(t) = 4 + 3\cos(5t + \theta)$, where θ is uniform on $[0, 2\pi]$.

(a) Compute the autocorrelation of X.

(b) Compute the statistical and time averages of X.

(c) Is X ergodic? Is X wide-sense stationary?

3.19 Consider the random process $X(t) = \cos(2t + \theta)$, where θ is a random variable exponentially distributed with mean $\mu = 1$.

(a) Find the autocorrelation.

(b) Is $X(t)$ stationary?

(c) Using part (a), find the statistical mean.

(d) Using part (a), find the statistical variance.

3.20 Show that the autocorrelation of the random process $X(t) = A\cos(\omega t + \theta)$, where θ is uniform on $[0, 2\pi]$, is $X(t) = \frac{1}{2}A^2 \cos \omega \tau$.

3.21 Show that the variance of a random walk with probability of advance p at time k is $\sigma^2 = 4kp(1 - p)$.

3.22 Show that the autocorrelation of a synchronous telegraph process with period T and amplitude A is given by

$$R_{xx}(\tau) = \begin{cases} A^2\left(1 - \dfrac{|\tau|}{T}\right), & |\tau| < T, \\ 0, & |\tau| \geqslant T. \end{cases}$$

3.23 Consider the random process defined recursively in Example 3.12 by $x(k) = w(k) - \frac{1}{2}x(k - 1) - \frac{1}{4}x(k - 2)$, where $w(k)$ is uniformly distributed white noise with unit variance and mean 5. Assume $x(0) = x(1) = 0$. Reproduce a table similar to Table 3.5.

(a) Generate 10 instances, each with 26 time slices of this process, and tabulate.
(b) Calculate the time average of each instance, and graph.
(c) Calculate the time average of the stationary phase, in each instance.
(d) Calculate the statistical averages of each time slice and graph.
(e) Calculate the time-average autocorrelation, and graph.

3.24 Consider a random synchronous signal with period 10 milliseconds and with the initial time t_0 uniformly distributed on $[0, 10]$. It is required to find the autocorrelation of this process.
(a) Assume that the signal $X(t)$ is bipolar with amplitude 5.
(b) Assume that the signal $Y(t)$ is binary, either 0 or 5. Note that $Y(t)$ is related to $X(t)$ by a linear transform that maps -5 into 0 and 5 into 5.

3.25 Show that the autocorrelation of an asynchronous telegraph process with event rate λ and amplitude A is given by $R_{xx}(\tau) = e^{-2\lambda|\tau|}$.

3.26 Consider a random process $X(t)$ that has autocorrelation $R_{xx}(\tau) = 4e^{-|\tau|} \cos 2\tau + 7$.
(a) Sketch this function, being careful to include critical points and features.
(b) Find the mean μ of $X(t)$.
(c) Find the variance σ^2 of $X(t)$.

3.27 Suppose that a random process $X(t)$ is wide-sense stationary.
(a) Prove that the maximum value of the autocorrelation $R_{XX}(\tau)$ is given by $R_{XX}(0)$.
(b) Give an example to show that the lag time τ for which a maximum is achieved is not necessarily unique.
(c) Under what conditions would you expect that the maximum value of the autocorrelation would/would not be unique?

3.28 The autocovariance of a random process $X(t)$ is defined as

$$C_{xx}(t, \tau) = E[\{X(t) - \mu_{X(t)}\}\{X(t + \tau) - \mu_{X(t+\tau)}\}],$$

where the means are defined as $\mu_{X(t)} = E[X(t)]$ and $\mu_{X(t+\tau)} = E[X(t + \tau)]$.
(a) Prove that the autocorrelation $R_{xx}(t, \tau) = C_{xx}(t, \tau) + E[X(t)]E[X(t + \tau)]$.

(b) Find the autocovariance for the random process $x(t) = 4 + 3 \cos(5t + \theta)$, where θ is a random variable uniformly distributed on $[0, 2\pi]$.

3.29 Show that the autocovariance for a Poisson random process with mean arrival rate λ is $C_{xx}(t, \tau) = \lambda \min(t, t + \tau)$.

3.30 Write a program that produces m instances of a simulation of a synchronous telegraph process producing outputs as both (i) event sequences and (ii) continuous-time sequences. The program should accept the period parameter T and run from time $t = 0$ to $t = L$ using integration step size $h = 0.1$.
(a) Execute this program for $m = 5$, $T = 2$, and $L = 30$.
(b) Compute the autocorrelation for $\tau = 0$ to 10 in steps of 0.5.
(c) Graph the results of parts (a) and (b).
(d) Compare the results of part (b) against the ideal autocorrelation.

3.31 Consider a two-dimensional random walk model in which the direction angle is a random variable uniformly distributed on $[0, 2\pi]$ and the distance traveled at each time-step is fixed at unity.
(a) Create a simulation for an arbitrary number of instances over 20 steps.
(b) Execute a single instance of this model and create graphs of
   (i) the psotion after each step in the x-y plane.
   (ii) the distance from the origin as a function of time.

(c) Compile the results of 20 instances of the process and find the distance of each step from the origin.

(d) Find the time average of the distances calculated in part (c) and graph the distance as a function of time.

(e) Make appropriate conjectures.

3.32 Repeat problem 3.31, but this time allow the distance traveled at each step to be a random variable that is exponentially distributed with unity mean, $\mu_D = 1$.

3.33 Consider a random walk in which a walker takes unit steps in random directions θ, where θ is uniformly distributed on $[0, 2\pi]$. Write a program to generate m instances of such a walk of n steps. The output should give the point (x, y) of each step in the sequence.

(a) Using the above program, create one such walk of length 20. Graph this walk.

(b) Create 10 instances of 1000 step walks, and record the radial r distance at each time step k. Graph r as a function of k for each instance along with the statistical average $\mu_R = E[R(k)]$.

(c) Assuming this process to be ergodic, compute time averages after a suitable steady state has been achieved.

(d) Select a single instance of this random process, and calculate the autocorrelation using reference times $k = 100, 150, 200, 250, 300$. Use lag times from $\tau = -50$ to 50 in steps of unity. Graph each.

(e) In light of your results for part (c), discuss the ergodic hypothesis.

(f) In light of your results for parts (c) and (d), discuss whether this process is wide-sense stationary or not.

3.34 Create a two-dimensional Brownian motion model in which the Gaussian standard deviation in the horizontal direction is σ_1 and that in the vertical direction is σ_2. Execute the simulation for $\sigma_1 = 1$ and $\sigma_2 = 2$, showing your results graphically.

3.35 Consider the two following simple ways to create white noise: a uniformly distributed random process and a normalized Gaussian random process.

(a) Create 26 time slices of a uniformly distributed random process with variance 10. Estimate its autocorrelation function using time averaging.

(b) Create 26 time slices of a Gaussian-distributed random process with variance 10. Estimate its autocorrelation function using time averaging.

(c) Graph the two methods of finding the autocorrelation of white noise with unit variance. Compare each against the theoretical white noise characteristic.

3.36 Find the spectral density of a random process with autocorrelation

$$R_{xx}(\tau) = \begin{cases} 1 - |\tau|, & |\tau| \leqslant 1, \\ 0, & |\tau| > 1. \end{cases}$$

3.37 Find the autocorrelation of a random process with spectral density

$$S_{xx}(\omega) = \delta(\omega - 12) + 4\delta(\omega - 5) + 6\delta(\omega) + 4\delta(\omega + 5) + \delta(\omega + 12).$$

3.38 Use the Shaping Theorem to implement an algorithm that generates a random process $x(k)$ with spectral density $S_{xx}(\omega) = 4/(\omega^2 + 9)$.

(a) Use this algorithm to generate 26 time slices of the required process, and graph.

(b) Using the data generated in part (a), estimate the autocorrelation function.

(c) Apply a Fourier transform to calculate the spectral density of the autocorrelation found in part (b).

(d) Compare the results of part (b) with the original $S_{xx}(\omega)$.

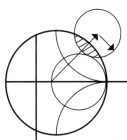

CHAPTER 4

Discrete Systems

4.1   SAMPLED SYSTEMS

In principle, models of natural physical systems assume time to be a continuous deterministic variable. On the other hand, engineered systems usually have to deal with discrete noisy signals, events, and measurements, and therefore see time differently. Designed systems are often built around microprocessors that deal with sampled measured signals produced by a transducer. This leads to models that have random errors associated and artificial time bases. Since microprocessors can only measure the signal (usually a voltage produced by the transducer) at regular discrete time instances, the resulting signal is discrete.

By convention, the variable t represents real, physical chronological time and k represents discrete time. This, combined with the fact that modelers often need to keep multiple clocks, imply that the concepts of real and physical time are somewhat confusing. This is because human notions of time are really just as artificial as those produced by a microprocessor or perceived by another biological organism. In this section, we consider two fundamental time bases and their relationship. The first is continuous time, which cannot exist in a simulation sense because computers are innately discrete. Thus, we characterize computer simulation time by a constant h, which is called the *integration step size* or *simulation granularity*. Unless integration is handled using an adaptive algorithm, h is almost always constant, and its relation to chronological time is fundamental:

$$t = hk + t_0. \tag{4.1}$$

This is not the only reason for discretizing time. For instance, if the model is of a computer system and its relationship to an analog device, the modeled computer must sample the system variables and thereby create another discrete time variable. Assuming that sampling occurs every δ time units, the fundamental relationship used to model a sampled signal is

$$t = \delta k + t_0, \tag{4.2}$$

where t_0 is the initial time. Even though Equation (4.1) looks very similar to Equation (4.2), they have quite different meanings and interpretations. In general, the integration step size h is not the same as the *sampling interval* δ. Often we speak of the *sampling frequency* f rather than the sampling interval. The sampling interval is the time it takes to take one sample, and the sampling frequency gives the number of samples per unit time. Therefore, they are related by $f = 1/\delta$.

The inverse of a sampled signal is a continuous signal that collocates at all the sampled times. This process, called *desampling*, should ideally result in the original signal. However, since the sampling process loses information, desampling can only approximate retrieval of the original continuous unsampled signal. This is illustrated in Figure 4.1, where a continuous signal $x(t)$ is sampled at frequency $f_1 = 1/\delta_1$. The sampled signal $x(k)$ is only defined at each of the integral times $k = 0, 1, 2, \ldots$. However, retaining the value $x(k)$ until the next sample occurs creates a new (discontinuous) function $x^*(t)$ defined for continuous time. This signal is called a *zero-order hold*, since the sampled value is held constant (a zeroth-order polynomial) until the next sampled value δ_1 time units later.

Since simulations are performed digitally on a computer, the actual process of desampling is also discrete. In general, desampling is done at a rate $f_2 = 1/\delta_2$ that is less frequent than the sampling rate and $f_1 \geqslant f_2$, where δ_2 is the desampling interval. This presents three specific cases of interest.

(i) The sampling frequency and the sampling frequency are the same: $f_1 = f_2$. Equivalently, $\delta_1 = \delta_2$.

(ii) The sampling frequency is an integer multiple of the desampling frequency: $f_1 = nf_2$. In other words, there are n desamples made for each sample. Equivalently, $\delta_2 = n\delta_1$.

(iii) The sampling frequency is smaller than the desampling frequency, but it is not an integral multiple of it. Thus, $\delta_2 > \delta_1$.

Each case is considered separately below.

Case (i) Let us assume that t is free-running continuous time over the interval $[0, T]$ as shown in Figure 4.2. Considering case (i), let $f_1 = f_2 = 1/\delta$ be the common sampling and desampling frequencies and suppose there are $m + 1$ samples taken and $m + 1$

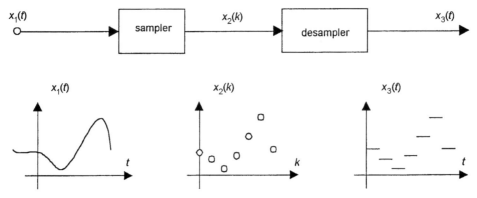

FIGURE 4.1   Signals created by a sampler with a zero-order hold.

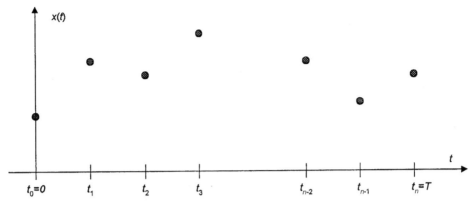

FIGURE 4.2   Desampled signal for case (i): $f_1 = f_2$.

desamples made over the entire interval. It follows that $m = T/\delta = Tf$. In this case, the sampling interval and the desampling interval both match the granularity of continuous time, and the following code sequence samples the signal $x(t)$ $m + 1$ times and prints the $m + 1$ desamples over the interval $[0, T]$ as required. The code in this case is as follows:

```
k=0
t=0
x(k)=x(t)
print t, x(0)
for k=1 to Tf
        t=t+1/f
        x(k)=x(t)
        print t, x(k)
next k
```

Case (ii) In the case where the sampling frequency is an integer multiple of the desampling frequency, $f_1 = nf_2$. This is shown in Figure 4.3, where there are $n = 4$

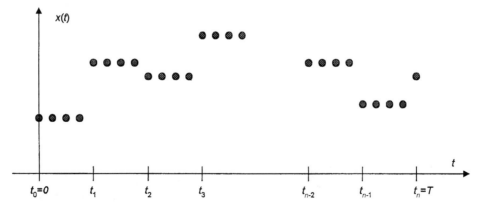

FIGURE 4.3   Desampled signal for case (ii): $f_2 = nf_1$.

desamples per sample and $m + 1$ samples on the interval $[0, T]$. Thus, there are $mn + 1$ desamples in total. In this case, the two frequencies are in synch, and the implementation can be handled using nested loops, the outside being the sampling loop and the inside the desampling loop. This means that there are Tf_1 iterations outside, $n = f_1/f_2$ iterations inside, and 1 initiating sample to begin the simulation. The time step size is $\delta_2 = 1/f_2$. The code in this case is as follows:

```
k=0
t=0
x(k)=x(t)
print t, x(0)
for i=1 to Tf₁
        for j=1 to f₁/f₂
                t=t+1/f₂
                x(k)=x(t)
                print t, x(k)
        next j
next i
```

Case (iii) This last case, where $f_2 < f_1$, is the most general. It is similar to case (ii), but the sample and desampler are not in synch. Therefore, nested loops are not the answer. Instead there is a structure clash and the two operations need to be handled separately. Again assuming a time interval of $[0, T]$ and a sampling frequency of f_1, the following module samples the signal $x(t)$ and forms an array $x(k)$:

```
k=0
for t=0 to T step 1/f₁
        t(k)=t
        x(k)=x(t)
        k=k+1
next t
```

Once the sampling has taken place and an array $x(k)$ has been formed, desampling requires a separate module that is independent of the sampler. Assuming a desampling rate of f_2, the module scans the interval $[0, T]$ at each point to find exactly where desampling should take place. While this is considerably slower than case (ii), it proceeds as follows:

```
for t=0 to T step 1/f₂
        for k=0 to Tf₁
                if t>t(k) and t<t(k+1) then print  t, x(k)
        next k
next t
```

EXAMPLE 4.1

Consider the sinusoidal signal $x(t) = \sin t$ over the interval $[0,10]$ with a sampling frequency of $f_1 = 5$ samples per second and a desampling frequency of $f_2 = 12$ samples per second. Show the results of sampling the continuous signal and then trying to invert the sampling process.

Solution
The original continuous signal is shown in Figure 4.4. The results of sampling using case (iii) are given in Figure 4.5. Since the frequencies of this example are non-conformal (that is, f_1/f_2 is not an integer), the desampler will not provide an equal number of points on each interval. This is evident in Figure 4.6 where it will be noted that the desampling intervals have an unequal number of points. Execution of the algorithms produces results consistent with the above explanations. ○

In practice the process of desampling is an extrapolation process. That is, given earlier points, an attempt is made at predicting the future. This is different from the interpolation process outlined in Section 2.5, where a look-up procedure was outlined in which approximated values were found *between* known points. When sampling occurs and the sampled value is retained until the next sampling period, the extrapolated signal is called a *zero-order hold (ZOH)*, since the approximating polynomial is a constant, which implies that the polynomial is of degree zero.

This extrapolation process can be extended to higher-degree polynomials as well. By retaining both the last sampled value x_k and the previous x_{k-1}, one can create a linear projection of anticipated values before the next sample is taken. Similarly, by retaining x_{k-2}, a quadratic projection is possible. In general, by retaining the previous $s + 1$ sampled signal values $x_k, x_{k-1}, \ldots, x_{k-s}$, an sth-degree polynomial extrapolation is in order. Suppose the signal value at present, as well as the previous s times is summarized by the following table.

k	t_k	x_k
⋮	⋮	⋮
$k - s$	t_{k-s}	x_{k-s}
⋮	⋮	⋮
$k - 2$	t_{k-2}	x_{k-2}
$k - 1$	t_{k-1}	x_{k-1}
k	t_k	x_k

Even though $k + 1$ samples are taken, only the $s + 1$ most recent samples are needed to extrapolate the signal value.

First notice that the following so-called *constituent polynomial* $L_k(t)$ is zero at each sampled data point $k - 1, k - 2, \ldots, k - s$, except $k = 0$, at which time it is the known value $L_0(t) = x_0$:

$$L_k(t) = x_k \frac{(t - t_{k-1})(t - t_{k-2})\ldots(t - t_{k-s})}{(t_k - t_{k-1})(t_k - t_{k-2})\ldots(t_k - t_{k-s})}.$$

In a similar manner, a constituent polynomial $L_{k-1}(t)$ can be created that is zero at all times except t_{k-1}, at which time $L_{k-1}(t) = x_{k-1}$:

$$L_{k-1}(t) = x_{k-1} \frac{(t - t_k)(t - t_{k-2})\ldots(t - t_{k-s})}{(t_{k-1} - t_k)(t_{k-1} - t_{k-2})\ldots(t_{k-1} - t_{k-s})}.$$

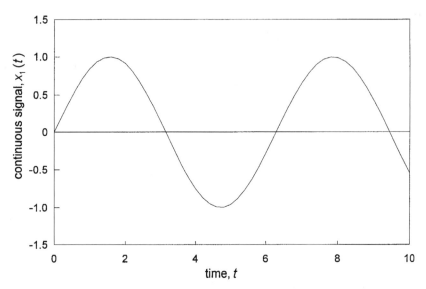

FIGURE 4.4   Graph of the unsampled continuous signal $x(t) = \sin t$ for Example 4.1.

These features will always be exhibited in the $L_{k-1}(t)$ polynomial, since the factors $t - t_{k-1}$ (in the numerator) and $t_{k-1} - t_{k-1}$ (in the denominator) are "missing", so that once again it is zero everywhere except at t_k. In general, the entire set of constituent polynomials is defined by

$$L_i(t) = x_i \prod_{j=0, j \neq i}^{s} \frac{t - t_j}{t_i - t_j}. \tag{4.3}$$

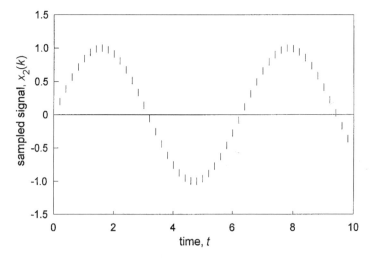

FIGURE 4.5   Sample signal for Example 4.1.

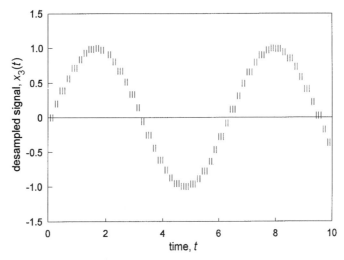

FIGURE 4.6   Desampled signal for Example 4.1.

This set of polynomials has the useful property that at each of the sample points t_k,

$$L_i(t_k) = \begin{cases} 0, & i \neq k, \\ x_k, & i = k. \end{cases}$$

Since each constituent contributes exactly one of the required collocation points (t_k, x_k) from the data set, the sum of all of the $L_i(t)$ forms the required so-called *Lagrangian extrapolation polynomial*. It follows from this that the approximating polynomial $\hat{x}(t)$ is

$$\hat{x}(t) = \sum_{i=0}^{s} \left(x_i \prod_{j=0, j \neq i}^{s} \frac{t_k - t_j}{t_i - t_j} \right). \tag{4.4}$$

EXAMPLE 4.2

Suppose that data is sampled at a rate of 2 samples per minute. Find the third-degree extrapolation polynomial at $t_k = 6.5$ minutes for the following data values. Predict data values for $t = 6$ minutes 50 seconds.

k	t_k	x_k
⋮	⋮	⋮
5	4.0	2
6	4.5	3
7	5.0	1
8	5.5	0
9	6.0	2
10	6.5	2

Solution
It is evident from the first two columns that $t_k = \frac{1}{2}k + \frac{3}{2}$, so the sampling time $\delta = 0.5$ minutes $= 30$ seconds, and $t_0 = 1.5$ seconds. Since we are looking for a third-degree polynomial at $k = 10$, only the $s + 1 = 4$ most recent samples ($k = 7, 8, 9, 10$) need be used. Computing each in turn by Equation (4.3),

$$L_{10}(t) = 2\frac{(t - 6.0)(t - 5.5)(t - 5.0)}{(6.5 - 6.0)(6.5 - 5.5)(6.5 - 5)}$$

$$= \tfrac{8}{3}(t - 6)(t - 5.5)(t - 5)$$

$$= \tfrac{8}{3}t^3 - 44t^2 + \tfrac{724}{3}t - 440,$$

$$L_9(t) = 2\frac{(t - 6.5)(t - 5.5)(t - 5.0)}{(6.0 - 6.5)(6.0 - 5.5)(6.0 - 5.0)}$$

$$= -8(t - 6.5)(t - 5.5)(t - 5)$$

$$= -8t^3 + 136t^2 - 766t - 1430,$$

$$L_7(t) = 1\frac{(t - 6.5)(t - 6.0)(t - 5.5)}{(5.0 - 6.5)(5.0 - 6.0)(5.0 - 5.5)}$$

$$= -\tfrac{4}{3}(t - 6.5)(t - 6)(t - 5.5)$$

$$= -\tfrac{4}{3}t^3 + 24t^2 - \tfrac{431}{3}t + 286.$$

$L_8(t) = 0$ is trivial, since $x_8 = 0$. Summing the four individual polynomials and simplifying gives the Lagrange extrapolation polynomial:

$$\hat{x}(t) = \tfrac{2}{3}t^3 + 116t^2 - \tfrac{2005}{3}t + 1276.$$

As a check, substituting the sampling times into Equation (4.4) produces $\hat{x}(5.0) = 1$, $\hat{x}(5.5) = 0$, $\hat{x}(6.0) = 2$, and $\hat{x}(6.5) = 2$, as expected. The anticipated value at $t = 6.83$ minutes is $\hat{x}(6.83) \approx -1.58$. ○

Lagrange's formula is very general in that there is no requirement for the data points to be at equally spaced times. However, since sampling a signal occurs at regular fixed time intervals, $\delta = t_k - t_{k-1} = 1/f$. This special case simplifies the Lagrangian formulation somewhat. To this end, consider the constituent polynomials:

$$L_k(t) = \frac{x_k(t - t_k)(t - t_{k-2})\ldots(t - t_{k-s})}{(t_k - t_{k-1})(t_k - t_{k-1})\ldots(t_k - t_{k-s})}$$

$$= \frac{x_k(t - t_k)(t - t_{k-2})\ldots(t - t_{k-s})}{(\delta)(2\delta)\ldots(s\delta)}$$

$$= \frac{1}{s!}f^s x_k(t - t_{k-1})(t - t_{k-2})\ldots(t - t_{k-s}),$$

$$L_{k-1}(t) = \frac{(t - t_k)x_{k-1}(t - t_{k-2})\ldots(t - t_{k-s})}{(t_{k-1} - t_k)(t_{k-1} - t_{k-2})\ldots(t_{k-1} - t_{k-s})}$$

$$= \frac{x_{k-1}(t - t_k)(t - t_{k-2})\ldots(t - t_{k-s})}{(-\delta)(\delta)(2\delta)\ldots[(s-1)\delta]}$$

$$= \frac{(-1)^1}{1!(s-1)!}f^s x_{k-1}(t - t_k)(t - t_{k-2})\ldots(t - t_{k-s}),$$

$$L_{k-2}(t) = \frac{(t - t_k)(t - t_{k-1})x_{k-2}\ldots(t - t_{k-s})}{(t_{k-2} - t_k)(t_{k-2} - t_{k-1})\ldots(t_{k-2} - t_{k-s})}$$

$$= \frac{x_{k-2}(t - t_k)(t - t_{k-1})\ldots(t - t_{k-s})}{(-2\delta)(-\delta)(2\delta)\ldots[(s-2)\delta]}$$

$$= \frac{(-1)^2}{2!(s-2)!}f^s x_{k-2}(t - t_k)(t - t_{k-1})\ldots(t - t_{k-s}).$$

Continuing in the above fashion for all constituents $L_k(t), L_{k-1}(t), \ldots, L_{k-s}(t)$, it is evident that the general term $L_j(t)$ is

$$L_{k-i}(t) = \frac{(-1)^i}{i!(s-i)!}f^s(t - t_k)(t - t_{k-1})\ldots(t - t_{k-i+1})x_{k-i}(t - t_{k-i-1})\ldots(t - t_{k-s})$$

$$= \frac{(-1)^i}{i!(s-i)!}f^s x_{k-i}\prod_{j=0,j\neq i}^{s}(t - t_{k-j}).$$

Summing over index i gives the Lagrange extrapolation for sampled signals:

$$\hat{x}(t) = f^s \sum_{i=0}^{s}\frac{(-1)^i}{i!(s-i)!}\left[x_{k-i}\prod_{j=0,j\neq i}^{s}(t - t_{k-j})\right]. \qquad (4.5)$$

In general, it is unnecessary, and certainly undesirable, to compute the explicit Lagrange polynomial, since this can be tedious at best. It is really only required that the evaluation can be done using a computer. This is done in practice, both in data acquisition applications and simulations. It is clear from Equation (4.5) that there are two different times. The first set, $t_0, t_1, \ldots, t_{k-s}, \ldots, t_{k-1}, t_k$, are the sample times that occur at time $t_k = \delta k + t_0$. The second time is the desampling time sequence characterizing continuous time t, and specified by the granularity h. Put another way, within each sampling interval there are n sub-intervals of size h at which computations can be made. Thus, in general,

$$t = \delta i + hj + t_0$$

```
h=δ/n
t=t₀
i=0
while ‾EOF‾ do
        get xᵢ
        for j=0 to n-1
                t=δi+hj
                x̂=lagrange(t, k)
                put x̂
        next j
        i=i+1
endo
```

LISTING 4.1   Desampling using Lagrange extrapolation for an sth-order hold.

for $i = 0, 1, \ldots, m$ and $j = 0, 1, \ldots, n - 1$. While the sampling frequency is $f_{\text{sample}} = \delta^{-1}$, the computation frequency is $f_{\text{compute}} = h^{-1}$. Since there are n computation intervals for every sampling interval, $nh = \delta$. Thus, the basic structure for a simulation routine is given in Listing 4.1, where the outside loop functions as a sampler and the inside loop functions as a desampler by putting the results after each computation.

Computation of the Lagrange extrapolation polynomial is accomplished by noting that Equation (4.5) is a sum from $i = 0$ to s data points over a product of $j = 0$ to s factors. A set of nested loops does the trick, but to do this requires the extrapolation time t and the index k of the most recent sample taken x_k. In performing the sum, the *sum* variable is initialized at *sum* $= 0$. In computing the product, the *prod* variable is initialized at *prod* $= 1$. If $i \neq j$, prod accumulates the $t - t_j$, factors and when $i = j$, prod tallies the x_i value. The extrapolation algorithm is shown as a function, and is given in Listing 4.2. Notice that the inputs include the extrapolation time t and the index of the most recent sample taken.

```
function lagrange(t, k)
        sum=0
        for i=0 to s
                prod=1
                for j=0 to s
                        if i=j then
                                prod=prod* xₖ₋ⱼ
                        else
                                prod=prod*(t-tₖ₋ⱼ)
                        endif
                next j
                sum=sum+(-1)ⁱ/i!(s - i)!*prod
        next i
        lagrange =sum*fˢ
return
```

LISTING 4.2   The function *lagrange* that performs an sth-order extrapolation for a sampled signal.

EXAMPLE 4.3

Using a second-order hold, estimate the sampled sinusoid $x(t) = \sin t$ if the sampling frequency $f_{sample} = 1$ samples per unit time and the computation frequency $f_{comput} = 4$ computations per unit time over the interval [0,20].

Solution
From the given frequencies, $\delta = 1$ time unit and $h = 0.25$ time units. It follows that there will be $n = 4$ computations for every sample taken and that only on the first computation will the computed estimate $\hat{x}(t_k)$ match the sampled value x_k. The remaining computations will estimate the sinusoid by means of a quadratic polynomial, since $s = 2$. It also follows that $m = 20$ and that there will be 21 samples taken, two of which will be needed to *prime* the array x_k. Thus, for this example, $t = k + \frac{1}{4}l$ and the pseudocode for Lagrange's quadratic (second-order-hold) proceeds as follows:

```
x₀=sin(0)
x₁=sin(1)
k=2
for i=2 to m
        xₖ=sin(k)
        for j=0 to 3
                t=i+¼j
                x=lagrange (t, k)
                k=k+1
                print t, x
        next j
next i
```

Since the first estimator requires two previous sampled values, the actual running of this simulation requires a start-up initializing loop of $s = 0$ and $s = 1$ before the main program in which estimations are performed. The results of this simulation for $s = 0$ (zero-order hold) and $s = 2$ (second-order hold) are shown in Figures 4.7 and 4.8. Figure 4.7 shows the characteristic ZOH, where the sampled signal is held constant until the next sample is taken. For this reason, the ZOH is sometimes called a *sample and hold*. Notice that in addition to the inaccuracies, the resulting signal appears to be slightly shifted to the right.

Figure 4.8 compares the second-order hold against the ideal signal. Clearly, there is a better fit than in the ZOH case, especially where the slope does not change orientation (from positive to negative or vice versa). However, in the regions where the slope changes abruptly, the propensity of this technique to anticipate causes problems in that it continues even after the ideal signal has changed direction. Notice that in addition to the inaccuracies, the resulting signal appears to have a larger amplitude than it should.   ○

4.2   SPATIAL SYSTEMS

Physical systems are the source of a number of rich models. For instance, in mechanical systems, if $x(t)$ is the displacement (position) of a mass at time t, then

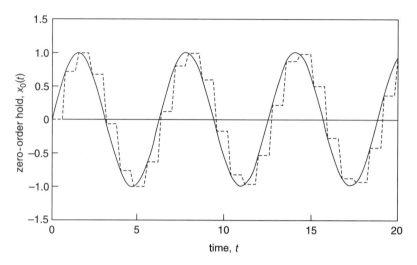

FIGURE 4.7   Sampling the signal using a zero order hold.

$\dot{x}(t) = dx/dt$ is the velocity and $\ddot{x}(t) = d^2x/dt^2$ is the acceleration of that mass at time t. In a specific well-posed problem, these are related by Newton's laws of mechanics. For example, assuming a constant mass attached to a spring in a viscous medium, as shown in Figure 4.9, the forces are as follows:

inertial force: $F_{\text{inert}} = M\ddot{x}$,
dissipative force: $F_{\text{dis}} = B\dot{x}$,
restoration force: $F_{\text{rest}} = Kx$,

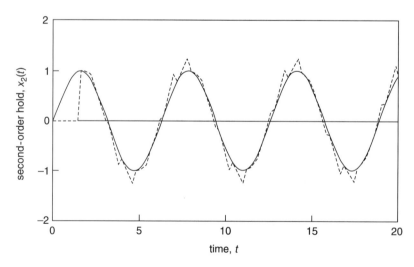

FIGURE 4.8   Sampling the signal using a second order hold.

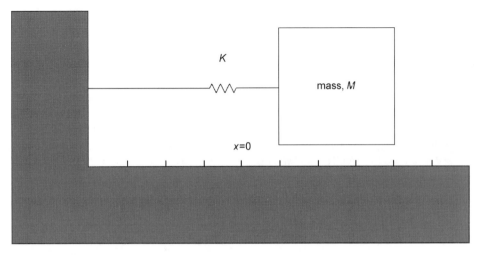

FIGURE 4.9   A one-dimensional Newtonian system.

where M is the mass, B is the damping constant, and K is the spring constant of the system.

In a general multidimensional system, Newton's fundamental law states that the vector sum of the forces must equal the external force applied to the system. However, since this is a one-dimensional system, the algebraic sum of these forces must be $f(t)$, where $f(t)$ is the external "driving" force. Thus,

$$M\ddot{x} = B\dot{x} + Kx = f(t). \tag{4.6}$$

Since this is a second-order linear system, two initial conditions are required – one for the initial position $x(t_0)$ and one for the initial velocity $\dot{x}(t_0)$. It follows that two states are required to characterize this system. The most straightforward approach is to simply define the position and velocity as the state variables, although there are infinitely many other possibilities as well:

$$\text{state variable one, position:} \quad x(t),$$
$$\text{state variable two, velocity:} \quad v(t) = \dot{x}(t).$$

From this, Equation (4.6) can be rewritten as two first-order differential equations, for which the techniques of Chapter 2 can immediately be employed:

$$\dot{x} = v,$$
$$\dot{v} = -\frac{K}{M}x - \frac{B}{M}v + \frac{1}{M}f(t), \tag{4.7}$$
$$x(t_0) = x_0,$$
$$v(t_0) = v_0.$$

If constants M, B, and K, initial states x_0 and v_0, and the driving force are given, this is a well-posed problem and the solution is straightforward. This is best illustrated by an example.

EXAMPLE 4.4

Consider a mechanical system whose dynamics are described by the following differential equation:

$$\ddot{x} + 2\dot{x} + x = \sin t + w(t), \tag{4.8}$$

where $w(t)$ is a random shock (in the form of white noise) that is Gaussian-distributed with mean zero and a standard deviation of 2. Initially, the mass is at rest and positioned at $x = 0.5$. Simulate the motion of this system over [0,50] using an integration frequency of 50 integrations per second, a print frequency of 10 prints per second, and a shock frequency of 1 shock per second.

Solution
Using the position and velocity as state variables, this problem can be reformulated as

$$\begin{aligned} \dot{x} &= v, \\ \dot{v} &= -x - 2v + \tfrac{1}{2}\sin t + w(t), \\ x(0) &= 0.5, \\ v(0) &= 0. \end{aligned} \tag{4.9}$$

Since $f_{\text{shock}} = 1$, $f_{\text{print}} = 10$, and $f_{\text{integrate}} = 50$, this simulation is conformal and nested looping is in order. Since time goes from 0 to 50 seconds, there are 51 seconds in all. As usual, one will be the initial print, followed by a loop from 1 to 50. Since the shocks must come each second, the shock will be introduced here. Within this outer loop will be the print loop, which will require 10 cycles, since there are 10 prints for each shock. Within this loop is another of 5 cycles to perform the integration, since there are $5 \times 10 = 50$ per second, each requiring an integration step size of $h = 0.02$ seconds. This is all summarized in Listing 4.3.

It is worth noting the need for the variables x_1 and v_1 in Listing 4.3. This is to ensure the proper update of position x and velocity v each integration cycle. Also, since the shock is Gaussian, it can be generated by the formula method outlined earlier. Since the noise is generated each 50 integration cycles, it must be turned off immediately after use, since it is a "one shot".

The graphical results are shown in Figures 4.10 and 4.11. Figure 4.10 plots both states x and v against time t. It should be noticed that while v appears quite noisy, x is much smoother, as should be expected, since integration is a smoothing operation and x is the integral of v. Even so, after the transient phase, x appears periodic even though the amplitude varies. The phase plot shown in Figure 4.11 is even more interesting. If there were no shocks to the system, the phase plot would degenerate to a continual cycle after the transient phase. However, shocks have created an almost galaxy-like picture in which arms sweep in and out of the limit cycle. Clearly, this is an interesting system with much room for interpretation.

○

Example 4.4 illustrates the concept of additive noise. In addition to a driving function $f(t) = \tfrac{1}{2}\sin t$, there is a periodic random shock described as "white noise". Loosely

```
h=.02
t=0
x=.5
v=0
print t , x , v
for i=0 to 50
        w=5√(-2 * ln(RND))cos(2π*RND)
        for j=1 to 10
            for k=1 to 5
                x₁=x+hv
                v₁=v+h(-2v-x+½sin(t)+w)
                w=0
                x=x₁
                v=v₁
                t=t+h
            next k
            print t, x, v
        next j
next i
```

LISTING 4.3   Simulation for Example 4.4.

speaking, white noise is a random process that is uncorrelated with any preceding or succeeding events, and has zero mean and a constant standard deviation. This is clearly the case in Example 4.4.

Such shocks are actually very practical in modeling natural systems. First, nature simply seems to act randomly at times. Even though we might not understand the root of

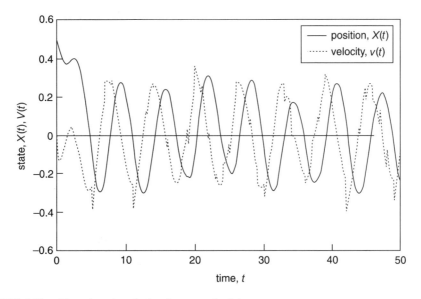

FIGURE 4.10   Time-domain solution for example 4.4.

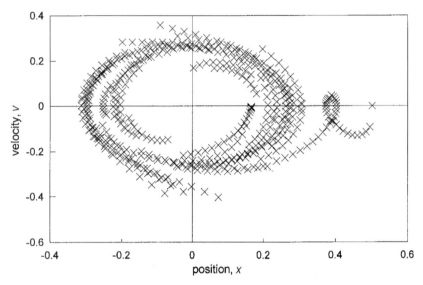

FIGURE 4.11   State-space solution for Example 4.4.

an underlying process, we can make use of statistics collected on the process. For instance, the weather certainly has a random appearance, even though its mathematical character is well understood. Thus, in modeling weather phenomena, we usually ignore the Navier–Stokes equations of fluid dynamics and go right to historical statistics. Also, the very act of taking measurements can create a disturbance due to the sensor or actuator of a physical system.

Additive noise is shown schematically in Figure 4.12. However, note that the character of the noise – be it white, Gaussian, continuous, or a shock – can vary considerably. If it is a shock, then the shock frequency – be it regular as in Example 4.4, or probabilistic such as arrivals in a supermarket queue – must be scheduled and characterized as well.

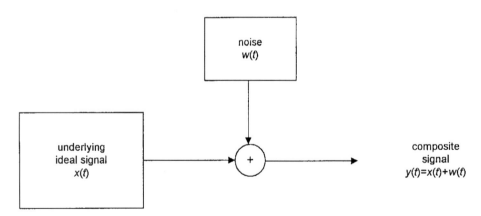

FIGURE 4.12   A signal with additive noise.

Up to now, we have considered only one-dimensional signals. Since signals are nothing more than state indicators, there is every reason to treat spatial dimensions using the same techniques. Traditionally, the notation used is index k for discrete time and i and j for discretized space. Continuous time is the variable t, while x, y, and z represent spatial coordinates. Usually spatial indices are given as subscripts, while time (either continuous or discrete) is given as a parenthetical argument. For example, the variable $\phi_{ij}(t)$ is a function of discretized position and continuous time. Presumably, $x = \delta_1 i + x_0$ and $y = \delta_2 j + y_0$ for appropriate spatial increments δ_1 and δ_2.

EXAMPLE 4.5

Consider the mechanical system shown in Figure 4.13, in which four equal masses are attached with rubber bands and the two ends are fixed to the reference base. Assume that the rubber bands obey Hooke's law in that they are linear and there are no frictional force or external forces of any kind. That is, ignore the mass of the rubber bands and the force of gravity.

Assume that initially the masses are held at the points shown in Figure 4.13 and that they are released at time $t = 0$. Write the equations of motion, assuming that there is no lateral movement of any spring. Simulate the motion of the system, assuming $K/M = 1$.

Solution
Using the conventions outlined above, we let $\phi_i(t)$ be the amplitude of the mass at position x_i on the x-axis at time t. We also take the common mass to be M and the spring constant of each rubber band to be K. The amplitudes at the two endpoints are each zero: $\phi_0 = 0$ and $\phi_5 = 0$. To write the equations of motion, consider a typical interior point x_i, where $i = 1, 2, 3$, or 4. Mass i has both an inertial force and the following restoring forces due to adjacent masses at points x_{i-1} and x_{i+1}:

inertial force: $\qquad F_{\text{inertia}} = M\dfrac{d^2\phi_i}{dt^2}$,

restoring force: $\qquad F_{\text{restore}} = K(\phi_i - \Phi_{i-1}) + K(\phi_i - \phi_{i+1})$.

Notice that this notation uses both subscripts for spatial coordinates and parenthetical arguments for time. This should not disguise the fact that we are

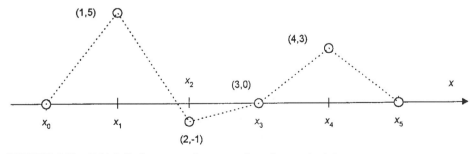

FIGURE 4.13   Initial displacement of wave motion of example 4.5.

really using partial derivatives for time and partial differences for the spatial coordinates. Since we have assumed no external driving forces, Newton's law states that

$$M\frac{d^2\phi_i}{dt^2} + K(-\phi_{i-1} + 2\phi_i - \phi_{i+1}) = 0. \tag{4.10}$$

Equation (4.10) is really four equations in disguise. Combined with the fact that $\phi_0(t) = 0$ and that $\phi_5(t) = 0$ for all time t, there is an equation for each node. Recalling that the problem stated that $K/M = 1$,

$$\ddot{\phi}_1 = -2\phi_1 + \phi_2,$$
$$\ddot{\phi}_2 = \phi_1 - 2\phi_2 + \phi_3,$$
$$\ddot{\phi}_3 = \phi_2 - 2\phi_3 + \phi_4,$$
$$\ddot{\phi}_4 = \phi_3 - 2\phi_4.$$

Since the application of Newton's law will always give a set of second-order differential equations, there are really twice as many states as equations. Let us define the velocity as $\theta_i(t) = \partial\phi/\partial t$. Since the system is released from at rest, the initial velocity is zero. This leads to eight first-order differential equations and eight initial conditions:

$$\begin{aligned}
\dot{\theta}_1 &= -2\phi_1 + \phi_2, & \theta_1(0) &= 0, \\
\dot{\theta}_2 &= \phi_1 - 2\phi_2 + \phi_3, & \theta_2(0) &= 0, \\
\dot{\theta}_3 &= \phi_2 - 2\phi_3 + \phi_4, & \theta_3(0) &= 0, \\
\dot{\theta}_4 &= \phi_3 - 2\phi_4, & \theta_4(0) &= 0, \\
\dot{\phi}_1 &= \theta_1, & \phi_1(0) &= 5, \\
\dot{\phi}_2 &= \theta_2, & \phi_2(0) &= -1, \\
\dot{\phi}_3 &= \theta_3, & \phi_3(0) &= 0, \\
\dot{\phi}_4 &= \theta_4, & \phi_4(0) &= 3.
\end{aligned}$$

The simulation using Euler integration follows the same procedure as outlined in Chapter 2. Since the six amplitudes are printed for each position in successive times, this produces results similar to those in Table 4.1.

   Notice that since there are two independent variables, there are two important views of the results. These are summarized in Figures 4.14 and 4.15. From a temporal perspective, each node oscillates according to the graphs presented in

TABLE 4.1   Amplitude as a Function of Time for Example 4.5

t	ϕ_0	ϕ_1	ϕ_2	ϕ_3	ϕ_4	ϕ_5
0.00	0	5.00	−1.00	0.00	3.00	0
0.25	0	4.66	−0.79	0.06	2.81	0
0.50	0	3.70	−0.18	0.24	2.29	0
0.75	0	2.27	0.68	0.52	1.49	0
⋮	⋮	⋮	⋮	⋮	⋮	⋮

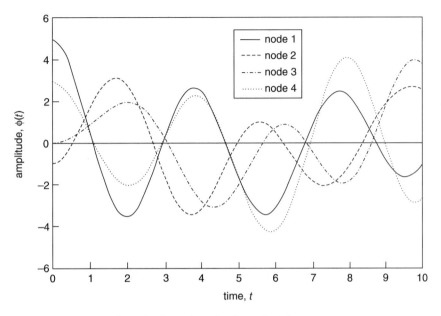

FIGURE 4.14   Temporal results for each node, Example 4.5.

Figure 4.14, where each node's amplitude is given as function of time. Figure 4.14 illustrates the motion of each node for successive time slices. It should also be noted that fundamental nodes occur at $x = \frac{5}{3}$ and $x = \frac{10}{3}$, even though no mass actually exists at that point.

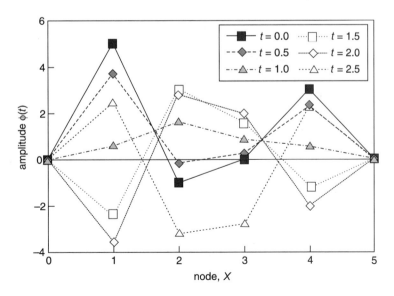

FIGURE 4.15   Space results from various time slices, Example 4.5.

Figure 4.14 is created by plotting each node's column from Table 4.1 against time, producing temporal results. Figure 4.15 is made by plotting several rows from Table 4.1 against the node number for various times, giving spatial results. Summarizing,

temporal results: vertical graph;
spatial results: horizontal graph.

Example (4.5) can be generalized to $m + 1$ nodes, $i = 0, 1, \ldots, m$. Taking the limiting case of $m \to \infty$ results in a continuous definition. In this case, Equation (4.10) changes from a difference equation to a differential equation. It can be shown that for continuous space and time,

$$\frac{\partial^2 \phi}{\partial x^2} = \frac{1}{\rho^2} \frac{\partial^2 \phi}{\partial t^2}, \tag{4.11}$$

where ρ is a constant defined by the physical circumstance. This equation can be shown to reduce to Equation (4.10) derived in Example 4.3 for the case of a discrete spatial variable. In the event of more than one spatial variable, the left-hand side of Equation (4.11) generalizes to the classical *wave equation*

$$\nabla^2 \phi = \frac{1}{\rho^2} \frac{\partial^2 \phi}{\partial t^2},$$

where $\nabla^2 \phi$ is the Laplacian of ϕ.

4.3   FINITE-DIFFERENCE FORMULAE

Example 4.5 illustrates two fundamental kinds of continuous-time modeling problems. Determination of the amplitude ϕ with respect to t is described as an *initial-value problem*. That is, $\phi(t_0)$ and $\dot{\phi}(t_0)$ are known a priori and it remains to describe $\phi(t, x)$ for all succeeding times $t \geqslant 0$. This is in contrast to the spatial variable x, where the value of ϕ is known at two points x_0 and x_5 but there is no derivative information. Thus, we call this a *boundary-value problem*. It remains to find the value of $\phi(t, x)$ at all points x such that $x_0 \leqslant x \leqslant x_5$. Since in this case there is both an initial-value problem and a boundary-value problem, Example 4.5 exhibits a hybrid system.

In general, it is not always possible to find explicit analytical solutions for a given boundary-value problem. At the same time, the Euler, Taylor, and Runge–Kutta methods are inappropriate, since they rely on knowledge of all state information at the initial time. Boundary-value problems only provide partial information on the state, but at multiple points. Accordingly, we need an approach giving derivative formulas in terms of spatially dispersed points. The most straightforward of these is the finite-difference method, which is illustrated here.

For sufficiently well-behaved functions $x(t)$, that is, those that are repeatedly differentiable over the domain of reference, Taylor's formula states that

$$x(t+h) = \sum_{k=0}^{\infty} \frac{h^k}{k!} x^{(k)}(t)$$

$$= x(t) + h\dot{x}(t) + \frac{1}{2}h^2\ddot{x}(t) + \frac{1}{6}h^3\dddot{x}(t) + \cdots \quad . \tag{4.12}$$

Replacing h by $-h$ in Equation (4.12) yields

$$x(t-h) = x(t) - h\dot{x}(t) + \frac{1}{2}h^2\ddot{x}(t) - \frac{1}{6}h^3\dddot{x}(t) + \cdots \quad . \tag{4.13}$$

For small h, Equations (4.12) and (4.13) are fairly accurate, even if truncated to the second order. Subtracting Equation (4.13) from (4.12) and ignoring higher-order terms gives $x(t+h) - x(t-h) \approx 2h\dot{x}(t)$. Solving for $\dot{x}(t)$,

$$\dot{x}(t) \approx \frac{x(t+h) - x(t-h)}{2h}. \tag{4.14}$$

Equation (4.14) is called the *centered-difference equation for the first derivative*, since it expresses the derivative in terms of the functional value on either side of time t. On the other hand, by adding Equations (4.12) and (4.13) gives

$$\ddot{x}(t) \approx \frac{x(t+h) - 2x(t) + x(t-h)}{h^2}, \tag{4.15}$$

which is the *centered-difference equation for the second derivative*. Higher-order derivative formulas are found in a similar manner.

In practice, Equations (4.14), (4.15), and their like are usually presented as true difference equations rather than approximations to differential equations. Using the granularity Equation (4.1), the continuous signal $x(t)$ defines a discrete counterpart x_k. Formally, the first-derivative approximation for continuous time becomes

$$\dot{x}(t) \approx \frac{x(t+h) - x(t-h)}{2h}$$

$$= \frac{x(t_0 + hk + h) - x(t_0 + hk - h)}{2h}$$

$$= \frac{x(t_0 + h(k+1)) - x(t_0 + h(k-1))}{2h}.$$

Converting to the discrete counterpart of $x(t)$,

$$\dot{x}(k) = \frac{x(k+1) - x(k-1)}{2h}. \tag{4.16}$$

Similarly, it can be shown that

$$\ddot{x}(k) = \frac{x(k+1) - 2x(k) + x(k-1)}{h^2}. \tag{4.17}$$

Using Equations (4.16) and (4.17), this discretization process replaces a signal defined in the continuous domain $t_0 \leqslant t \leqslant t_n$ with an analogous discrete-domain description over $k = 0, 1, 2, \ldots, n$. This is illustrated in Example 4.6.

EXAMPLE 4.6

Consider the following boundary-value problem defining a signal $x(t)$ on the time interval $[1,3]$:

$$t^2\ddot{x} - 2t\dot{x} + 2x + 4 = 0,$$
$$x(1) = 0, \tag{4.18}$$
$$x(3) = 0.$$

(a) Find the unique analytical solution $x(t)$ that satisfies both the differential equation and the boundary conditions.
(b) Using the finite-difference method, find an approximate numerical solution to this system.

Solution
Since the differential equation is linear, the general solution is the sum of the homogenous and particular solutions. Consider first the homogenous equation, $t^2\ddot{x} - 2t\dot{x} + 2x = 0$. The degree of the coefficient of each derivative equals the order of the derivative, so the equation is of Euler type. Therefore, a suitable solution is of the form $x(t) = t^\alpha$, for constant α. Substitution into the homogenous equation results in

$$t^2\ddot{x} - 2t\dot{x} + 2x = t^2\alpha(\alpha-1)t^{\alpha-2} - 2t\alpha t^{\alpha-1} + 2t^\alpha$$
$$= t^\alpha(\alpha-1)(\alpha-2) = 0.$$

From this, there are apparently two solutions: one for $\alpha = 1$ and one for $= 2$. The corresponding solutions are $x_1(t) = t$ and $x_2(t) = t^2$.

The only difference between the original differential equation and its associated homogenous equation is the constant 4. In analogy with Equation (4.6), this corresponds to the input to the system. Since this input is a constant, it will create a steady-state output, which is the particular solution, that is also constant. Clearly, the particular solution is $x_p(t) = -2$, since that is the only constant that satisfies Equation (4.18). The *general solution* is the following sum of a linear combination of the homogeneous solutions and the particular solution

$$x(t) = At + Bt^2 - 2, \tag{4.19}$$

for constants A and B to be determined using the boundary conditions given in Equation (4.18). Substituting each condition into Equation (4.19) produces

$$x(1) = A + B - 2 = 0,$$
$$x(3) = 3A + 9B - 2 = 0.$$

Solving simultaneously, $A = \frac{8}{3}$ and $B = -\frac{2}{3}$, and the specific solution of the system (4.18) is

$$x(t) = \frac{2}{3}(t-1)(3-t). \tag{4.20}$$

Unfortunately, only in very idealized systems can an analytical solution even be found – let alone so easily. Even though the system dynamics of Equation (4.18) might be realistic, the input (the constant 4) is not. More than likely, the driving input will be a function of time involving noise, such as that in Example 4.4. Therefore, we now consider a numerical solution to the same system.

Using the granularity Equation (4.1) with $t_0 = 1$ and $t_n = 3$, if there are n intervals on $[1,3]$, it follows that $h = 2/n$ and $t = 2k/n + 1$. Combining this with the difference formulas (4.16) and (4.17) yields

$$\left(1 + \frac{2k}{n}\right)^2 \frac{x(k+1) - 2x(k) + x(k-1)}{(2/n)^2}$$
$$- 2\left(1 + \frac{2k}{n}\right)\frac{x(k+1) - x(k-1)}{2(2/n)} + 2x(k) + 4 = 0.$$

Simplifying by first clearing the fractions then writing in descending difference orders,

$$(n + 2k)(n + 2k - 2)x(k+1) - 2(n + 2k + 2)(n + 2k - 2)x(k)$$
$$+ (n + 2k)(n + 2k + 2)x(k-1) + 16 = 0. \qquad (4.21)$$

Recalling that the boundary points are specified by Equation (4.18), Equation (4.21) actually applies only to points in the interior region, that is, for $k = 1, 2, \ldots, n - 1$. At $k = 0$ and $k = n$, $x(0) = 0$ and $x(n) = 0$ owing to the boundary conditions imposed in the problem statement. For instance if $n = 2$, then $h = 1$, and there are three equations:

$$\begin{aligned} k = 0: &\quad x(0) = 0, \\ k = 1: &\quad 24x(0) - 24x(1) + 8x(2) + 16 = 0, \\ k = 2: &\quad x(2) = 0, \end{aligned}$$

from which the solution is $x(0) = 0$, $x(1) = \frac{2}{3}$, and $x(2) = 0$.

For larger values of n, a finer *mesh* is defined over the domain, and is thus more accurate, but there is also a larger set of equations to be solved. For instance, if $n = 4$,

$$x(0) = 0,$$
$$48x(0) - 64x(1) + 24x(2) + 16 = 0,$$
$$80x(1) - 120x(2) + 48x(3) + 16 = 0,$$
$$120x(2) - 192x(3) + 80x(4) + 16 = 0,$$
$$x(4) = 0.$$

This set of equations may be written as a single matrix equation in the form $\mathbf{AX} = \mathbf{B}$, namely

$$\begin{bmatrix} 1 & 0 & 0 & 0 & 0 \\ 48 & -64 & 24 & 0 & 0 \\ 0 & 80 & -120 & 48 & 0 \\ 0 & 0 & 120 & -192 & 80 \\ 0 & 0 & 0 & 0 & 1 \end{bmatrix} \begin{bmatrix} x(0) \\ x(1) \\ x(2) \\ x(3) \\ x(4) \end{bmatrix} = \begin{bmatrix} 0 \\ -16 \\ -16 \\ -16 \\ 0 \end{bmatrix}.$$

Inverting the matrix $\mathbf{A}$ and solving for $\mathbf{x}$ produces the solution $x(0) = 0$, $x(1) = \frac{1}{2}$, $x(2) = \frac{2}{3}$, $x(3) = \frac{1}{2}$, and $x(4) = 0$.

TABLE 4.2   Comparison of Numerical Solutions of the System Described by Equation (4.18) of Example 4.6 Using Finite Differences

k	t	n = 2	n = 4	n = 8	n = 16	Exact solution
0	1.000	0.000	0.000	0.000	0.000	0.000
1	1.125				0.156	0.156
2	1.250			0.292	0.292	0.292
3	1.375				0.406	0.406
4	1.500		0.500	0.500	0.500	0.500
5	1.625				0.573	0.573
6	1.750			0.625	0.625	0.625
7	1.875				0.656	0.656
8	2.000	0.667	0.667	0.667	0.667	0.667
9	2.125				0.656	0.656
10	2.250			0.625	0.625	0.625
11	2.375				0.573	0.573
12	2.500		0.500	0.500	0.500	0.500
13	2.675				0.406	0.406
14	2.750			0.292	0.292	0.292
15	2.875				0.156	0.156
16	3.000	0.000	0.000	0.000	0.000	0.000

This procedure is straightforward, and will always result in a banded matrix, for which there are more efficient solution methods than inversion. Even so, the conceptually simple but computationally hard inversion is satisfactory for $n = 2, 4, 8$, and 16. It will be noticed that in each case the exact analytical solution is identical to the approximate numerical solution. This is not true in general, but in this example is due to the fact that the analytical result and difference methods are both of the second order in this example. Table 4.2 also illustrates how the mesh size grows geometrically with interval number n.   ○

It is not always so easy to solve a large set of equations simultaneously. Even so, for linear systems such as that in Example 4.6, an iterative method can be employed effectively. By enforcing the boundary constraints on $x(0)$ and $x(n)$, then repeatedly computing $x(k)$ for $k = 1, 2, \ldots, n - 1$ on the interior points, the sequence $x(k)$ will converge to the exact solution. This is illustrated in Example 4.7.

EXAMPLE 4.7

Again consider the boundary-value problem of Example 4.6. Solving Equation (4.21) for $x(k)$ in terms of $x(k + 1)$ and $x(k - 1)$,

$$x(k) = \frac{(n + 2k)(n + 2k - 2)x(k + 1) + (n + 2k)(n + 2k + 2)x(k - 1) + 16}{2(n + 2k + 2)(n + 2k - 2)},$$

(4.22)

which holds for $k = 1, 2, \ldots, n - 1$. Implement this as an iterative algorithm to find the numerical solution to the system specified by Equation (4.18).

Solution

On the boundary, $t = 1$ and $t = 3$, so, using a mesh of n intervals, $h = 2/n$ and $t = 1 + 2k/n$. Using discrete time k, $x(0) = 0$ and $x(n) = 0$. From this, an appropriate algorithm proceeds as follows:

(i) Set the interior x-values to arbitrary initial values; fix the boundary values as specified:

```
for k=1 to n-1
        x(k)=0 (or some other arbitrary values)
next k
x(0)=0
x(n)=0
```

(ii) Iteratively update the interior points as per Equation (4.22):

```
for k=1 to n-1
        x(k) as given by Equation (4.22)
next k
```

(iii) Repeat step (ii) until convergence is achieved. ○

Convergence of the sequence $x(k)$ is tricky, since the question is really how many iterations are required to be close enough to the exact solution. This presents two problems. First, since $x(k)$ isn't a single number, but a vector of numbers, in reality all components must be close enough. The second problem is that since we don't know the answer, how can we possibly know when we are close to it? A common resolution of this question lies in the use of the so-called *Cauchy criterion*. The Cauchy criterion states that when all components $x_0(k)$, $x_1(k)$, $x_2(k)$, . . . , $x_n(k)$ of the estimated vector are within a given tolerance of the previous estimate $x_0(k-1)$, $x_1(k-1)$, $x_2(k-1)$, . . . , $x_n(k-1)$, convergence is close enough. Specifically, using the traditional Euclidean norm $\|x(k)\| = \sqrt{\sum_{i=1}^{n} x_i^2}$ of the vector $x(k) = [x_1(k), . . . , x_n(k)]$, the Cauchy criterion states that for a tolerance ϵ,

$$\|x(k) - x(k-1)\| < \epsilon.$$

Use of the Cauchy convergence criterion requires that the previous **x**-vector be remembered and that the Euclidean norm be computed. Comparing this with the specified tolerance, another iteration is taken, otherwise termination ensues. This is shown in Listing 4.4, where the iteration loop is broken into two parts: the update portion and the norm calculation. Notice that the previous **x**-vector $xp(k)$ is compared against $x(k)$, the square of the norm is the variable *diff*, and the tolerance is ϵ.

There are good theoretical reasons for using the Cauchy criterion, but there are also many fine reasons to be wary as well. However, in the case of linear systems, there are no real problems. Eventually, the signal values repeat and convergence is assured. Unfortunately, this is often quite a while! Even though it can be shown that for linear equations there will be convergence to the correct solution independently of the initial conditions, this is not a fast technique. For the specific case of $n = 16$ and beginning at the specific initial "guess" of $x(k) = 0$ for $k = 1, 2, . . . , n - 1$, some of the iterates are shown in Table 4.3. The graph

```
for k=1 to n-1
        x(k)=0
next k
x(0)=0
x(n)=0
do until diff<ε
        print 0, x(0)
        for k=1 to n-1
                t=1+2k/n
                xp(k)=x(k)
                compute x(k) using Equation (4.22)
                print k, x(k)
        next k
        print n, x(n)
        diff=0
        for k=1 to n-1
                diff=diff+[x(k)-xp(k)]²
        next k
end do
```

LISTING 4.4   Solution to the system described by Equation (4.18) using iteration.

of each iterate is also shown in Figure 4.16. Figure 4.17 presents the same result, but starting from the initial guess of $x(k) = 1$ in the interior. Clearly, the iterative approach is not efficient, but it works. Recall that the exact solution is given by $x(t) = \frac{2}{3}(t - 1)(3 - t)$. Figure 4.16 shows that each iterate builds up to the correct answer, while Figure 4.17 builds down. Regardless, convergence is assured.

TABLE 4.3   Comparative Iterates in the Iterative Solution to the System (4.18), Starting with $x(k) = 0$ in the Interior

k	t	Initial	50 tries	100 tries	250 tries	500 tries	Actual solution
0	1.000	0	0.000	0.000	0.000	0.000	0.000
1	1.125	0	0.126	0.145	0.156	0.156	0.156
2	1.250	0	0.226	0.267	0.290	0.292	0.292
3	1.375	0	0.301	0.367	0.404	0.406	0.406
4	1.500	0	0.355	0.446	0.497	0.500	0.500
5	1.675	0	0.388	0.504	0.569	0.573	0.573
6	1.750	0	0.405	0.542	0.620	0.625	0.625
7	1.875	0	0.406	0.562	0.651	0.656	0.656
8	2.000	0	0.396	0.564	0.661	0.667	0.667
9	2.125	0	0.375	0.549	0.650	0.656	0.656
10	2.250	0	0.346	0.518	0.619	0.625	0.625
11	2.375	0	0.309	0.472	0.567	0.573	0.573
12	2.500	0	0.264	0.409	0.495	0.500	0.500
13	2.675	0	0.212	0.332	0.402	0.406	0.406
14	2.750	0	0.152	0.238	0.289	0.292	0.292
15	2.875	0	0.082	0.128	0.155	0.156	0.156
16	3.000	0	0.000	0.000	0.000	0.000	0.000

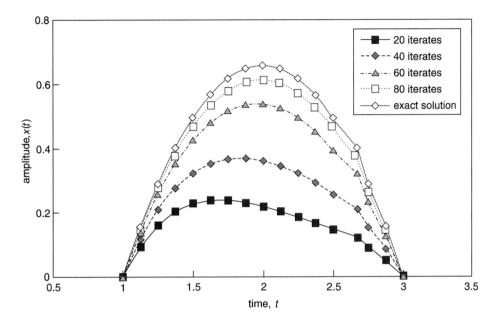

FIGURE 4.16 Successive iterates in the iterative solution to the system (4.18), starting with $x(k) = 0$ in the interior

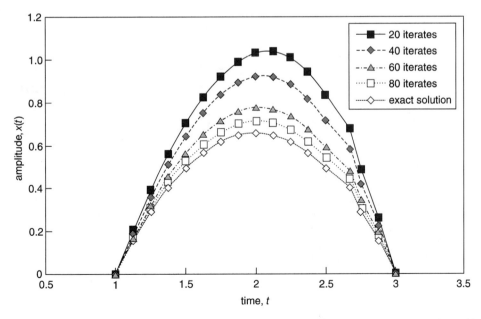

FIGURE 4.17 Successive iterates inthe iterative solution to the system (4.18), starting with $x(k) = 1$ in the interior.

4.4 ___ PARTIAL DIFFERENTIAL EQUATIONS

As the dimension of the system domain increases, the modeling situation becomes more complicated. For instance, a typical domain in two dimensions is shown in Figure 4.18, where a boundary $\mathcal{B}$ encloses a domain $\mathcal{D}$. Unlike the one-dimensional domain of Example 4.6, where the boundary was just two points, the boundary is now defined by planar curves rather than simply points defining an interval.

Typical *boundary-value problems* prescribe a system dynamics over $\mathcal{D}$ (usually described by a partial differential equation), together with a constraint over $\mathcal{B}$. If a unique solution is known to exist that satisfies both the dynamics and the boundary conditions, the problem is said to be *well posed*. Let ϕ be the function in question for a continuous model with spatial coordinates (x, y). The following system descriptions are known to be well-posed boundary-value problems:

(i) *heat equation* (also called the *diffusion equation*)

$$\frac{\partial^2 \phi}{\partial x^2} = \frac{1}{\rho^2} \frac{\partial \phi}{\partial t}, \qquad a < x < b, \qquad t > t_0,$$

known a priori: $\begin{cases} \phi(t_0, x), \\ \phi(t, a), \\ \phi(t, b); \end{cases}$ (4.23)

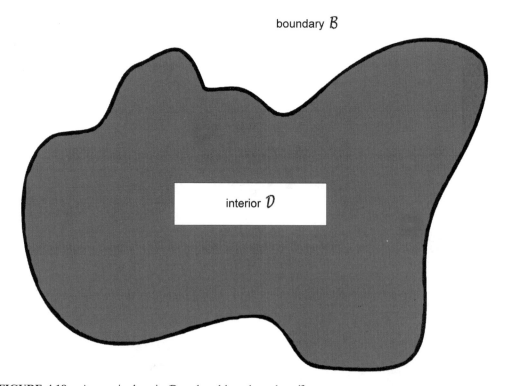

boundary $\mathcal{B}$

interior $\mathcal{D}$

FIGURE 4.18   A generic domain $\mathcal{D}$ enclosed by a boundary $\mathcal{B}$.

(ii) *wave equation*

$$\frac{\partial^2 \phi}{\partial x^2} = \frac{1}{\rho^2} \frac{\partial^2 \phi}{\partial t^2}, \qquad a < x < b, \qquad t > t_0,$$

$$\text{known a priori:} \quad \begin{cases} \phi(t_0, x), \\ \dot{\phi}(t_0, x), \\ \phi(t, a), \\ \phi(t, b). \end{cases} \qquad (4.24)$$

(iii) *Laplace's equation*

$$\frac{\partial^2 \phi}{\partial x^2} + \frac{\partial^2 \phi}{\partial y^2} = 0, \qquad a < x < b, \qquad c < y < d,$$

$$\text{known a priori:} \quad \begin{cases} \phi(a, y) \\ \phi(b, y), \\ \phi(x, c), \\ \phi(x, d). \end{cases} \qquad (4.25)$$

Each of the above well-posed problems can also be extended to higher dimensions by simply replacing the left-hand-side of each partial differential equation by the Laplacian,

$$\nabla^2 \phi = \frac{\partial^2 \phi}{\partial x^2} + \frac{\partial^2 \phi}{\partial y^2} + \frac{\partial^2 \phi}{\partial z^2}.$$

Also, the boundary region of interest can be adjusted by using different coordinate systems. Regardless, models (i), (ii), and (iii) are classical and well understood. Even so, their analytical solution is non-trivial, as illustrated by the following example.

EXAMPLE 4.8

Consider the boundary-value problem defined by the following partial differential equation and boundary conditions:

$$\frac{\partial^2 \phi}{\partial x^2} + \frac{\partial^2 \phi}{\partial y^2} = 0, \qquad (4.26)$$

condition (*a*):	$\phi(0, y) = 1,$	$0 < y < 1,$	(4.27a)
condition (*b*):	$\phi(1, y) = 1,$	$0 < y < 1,$	(4.27b)
condition (*c*):	$\phi(x, 0) = 0,$	$0 < x < 1,$	(4.27c)
condition (*d*):	$\phi(x, 1) = 0,$	$0 < x < 1,$	(4.27d)

Equation (4.26) is the classical Laplace equation, and is often used to model the steady-state heat distribution over a thermally conducting plate of domain $\{(x, y)|0 < x < 1, 0 < y < 1\}$ with boundary values as specified by Equations (4.27). This is shown geometrically in Figure 4.19. Find an explicit analytical solution to this problem.

Solution
We assume a variable-separable solution of the form $\phi(x, y) = X(x)Y(y)$, where X

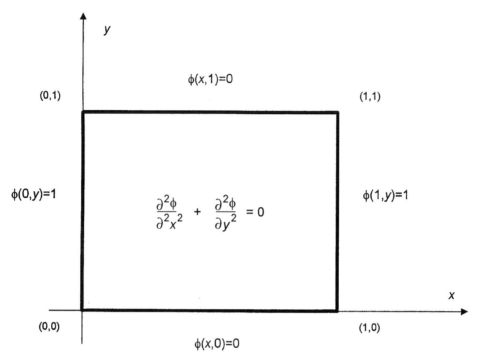

FIGURE 4.19   Domain and boundary conditions prescribed by example 4.8.

and Y are functions of only x and y respectively. It follows that

$$\frac{\partial^2 \phi}{\partial x^2} = X''(x)Y(y),$$

$$\frac{\partial^2 \phi}{\partial y^2} = X(x)Y''(y). \tag{4.28}$$

Thus, $X''Y + XY'' = 0$ and

$$\frac{X''}{X} = \frac{Y''}{Y} = \text{constant.} \tag{4.29}$$

The ratios X''/X and Y''/Y are constant, since they must be independent of each other because as they are functions of different variables. The constant per se is yet to be determined. There are three cases to be considered:

(i) Assume that the constant is zero. Therefore, $X'' = 0$, and it follows that $X(x) = Ax + B$. Similarly, $Y(y) = Cy + D$, where $A, B, C,$ and D are constants to be determined. Furthermore, $\phi(x, y) = (Ax + B)(Cy + D)$.

Applying the boundary conditions (4.27) results in $A = B = 0$, leading to the conclusion that the constant cannot be zero, since this would require that $\phi(x, y) = 0$. This cannot be the case, since this solution does not satisfy all of the boundary conditions. Since this leads to a contradiction, case (i) must be rejected as untenable.

(ii) Assume that the constant is $-\lambda^2$, a negative number. Equation (4.26) now becomes

$$X'' + \lambda^2 X = 0,$$
$$Y'' - \lambda^2 Y = 0,$$

from which it follows that

$$X(x) = A \cos \lambda x + B \sin \lambda x, \qquad Y(y) = Ce^{\lambda y} + De^{-\lambda y}.$$

Thus,

$$\phi(x, y) = (A \cos \lambda x + B \sin \lambda x)(Ce^{\lambda y} + De^{-\lambda y}).$$

Applying the boundary conditions (4.27) again produces an inconsistency. Therefore, we exclude this case too.

(iii) Assume that the constant is λ^2, a positive number. Equation (4.26) now becomes

$$X'' - \lambda^2 X = 0,$$
$$Y'' + \lambda^2 Y = 0.$$

Here the implication is similar to that in case (ii), but this time

$$X(x) = Ae^{\lambda x} + Be^{-\lambda x}, \qquad Y(y) = C \cos \lambda y + D \sin \lambda y.$$

It follows that

$$\phi(x, y) = (Ae^{\lambda x} + Be^{-\lambda x})(C \cos \lambda y + D \sin \lambda y),$$

and this time there are no inconsistencies in the boundary conditions.

Staying with case (iii), using Equation (4.27c) implies $C = 0$, so we incorporate the constant D into A and B. Applying the conditions (4.27a) and (4.27b) and then eliminating the factor $\sin y$ implies that $B = Ae^{\lambda}$. Thus,

$$\phi(x, y) = A(e^{\lambda x} + e^{\lambda(1-x)}) \sin \lambda y.$$

Now using Equation (4.27d),

$$\phi(x, 1) = A(e^{\lambda x} + e^{-\lambda x}) \sin \lambda = 0,$$

which requires that λ be a multiple of π. That is, $\lambda = k\pi$ for $k = 0, 1, 2, \ldots$. Thus, there are infinitely many solutions, one for each integer k:

$$\phi_k(x, y) = A_k(e^{k\pi x} + e^{-k\pi x}) \sin k\pi y.$$

By superposition, the general solution is the sum

$$\phi(x, y) = \sum_{k=0}^{\infty} A_k(e^{k\pi x} + e^{-k\pi x}) \sin k\pi y.$$

Imposing the condition (4.27b),

$$\phi(1, y) = \sum_{k=0}^{\infty} A_k(e^{k\pi} + 1) \sin k\pi y = 1.$$

We recognize the Fourier sine series for unity, with coefficient $A_k(e^{k\pi} + 1)$. This coefficient can be found by applying Fourier's formula to solve for the Fourier coefficients:

$$A_k(e^{k\pi} + 1) = 2 \int_0^1 \sin k\pi y \, dy$$
$$= \frac{2[1 + (-1)^{k+1}]}{k\pi}.$$

Therefore,

$$\phi(x, y) = \sum_{k=0}^{\infty} \frac{2[1 + (-1)^{k+1}](e^{k\pi x} + e^{k\pi(1-x)})}{k\pi(1 + e^{k\pi})} \sin k\pi y,$$

which is sometimes more conveniently written as

$$\phi(x, y) = \frac{4}{\pi} \sum_{k=1}^{\infty} \frac{e^{(2k-1)\pi x} + e^{(2k-1)\pi(1-x)}}{(2k - 1)(1 + e^{(2k-1)\pi})} \sin[(2k - 1)\pi y]. \qquad (4.30)$$

○

Equation (4.30) presents an analytical solution to the classical Laplace equation on a square plate. It can be generalized, but this will not be done here. However, it includes an infinite series and is therefore not in *closed* form. Even so, it can be computed to any desired degree of accuracy in a straightforward manner. Using discrete variables i and j for the continuous x and y and noting that the step size is $h = 1/n$,

$$x_i = \frac{i}{n}, \quad i = 0, 1, \ldots, n,$$

$$y_j = \frac{j}{n}, \quad j = 0, 1, \ldots, n.$$

Code that sums the first m terms of the Fourier series computing points within the domain for this problem is shown in Listing 4.5. The output for 81 cells ($n = 8$) is shown in Table 4.4.

4.5   FINITE DIFFERENCES FOR PARTIAL DERIVATIVES

The finite-difference Equations (4.14) and (4.15) are easily extended to multiple dimensions. If $\phi(x, y)$ is a function of the two independent variables x and y, partial

```
for i=0 to n
    for j=0 to n
        x=i/n
        y=j/n
        φ=0
        for k=1 to m
```

$$\phi = \phi + \frac{\exp[(2k-1)\pi x] + \exp[(2k-1)\pi(1-x)]}{(2k-1)(1+\exp[(2k-1)\pi])} \sin[(2k-1)\pi y]$$

```
        next k
        φ=4φ/π
        print x, y, φ
    next j
next i
```

LISTING 4.5   Solution to the system specified in Example 4.8 given by the infinite series in Equation (4.30).

TABLE 4.4   Tabulation of Equation (4.30) to 10 Terms

8	1.000	0.000	0.000	0.000	0.000	0.000	0.000	0.000	0.000	0.000
7	0.875	1.068	0.500	0.300	0.223	0.201	0.223	0.300	0.500	1.068
6	0.750	0.963	0.700	0.500	0.396	0.364	0.396	0.500	0.700	0.963
5	0.675	1.029	0.777	0.604	0.500	0.466	0.500	0.604	0.777	1.029
4	0.500	0.974	0.799	0.636	0.534	0.500	0.534	0.636	0.799	0.974
3	0.375	1.029	0.777	0.604	0.500	0.466	0.500	0.604	0.777	1.029
2	0.250	0.963	0.700	0.500	0.396	0.364	0.396	0.500	0.700	0.963
1	0.125	1.068	0.500	0.300	0.223	0.201	0.223	0.300	0.500	1.068
0	0.000	0.000	0.000	0.000	0.000	0.000	0.000	0.000	0.000	0.000
j	y/x	0.000	0.125	0.250	0.375	0.500	0.675	0.750	0.875	1.000
	i	0	1	2	3	4	5	6	7	8

derivatives are computed by differentiating as usual by differentiating one variable while holding the other variable fixed. Thus,

$$\frac{\partial}{\partial x}\phi(x,y) \approx \frac{\phi(x+h,y) - \phi(x-h,y)}{2h},$$

$$\frac{\partial}{\partial y}\phi(x,y) \approx \frac{\phi(x,y+h) - \phi(x,y-h)}{2h}$$

and

$$\frac{\partial^2}{\partial x^2}\phi(x,y) \approx \frac{\phi(x+h,y) - 2\phi(x,y) + \phi(x-h,y)}{h^2},$$

$$\frac{\partial^2}{\partial y^2}\phi(x,y) \approx \frac{\phi(x,y+h) - 2\phi(x,y) + \phi(x,y-h)}{h^2}.$$

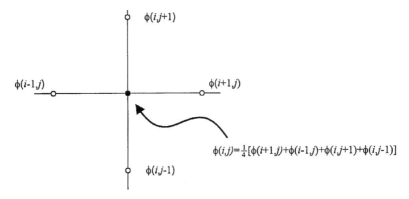

$$\phi(i,j)=\tfrac{1}{4}[\phi(i+1,j)+\phi(i-1,j)+\phi(i,j+1)+\phi(i,j-1)]$$

FIGURE 4.20 Discrete form of Laplace's steady-state heat conduction model.

The corresponding discrete difference equation formulas are

$$\frac{\partial}{\partial x}\phi(x, y) = \frac{\phi(i+1,j) - \phi(i-1,j)}{2h},$$

$$\frac{\partial}{\partial y}\phi(x, y) = \frac{\phi(i,j+1) - \phi(i,j-1)}{2h}$$

and

$$\frac{\partial^2}{\partial x^2}\phi(x, y) = \frac{\phi(i+1,j) - 2\phi(i,j) + \phi(i-1,j)}{h^2},$$

$$\frac{\partial^2}{\partial y^2}\phi(x, y) = \frac{\phi(i,j+1) - 2\phi(i,j) + \phi(i,j-1)}{h^2}$$

Application of the finite-difference formula for partial differential equations proceeds in the same way as for ordinary differential equations. For instance, consider Laplace's equation (4.26) from Example 4.8. Upon direct substitution into Equation (4.26),

$$\frac{\phi(i+1,j) - 2\phi(i,j) + \phi(i-1,j)}{h^2} + \frac{\phi(i,j+1) - 2\phi(i,j) + \phi(i,j-1)}{h^2} = 0.$$

Solving for $\phi(i,j)$ gives the following very intuitive relationship:

$$\phi(i,j) = \tfrac{1}{4}[\phi(i+1,j) + \phi(i-1,j) + \phi(i,j+1) + \phi(i,j-1)]. \qquad (4.31)$$

This equation possesses a certain elegance. It states that the temperature ϕ at any given point (i,j) is simply the average of the temperatures at the adjacent points. Recalling that Laplace's equation is applied in heat-distribution models, this makes a great deal of sense, and is illustrated in Figure 4.20.

EXAMPLE 4.9

Consider the application of Laplace's equation posed as the boundary-value problem described in Example 4.8. Apply the finite-difference method to approximate the actual solution.

Solution
The discrete form of Equation (4.26) is given by using Equation (4.31). Also, the boundary points have the following values:

$$\begin{aligned}
\phi(i, 0) &= 0, & i &= 1, 2, \ldots, n - 1, \\
\phi(i, 1) &= 0, & i &= 1, 2, \ldots, n - 1, \\
\phi(0, j) &= 1, & j &= 1, 2, \ldots, n - 1, \\
\phi(1, j) &= 1, & j &= 1, 2, \ldots, n - 1.
\end{aligned}$$ (4.32)

In general, there will be $n - 1$ boundary points for each of the four sides and $(n - 1)^2$ interior points. For $n = 4$, there are 12 boundary points and 9 interior points, as illustrated in Figure 4.21.

Using Equation (4.31), the nine independent equations are as follows:

$$\begin{aligned}
4\phi(1,1) &= 1 + \phi(1,2) + \phi(2,1), \\
4\phi(2,1) &= 1 + \phi(2,2) + \phi(1,1) + \phi(3,1), \\
4\phi(3,1) &= 1 + \phi(3,2) + \phi(2,1), \\
4\phi(1,2) &= \phi(2,2) + \phi(1,1) + \phi(1,3), \\
4\phi(2,2) &= \phi(1,2) + \phi(3,2) + \phi(2,1) + \phi(2,3), \\
4\phi(3,2) &= \phi(2,2) + \phi(3,1) + \phi(3,3), \\
4\phi(1,3) &= 1 + \phi(1,2) + \phi(2,3), \\
4\phi(2,3) &= 1 + \phi(1,3) + \phi(2,2) + \phi(3,3), \\
4\phi(3,3) &= 1 + \phi(2,3) + \phi(3,2).
\end{aligned}$$

These may be represented in linear form $\mathbf{A}\boldsymbol{\phi} = \mathbf{B}$ with the following banded matrix equation:

$$
\begin{bmatrix}
-4 & 1 & 0 & 1 & 0 & 0 & 0 & 0 & 0 \\
1 & -4 & 1 & 0 & 1 & 0 & 0 & 0 & 0 \\
0 & 1 & -4 & 0 & 0 & 1 & 0 & 0 & 0 \\
1 & 0 & 0 & -4 & 1 & 0 & 1 & 0 & 0 \\
0 & 1 & 0 & 1 & -4 & 1 & 0 & 1 & 0 \\
0 & 0 & 1 & 0 & 1 & -4 & 0 & 0 & 1 \\
0 & 0 & 0 & 1 & 0 & 0 & -4 & 1 & 0 \\
0 & 0 & 0 & 0 & 1 & 0 & 1 & -4 & 1 \\
0 & 0 & 0 & 0 & 0 & 1 & 0 & 1 & -4
\end{bmatrix}
\begin{bmatrix}
\phi(1,1) \\
\phi(2,1) \\
\phi(3,1) \\
\phi(1,2) \\
\phi(2,2) \\
\phi(3,2) \\
\phi(3,1) \\
\phi(3,2) \\
\phi(3,3)
\end{bmatrix}
=
\begin{bmatrix}
-1 \\
-1 \\
-1 \\
0 \\
0 \\
0 \\
-1 \\
-1 \\
-1
\end{bmatrix}.
$$

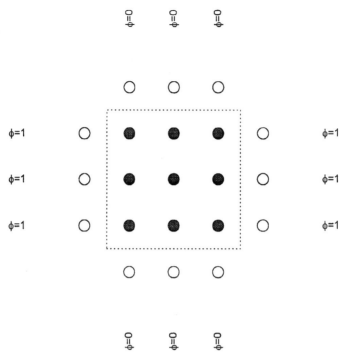

FIGURE 4.21   Geometry for $n = 4$ of the 5-by-5 grid defined by Equation (4.32).

Inverting the matrix **A** and solving for ϕ produces

$$\Phi = \begin{bmatrix} \phi(1,1) \\ \phi(2,1) \\ \phi(3,1) \\ \phi(2,1) \\ \phi(2,2) \\ \phi(2,3) \\ \phi(3,1) \\ \phi(3,2) \\ \phi(3,3) \end{bmatrix} = \begin{bmatrix} \frac{1}{2} \\ \frac{5}{8} \\ \frac{1}{2} \\ \frac{3}{8} \\ \frac{1}{2} \\ \frac{3}{8} \\ \frac{1}{2} \\ \frac{5}{8} \\ \frac{1}{2} \end{bmatrix}.$$

The results, shown in Table 4.5 on a grid similar to that in Table 4.4, are good. It should be noted that they do not reproduce the exact solution – but even with an $n = 4$ sized mesh there is fairly good accuracy, nonetheless.   ○

The obvious problem with this method is the large number of simultaneous equations that must be handled. Even so, it should be remembered that in practice the resulting banded matrices are handled by special (non-inversion) means that are far more efficient.

TABLE 4.5 Solution of Laplace's Boundary-Value Problem (4.31), (4.32) of Example 4.9 Using Finite Differences with $n = 4$

j	y	Temperature				
4	1.00		0.000	0.000	0.000	
3	0.75	1.000	0.500	0.375	0.500	1.000
2	0.50	1.000	0.625	0.500	0.625	1.000
1	0.25	1.000	0.500	0.375	0.500	1.000
0	0.00		0.000	0.000	0.000	
	x	0.00	0.25	0.50	0.75	1.00
	i	0	1	2	3	4

An alternative is to proceed iteratively as in Example 4.7. This is straightforward, but, as was illustrated earlier, convergence can be exceedingly slow.

4.6 CONSTRAINT PROPAGATION

There are a number of applications involving difference equations that have nothing whatsoever to do with heat transfer, temperature distributions, vibrating membranes, or mechanical systems. Applications in image and signal processing and data smoothing are also commonplace. These may often be thought of as generalizations of the boundary-value problems of difference equations in that the problem constraints may be scattered throughout the entire domain and not just on the domain boundary. These constraints are then propagated throughout the system.

Rather than developing a general theory, the data-smoothing technique described here, called *relaxation*, will be illustrated by example. Roughly speaking, relaxation is a heuristic by which imposed constraints are set a priori, then iteratively propagated throughout the system to a solution. These constraints need not even be on the boundary. As long as the problem is well posed, a unique solution exists, and relaxation will find it.

Consider the following problem:

> It is required to draw a topographic map of land on a straight line joining two points. The heights at the two endpoints are known accurately, and selected points in the interior are estimated using barometric pressure readings. However, the confidence in many of the barometric readings is less than ideal, and several points are not even estimated. It is desired to find the height at all points along the line without making direct measurements.

At first, this problem seems untenable. However, by assuming that intermediate points tend to be the average of adjacent points (a fairly plausible assumption), a reasonable solution may be approximated.

More precisely, consider a one-dimensional grid of $n + 1$ points at positions x_0, x_1, $x_2, \ldots, x_n$. The actual height (which we wish to determine) is taken to be $a(i)$ for $i = 0, 1, \ldots, n$. Similarly, the measured barometric height is $b(i)$ and the corresponding

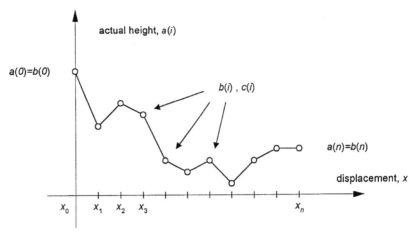

FIGURE 4.22   Rough topographic map with estimated heights $b(i)$, confidences $c(i)$ and actual heights $a(i)$ at position $x(i)$.

confidence in that reading is $c(i)$. We assume that a confidence of unity is perfectly accurate, a confidence of zero is not at all to be trusted, and intermediate values are somewhere in between. It is desired to find the actual height $a(i)$. See Figure 4.22. Using the averaging assumption for interior points, an appropriate difference equation is

$$a(i) = b(i)c(i) + \tfrac{1}{2}[a(i+1) + a(i-1)][1 - c(i)], \qquad i = 2, 3, \ldots, n-1. \qquad (4.33)$$

Notice that if there is high confidence in $b(i)$, then $a(i)$ tends to match. On the other hand, low confidence in $b(i)$ tends to weight the averaging term most heavily. For each $i = 0, 1, \ldots, n$, $b(i)$ and $c(i)$ are known. (If $b(i)$ is unknown, simply assume $b(i) = 0$ with corresponding confidence $c(i) = 0$.) The result is a well-posed boundary-value problem with additional interior constraints. We will use the so-called relaxation method to exercise constraint propagation and therefore estimate the heights of each interior point. Since the difference Equation (4.33) is linear, the iterative method outlined in Example 4.7 will suffice. Speed of convergence will not be as much of a problem here, since the internal interval constraints provide more information than for a pure boundary-value problem.

EXAMPLE 4.10

Consider the topographical map problem defined above for the boundary and interior constraints defined in the following table.

Index, i	0	1	2	3	4	5	6	7	8
Position, $x(i)$	0.0	1.5	3.0	4.5	6.0	7.5	9.0	10.5	12.0
Barometric height, $b(i)$	10	0	0	4	7	0	1	6	3
Confidence, $c(i)$	1.0	0.0	0.0	0.5	0.3	0.0	0.7	0.4	1.0

Notice that at the two endpoints, the barometric pressure is known with certainty, but at points interior to the interval, heights are less certain. Generate the

```
read a(0), a(n)
for i=1 to n-1
        read b(i),c(i)
        a(i)=b(i)
next i
k=1
do until diff<ε
        print k, a(0)
        for i=1 to n-1
                x=x₀+hi
                ap(i)=a(i)
                a(i)=b(i)c(i)+½[a(i+1)+a(i-1)][1-c(i)]
                print a(i),
        next i
        print a(n)
        diff=0
        for i=1 to n-1
                diff=diff+[a(i)-ap(i)]²
        next i
        k=k+1
end do
```

LISTING 4.6   Solution to Example 4.10 using iteration applied to Equation (4.33).

sequence of iterates, thus finding the resulting apparent heights using the relaxation method.

Solution
We will apply Equation (4.33) iteratively. A natural starting point is to select each of the interior $a_0(i)$ heights to match the best estimate $b(i)$. Using k as the index of iteration,

$$a_{k+1}(i) = b(i)c(i) + \tfrac{1}{2}[a_k(i+1) + a_k(i-1)][1 - c(i)], \qquad i = 1, 2, \ldots, n-1,$$

where $k = 0, 1, \ldots$ until convergence. A program segment performing this is given in Listing 4.6, where the Cauchy convergence criterion is used. As in Listing 4.4, note the use of a temporary vector $ap(i)$ to apply the Cauchy condition. The resulting sequence of heights are given in Table 4.6. The clearly rapid convergence is due to the additional interior constraints that are propagated throughout the interval by the relaxation algorithm.   ○

There is no real need to treat problems of this type as strict boundary-value problems. From a difference-equation point of view, constraint propagation problems are a more general class than boundary-value problems. Therefore, non-boundary conditions as well as derivative conditions (recall Chapter 2) can be imposed relatively easily. Solution by relaxation simply propagates the constraints. However, in general, this is an application-specific endeavor, and so will be left at this point. It should be clear that modeling systems in this way is a fruitful enterprise.

TABLE 4.6   Results of the Relaxation Method Applied to Example 4.10

Iteration	0	1	2	3	4	5	6	7	8
0	10.00	5.00	4.50	4.86	3.81	2.40	1.96	3.89	3.00
1	10.00	7.25	6.06	4.47	4.51	3.23	1.77	3.83	3.00
2	10.00	8.03	6.41	4.69	4.87	3.32	1.77	3.83	3.00
3	10.00	8.20	6.51	4.82	4.95	3.36	1.89	3.83	3.00
4	10.00	8.26	6.56	4.87	4.98	3.38	1.78	3.84	3.00
5	10.00	8.28	6.58	4.88	4.99	3.39	1.78	3.84	3.00
6	10.00	8.29	6.59	4.89	5.00	3.39	1.78	3.84	3.00

BIBLIOGRAPHY

Bennett, W. R., *Scientific and Engineering Problem-Solving with the Computer.* Prentice-Hall, 1976.

Bronson, R., *Theory and Problems of Modern Introductory Differential Equations* [Schaum's Outline Series] McGraw-Hill, 1973.

Burden, R. L., J. D. Faires and S. Reynolds. *Numerical Analysis*, 2nd edn. Prindle, Weber & Schmidt, 1981.

Conte, C. M., S. K. Dean, and C. de Boor, *Elementary Numerical Analysis*, 3rd edn. McGraw-Hill, 1980.

Fausett, L. V., *Applied Numerical Analysis Using MATLAB*. Prentice-Hall, 1999.

Press, W. H., B. P. Flannery, S. A. Teukolsky, and W. T. Vettering, *Numerical Recipes: The Art of Scientific Computing*. Cambridge University Press, 1986.

Richardson, C. H., *An Introduction to the Calculus of Finite Differences*. Van Nostrand-Reinhold, 1954.

Scheid, F., *Theory and Problems of Numerical Analysis* [Schaum's Outline Series]. McGraw-Hill, 1968.

Spiegel, M. R., *Theory and Problems of Fourier Analysis with Boundary Value Problems*. McGraw-Hill, 1968.

Winston, P. H., *Artificial Intelligence*. McGraw-Hill, 1992.

Yakowitz, S. and F. Szidarovszky, *An Introduction to Numerical Computations*. MacMillan, 1986.

EXERCISES

4.1 Suppose that $w_1(t)$ and $w_2(t)$ are two shock signals with the following definitions:

- $w_1(t)$ delivers Gaussian white noise with standard deviation 2 each $\frac{1}{2}$ second;
- $w_2(t)$ delivers Gaussian white noise with standard deviation 1 each $\frac{1}{3}$ second.

Create a signal $x(t)$ defined by the equation $x(t) = x(t - \frac{1}{7}) + w_1(t) + w_2(t)$.
(a) Print the signal each $\frac{1}{42}$ seconds from 0 to $\frac{1}{6}$.
(b) Print and graph the signal each 0.001 seconds over the interval [0,2].

4.2 Let $x(t)$ be a periodic signal with one cycle defined by

$$x_1(t) = \begin{cases} 5, & 0 \leqslant t < 2, \\ 0, & 2 \leqslant t < 3, \\ 2, & 3 \leqslant t < 5. \end{cases}$$

(a) Write a function to return the value of x for any specified t.
(b) Using the function defined in part (a), graph $x(t)$ on the interval [0,30].

4.3 Consider two periodic signals $x(t)$ with period 3 and $y(t)$ with period 4. A single cycle of each is defined as follows:

$$x_1(t) = \begin{cases} \frac{5}{2}t, & 0 \leqslant t < \frac{3}{2}, \\ \frac{5}{2}(3-t), & \frac{3}{2} \leqslant t < 3, \end{cases}$$

$$y_1(t) = \begin{cases} 2, & 0 \leqslant t < 3, \\ 0, & 3 \leqslant t < 4. \end{cases}$$

(a) Write functions to evaluate $x(t)$ and $y(t)$ for arbitrary true t.
(b) Write a program to graph the composite signal $f(t) = 5x(t) - 3y(t)$ over [0,24].
(c) Graph $x(t)$, $(y(t)$ and $f(t)$ as an overlay.

4.4 Consider the following data. Using this data, derive Lagrange's collocation fourth-degree polynomial that goes through each of the tabulated points.

t	$x(t)$
1.0	4
2.5	6
3.0	3
4.5	1
6.0	2

(a) Derive each of the required constituent polynomials.
(b) Derive Lagrange's collocation polynomial.
(c) On the basis of Lagrange's polynomial, part (b), extrapolate an estimate of the functional value at $t = 8.5$.

4.5 Consider the continuous signal $x(t) = 2\cos 5t + 7\sin 10t$, which is sampled at frequency $f_s = 20$ samples per second. Assuming a granularity of $h = 0.01$ seconds:
(a) Graph the continuous and zero-order-hold signals.
(b) Write a program to extrapolate the second-order-hold (quadratic-hold) signal.
(c) Graph the continuous and quadratic-hold signals.

4.6 Consider a continuous signal $x(t)$ that is sampled at f_s samples per second. Assuming a generic granularity of h seconds:
(a) Write a program to sample and extrapolate the signal using the Lagrange extrapolation polynomial.
(b) Execute the program, and compare $x(t)$ against the extrapolated signal $\hat{x}(t)$.
(c) On the basis of part (b), make an intelligent guess at optimizing the optimal extrapolation polynomial.

Make the comparison by performing a computer simulation and computing the sum of the differences between $x(t)$ and $\hat{x}(t)$ at each point from $t = t_0$ to $t = t_n$ in the time interval:

$$\sum_{t=t_0, \text{step } h}^{t_n} |x(t) - \hat{x}(t)|.$$

Form this sum for different degrees s of $\hat{x}(t)$, and complete a table such as

Degree of $\hat{x}(t)$	Error sum
0 (ZOH)	
1 (linear)	
2 (quadratic)	
3	
4	
5	

Do this for the following signals:
(d) $x(t) = 2 \sin 7t + 5 \cos 12t$;
(e) $x(t) = 2 \sin 7t + 5 \cos 12t + w(t)$, where $w(t)$ is Gaussian white noise of variance 2.
(f) $x(t) = x(t-1) + 2 \sin 7t + 5 \cos 12t + w(t)$, where $w(t)$ is Gaussian white noise of variance 1.

4.7 Consider the boundary-value problem

$$\frac{d^2}{dx^2} x(t) = C,$$

where C is a constant over the interval $t_0 \leqslant t \leqslant t_n$ with $x(t_0) = a$ and $x(t_n) = b$. Derive an explicit solution for $x(t)$.

4.8 Derive an analytical solution of the following boundary-value problem:

$$\ddot{x}(t) = 4x(t), \qquad 1 \leqslant t \leqslant 5,$$

subject to $x(1) = 3$ and $x(5) = 1$ at the boundaries.

4.9 Solve the following systems using the method of finite differences with mesh $n = 4$:
(a) $\ddot{x}(t) + 5tx(t) = t^2$, $\quad 0 \leqslant t \leqslant 4$,
  subject to $x(0) = 1$ and $x(4) = 1$;
(b) $\ddot{x}(t) + 3\dot{x}(t) + 7x(t) = 7$, $\quad 0 \leqslant t \leqslant 8$,
  subject to $x(0) = 1$ and $x(8) = 0$;
(c) $\ddot{x}(t) + x(t) = 2t$, $\quad 0 \leqslant t \leqslant 8$,
  subject to $x(0) = 0$ and $x(8) = 1$.

4.10 Repeat Exercise 4.9 using a mesh of $n = 8$.

4.11 **(a)** Write a computer program to solve the following system for general values of the mesh size n, boundary times t_0 and t_n, and boundary values x_0 and x_n:

$$\ddot{x}(t) + \dot{x}(t) - x(t) = -t, \qquad t_0 \leqslant t \leqslant t_n,$$
$$\text{subject to } x(t_0) = x_0 \text{ and } x(t_n) = x_n.$$

(b) Apply the program in part (a) to the system, using $n = 20$, $h = 1$, $x_0 = 1$, $x_{20} = 3$, and $t_{20} = 10$.

4.12 Consider the following table of estimated barometric data along with confidence in each of the corresponding measurements. Apply the method of constraint propagation to estimate the actual heights using the iterative method outlined in Example 4.10.

Index, i	0	1	2	3	4	5	6	7	8	9	10	11	12
Position, $x(i)$	0	3	6	9	12	15	18	21	24	27	30	33	36
Barometric height, $b(i)$	4	8	0	5	6	0	0	5	8	9	0	4	7
Confidence, $c(i)$	1.0	0.6	0.0	0.5	0.3	0.0	0.0	0.9	0.6	0.2	0.7	0.1	1.0

(a) Show the results of your solution graphically after each of the first five iterations.
(b) Using the Cauchy convergence criterion and a tolerance of $\epsilon = 0.0001$, find the solution to this problem.
(c) Graph the results of parts (a) and (b) as an overlay.

4.13 Extend the constraint propagation technique to cover a rectangular grid of $m + 1$ by $n + 1$ mesh points. The actual height $a(i, j)$ should weight the barometric height $b(i, j)$ with confidence $c(i, j)$ at the point (i, j), and also weight the average of adjacent points with the complement of $c(i, j)$. *Note:* be careful at the corners and points along the edge.
(a) Express this specification algebraically.

(b) Apply your result to the rectangular region shown below,

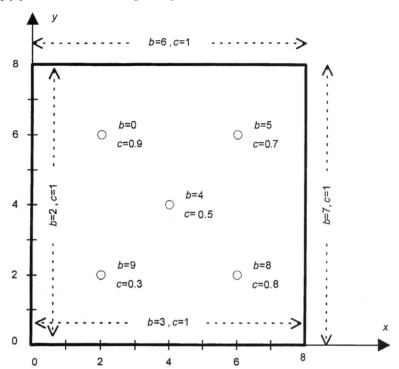

(c) Show the results at the end of each iteration until convergence.
(d) Plot the final results as a contour plot.

4.14 Consider the following version of Laplace's equation in two dimensions, where it is required to find a function $\phi(x, y)$ subject to the following partial differential equation and boundary conditions:

$$\frac{\partial^2 \phi}{\partial x^2} + \frac{\partial^2 \phi}{\partial x^2} = 0, \qquad 0 < x < 1, \qquad 0 < y < 1,$$

subject to

$$\phi(x, 0) = \begin{cases} 2x, & 0 \leqslant x \leqslant \frac{1}{2}, \\ 2(1 - x), & \frac{1}{2} < x \leqslant 1, \end{cases}$$

$$\phi(x, 1) = \begin{cases} 2x, & 0 \leqslant x \leqslant \frac{1}{2}, \\ 2(1 - x), & \frac{1}{2} < x \leqslant 1, \end{cases}$$

$$\phi(0, y) = 0, \qquad 0 \leqslant y \leqslant 1,$$

$$\phi(1, y) = 0, \qquad 0 \leqslant y \leqslant 1.$$

(a) Determine the discrete initial conditions, assuming a boundary of five mesh points on each side.
(b) Write a computer program that finds a numerical solution to this system, assuming the Cauchy convergence criterion.
(c) Execute the program of part (b) to find the solution. Plot the corresponding contour plot.

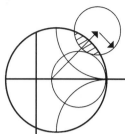

CHAPTER 5

Stochastic Data Representation

5.1 — RANDOM PROCESS MODELS

In most stochastic models, the system per se is fixed and deterministic. Even though they might vary with time, systems are usually well behaved and not subject to erratic fluctuation. In contrast, the signals that drive these systems often appear somewhat random and noisy. Thus, modeling signals is fundamentally different than modeling systems. Signals are in essence time series of statistically related processes, while systems are simply devices that transform those processes.

On the surface, a random signal would seem to be of little use, and it is rare that one is used in isolation as a single instance. This is because, as we have learned earlier, random signals come from ensembles of random processes. What is really important is to discover how the statistical characteristics of the random process inputs relate to those of the random process outputs. Thus, stochastic models are very similar to deterministic models except that a random process (in practice, a statistically large number of random signals) is input to the system and their effects analyzed statistically.

Table 5.1 summarizes the salient differences between deterministic and stochastic models. Deterministic models have finite sets of input and output signal vectors. In practice, each signal is simulated by an explicit formula expressed as a function of time. In contrast, stochastic signals do not have explicit formulas. Instead, they are characterized by defining statistical time series descriptions such as the *autocorrelation* or *spectral density*. Another major difference between deterministic and stochastic systems lies in their long-term behavior. After an initial transient phase, deterministic systems tend to produce output signals with regular steady-state qualities such as constant, periodic, or chaotic states. On the other hand, stochastic systems are usually random, even in the long run. Although the randomness still has no explicit definition, the output can still be characterized by autocorrelation and spectral density functions.

Figure 5.1 illustrates several of the characteristic differences between deterministic and stochastic systems. In the deterministic case, the output can be a constant steady state or periodic. One of these two cases will always occur in a stable linear system. In a nonlinear system, the behavior can be more exotic and perhaps even chaotic. Regardless, a

TABLE 5.1 Comparison of Deterministic and Stochastic Signals

Signal	Deterministic	Stochastic
Input	Single input vector	Ensemble of input vectors
Output	Single output vector	Ensemble of output vectors
Analysis	Transient phase	Initial non-stationary phase
	Steady-state phase	Stationary phase
Defining input	Impulse	White noise
Signal descriptor	Explicit formula	Autocorrelation
		Spectral density

deterministic system can, in principle, be described using formulas that are explicit functions of time. On the other hand, stochastic systems seem to have an underlying deterministic behavior with randomness superimposed, as illustrated in Figure 5.1. In the case of linear systems, output signals will have steady states analogous to those of deterministic systems. If the steady state appears essentially constant with a fixed amount of randomness superimposed, the signal has achieved stationarity, since all statistics are the same for each time slice. However, in the case of a periodic signal with superimposed randomness, there is no stationarity, since differing time slices have different averages. This is illustrated in Figure 5.1.

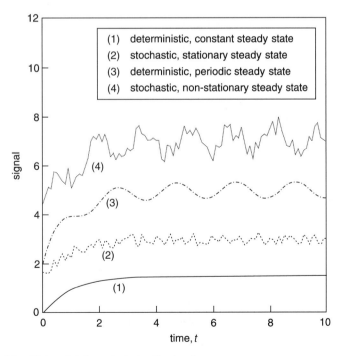

FIGURE 5.1   Deterministic and stochastic signals.

In order for a model to be useful, it must first be analyzed and validated before it can be used as a predicator. To do this requires that the statistical characteristics of all input signals be well defined. These characteristics, usually the statistical mean, autocorrelations, and spectral density functions, form the basis of random models. Thus, knowledge of these statistics can be used for system identification and prediction.

In the case of linear systems, there are a number of extremely useful results. The primary one involves the spectral density $S_{xx}(\omega)$, which is the Fourier transform of the autocorrelation $R_{xx}(\tau)$. It turns out that $S_{xx}(\omega)$ and $R_{xx}(\tau)$ are equivalent in that knowing one is the same as knowing the other. Recalling that the autocorrelation is an even function, the inter-relationship between the two is given by the following *Wiener–Khinchine relations*:

$$S_{xx}(\omega) = \int_{-\infty}^{\infty} R_{xx}(\tau) \cos \omega t \, dt, \tag{5.1a}$$

$$R_{xx}(\tau) = \frac{1}{2\pi} \int_{-\infty}^{\infty} S_{xx}(\omega) \cos \omega \tau \, d\omega. \tag{5.1b}$$

In a deterministic linear system, there is a fundamental relationship between the input and output spectral densities. If the input and output signals are $x(t)$ and $y(t)$ respectively, and $X(\omega)$ and $Y(\omega)$ are the corresponding Fourier transforms, then

$$Y(\omega) = H(j\omega)X(\omega),$$
$$S_{yy}(\omega) = |H(j\omega)|^2 S_{xx}(\omega), \tag{5.2}$$

where $H(s)$ is the system transfer function, $s = j\omega$ is the complex argument, and $S_{xx}(\omega)$ and $S_{yy}(\omega)$ are the input and output spectral densities. Thus, under ideal linear conditions where the input statistical characteristics are well understood, the output characteristics can be inferred. Of course, real systems are usually too complicated or nonlinear, and this motivates the use of simulation as a more realistic approach to modeling. Even so, Equation (5.2) is so basic that we often fall back on it when the need arises to validate system models.

EXAMPLE 5.1

Consider the linear system with input $x(t)$ and output $y(t)$ described by the following block diagram:

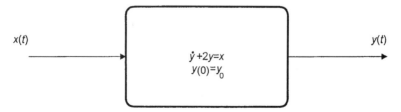

Since the differential equation describing this system is linear, its Laplace transform is straightforward. Taking the Laplace transform of the differential equation, $sY(s) - y(0) + 2Y(s) = X(s)$. Thus, the transfer function, which by definition ignores the initial condition, is

$$H(s) = \frac{Y(s)}{X(s)} = \frac{1}{s+2},$$

and the equivalent block diagram in the Laplace domain is

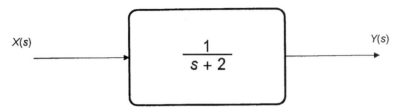

Assuming that the input is white noise with variance $\sigma_x^2 = 4$, characterize the output.

Solution
Since the input is white noise, the mean $\mu_x = 0$ and the autocorrelation is $R_{xx}(\tau) = 4\delta(\tau)$. The input spectral density is found using Equation (5.1):

$$S_{xx}(\omega) = \int_{-\infty}^{\infty} 4\delta(\tau) \cos \omega\tau \, d\tau = 4.$$

As should be expected, the *uncorrelatedness* of white noise in the time domain is equivalent to the spectral density being constant. It follows that the physical interpretation of white noise in the frequency domain is a signal that is independent of frequency. In other words, the signal power is the same at all frequencies. See Figures 5.2(a) and 5.3(a), where the input $x(t)$ is white noise.
Since $s = j\omega$, the transformation of $H(s)$ to the frequency domain gives

$$H(j\omega) = \frac{1}{j\omega + 2},$$

and the square of the magnitude is

$$|H(j\omega)|^2 = \frac{1}{\omega^2 + 4}.$$

Thus, by Equation (5.2), the output spectral density is

$$S_{yy}(\omega) = \frac{4}{\omega^2 + 4},$$

and it remains only to return this to the time domain using the second Wiener–Khinchine relation, Equation (5.1b):

$$\begin{aligned}
R_{yy}(\tau) &= \frac{1}{2\pi} \int_{-\infty}^{\infty} \frac{4}{\omega^2 + 4} \cos \omega\tau \, d\omega \\
&= \frac{2}{\pi} \int_{-\infty}^{\infty} \frac{e^{j\omega\tau}}{(\omega + 2j)(\omega - 2j)} \qquad (5.3) \\
&= \frac{2}{\pi} 2\pi j \, \mathrm{Res}(2j),
\end{aligned}$$

where Res(2j) is the residue associated with the contour integral

$$\oint \frac{e^{jst}}{s^2 + 4}$$

over the infinite semicircle in the top half-plane:

$$\text{Res}(2j) = \lim_{s \to 2j} \frac{e^{jst}}{s^2 + 4}(s - 2j)$$

$$= \lim_{s \to 2j} \frac{e^{jst}}{s + 2j}$$

$$= \frac{e^{-2\tau}}{4j}. \tag{5.4}$$

Combining Equations (5.3) and (5.4) and noting that the autocorrelation is always even, $R_{yy}(\tau) = e^{-2|\tau|}$. It will be noted that the spectral density of the output, unlike that of the input, is not constant, as is shown in Figure 5.3(b). Similarly, the output signal is no longer white noise, as is indicated by the autocorrelation of the output plotted in Figure 5.2(b). It follows that the output mean $\mu_y = \sqrt{\lim_{\tau \to \infty} R_{yy}(\tau)} = 0$ and the variance $\sigma_y^2 = R_{yy}(0) = 1$.   ○

In practice, linear theoretical models have limited utility. In order to obtain a reasonable degree of realism, numerical integrations and simulation runs are the norm. One technique is to use a large number instances (from 100 to 1000 is not unusual) of random process input. Since getting statistically meaningful results by this method is clearly a time-consuming operation, time averaging for a single instance is far more preferable. This requires an ergodic random process, and while it may be possible to demonstrate this for the input, doing so for the output is generally impossible. The

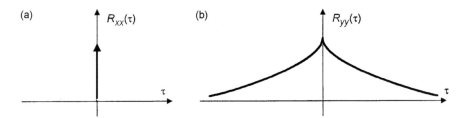

FIGURE 5.2   Input and output autocorrelation functions, Example 5.1.

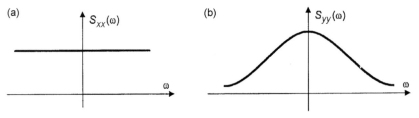

FIGURE 5.3   Input and output spectral density functions, Example 5.1.

following procedure is often used in rigorous studies. Although it avoids the really desirable absolute guarantee of ergodicity, it tends to work in practice.

(i) Create a large number of instances (100 or so) of the random process over the time horizon of the model (at least 1000 time steps is best). Place in a rectangular array in a spreadsheet.

(ii) From the simulations, calculate the statistical averages of each time slice.

(iii) After eliminating the transient phase, calculate the time averages of each instance. An ergodic random process will have essentially constant statistical time averages and time averages for the mean.

(iv) Calculate the autocorrelation using the time average of each instance. An ergodic random process will have essentially the same time average for each instance.

(v) If both tests (iii) and (iv) are positive, there is good reason to suspect ergodicity. Therefore, use the results of (iv) to estimate the autocorrelation.

An example of this procedure is given as Example 3.12 in Chapter 3. Here we show how such an example applies to a total system rather than just a signal.

EXAMPLE 5.2

Repeat the statistical analysis of Example 5.1 using simulation techniques.

Solution

The simulation alternative requires two steps. First, an input signal must be generated. Assuming ergodicity, it is necessary to generate only a single instance, from which the output can be produced. Since the only input characteristic specified in the problem statement is "white noise with variance 4", there are a number of potential random processes that are sufficient.

As a first candidate input signal, consider an uncorrelated signal with uniform random variable. Since RND is uniform on [0,1], $2\,\text{RND} - 1$ is uniform on $[-1, 1]$ with variance $\sigma^2 = \frac{1}{3}$. It follows that $x = 2\sqrt{3}(2\,\text{RND} - 1)$ is uniform on $[-2\sqrt{3}, 3\sqrt{3}]$ with variance $\sigma_x^2 = 4$. An algorithm producing n signal inputs from $t = h$ to $t = hn$ in increments of h proceeds as follows:

```
for k=1 to n
        x-2√3(2*RND-1)
        t=hk
     next k
```

In order to produce the output, the differential equation specifying the system can be integrated using the Euler approximation as $y_{\text{new}} = y + h(x - 2y)$. From this, an algorithm producing the output $y(k)$ over $[0, nh]$ is given in Listing 5.1.

As a second candidate input signal, generation of random variates $x(t)$ can be made using a Gaussian rather than uniform distribution by replacing the line `x=2√3(2*RND-1)` with `x=2√-2*ln(RND)*cos(2π*RND)`, since both the uniform and Gaussian signals satisfy the requirements of mean zero and variance four. A single instance of each of these random process inputs is shown in Figure 5.4.

```
x=2√3 (2*RND-1)
y(0)=y₀
k=0
print k, t, x, y(k)
for k=1 to n
        x=2√3 (2*RND-1)
        y(k)=y(k-1)+h[x-2y(k-1)]
        t=hk
        print k, t, x, y(k)
    next k
```

LISTING 5.1   Random process system output, Example 5.2.

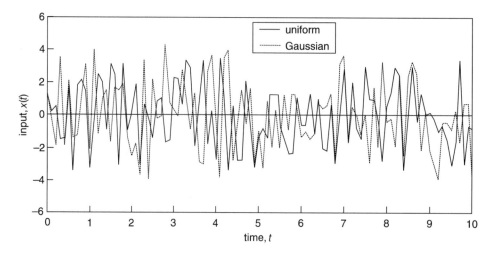

FIGURE 5.4   Single instance of uniform and Gaussian random signal inputs.

Generation of the output can be accomplished by using the Eulerian integration shown in Listing 5.1, which produces the input x and output $y(k)$, where k is the discrete lag time and $\tau = hk$ is the continuous lag time. From this, the autocorrelation can be computed using the algorithm shown in Listing 5.2.

It is important to note that the output results are similar for both of these random signals as well. Outputs $y(t)$ for both the uniform and Gaussian white noises are plotted in Figure 5.5. A comparison of Figures 5.4 and 5.5 shows the increase in frequency and decrease in amplitude predicted by Example 5.1.

Finally, it is important to note the number of time steps required in order to obtain good statistics. Figure 5.6 shows how the accuracy improves as a function of the number of time steps n. Even so, it is evident that the ergodic hypothesis is appropriate here, since the results for ever-larger n tend to the expected autocorrelation as found in Example 5.1.   ○

```
for k=0 to m
    tot=0
    for i=0 to n
        tot=tot+y(k)*y(k+i)
    Next i
    R=tot/(n-k+1)
    τ=kh
    print  τ, R
next k
```

LISTING 5.2   Computation of the autocorrelation estimate: continuous time τ

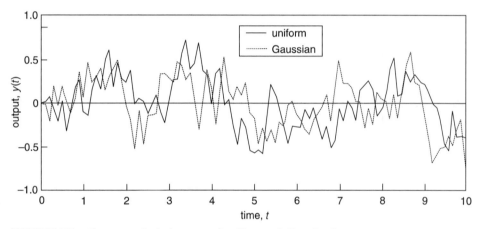

FIGURE 5.5   Output to single instance of uniform and Gaussian inputs.

As a final illustration of the power of stochastic simulation techniques, let us consider the phenomenon of so-called *shot noise*. Shot noise can be thought of as isolated instances of a random positive signal in a sea of zeros. That is, it occurs randomly and infrequently, and its amplitude is positive but random. Rather than develop a theory, let us progress directly to the simulation stage. From the description, a reasonable model of shot noise is that of a Poisson process with event rate λ events per unit time and with amplitudes uniform on $[0,a]$, where a is a known positive constant. Such a model can be implemented and analyzed in three stages:

- Stage 1: determine shot event time sequence and amplitudes.
- Stage 2: determine the continuous time sequence.
- Stage 3: determine the autocorrelation from Stage 2.

Recall from Sections 1.2 and 3.2 just how random scheduling times and random signal amplitudes are simulated. Poisson processes are equivalent to exponential inter-event times: $t_k - t_{k-1} = \mu \ln(\text{RND})$, where $1/\lambda = \mu$ is the average inter-event time. The required uniform amplitudes can be generated by $x_k = a * \text{RND}$, where a is the maximum

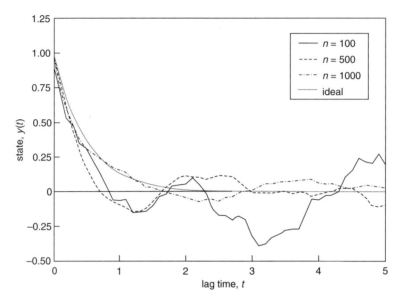

FIGURE 5.6   White noise reponse to simulations of Example 5.2 using $n = 100$, 500, and 1000 integration time steps.

amplitude. These two facts can be combined to generate the required sequence by

```
t=t₀
print t, x
for k=1 to n
            tₖ=tₖ₋₁-μ*ln(RND)
            xₖ=a*RND
            print t, x
    next k
```

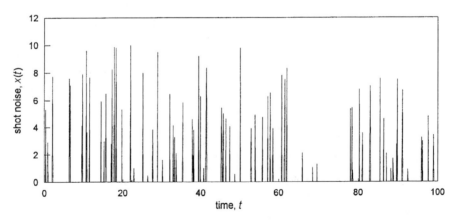

FIGURE 5.7   Shot noise signal: exponential inter-event times with $\lambda = 1$ and uniform amplitudes on $[0, 10]$.

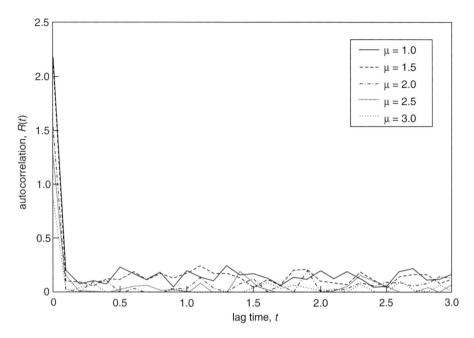

FIGURE 5.8   Autocorrelation of shot noise varied with event rate λ.

A graph showing a signal of pure shot noise is shown in Figure 5.7 for $a = 10$ and $\lambda = 1$, which implies that the mean is $\mu = 1/\lambda = 1$. By fixing the maximum amplitude at $a = 10$ and varying the mean inter-event time $\mu = 1.0$, 1.5 , 2.0 , 2.5 , 3.0, the family of autocorrelation graphs shown in Figure 5.8 is produced for $\tau = 0$ to 3. A quick review of this family of results shows two characteristics of shot noise. First, the autocorrelation is a delta function; that is, the signal is uncorrelated with itself unless the lag is zero. This is not surprising – but a fact worth verifying, nonetheless! Also, the variance, which is the magnitude of the autocorrelation σ^2 at the origin, is inversely related to the mean event time μ, or, equivalently, it is directly related to the event rate λ.

5.2   MOVING-AVERAGE (MA) PROCESSES

There are two reasons that random signals are used in simulations. First, there is often noise that tends to contaminate the underlying basic process. By adding a noise component to a deterministic signal, the characteristics of realistic processes can be modeled most effectively. In this case, the noise is often of the form $x(k) = w(k)$, where $w(k)$ is white noise. Because each time slice is independent, the autocorrelation $R_{xx}(\tau) = \sigma_w^2 \delta(\tau)$, where σ_w^2 is the variance associated with the white noise.

The other reason that random signals are used is because of ignorance. For instance, it can happen that the signal $x(k)$ depends on the two most recent white noise events (sometimes called *shocks*) $w(k)$ and $w(k-1)$. This suggests the linear model

$x(k) = b_0 w(k) + b_1 w(w - 1)$, where the constants b_0 and b_1 are parameters yet to be specified since the process depends on previous shock events, even though neither the current or previous input is actually known. This can be generalized so as to make the following remarkably useful model for random signals, known as the *moving-average* (*MA*) process:

$$x(k) = \sum_{i=0}^{q} b_i w(k - i). \tag{5.5}$$

The reason that this is called a moving-average process is that Equation (5.5) actually acts like a low-pass filter. For instance, by taking $b_0 = b_1 = b_2 = \frac{1}{3}$ and $q = 2$, the signal becomes $x(k) = \frac{1}{3}[w(k) + w(k - 1) + w(k - 2)]$. Thus, $x(k)$ is the average of the most recent white noise contributions. Being an average, it has the effect of softening the peaks and valleys of the white noise environment. This in turn reduces the erratic behavior of the signal, and a lower frequency results. It is important to note that even though $x(k)$ is defined in terms of white noise, the autocorrelation $R_{xx}(\tau)$ will not be a single delta function. Since $x(k)$ always has some residual memory of $w(k - i)$, a multitude of delta functions is inevitable.

There is actually another possibility as well. Rather than retaining memory through delayed inputs $w(k)$, it can be retained through delays in the output. This model, known as the *autoregressive* (*AR*) process, is defined as follows:

$$x(k) = w(k) - \sum_{i=1}^{p} a_i x(k - i). \tag{5.6}$$

The autoregressive process not only observes the current white noise input, but also has access to previous states as well. Thus, there is memory and non-trivial autocorrelation.

Finally, both the autoregressive and moving-average processes can be combined into one. In so doing, the most general of all linear random processes is the *autoregressive moving-average* (*ARMA*) model, defined as

$$x(k) = -\sum_{i=1}^{p} a_i x(k - i) + \sum_{i=0}^{q} b_i w(k - i). \tag{5.7}$$

This general ARMA model retains memory through both the white noise shock inputs and the signal value at sampling times.

Each of these models is driven by the white noise input $w(k)$, which produces the signal $x(k)$ as output in the diagram shown in Figure 5.9. In the case where there is no signal feedback $x(k - i)$, Figure 5.9 corresponds to the MA model, and when there is shock feedback $w(k - i)$, it corresponds to an AR model. If both feedbacks exist, it is a general ARMA model. Since the system is driven by white noise in every case, modeling $w(k)$ is quite important. It will be recalled that as far as autocorrelation is concerned, only the variance σ_w^2 is needed. Even so, at each time slice, the statistical distribution is important, just so long as the mean is zero and the variance as given.

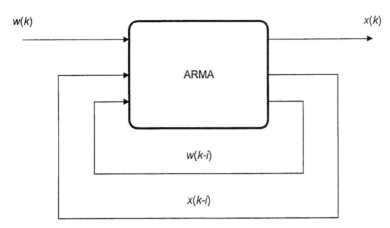

FIGURE 5.9 General ARMA model.

EXAMPLE 5.3

Consider the moving-average process defined by

$$x(k) = w(k) + 3w(k-12) + 5w(k-15).$$

The sequence $x(k)$ is graphed in Figure 5.10 using two different unit-variance white noise inputs:

uniform: $\quad w(k) = \sqrt{3}(2\,\text{RND} - 1)$

Gaussian: $\quad w(k) = \sqrt{-2\ln(\text{RND})}\cos(2\pi\,\text{RND})$

In both input processes the mean is $\mu_w = 0$ with variance $\sigma_w^2 = 1$. Therefore, $R_{ww}(\tau) = \delta(\tau)$,

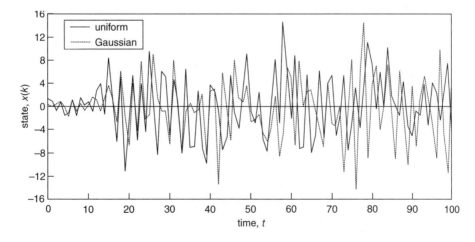

FIGURE 5.10 Generation of MA signals using uniform and Gaussian input, Example 5.3.

It will be noted that there are two rather distinct phases. From $k = 0$ to 15, the system seems to be building up in a transient phase. Beyond that point, it falls into a steady fluctuation centered about the mean $\mu_x = 0$ and a seemingly constant standard deviation $\sigma_x \approx 6$. Thus, even though the signal is not a fixed steady state, its statistics are such that the process appears to be stationary. However, there are no obvious visual signs of how the process is autocorrelated; this requires computation of the autocorrelation itself.

The autocorrelation is the statistic of choice for random processes. To see this, let us derive $R_{xx}(\tau)$:

$$
\begin{aligned}
R_{xx} &= E[x(k)x(k+\tau)] \\
&= E[\{w(k) + 3w(k-12) + 5w(k-15)\}\{w(k+\tau) + 3w(k+\tau-12) \\
&\quad + 5w(k+\tau-15)\}] \\
&= E[w(k)w(k+\tau+2) + 3w(k)w(k+\tau-12) + 5w(k)w(k+\tau-15) \\
&\quad + 3w(k-12)w(k+\tau+2) + 9w(k-12)w(k+\tau-12) \\
&\quad + 15w(k-12)w(k+\tau-15) + 5w(k-15)w(k+\tau+2) \\
&\quad + 15w(k-15)w(k+\tau-12) + 25w(k-15)w(k+\tau-15)] \\
&= 35R_{ww}(\tau) + 3\{R_{ww}(\tau-12) + R_{ww}(\tau+12)\} \\
&\quad + 5\{R_{ww}(\tau-15) + R_{ww}(\tau+15)\} + 15\{R_{ww}(\tau-3) + R_{ww}(\tau+3)\} \\
&= 35\delta(\tau) + 15\{\delta(\tau-3) + \delta(\tau+3)\} + 5\{\delta(\tau-12) + \delta(\tau+12)\} \\
&\quad + 3\{\delta(\tau-15) + \delta(\tau+15)\}.
\end{aligned}
$$

As expected, $R_{xx}(\tau)$ is highest at $t = 0$. Also, there are non-zero contributions at $\tau = \pm 12$ and $\tau = \pm 15$. However, what might not have been expected is the strong component at $\tau = \pm 3$. As seen in the derivation, this is due to the difference $15 - 12 = 3$ of the two lag times.   ◯

It can be shown that the autocorrelation of the MA process defined in Equation (5.5) is, in general,

$$
R_{xx}(\tau) = \sigma_w^2 \sum_{k=0}^{q-|\tau|} b_k b_{k+|\tau|}, \tag{5.8}
$$

where it will be noted that the signal of Example 5.3 is the special case with $b_k = 0$ except for $b_0 = 1$, $b_{12} = 3$, $b_{15} = 5$, $q = 15$, and $\sigma_w = 1$. Also, the mean and variance of the process are $\mu_x = 0$ and $R_{xx}(0) = \sigma_x^2 = 35$.

There is an especially useful interpretation of Equation (5.8). Since $R_{xx}(\tau)$ is the product of coefficient pairs $b_k b_{k-|\tau|}$, the only non-zero terms in the autocorrelation are those in which the difference of the subscripts is the argument τ. In other words, if b_i and b_j are non-zero MA terms, the arguments that express themselves are those in which $i - j = |\tau|$. Equivalently, τ can be found by taking the set of differences as either $\tau = i - j$ or $\tau = j - i$. For example, the set of non-zero terms in Example 5.3 is $\{0, 12, 15\}$. From this, the set of all possible differences is $\{-15, -12, -3, 0, 3, 12, 15\}$. Accordingly, these should be the only terms in the autocorrelation. These findings can be verified, as shown in the following example.

EXAMPLE 5.4

Once in the steady-state phase, the moving-average process is wide-sense stationary, as implied by the graph shown in Figure 5.10. In fact, the MA process is also ergodic. Accordingly, the autocorrelation can be computed by taking time averages of a single instance. Since the MA is a discrete random process, its calculation is the same as that in Listing 5.2, but this time does not involve the simulation granularity h. The calculation of $R_{xx}(\tau)$ for a discrete ergodic random process is given in Listing 5.3.

In order to both demonstrate this and obtain reasonable results, it is necessary to use a large number of time slices. For instance, let us once again consider the random process $x(k) = w(k) + 3w(k - 12) + 5w(k - 15)$ of Example 5.3. Using $m = 500$, 1000, and 10 000 times k, the autocorrelation $R_{xx}(\tau)$ is computed using time averages by the algorithm given in Listing 5.3 for discrete autocorrelation.

```
for τ=1 to m
     tot=0
     for k=0 to m-τ
               tot=tot+y(k)*y(k+τ)
     next k
     R=tot/(n-τ+1)
     print τ, R
next τ
```

LISTING 5.3   Computation of the autocorrelation estimate: discrete time k.

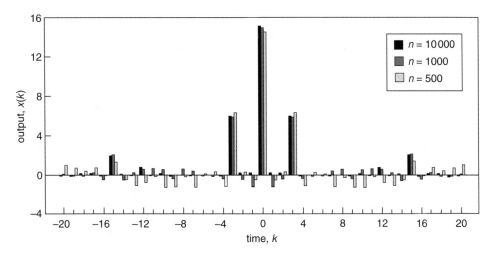

FIGURE 5.11   Autocorrelation for the MA process.

This is comparable to the continuous-time algorithm of Listing 5.2. The results of this are shown in Figure 5.11.

A quick look at Figure 5.11 shows that the random process behaves just as predicted. There are non-zero components for $R_{xx}(\tau)$ only at $\tau = -15, -12, -3, 0, 3, 12,$ and 15. Furthermore, the relative heights are all more or less as expected too: $R_{xx}(0) = 35$, $R_{xx}(\pm 3) = 15$, $R_{xx}(\pm 12) = 3$, and $R_{xx}(\pm 15) = 5$.

However, it should also be noted that the cost in simulation times is rather high. In order to get these really quality results requires approximately $n = 10\,000$ times, and without this accuracy, spurious correlations are inevitable.

The correlations seen at $\tau = 0, \pm 12,$ and ± 15 are not surprising, since they have counterparts in the original MA signal. However, the correlation at $\tau = \pm 3$ and its strength of 15 might be somewhat unexpected. This signal, which is the result of the $15 - 12 = 3$ contribution, should be anticipated when modeling with MA processes.   ○

5.3 ___ AUTOREGRESSIVE (AR) PROCESSES

The moving-average process expresses knowledge of the past in that it remembers previous (lagged) white noise inputs $w(k)$. The autoregressive process (AR) model is different. It is also driven by white noise input, but this is not explicitly recalled. Instead, it remembers past signal values. Mathematically, the autoregressive process takes the form

$$x(k) = w(k) - \sum_{i=1}^{p} a_i x(k-i),$$

where memory is p time units distant and a_i is the strength associated with each lag time. The minus sign is simply a mathematical convenience, and is otherwise just absorbed into the constant a_i.

By thinking of the signal value as analogous to the system state, an AR process is similar to a random walk in that its next state (signal value) depends directly on the current state (signal value). Actually, the AR process is a generalization of the random walk, since it depends on any of a number of past signal values and it is different in that it has a white noise input rather than a Bernoulli bipolar input.

The AR process is similar to the MA process in that it usually has two phases: transient and post-transient, when the process is stationary. However, unlike the MA process, an AR process can also be unstable. By this we mean that the post-transient signal values can go to infinity. This causes the signal to fluctuate wildly, and therefore to be of little practical value in system simulation. Once the process has achieved steady state, the autoregressive process is *wide-sense stationary* (recall Chapter 3), and therefore each time slice has identical statistics, as is shown in the following example.

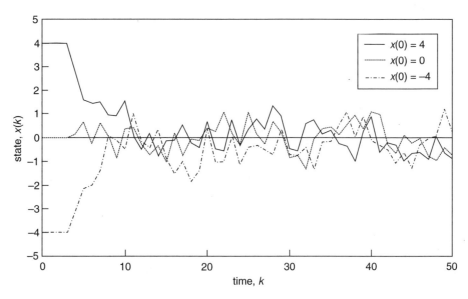

FIGURE 5.12   Autoregressive process for Example 5.5.

EXAMPLE 5.5

Consider the autoregressive signal

$$x(k) = w(k) + \tfrac{1}{2}x(k-1) + \tfrac{1}{3}x(k-3) - \tfrac{1}{4}x(k-4),$$

where $w(k)$ is white noise uniformly distributed on $[-1, 1]$. Thus, the white noise variance $\sigma_w^2 = \tfrac{1}{3}$ and its autocorrelation $R_{ww}(\tau) = \tfrac{1}{3}\delta(\tau)$.

Three autoregressive (AR) sequences are generated and graphed for three initial states (outputs) in Figure 5.12. Note that after the initial phase of $k = 0, 1, 2, 3, 4$, the process becomes stationary regardless of the initial state. As with the MA process, the AR process appears random, even though it appears to be confined to an interval centered about $x(k) = 0$ once in steady state.   ○

Autoregressive processes have one important consideration that is not a concern for moving averages. For MA processes, the signal sequence always becomes stationary after the transient phase, regardless of the coefficients. However, in AR sequences, not all coefficient sets lead to a steady state. If the coefficients are chosen badly, the sequence will not remain bounded and goes to infinity. In this case, the process is nonstationary.

The stationarity of random processes is an important consideration, and there exists a significant theory concerning it, especially for the case of linear systems. For AR processes, which are inherently linear, there is an especially important result that serves as a useful stability test. This is summarized by the Autoregressive Stationarity Theorem:

Autoregressive Stationarity Theorem *Let the general autoregressive process $x(k)$ be defined by $x(k) + \sum_{i=1}^{p} a_i x(k - i) = w(k)$. Further define the associated complex polynomial $C(z) = z^p + \sum_{i=1}^{p} a_i z^{p-i} = 0$, where z is a complex number with real and imaginary parts $\mathrm{Re}(z)$ and $\mathrm{Im}(z)$ respectively. Then the process $x(k)$ is stable if and only if all roots of $C(z)$ are confined to the interior of the unit circle; that is, if $[\mathrm{Re}(z)]^2 + [\mathrm{Im}(z)]^2 < 1$.*

The polynomial $C(z)$ is called the *characteristic polynomial*, and solutions to the *characteristic equation* $C(z) = 0$ can be solved with the aid of a good polynomial root-finder computer program. Equivalently, once the roots have been found, if each of their magnitudes is less than unity, the roots are interior to the unit circle and the process is stationary. Since the roots of z are in general complex, their magnitudes satisfy the stationarity requirement only when $[\mathrm{Re}(z)]^2 + [\mathrm{Im}(z)]^2 < 1$.

As a specific example, consider the AR process of Example 5.5. The characteristic polynomial is $C(z) = z^4 - \frac{1}{2}z^3 - \frac{1}{3}z + \frac{1}{4} = 0$, whose roots are $z_{1,2} \approx +0.3912 \pm 0.6233j$ and $z_{3,4} \approx 0.6412 \pm 0.2248j$. It remains to test whether or not these roots fall inside the unit circle. In the example, since $(0.3912)^2 + (0.6233)^2 = 0.5374 < 1$ and $(0.6412)^2 + (0.2248)^2 = 0.6359 < 1$, the process must be stationary, as is confirmed by the stable simulation results of Example 5.5. The geometric interpretation is best seen from the diagram in Figure 5.13.

EXAMPLE 5.6

Consider the following three autoregressive processes:

(a) $x(k) = w(k) + x(k - 1) - \frac{1}{4}x(k - 2)$;

(b) $x(k) = w(k) - \frac{1}{4}x(k - 1) - \frac{1}{4}x(k - 2) + \frac{3}{4}x(k - 3)$;

(c) $x(k) = w(k) + 2x(k - 1) - \frac{1}{4}x(k - 2) + \frac{1}{2}x(k - 3)$.

Analyze the stability of each, and observe the resulting time sequences, assuming zero initial states and Gaussian white noise with variance $\sigma_w^2 = 1$.

Solution

(a) The characteristic equation of the first process is

$$C_1(z) = z^2 - z + \frac{1}{4}$$
$$= \left(z - \frac{1}{2}\right)^2 = 0.$$

Clearly, there are two roots at $z_{1,2} = \frac{1}{2}$, both of which are inside the unit circle. Since $C(z)$ is of second degree, $x(k)$ is a second-order process, and it follows that for $k = 0$, 1, and 2, the sequence is in the transient phase. Beyond that time (after $k = 2$), the process becomes stationary. One instance of this process is simulated in Figure 5.14, which demonstrates that this is indeed the case.

(b) The second process has characteristic equation

$$C_2(z) = z^3 + \frac{1}{4}z^2 + \frac{1}{4}z - \frac{3}{4}$$
$$= (z^2 + z + 1)\left(z - \frac{3}{4}\right) = 0$$

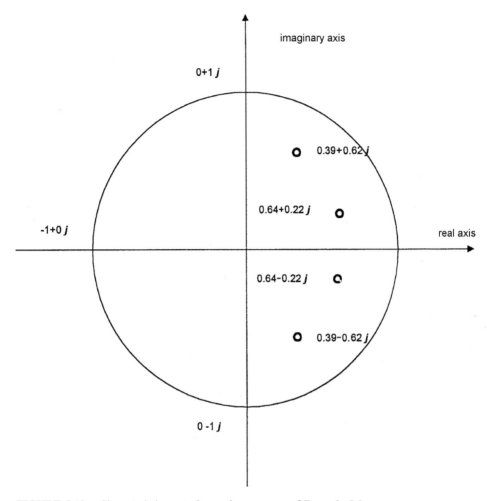

FIGURE 5.13   Characteristic roots for random process of Example 5.5.

The root $z = \frac{3}{4}$ is obviously inside the unit circle, but the two roots of the quadratic factor are $z_{2,3} = -\frac{1}{2} \pm \frac{1}{2}\sqrt{3}j$ and lie exactly on the unit circle since $(-\frac{1}{2})^2 + (\pm\frac{1}{2}\sqrt{3})^2 = 1$. Since the circle itself is the border between stationary and non-stationary, this is called a *marginally* stationary or an ARIMA process. As seen in Figure 5.14, the time-domain behavior of such a process no longer has a stationary region, even though the signal does not shoot off to infinity.

   (c) The third process has characteristic equation

$$C_3(z) = z^3 - 2z^2 + \tfrac{1}{4}z - \tfrac{1}{2}$$
$$= (z^2 + \tfrac{1}{4})(z - 2) = 0.$$

Since the root $z = 2$ is outside the unit circle, it is not necessary to look into the effect of the imaginary roots $z = \pm\frac{1}{2}j$, even though they are inside the unit circle.

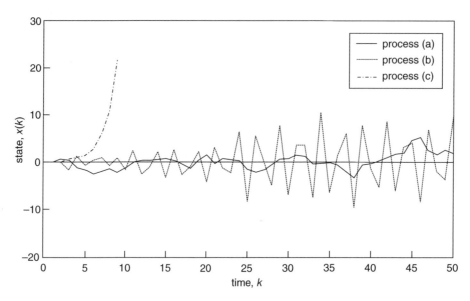

FIGURE 5.14   Time domain behavior of random processes in Example 5.6.

The time-domain behavior seen in Figure 5.14 shows that this signal immediately drifts off to infinity, presumably never to return.   ○

As in the MA process, understanding an AR sequence requires the autocorrelation statistic. If the process is stationary, the AR process is ergodic and $R_{xx}(\tau)$ can be found experimentally using time averages. However, if an analytical expression is required, the so-called *Yule–Walker equations* must be used. These are found by using Equation (5.6), as shown in the following derivation:

$$R_{xx}(\tau) = E[x(k)x(k+\tau)]$$
$$= E\left[x(k)\left\{w(k+\tau) - \sum_{i=1}^{p} a_i x(k+\tau - i)\right\}\right]$$
$$= E[x(k)w(k+\tau)] - \sum_{i=1}^{p} a_i E[x(k)(x(k+\tau - i)]$$
$$= R_{xw}(\tau) - \sum_{i=1}^{p} a_i R_{xx}(\tau - i).$$

The cross-correlation $R_{xw}(\tau)$ can be found by noting that in Equation (5.6), there is no relationship between $x(k)$ and $w(k+\tau)$ unless $\tau = 0$. In this case, $R_{xw}(0) = E[w(k)w(k)] = \sigma_w^2$. Therefore, in general,

$$R_{xx}(\tau) + \sum_{i=1}^{p} a_i R_{xx}(\tau - i) = \sigma_w^2 \sigma(\tau). \tag{5.9}$$

Equation (5.9) is actually a system of linear equations. By taking $\tau = 0, 1, \ldots, p$ and writing the resultant equations in matrix form, we have

$$
\begin{bmatrix}
R_{xx}(0) & R_{xx}(-1) & \cdots & R_{xx}(-p) \\
R_{xx}(1) & R_{xx}(0) & \cdots & R_{xx}(1-p) \\
\vdots & \vdots & \cdots & \vdots \\
R_{xx}(p) & R_{xx}(p-1) & \cdots & R_{xx}(0)
\end{bmatrix}
\begin{bmatrix}
1 \\
a_1 \\
\vdots \\
a_p
\end{bmatrix}
=
\begin{bmatrix}
\sigma_w^2 \\
0 \\
\vdots \\
0
\end{bmatrix}.
\tag{5.10}
$$

Since $R_{xx}(\tau)$ is an even function, there are actually only p unknown autocorrelations rather than the $2p - 1$ displayed in the matrix. Thus, if the coefficients $a_1, a_2, \ldots, a_p$ and the variance σ_w^2 are known, the autocorrelations $R_{xx}(0)$, $R_{xx}(\pm 1), \ldots, R_{xx}(\pm p)$ can be determined. Similarly, knowing $R_{xx}(0)$, $R_{xx}(\pm 1), \ldots, R_{xx}(\pm p)$ leads directly to the computation of $a_1, a_2, \ldots, a_p$ and σ_w^2.

EXAMPLE 5.7

Calculate the theoretical autocorrelations $R_{xx}(\tau)$ for the AR process given in Example 5.5:

$$
x(k) = w(k) + \tfrac{1}{2}x(k-1) + \tfrac{1}{3}x(k-3) - \tfrac{1}{4}x(k-4),
$$

where $w(k)$ is white noise uniform on $[-1, 1]$. Therefore, $\sigma_w^2 = \tfrac{1}{3}$.

Solution

According to the Yule–Walker equations and noting that $R_{xx}(\tau)$ is even,

$$
\begin{bmatrix}
R_{xx}(0) & R_{xx}(1) & R_{xx}(2) & R_{xx}(3) & R_{xx}(4) \\
R_{xx}(1) & R_{xx}(0) & R_{xx}(1) & R_{xx}(2) & R_{xx}(3) \\
R_{xx}(2) & R_{xx}(1) & R_{xx}(0) & R_{xx}(2) & R_{xx}(2) \\
R_{xx}(3) & R_{xx}(2) & R_{xx}(1) & R_{xx}(0) & R_{xx}(1) \\
R_{xx}(4) & R_{xx}(3) & R_{xx}(2) & R_{xx}(1) & R_{xx}(0)
\end{bmatrix}
\begin{bmatrix}
1 \\
-\tfrac{1}{2} \\
0 \\
-\tfrac{1}{3} \\
\tfrac{1}{4}
\end{bmatrix}
=
\begin{bmatrix}
\tfrac{1}{3} \\
0 \\
0 \\
0 \\
0
\end{bmatrix}.
$$

Expanding this matrix equation and creating a new equation linear in $R_{xx}(\tau)$ gives

$$
\begin{bmatrix}
1 & -\tfrac{1}{2} & 0 & -\tfrac{1}{3} & \tfrac{1}{4} \\
-\tfrac{1}{2} & 1 & -\tfrac{1}{3} & \tfrac{1}{4} & 0 \\
0 & -\tfrac{5}{6} & \tfrac{3}{4} & 0 & 0 \\
\tfrac{1}{3} & \tfrac{1}{4} & -\tfrac{1}{2} & 1 & 0 \\
\tfrac{1}{4} & -\tfrac{1}{3} & 0 & -\tfrac{1}{2} & 1
\end{bmatrix}
\begin{bmatrix}
R_{xx}(0) \\
R_{xx}(1) \\
R_{xx}(2) \\
R_{xx}(3) \\
R_{xx}(4)
\end{bmatrix}
=
\begin{bmatrix}
\tfrac{1}{3} \\
0 \\
0 \\
0 \\
0
\end{bmatrix}.
$$

By inverting the left-hand matrix, and recalling that $R_{xx}(\tau)$ is even the autocorrelations can be found for $\tau = -4, -3, \ldots, 4$. By using the recursion $R_{xx}(\tau) = -\tfrac{1}{2}R_{xx}(\tau - 1) - \tfrac{1}{3}R_{xx}(\tau - 3) + \tfrac{1}{4}R_{xx}(\tau - 4)$ for $\tau > 4$ and the fact that $R_{xx}(\tau)$ is even for $\tau < -4$, as many theoretical values as needed can be computed. These, along with a simulation using $n = 15\,000$ times for time averaging, are calculated, and are shown in Table 5.2 and Figure 5.15. $\bigcirc$

The Yule–Walker equations are even more useful in signal identification. In this case, Equation (5.10) reduces to finding the coefficients a_i and the variance of the white noise. Suppose that a simulation is performed in which $n + 1$ autocorrelations are calculated

TABLE 5.2   Example 5.7 Autocorrelation

k	Theory	Experiment
−10	−0.014	−0.002
−9	−0.002	0.007
−8	−0.014	0.005
−7	−0.014	0.014
−6	0.031	0.058
−5	0.022	0.055
−4	0.058	0.092
−3	0.196	0.214
−2	0.181	0.198
−1	0.271	0.287
0	0.519	0.527
1	0.271	0.287
2	0.181	0.198
3	0.196	0.214
4	0.058	0.092
5	0.022	0.055
6	0.031	0.058
7	−0.014	0.014
8	−0.014	0.005
9	−0.002	0.007
10	−0.014	−0.002

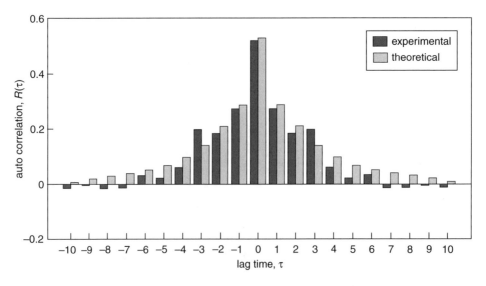

FIGURE 5.15   Example 5.7 autocorrelation using Yule–Walker equations.

using time averaging: $R_{xx}(0), R_{xx}(1), \ldots, R_{xx}(p)$. We wish to use these empirical calculations to estimate a pth-order AR model. That is, we need to calculate the parameters
$a_1, a_2, \ldots, a_p$ and σ_w^2. By again rearranging the Yule–Walker equations into an over-
determined set of equations, it is possible to find the best-fitting coefficients in the least
squares sense. Specifically, suppose that we have a model with p coefficients
$a_1, a_2, \ldots, a_p$, assuming $n+1$ observations at lag times $\tau = 0, 1, \ldots, n$, where $n > p$.
The Yule–Walker equations now take the form

$$
\begin{bmatrix}
R_{xx}(0) & R_{xx}(-1) & \cdots & R_{xx}(-p) \\
R_{xx}(1) & R_{xx}(0) & \cdots & R_{xx}(1-p) \\
\vdots & \vdots & \cdots & \vdots \\
R_{xx}(p) & R_{xx}(p-1) & \cdots & R_{xx}(0) \\
\vdots & \vdots & \cdots & \vdots \\
R_{xx}(n) & R_{xx}(n-1) & \cdots & R_{xx}(n-p)
\end{bmatrix}
\begin{bmatrix}
1 \\ a_1 \\ \vdots \\ a_p
\end{bmatrix}
=
\begin{bmatrix}
\sigma_w^2 \\ 0 \\ \vdots \\ 0
\end{bmatrix}.
\tag{5.11}
$$

Notice that unknowns are on both the left- and right-hand sides of Equation (5.11).
Equivalently, this can be put in the form

$$
\begin{bmatrix}
R_{xx}(-1) & R_{xx}(-2) & \cdots & R_{xx}(-p) & -1 \\
R_{xx}(0) & R_{xx}(-1) & \cdots & R_{xx}(1-p) & 0 \\
\vdots & \vdots & \cdots & \vdots & \\
R_{xx}(p-1) & R_{xx}(p-2) & \cdots & R_{xx}(0) & 0 \\
\vdots & \vdots & \cdots & \vdots & \\
R_{xx}(n-1) & R_{xx}(n-2) & \cdots & R_{xx}(n-p) & 0
\end{bmatrix}
\begin{bmatrix}
a_1 \\ \vdots \\ a_p \\ \sigma_w^2
\end{bmatrix}
=
\begin{bmatrix}
-R_{xx}(0) \\ -R_{xx}(1) \\ \vdots \\ -R_{xx}(n)
\end{bmatrix}.
\tag{5.12}
$$

This system takes the form $\mathbf{Ra} = \boldsymbol{\rho}$, where $\mathbf{R}$ is the $(n+1) \times (p+1)$ matrix of $n+1$ data
points of the p modeling parameters, $\boldsymbol{\rho}$ is an $(n+1) \times 1$ data vector, and $\mathbf{a}$ is a $(p+1) \times 1$
vector of unknowns. Using the left-hand pseudo-inverse, a solution for $\mathbf{a}$ can be found that
is the best fit in a least squares sense:

$$
\mathbf{a} = (\mathbf{R}^T\mathbf{R})^{-1}\mathbf{R}^T\boldsymbol{\rho}.
\tag{5.13}
$$

The following example shows the power of this technique.

EXAMPLE 5.8

Model the random signal of Example 5.5 as an autoregressive process of the form
$x(k) = w(k) - a_1 x(k-1) - a_2 x(k-2)$, where $w(k)$ is white noise with unknown
variance σ_w^2. Use the simulation results for $\tau = -5, -4, \ldots, 4, 5$.

Solution
We should not expect this model to work well, since the AR process for
generating it was of fourth order and the hypothesized model in this exercise is
of only second order. Even so, trying to fit such an inferior model to the data
should prove instructive.

The data described above, which is taken directly from Table 5.2, is as follows:

τ	$R(\tau)$
0	0.527
1	0.287
2	0.198
3	0.214
4	0.092
5	0.055

Since $n = 5$ and $p = 2$, the matrix $\mathbf{R}$ is given by

$$\mathbf{R} = \begin{bmatrix} R_{xx}(1) & R_{xx}(2) & -1 \\ R_{xx}(0) & R_{xx}(1) & 0 \\ R_{xx}(1) & R_{xx}(0) & 0 \\ R_{xx}(2) & R_{xx}(1) & 0 \\ R_{xx}(3) & R_{xx}(2) & 0 \\ R_{xx}(4) & R_{xx}(3) & 0 \end{bmatrix} = \begin{bmatrix} 0.287 & 0.198 & -1 \\ 0.527 & 0.287 & 0 \\ 0.287 & 0.527 & 0 \\ 0.198 & 0.287 & 0 \\ 0.214 & 0.198 & 0 \\ 0.092 & 0.214 & 0 \end{bmatrix}$$

Likewise, the vector $\boldsymbol{\rho}$ is given by

$$\boldsymbol{\rho} = \begin{bmatrix} -R_{xx}(0) \\ -R_{xx}(1) \\ -R_{xx}(2) \\ -R_{xx}(3) \\ -R_{xx}(4) \\ -R_{xx}(5) \end{bmatrix} = \begin{bmatrix} -0.527 \\ -0.287 \\ -0.198 \\ -0.214 \\ -0.092 \\ -0.055 \end{bmatrix}.$$

Solving for the vector $\mathbf{a} = [a_1, a_2, \sigma_w^2]^T$ using Equation (5.13) results in $a_1 = -0.452$, $a_2 = -0.164$, and $\sigma_w^2 = 0.365$. Thus, the second-order model prescribed is $x(k) = w(k) + (0.452)x(k-1) + (0.164)x(k-2)$, where $w(k)$ is white noise of variance $\sigma_w^2 = 0.365$.

The autocorrelation of this second-order model is compared in Figure 5.16 against the actual fourth order AR model. It should be noted that there is a rather good fit in the center, but it becomes far less than ideal toward the "wings", where no data were fitted. In particular, $R_{xx}(\pm 3)$ deviates significantly from the ideal. Evidently, this is an inferior model, and it is best to try a higher-order model instead. ○

5.4 BIG-Z NOTATION

It is useful to represent discrete signals using a special operator notation called "big-Z". The big-Z operator can be thought of as a signal advance where $\mathbf{Z}[x(k)]$ is expressed one time unit sooner than $x(k)$. Thus, $\mathbf{Z}[x(k)] = x(k+1)$, as illustrated in the following table:

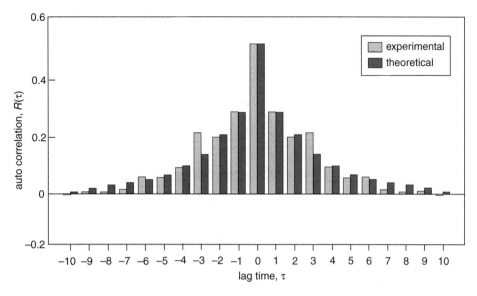

FIGURE 5.16   Comparison of fourth-order system modeled with a second-order system.

k	$x(k)$	$x(k+1)$	$x(k-1)$
0	5	3	—
1	3	4	5
2	4	7	3
3	7	—	4

Higher-order operators can be defined similarly, such as $\mathbf{Z}^2 = \mathbf{Z}[\mathbf{Z}x(k)] = x(k+2)$, which is the two-time-step advance. The inverse operator $\mathbf{Z}^{-1}$, called the *signal delay*, is defined in an analogous manner: $\mathbf{Z}^{-1}[x(k)] = x(k-1)$.

It is important to note that even though $\mathbf{Z}$ is written as an algebraic symbol, it is meaningless in isolation; it must always *act* on a discrete signal $x(k)$. Even so, $\mathbf{Z}$ is a powerful shorthand notation that behaves algebraically in practical situations. The following is a list of useful properties, assuming a linear discrete signal:

Property	Interpretation
$\mathbf{Z}$ is linear	
$\mathbf{Z}^n$	$x(k+n)$
$\mathbf{Z}^{-n}$	$x(k-n)$
$\mathbf{Z}^n\mathbf{Z}^{-n} = \mathbf{Z}^0 = 1$	$x(k)$
$\mathbf{Z}^{m+n} = \mathbf{Z}^m\mathbf{Z}^n$	$x(k+m+n)$
$a\mathbf{Z} = b\mathbf{Z} = (a+b)\mathbf{Z}$	$(a+b)x(k+1)$
$\mathbf{Z}a = a\mathbf{Z}$	$ax(k+1)$

Consider the specific example of a discrete signal $x(k)$ defined recursively by

$$x(k) = w(k) - \tfrac{1}{4}x(k-1)\tfrac{1}{4}x(k-2) + \tfrac{3}{4}x(k-3),$$

where $w(k)$ is unit white noise. Expressing each term using the big-$\mathbf{Z}$ notation,

$$\mathbf{Z}^0 x(k) = \mathbf{Z}^0 w(k) - \tfrac{1}{4}\mathbf{Z}^{-1}x(k) - \tfrac{1}{4}\mathbf{Z}^{-2}x(k) + \tfrac{3}{4}\mathbf{Z}^{-3}x(k).$$

Placing the x-terms on the left-hand side and the w-terms on the right-hand side of the equation,

$$(1 + \tfrac{1}{4}\mathbf{Z}^{-1} + \tfrac{1}{4}\mathbf{Z}^{-2} - \tfrac{3}{4}\mathbf{Z}^{-3})x(k) = w(k).$$

From this, the ratio of the signal $x(k)$ to the white noise driver $w(k)$ is formed:

$$\frac{x(k)}{w(k)} = \frac{1}{1 + \tfrac{1}{4}\mathbf{Z}^{-1} + \tfrac{1}{4}\mathbf{Z}^{-2} - \tfrac{3}{4}\mathbf{Z}^{-3}} = \frac{4\mathbf{Z}^3}{\mathbf{Z}^2 + \mathbf{Z}^2 + \mathbf{Z} - 3} \equiv H(\mathbf{Z}).$$

This ratio is known as the *signal transfer function*, since it relates the input driver to the input signal. Note the analogous definition of the *signal transfer function* to the *system transfer function*. For linear models, the transfer function is always independent of both the input and output. It should also be noted that this example is the same as Example 5.6 (b), where the denominator is the characteristic equation $C(\mathbf{Z}) = 0$ from that example. Transfer functions have a very important place in linear systems theory. They express the dynamic relationship between input and output in an algebraic form.

Since ARMA models are always linear, the ratio of output $x(k)$ to $w(k)$ can be simplified to a ratio of two polynomials $H(\mathbf{Z}) = B(\mathbf{Z})/C(\mathbf{Z})$. In general, $\deg[B(\mathbf{Z})] \leqslant \deg[C(\mathbf{Z})]$ because of causality. If this were not the case, the $x(k)$ would depend on future values $w(k)$. There is one especially important interpretation of $H(\mathbf{Z})$. If the input $w(k) = \delta(k)$ is the unit impulse function, then $h(k) = H(\mathbf{Z})\delta(k)$ is the impulse response. Therefore, if $C(\mathbf{Z})$ is formally divided into $B(\mathbf{Z})$, the resulting polynomial form for $H(\mathbf{Z})$ is the corresponding view for $h(k)$ in the
$\mathbf{Z}$-domain:

$$h(k) = [h(0) + h(1)\mathbf{Z}^{-1} + h(2)\mathbf{Z}^{-2} + \ldots]\delta(k)$$
$$= h(0)\delta(k) + h(1)\delta(k-1) + h(2)\delta(k-2) + \ldots \quad .$$

It follows that the functional values $h(k)$ are the signal values of the impulse response. In the case of $x(k) = w(k) - \tfrac{1}{4}x(k-1) - \tfrac{1}{4}x(k-2) + \tfrac{3}{4}x(k-3)$, the division leads to a divergent infinite series, since, as was found above, solutions to the characteristic equation lie on the unit circle:

$$h(k) = 1 - \tfrac{1}{4}\mathbf{Z}^{-1} - \tfrac{3}{16}\mathbf{Z}^{-2} + \tfrac{55}{64}\mathbf{Z}^{-3} + \ldots$$
$$= \delta(k) - \tfrac{1}{4}\delta(k-1) - \tfrac{3}{16}\delta(k-2) + \tfrac{55}{64}\delta(k-3) + \ldots \quad .$$

In summary, the big-$\mathbf{Z}$ notation is very closely related to the common Z-transform for discrete signals. It is also analogous to the Laplace transform for continuous signals. Table 5.3 summarizes these other signal views.

TABLE 5.3   Transforms from Continuous Time to the s-Domain (Laplace), and Transforms from Discrete Time to the z-Domain (**Z**-Transforms)

Laplace transforms:	$\mathscr{L}[\dot{x}(t)] = sX(s) - x(o)$	$\mathscr{L}[z \int x(t)\, dt] = \dfrac{1}{s} X(s)$
Z-transforms:	$Z[x(k+1)] = z[X(z) - x(o)]$	$Z[x(k-1)] = z^{-1} X(z)$

5.5 AUTOREGRESSIVE MOVING-AVERAGE (ARMA) MODELS

It is logical to generalize on the autoregressive and moving-average processes by defining an autoregressive moving-average process as follows:

$$x(k) = -\sum_{i=1}^{p} a_i x(k-i) + \sum_{i=0}^{q} b_i w(k-i). \tag{5.14}$$

If all the $a_1, \ldots, a_p$ are zero, this reduces to an MA process. If $q = 0$, this reduces to an AR process. In the more general case of Equation (5.14), the process is known as an ARMA random process. As with the AR and MA processes, the autocorrelation is the statistic of choice.

Since the ARMA equation (5.14) is linear and rewriting by using big-Z,

$$x(k) = -x(k)\sum_{i=1}^{p} a_i Z^{-i} + w(k)\sum_{i=0}^{q} b_i Z^{-i}$$

Equivalently, the transfer function is

$$H(Z) = \frac{x(k)}{w(k)} = \frac{\displaystyle\sum_{i=0}^{q} b_i Z^{-i}}{1 + \displaystyle\sum_{i=1}^{p} a_i Z^{-i}}. \tag{5.15}$$

The denominator of this ratio of polynomials is called the characteristic polynomial $C(Z)$, and solutions to the characteristic equation $C(Z) = 0$, called the system *poles*, determine the stability of the random process. As noted in the Autoregressive Statonarity Theorem (Section 5.3), if the poles are all inside the unit circle, then the system is stable, and if any are outside, then it is unstable. If the transfer function ratio is divided out so that it is a power series

$$H(Z) = \frac{x(k)}{w(k)} = h(0) + h(1)Z^{-1} + h(2)Z^{-2} + \ldots,$$

then the coefficient $h(i)$ of Z^{-i} is called the *signal impulse response*.

EXAMPLE 5.9

Characterize the ARMA signal defined by the equation

$$x(k) = x(k-1) - \tfrac{1}{4}x(k-2) + w(k) + 3w(k-1) + 2w(k-2).$$

Specifically:
(a) Find the signal transfer function.
(b) Find the characteristic equation and determine the system poles.
(c) Thus infer signal stability.
(d) Find the impulse response of the process.

Solution
The transfer function is found by first expressing the ARMA equation in the
z-domain by

$$x(k) = Z^{-1}x(k) - \frac{1}{4}Z^{-2}x(k) + w(k) + 3Z^{-1}w(k) + 2Z^{-2}w(k).$$

Collecting the $x(k)$ terms on the left and the $w(k)$ terms on the right, then finding
the ratio $x(k)/w(k)$, the transfer function can be written with either negative or
positive powers of Z as follows:

$$H(Z) = \frac{x(k)}{w(k)} = \frac{1 + 3Z^{-1} + 2Z^{-2}}{1 + Z^{-1} + \frac{1}{4}Z^{-2}}$$

$$= \frac{Z^2 + 3Z + 2}{Z^2 + Z + \frac{1}{4}}.$$

The denominator $C(Z) = Z^2 + Z + \frac{1}{4} = 0$ is the characteristic equation, and its
two roots at $Z = \frac{1}{2}$ are inside the unit circle. Therefore, this ARMA signal is
stable. By formally dividing the denominator into the numerator, $H(Z)$ can be
written as an infinite series: $H(Z) = 1 + 4Z^{-1} + \frac{23}{4}Z^{-2} + \frac{19}{8}Z^{-3} + \dots$ In
general, this expansion for $H(Z)$ will be infinite, but, for now, the first few
terms are sufficient. It follows from this that the impulse response has coefficients
$h(0) = 1$, $h(1) = 4$, $h(2) = 5.75$, $h(3) = 4.75$, $h(4) = 3.6875$, etc., and in general
for $k > 1$, $h(k) = (\frac{1}{2})^k(15k - 7)$.   ○

As the name implies, the impulse response should be thought of as the output
resulting from a single unit-impulse input. It is not random, and even though it is
deterministic, it completely defines this signal. Thus, in Example 5.9, the coefficients $h(k)$
can be computed using the ARMA equation directly. By assuming an initial quiescent state
(zero initial conditions) and a unit impulse input, the output $x(k) = h(k)$. A program with
which to compute the impulse response is given in Listing 5.4, where it should be noted

```
h(0)=1
print 0, h(0)
h(1)=h(0)+3
print 1, h(1)
h(2)=h(1)-¼h(0)+2
print 2, h(2)
for k=3 to n
         h(k)=h(k-1)-¼h(k-2)
print k, h(k)
next k
```

LISTING 5.4   Computing the impulse response, Example 5.9.

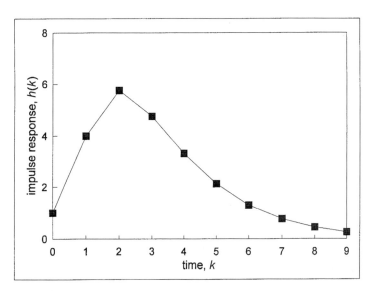

FIGURE 5.17   Impulse response for ARMA process.

three printing computations are needed to initialize the process loop. A graph of the impulse response is shown in Figure 5.17. As should be expected in a stable system, the output resulting from a single unit-impulse input will rise, then fall and die out as time k becomes large. In fact, one definition of a stable system is one in which the steady-state impulse response is zero, just as seen in Figure 5.17.

On the other hand, the response to a white noise input should be stationary in the steady state. That is, after a short transient, the random process should wander about its mean $\mu_x = 0$ with a constant variance forever. Listing 5.5 generates such a sequence, which is graphed in Figure 5.18, showing the results of a simulation with several instances of white noise with variance $\sigma_w^2 = 1$. Unlike the impulse response where non-zero times

```
w(-2)=√−2*ln(RND)*cos(2π*RND)
w(-1)=√−2*ln(RND)*cos(2π*RND)
w(0)=√−2*ln(RND)*cos(2π*RND)
x(0)=w(0)+3w(-1)+2w)(-2)
print 0, x(0)
w(1)=√−2*ln(RND)*cos(2π*RND)
x(1)=x(0)+w(1)+3w(0)+2w(-1)
print 1, x(1)
for k=2 to n
        w(k)=√−2*ln(RND)*cos(2π*RND)
        x(k)=x(k-1)−¼x(k-2)+w(k)+3w(k-1)+2w(k-2)
        print k, x(k)
next k
```

LISTING 5.5   Generating an ARMA sequence, Example 5.9.

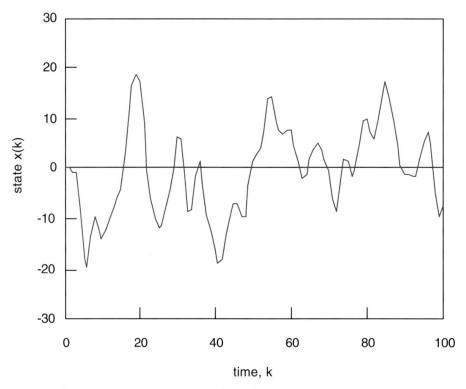

FIGURE 5.18   ARMA system response to white noise.

produced no effect, here we must assume shocks at all times, including those that are negative.

As with AR and MA random processes, the important statistic in the general ARMA process is the autocorrelation. However, in the case of an ARMA process, this requires knowledge of the impulse response. It can be shown that for the general ARMA process of Equation (5.14), the autocorrelation is found by the following equation:

$$
R_{xx}(\tau) = \begin{cases} -\displaystyle\sum_{i=1}^{p} a_i R_{xx}(|\tau| - i) + \displaystyle\sum_{i=0}^{q-|\tau|} h(i) b_{i+|\tau|}, & |\tau| \leqslant q, \\[2em] -\displaystyle\sum_{i=1}^{p} a_i R_{xx}(|\tau| - i), & |\tau| > q. \end{cases}
\tag{5.16}
$$

Equation (5.16) looks more intimidating than it really is if one remembers that the autocorrelation is an even function and that a sum in which the upper limit is smaller than the lower limit is zero. Regardless, this equation is straightforward to apply, as is shown in the following example:

EXAMPLE 5.10

Determine the autocorrelation of the random process

$$x(k) = x(k-1) - \tfrac{1}{4}x(k-2) + w(k) + 3w(k-1) + 2w(k-2),$$

where $w(k)$ is white noise with variance $\sigma_w^2 = 1$, using both theoretical and simulation techniques.

Solution
Comparing the general ARMA equation (5.14) against the problem at hand, $p = 2$, $q = 2$, $a_1 = 1$, $a_2 = \tfrac{1}{4}$, $b_0 = 1$, $b_1 = 3$, and $b_2 = 2$. Also, recalling the impulse response computed in Example 5.9, $h(0) = 1$, $h(1) = 4$, and $h(2) = 5.75$. Thus, for $\tau = 0$, $\tau = 1$, and $\tau = 2$, the first of Equations (5.16) reduces to three equations:

$$
\begin{aligned}
R_{xx}(0) &= -a_1 R_{xx}(-1) - a_2 R_{xx}(-2) + \sigma_w^2[h(0)b_0 + h(1)b_1 + h(2)b_2] \\
&= R_{xx}(1) - \tfrac{1}{4}R_{xx}(2) + 24.5, \\
R_{xx}(1) &= -a_1 R_{xx}(0) - a_2 R_{xx}(-1) + \sigma_w^2[h(0)b_1 + h(1)b_2] \\
&= R_{xx}(0) - \tfrac{1}{4}R_{xx}(1) + 11, \\
R_{xx}(2) &= -a_1 R_{xx}(1) - a_2 R_{xx}(0) + \sigma_w^2 h(0)b_2 \\
&= R_{xx}(1) - \tfrac{1}{4}R_{xx}(0) + 2.
\end{aligned}
\tag{5.17}
$$

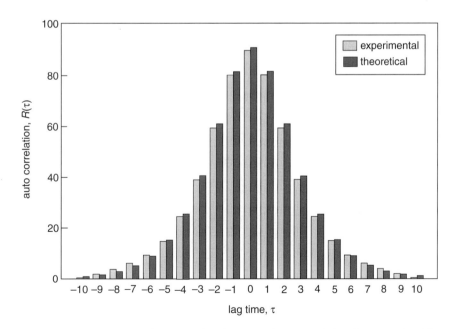

FIGURE 5.19   Comparison of calculated and computed autocorrelation for Example 5.10.

Equations (5.17) form a set of three linear equations with three unknowns, which can be written in matrix form as

$$\begin{bmatrix} 1 & -1 & \frac{1}{4} \\ -1 & \frac{5}{4} & 0 \\ \frac{1}{4} & -1 & 1 \end{bmatrix} \begin{bmatrix} R_{xx}(0) \\ R_{xx}(1) \\ R_{xx}(2) \end{bmatrix} = \begin{bmatrix} 24.5 \\ 11 \\ 2 \end{bmatrix}.$$

Solving these, $R_{xx}(0) = 90\frac{2}{3}$, $R_{xx}(1) = 81\frac{1}{3}$, and $R_{xx}(2) = 60\frac{2}{3}$. Since the auto-correlation is an even function, $R_{xx}(-1) = 81\frac{1}{3}$ and $R_{xx}(-2) = 60\frac{2}{3}$. If $\tau > 2$, the second of Equations (5.16) applies recursively: $R_{xx}(3) = -R_{xx}(2) - \frac{1}{4}R_{xx}(1) = -\frac{1}{3}$ and $R_{xx}(4) = -R_{xx}(3) - \frac{1}{4}R_{xx}(2) = -\frac{5}{12}$, and so forth.

ARMA processes are, in general, ergodic, so, after generating a single instance, time averaging provides the autocorrelation experimentally by the usual methods described earlier. Comparisons of these methods are shown in Figure 5.19, confirming the reliability of both.   ○

5.6   ADDITIVE NOISE

There are two reasons that statistical noise models are so important. First, in modeling signals, it is rarely enough to analyze the response to an ideal input. Even after accounting for all understood signals, there is inevitably a residual error between an ideal and a realistic input. Therefore, it is natural to add a random noise component to the ideal signal. This model, which is called the *additive noise model*, is used primarily because of the *superposition principle* in linear systems. That is, if the input can be additively decomposed, so can the output. This idea is illustrated in Figure 5.20.

The second reason that linear statistical models are so important is that to some extent all noise looks the same. A signal that contains an additive noise component will appear like high-frequency "static" overlaying an ideal carrier. The amplitude of the noise is characterized by the variance σ^2, but otherwise they look more or less the same. However, the signal statistics can be quite different, and, when acted on by a system, the response can be significantly different. That is why statistical measures such as the autocorrelation, or equivalently the spectral density, are so important.

Consider a daily temperature model in a climatological system viewed over a year's time. Clearly there will be an ideal sinusoid-shaped signal of frequency 1 cycle every 365 days because of the Earth's annual motion about the Sun. There will also be apparently random daily temperature fluctuations with frequency 1 cycle per day because of the

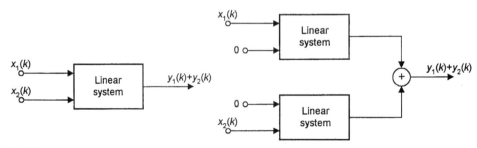

FIGURE 5.20   The superposition principle.

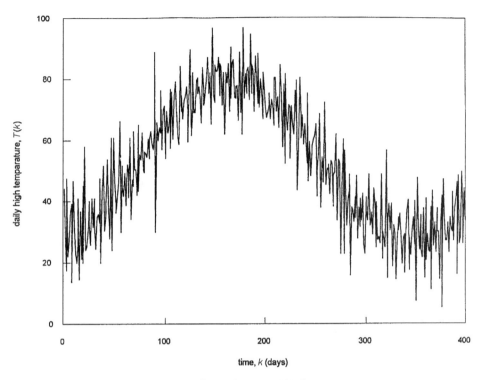

FIGURE 5.21   Empitical temperature fluctuation over 400 days.

Earth's daily rotation about its own axis. An example of this is the empirical data shown in Figure 5.21 of a year's daily high temperatures at a typical midwestern US city. Suppose that it is necessary to use data such as this to drive a simulation such as the temperature-driven population models of Sections 2.4 and 2.5. To find a suitable additive model, we must (1) estimate the ideal signal and (2) characterize the additive noise. The following examples demonstrate a practical means for finding such a model.

EXAMPLE 5.11

We know that an ideal signal representing daily temperature will be periodic with period $n = 365$ days and approximately sinusoidal:

$$x(k) = A + B\cos(\omega k + \theta), \tag{5.18}$$

where $\omega = 2\pi/365 \approx 0.017$ radians per day. Consider the empirical data shown in Figure 5.21 and partially tabulated in Table 5.3. Estimate the parameters A, B, and θ.

Solution
In order to remove what is apparently a very noisy signal, we will filter it with a so-called *moving-average low-pass filter*. (Another popular low-pass filter will be

TABLE 5.4   Signal Models for Figure 5.21 data, Example 5.11

Day	Empirical signal	MA(5)	MA(13)	Residual	Ideal signal
1	31.29				29.25
2	32.65				29.39
3	39.77	31.74			29.54
4	26.18	32.58			29.69
5	28.70	32.26			29.85
6	35.62	30.68			30.02
7	31.05	28.70	29.62	−1.42	30.20
8	31.88	26.98	29.15	−2.72	30.38
9	16.24	27.20	29.35	13.11	20.57
10	20.11	27.20	28.62	8.51	30.77

discussed in Chapter 6.) Moving average filters approximate one day's signal by averaging signal occurences of adjacent days, thereby obtaining a smoothing effect. Closely related to the MA random process, two such filters are as follows:

MA(5): $x_5(k) = \frac{1}{5}[x(k-2) + x(k-1) + x(k) + x(k+1) + x(k+2)]$;

MA(13): $x_{13}(k) = \frac{1}{13} \sum_{i=-6}^{6} x(k+i)$.

In the spreadsheet of Table 5.4, columns 3 and 4 show these filtered signals for the data $x(k)$. Notice that because of the definitions of x_5 and x_{13}, several of the

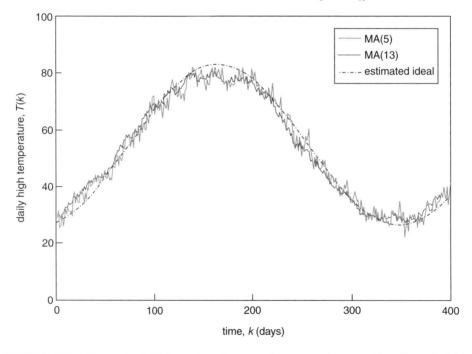

FIGURE 5.22   Removal of additive noise using a moving-average low-pass filter, Example 5.11.

initial (and final) rows are undefined. Both $x_5(k)$ and $x_{13}(k)$ are graphed in Figure 5.22, where they should be compared against $x(k)$ in Figure 5.21. Clearly, the filter MA(m): $x_m(k)$ becomes increasingly smooth as m increases.

Since $x_{13}(k)$ seems sufficiently smooth to be a sinusoid, we can estimate the parameters in Equation (5.18) as

$$
\begin{aligned}
A &= \tfrac{1}{2}[x_{13}(k_{\max}) + x_{13}(k_{\min})], \\
B &= \tfrac{1}{2}[x_{13}(k_{\max}) - x_{13}(k_{\min})], \\
\theta &= -\tfrac{2}{365}\pi k_{\max},
\end{aligned}
\tag{5.19}
$$

where $k_{\min}$ and $k_{\max}$ are the times at which the filtered signal achieves its minimum and maximum values. Equations (5.19) are immediate consequences of Equation 5.18. In this example, $k_{\max} = 166$, $k_{\min} = 347$, $x(k_{\max}) = 82.8$, and $x(k_{\min}) = 26.2$. Thus, $A = 54.5$, $B = 28.3$, and $\theta = -2.84789$, from which the ideal signal is estimated to be $x_{\text{ideal}}(k) = 54.5 + 28.3\cos(0.017t - 2.84789)$.

There is a question of how many terms to use to make the filter, but here the results show the way. Since the MA(13) is "close" to the desired sinusoid, we simply go with $x_{\text{ideal}}(k)$ as derived above. Thus, the difference $x(k) - x_{\text{ideal}}(k)$ amounts to the residual noise. With luck, this noise can be modeled by one of the ARMA models discussed earlier.

The simplest noise model is the zeroth-order or white noise model. Since it is of zeroth order, the signal cannot depend on previous observations, and each time slice is totally independent of other events. Still, the question remains as to which distribution is used for creating the white noise. This is illustrated below.

EXAMPLE 5.12

Create an additive white noise model to be superimposed onto the ideal filtered signal of Example 5.11. Test it for adequacy against the empirical temperature data given in Figure 5.21.

Solution
The residual error $x(k) - x_{\text{ideal}}(k)$ is tabulated in column 5 of Table 5.4. A quick calculation of the mean (0.0174) and standard deviation (8.6496) confirms that a white noise model is possible, since the mean is close to zero, as required for white noise. The actual distribution of the residual noise can be estimated by plotting a histogram of the residuals. This is done in Figure 5.23.

A quick look at this graph reveals that the distribution is much closer to a Gaussian than to, say, a uniform distribution. Thus, the white noise assumption implies a simulated noise signal $x_{\text{noise}}(k)$, which, when combined with the estimated ideal signal $x_{\text{ideal}}(k)$, produces the statistically appropriate modeled signal $\hat{x}(k)$:

$$
\begin{aligned}
x_{\text{noise}}(k) &= 8.65\sqrt{-2\ln(\text{RND})}\cos(2\pi\,\text{RND}), \\
x_{\text{ideal}}(k) &= 54.5 + 28.3\cos(0.017t - 2.84789), \\
\hat{x}(k) &= x_{\text{ideal}}(k) + x_{\text{noise}}(k).
\end{aligned}
\tag{5.20}
$$

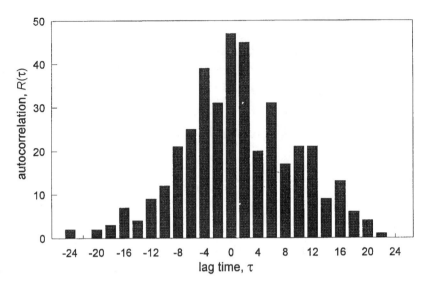

FIGURE 5.23   Residuals resulting from the additive white noise model, Example 5.12.

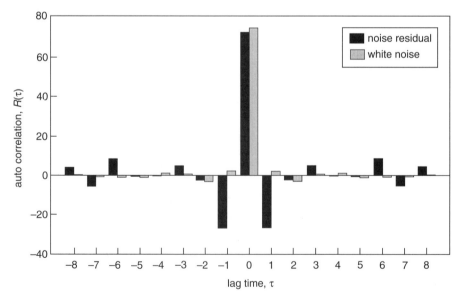

FIGURE 5.24   Comparison of noise residual with the white noise model using the autocorrelation statistic.

It remains to test the white noise assumption. A simulated signal $\hat{x}(k) = x_{\text{ideal}}(k) + x_{\text{noise}}(k)$ will undoubtably "look" like the original noisy signal $x(k)$, but it will never be the "same". This is because the signal is merely one of the entire process ensemble. The answer is to check the autocorrelation of the white noise simulation against that of the noise residual. The results of this calculation are shown in Figure 5.24. Notice that, as expected, the white noise is concentrated at $\tau = 0$, while the autocorrelation of the residual noise is more spread out. ○

As a practical matter, an estimate of the autocorrelation can be calculated from the single instance of the random process as originally described in Example 5.11. Assuming ergodic behavior for the residuals and using the algorithms of Chapter 3 to calculate the autocorrelations of both $R_{\text{residual}}(\tau)$ and $R_{\text{white}}(\tau)$, these are graphed in Figure 5.24. Clearly, there are significant components of $R_{\text{residual}}(\tau)$ that are unaccounted for in $R_{\text{white}}(\tau)$. Thus, a better noise model is called for.

EXAMPLE 5.13

The autocorrelation of the noise estimates Figure 5.24 indicates that there is a significant noise component at lag time $\tau = 1$ as well as one at $\tau = 0$. Therefore, we reject the simple white noise model of Example 5.12 graphed in Figure 5.24 are as follows:.

τ	$R_{\text{residual}}(\tau)$
0	72.50
1	−26.74
2	−2.17
3	4.71
4	−0.06
5	−0.36

Derive an AR estimator for the noise residual of Example 5.11. From this, create a simulated signal that is a faithful generator of the original input signal $\hat{x}(k)$.

Solution
This calls for an application of the Yule–Walker equation (5.12). From the above table,

$$
\begin{bmatrix}
-26.74 & -2.17 & 4.71 & -0.06 & -0.36 & -1 \\
72.50 & -26.74 & -2.17 & 4.71 & -0.06 & 0 \\
-26.74 & 72.50 & -26.74 & -2.17 & 4.71 & 0 \\
-2.17 & -26.74 & 72.50 & -26.74 & -2.17 & 0 \\
4.71 & -2.17 & 26.74 & 72.50 & -26.74 & 0 \\
-0.06 & 4.71 & -2.17 & -26.74 & 72.50 & 0
\end{bmatrix}
\begin{bmatrix}
a_1 \\ a_2 \\ a_3 \\ a_4 \\ a_5 \\ \sigma^2
\end{bmatrix}
=
\begin{bmatrix}
-72.50 \\ 26.74 \\ 2.17 \\ -4.71 \\ 0.06 \\ 0.36
\end{bmatrix}.
$$

This is a linear system that can be solved using elementary means, giving the solution as

$$\sigma_w^2 = 60.26,$$
$$a_1 = 0.444,$$
$$a_2 = 0.199,$$
$$a_3 = 0.013,$$
$$a_4 = -0.023,$$
$$a_5 = -0.016.$$

Since the terms a_3, a_4, and a_5 seem insignificantly small, a reasonable noise generator can be based on the terms a_1 and a_2 only, along with the variance white noise variance. Thus, an AR noise model leads to the following model for realistic signal generation:

$$x_{\text{noise}}(k) = 7.76\sqrt{-2 \ln(\text{RND})} \cos(2\pi\, \text{RND})$$
$$+ 0.44x_{\text{noise}}(k-1) + 0.2x_{\text{noise}}(k-2),$$
$$x_{\text{ideal}}(k) = 54.5 + 28.3 \cos(0.017t - 2.84789),$$
$$\hat{x}(k) = x_{\text{ideal}}(k) + x_{\text{noise}}(k).$$
$$(5.21)$$

The last step is to once again check and see if this signal produces an autocorrelation function similar to that of the original residual noise. By running a simulation of the first of Equations (5.21), it turns out that it is very similar to the original autocorrelation shown in Figure 5.24. Thus, we conclude that we have reached our goal of a faithful signal generator.   ○

Equations (5.21) present a reliable generator of realistic signals for the problem at hand. Not only does it produce signals that appear right, but it also produces signals that are statistically right. Only from analysis such as that described above can dependable system responses be obtained.

Additive noise makes a great deal of sense when dealing with linear systems, because of the superposition principle. This states that if the input signal can be decomposed into ideal and noisy parts, so can the output. Moreover, the sum of the ideal signal response and the output noise response is the total system response. Even though this does not hold true for nonlinear systems, the principle of statistically reliable signals in simulation is extremely important. Simply put, no matter how good the system model is, without statistically realistic inputs, no reliable predictions can be expected.

BIBLIOGRAPHY

Anderson, T. W., *The Statistical Analysis of Time Series*. Wiley, 1994.
Box, G. and G. Jenkins, *Time Series Analysis, Forecasting and Control*. Prentice-Hall, 1976.
Feller, W., *An Introduction to Probability Theory and Its Applications*. Wiley, 1971.

Fuller, W. A., *Introduction to Statistical Time Series*, 2nd edn. Wiley, 1995.

Gardener, W., *Introduction to Random Processes*, 2nd edn. McGraw-Hill, 1990.

Goodman, R., *Introduction to Stochastic Models*. Benjamin-Cummings, 1988.

Juang, J.-N., *Applied System Identification*. Prentice-Hall, 1994.

Ljung, L., *System Identification – Theory for the User*, 2nd edn. Prentice-Hall, 1999.

Masters, T., *Advanced Algorithms for Neutral Networks – aC++ Sourcebook*. Wiley, 1993.

Masters, T., *Neutral, Novel and Hybrid Algorithms for Time Series Prediction*. Wiley, 1995.

Proakis, J. G. and D. G. Manolakis, *Digital Signal Processing, Principles, Algorithms and Applications*, 3rd edn. Prentice-Hall, 1996.

Thompson, J. R., *Simulation: a Modeler's Approach*. Wiley-Interscience, 2000.

EXERCISES

5.1 Consider a continuous linear system with transfer function

$$H(s) = \frac{2}{s^2 + 4s + 3}$$

and white noise input $x(t)$ with variance $\sigma_x^2 = 5$.

(a) Calculate the spectral density of the output, $S_{yy}(\omega)$.

(b) Calculate the autocorrelation of the output, $R_{yy}(\tau)$.

5.2 Consider the system defined in Exercise 5.1. Using a step size $h = 0.1$ and a time horizon $[0,10]$:

(a) Perform a simulation producing the output $y(t)$. Assume a Gaussian-distributed input.

(b) Estimate the autocorrelation $\hat{R}_{yy}(\tau)$ of the output, assuming ergodicity.

(c) Compare the results of part (b) against that of Exercise 5.1(b) graphically.

5.3 Consider the continuous system with transfer function

$$H(s) = \frac{2}{s^2 + 4s + 3}$$

and shot noise input with exponential inter-event times with mean $\mu = 2$ seconds and maximum amplitude $a = 5$.

(a) Simulate the input over the interval $[0,10]$.

(b) Using $h = 0.1$, simulate the system response over the interval $[0,10]$.

(c) Graph the output.

(d) Estimate the autocorrelations of both inputs and outputs empirically.

5.4 Consider the MA process $x(k) = 2w(k) + 3w(k - 3) + 4w(k - 7)$.

(a) Generate several instances of the this process, and graph for $n = 100$ time steps.

(b) Consider one instance of the process for $n = 1000$ time steps. From this instance, find and graph the autocorrelation $R_{xx}(\tau)$.

(c) Find the theoretical values for the autocorrelation function $R_{xx}(\tau)$.

(d) Compare parts (b) and (c) graphically.

5.5 Consider the ARMA process $x(k) = -\frac{1}{4}x(k - 2) + w(k) + 4w(k - 2)$.

(a) Find the system function $H(\mathbf{Z})$ and characteristic polynomial $C(\mathbf{Z})$.

(b) Determine the system poles and discuss the system's stability.

(c) Find the autocorrelation function $R_{xx}(\tau)$.

(d) Compare the results of part (c) against those determined experimentally.

5.6 For each of the following ARMA random processes:

    (i) Find the transfer function $H(\mathbf{Z})$.

    (ii) Find the impulse response for $k = 0, \ldots, 5$.

    (iii) Discuss the stability.

    (a) $x(k) = -2x(k-1) - 2x(k-2) - x(k-3) + w(k) + 2w(k-2) + w(k-3)$.

    (b) $x(k) = \frac{1}{4}x(k-3) + w(k) + 2w(k-2) + w(k-3)$.

    (c) $x(k) = -\frac{13}{4}x(k-1) - \frac{11}{4}x(k-2) - \frac{1}{2}x(k-3) + w(k) + 2w(k-2) + w(k-3)$.

5.7 Consider the AR process $x(k) = -\frac{1}{6}x(k-1) + \frac{1}{6}x(k-2) + w(k)$, where $w(k)$ is Gaussian white noise with variance $\sigma_w = 3$.

    (a) Find the system function $H(Z)$ and characteristic polynomial $C(\mathbf{Z})$.

    (b) Determine the system poles and discuss the system's stability.

    (c) Find the autocorrelation function $R_{xx}(\tau)$.

    (d) Compare the results of part (c) against those determined experimentally.

5.8 Model the following autocorrelation data as an AR random process. After determining each model, compare the predicted autocorrelations against the original data as well as the different models. Do this by first modeling the process, then producing an instance of the simulation. From the simulation results, calculate the autocorrelation coefficients and variance. Compare against the original data as illustrated in the text and draw appropriate conclusions.

    (a) Model using a first-order process.

    (b) Model using a second-order process.

    (c) Model using a third-order process.

τ	$R_{xx}(\tau)$
0	10.0
1	7.0
2	6.5
3	5.5
4	6.0
5	4.0
6	2.0
7	1.0

5.9 Find the autocorrelation of each of the following random processes, where $w(k)$ is white noise with variance $\sigma_w^2 = 3$:

    (a) $x(k) = -\frac{13}{2}x(k-1) - \frac{17}{24}x(k-2) + \frac{1}{24}x(k-3) + w(k)$;

    (b) $x(k) = -\frac{13}{2}x(k-1) - \frac{17}{24}x(k-2) + \frac{1}{24}x(k-3) + w(k) + 3w(k-1) + 2w(k-2)$;

    (c) $x(k) = w(k) + 3w(k-1) + 2w(k-2)$.

5.10 Consider the sinusoidal random process defined by $x(t) = 3\cos(2t + \theta)$, where θ is uniformly distributed on $[0, 2\pi]$. Recalling that this process is ergodic, find the autocorrelation function and compare it graphically against the predicted result derived in Chapter 4.

5.11 Consider the nonlinear system where the output is the square of the input. Use as input uniform white noise with variance $\sigma_w^2 = 3$.

    (a) Compute an estimate of the autocorrelation of the output.

    (b) Model the output as a second-order AR random process.

    (c) Compare the model against actual autocorrelation data.

(d) What is the major flaw in making a model of the type described in part (c)?

5.12 Consider a event-driven random process in which the inter-event times are distributed in a 2-Erlang fashion and the amplitude is Gaussian white noise with variance $\sigma_w^2 = 3$.

(a) Create an event-scheduling algorithm by simulation accomplished in sequence time.

(b) Create a regular time sequence of the state, and graph several instances of this random process.

(c) Find both time averages and statistical averages of the mean to determine whether or not it is reasonable to assume ergodicity.

(d) Assuming ergodicity, compute the autocorrelation, and graph your results.

5.13 Consider the AR process $x(k) = ax(k-1) + w(k)$, where $w(k)$ is white noise with variance σ_w^2. Show that the autocorrelation is given by

$$R_{xx}(\tau) = \frac{\sigma_w^2}{1 - a^2}, \qquad \text{where} \qquad |a| < 1.$$

5.14 Explain why a discussion of stability is not required for an MA process.

5.15 Assume that an ideal signal has the form $x(k) = A + B\cos(\omega t + \theta)$ and period T. If k_{max} and k_{min} are the times at which the maximum and minimum x-values are achieved, show that

$$A = \tfrac{1}{2}[x(k_{max}) + x(k_{min})],$$
$$B = \tfrac{1}{2}[x(k_{max}) - x(k_{min})],$$
$$\theta = -\frac{2\pi k_{max}}{T}.$$

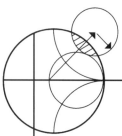

CHAPTER 6

Modeling Time-Driven Systems

Nature, be it physical or biological is the source of a great number of interesting, yet non-trivial models. The phenomena that form the basis of these models are not only important for engineered systems, but also fascinating in their own right. These natural models are characterized by several important properties:

(i) They are based on continuous time. As such, their mathematical description is usually expressed using partial or ordinary differential equations.
(ii) These systems are either autonomous or they have time-dependent inputs. Time-dependent inputs are regular and not event-based, and so they cannot be characterized by inter-arrival statistics.
(iii) They often have either transport delays (physical systems) or gestation delays (biological systems). These lead to difference or differential–difference equation descriptors.

While the system mechanism is often well behaved, the signals that drive the system are often not. This is partly because nature tends to be fickle and partly because the process of measurement itself introduces noise and complication. This presents a problem for both the scientist/engineer and the modeler/simulator in that there are really two interconnected modeling problems. The first goal is to simply model the system – how it transforms, manipulates, and combines signals. The second problem is to model the signal. Unless the signals are of the type the model expects, are realistic to the degree the model requires, and have consistent statistical properties, simulations will be totally meaningless.

The central idea in modeling natural systems is to formulate a precise mathematical description. For systems, this is usually a set of difference equations, differential equations, or differential–difference equations along with appropriate delays. For modularity, the system is often represented as a set of subsystems of components, each with its own input/output characteristics. This is, of course, the stuff of system design. However, the important thing is that each component is defined as a mathematical entity, and each can be modeled using the techniques outlined earlier in this text. The input signals are often random processes characterized by autocorrelation noise models superimposed on deterministic signals. This idea, called the *additive noise model*, has both a theoretical motivation and a practical rationale. Both will be exploited in a number of examples throughout this chapter. Once modeling has been completed, simulation in the form of computer programs can be accomplished. With simulation, system performance can be validated, and sensitivity and optimization can be demonstrated or enhanced. This understanding is fundamental to engineering and scientific enterprises.

6.1 MODELING INPUT SIGNALS

Ultimately, input drives the system. This is true for both event-driven and time-driven systems. In time-driven models, the input often shows itself on the right-hand side of a differential or difference equation. For example, each of the following equations describes an input–output relationship between input $x(t)$ and output $z(t)$:

$$\ddot{z} + 2\dot{z} + 4z = x, \tag{6.1a}$$

$$\ddot{z} + 2\dot{z} + 4z = x + 2\dot{x}, \tag{6.1b}$$

$$\ddot{z} + 2\dot{z} + 4z = x + 3tx. \tag{6.1c}$$

In each case, the left-hand sides exclusively describe critical output dynamics. For instance, in a mechanical system, this would relate fundamental mechanical variables:

$$\text{position}: \quad z(t),$$

$$\text{velocity}: \quad \dot{z}(t) = \frac{dz}{dt},$$

$$\text{acceleration}: \quad \ddot{z}(t) = \frac{d^2z}{dt^2}.$$

The right-hand sides incorporate the input in a number of ways. The simplest is Equation (6.1a), where the right-hand side is expressly the input, but there can also be derivatives of the input, as in Equation (6.1b). The description can also be time-dependent, as in Equation (6.1c). Regardless of the input–output relationship, it is essential to have a means for modeling $x(t)$. Only in the most idealistic of approximations is there an explicit formula for the input. Indeed, it is the heart and soul of simulation, where we must predict the output behavior of a system using a realistic input.

Since explicit formulas for input signals are usually unrealistic, it is often useful to use a relational notation for writing composite functions. For instance, the relation $(t > 1)$ is either true (1) or false (0). It follows that the function defined by $f(t) = (t > 0) - (t < 0)$ is $+1$ for all positive t and -1 for all negative t. In short, $f(t)$ is the *signum function*:

$$f(t) = (t \geqslant 0) - (t < 0) = \begin{cases} 1, & t > 0, \\ 0, & t = 0, \\ -1, & t < 0. \end{cases}$$

Similarly, the absolute value is given by $|t| = t[(t > 0) - (t < 0)]$. The technique is best shown by example.

EXAMPLE 6.1

Consider the idealized periodic signal shown in Figure 6.1. Using relational function notation, write a formula for this signal. Generalize this to a pulse train with arbitrary maximum f_{max}, minimum f_{min}, and duty cycle p.

Solution
Notice that, beginning with time $t = 1$, the signal is periodic with period 5. The signal is high 80% of the time and low 20% of the time. Thus, for the time interval $1 \leqslant t \leqslant 6$,

$$x_1(t) = \begin{cases} 5, & 1 \leqslant t < 5, \\ 2, & 5 \leqslant t < 6, \end{cases}$$
$$= [(t > 1) \text{ and } (t < 5)]3 + 2.$$

However, this signal must be defined for all time and not simply for one cycle of its existence. This can be done by referencing all evaluations back to the fundamental [1,6] time interval by creative use of the INT function, where $\text{INT}(t)$ is the integer part of variable t. It follows that an explicit formula for the signal of Figure 6.1 is given by

$$x_1(t) = 3\{([t > 1 + 5 * \text{INT}(\tfrac{1}{5}t)] \text{ and } [t < 5 + 5 * \text{INT}(\tfrac{1}{5}t)]\} + 2.$$

This can be verified using user-defined functions in any algorithmic computer programming language.

In general, if the signal starts at time $t = t_0$, is periodic with period T and is high a proportion p of every cycle, p is called the *duty cycle* of the signal. Assuming the signal alternates between a low of f_{min} and a high of f_{max}, the

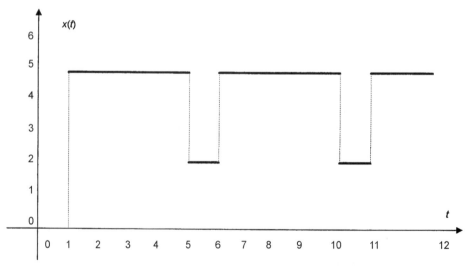

FIGURE 6.1   Idealized pulse-train signal with $f_{min} = 2$, $f_{max} = 5$, and a duty cycle $p = 80\%$.

explicit formula for this signal is

$$x_1(t) = (f_{max} - f_{min})\left\{\left[t > t_0 + T * \text{INT}\left(\frac{t}{T}\right)\right] \text{ and } \left[t < t_0 + pT * \text{INT}\left(\frac{t}{T}\right)\right]\right\} + f_{min}.$$

$$(6.2)$$

It is useful to be able to express the ideal signal by formulas such as Equation (6.2). However, to make signals more realistic, additive noise is often introduced, as shown in Figure 6.2, where $x_1(t)$ is an ideal signal and $w(t)$ is suitably defined noise. In general, $w(t)$ can be defined by any number of noise models, such as white noise or the ARMA models described in Chapter 5.

In the theoretically continuous case, white noise should be added at every instant of time. This, of course, is impossible, since we simulate our systems using computers, which are innately discrete devices. While adding noise to an ideal deterministic signal might make signals appear realistic, it cannot completely define the total behavior of the input. In particular, the frequency spectrum of the signal will be severely band-limited, even though this should not be the case. For example, by definition, white noise is frequency-independent and has an infinite bandwidth. Therefore, especially when dealing with non-simulated noisy signals, it is common to put them through a low-pass filter (LPF) before entering the system per se. This will remove much of the "wiggle" and make sampling the underlying non-noisy signal $x_1(t)$ more accurate.

Filtering in general is well beyond the scope of this text, and the low-pass "RC" filter described here is the simplest possible. However, being simple has its advantages; RC filters are used virtually everywhere. The basic idea is motivated by the basic electrical resistor–capacitor (RC) circuit shown in Figure 6.3. It is well known that by varying the

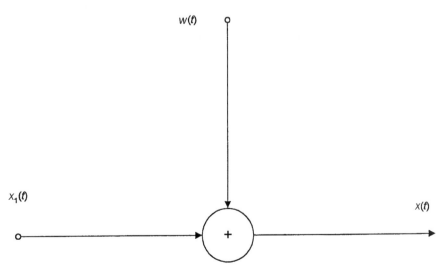

FIGURE 6.2   Additive noise.

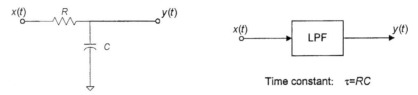

FIGURE 6.3   Basic *RC* low-pass filter.

values of R and C, high-frequency input signals are attenuated while low-frequency signals pass through the circuit unaltered. It is included in every electrical engineer's bag of tricks.

Referring to Figure 6.3, $x(t)$ and $y(t)$ are the source and output voltage respectively. The value of $\tau = RC$ is called the time constant, and can be adjusted so as to optimize the filtering effect. The differential equation defining the input–output relationship is

$$\tau \frac{dy}{dt} + y(t) = x(t). \tag{6.3}$$

This can be transformed into a digital filter by applying a numerical technique to solve for $y(t)$. For example, assuming an integration step size h, Euler's formula reduces to

$$y^{(\text{new})} = y + \frac{h}{\tau}[x(t) - y(t)].$$

Again, since the filter's input signal $x(t)$ is likely non-deterministic and without an explicit formula, numerical methods are essential here.

EXAMPLE 6.2

Consider an input $x(t)$ that is pre-processed using a low-pass filter producing filtered output $y(t)$ as shown in Figure 6.4. This ideal input $x_1(t)$ is the periodic signal given in Example 6.1 and $w(t)$ is Gaussian white noise with variance 4. Implement the low-pass filter of Equation (6.3) for various values of the time constant τ.

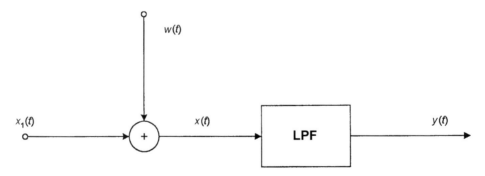

FIGURE 6.4   Additive noise signal with a low-pass filter.

Solution

Assuming the signal granularity to be the same as the integration step size h, the various signals defined in Figure 6.4 are given by

$$x_0(t) = [t > 1 + 5\,\mathrm{INT}(\tfrac{1}{5}t)] \text{ and } t < 5 + 5\,\mathrm{INT}(\tfrac{1}{5}t)],$$

$$x_1(t) = 3x_0(t) + 2,$$

$$w(t) = 2\sqrt{-2\ln(\mathrm{RND})}\cos(2\pi\,\mathrm{RND}),$$

$$x(t) = x_1(t) + w(t).$$

The graphs of the ideal signal $x_1(t)$ and the realistic signal $x(t)$ are shown in Figure 6.5. It should be obvious that on sampling $x(t)$, the chances of inferring the high or low ideal state are somewhat dubious at best. Thus, we try filtering.

Using the Euler integration method, the update equation for output $y(t)$ is

$$y^{(\mathrm{new})} = y + \frac{h}{\tau}(x - y).$$

A routine producing all input signals, both filtered and unfiltered is given in Listing 6.1. Note that the initial value of the filter was chosen arbitrarily to be $y(0) = 0$, since there is no way of knowing where to start. Thus, a short transient phase is expected before this filter becomes accurate.

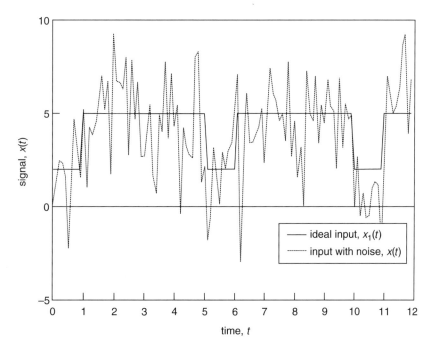

FIGURE 6.5   Ideal signal $x_1(t)$ superimposed on noise $w(t)$ to form an additive noise signal $x(t)$, Example 6.2.

```
y=0
for t=t₀ to tmax step h
        x₀=[t>1+5*INT (⅕t)]and[t<5+5*INT (⅕t)]
        x₁=3x₀+2
        w=2√(-2*ln(RND))cos(2π*RND)
        x=x₁+w
            h
        y=y+─(x-y)
            τ
        print t, x, y
next t
```

LISTING 6.1   Simulation of additive noise signal $x(t)$, Example 6.2.

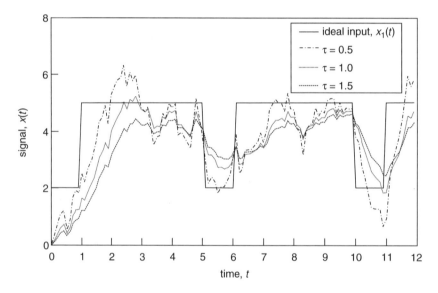

FIGURE 6.6   Filtered signal, Example 6.2.

Results for $\tau = 0.5$, $\tau = 1$, and $\tau = 1.5$ are given in Figure 6.6. Notice that as τ is increased, the filtered signal becomes smoother, as desired. However, it also becomes less responsive in that it fails to rise up to the ideal before it is time to go back down. Conversely, as τ decreases, the response is much better. But, now it is more erratic; that is, high frequencies become more prevalent.   ○

For convenience, it is desirable to define a set of common periodic signals. These signals can then be used to form the basis of the deterministic portion of the input signal. Each is defined in terms of characteristic quantities such as those listed here:

V_{min}:   minimum value of signal;
V_{max}:   maximum value of signal;
f: frequency (the reciprocal of the period) of the signal;
p: duty cycle (proportion of time pulse remains in the V_{max} mode).

A partial list is presented here, along with an explicit mathematical description over one cycle:

sinusoid: $\sin(V_{\min}, V_{\max}, f, t) = \dfrac{(V_{\min} + V_{\max})}{2} + \dfrac{(V_{\max} - V_{\min})}{2}\sin(2\pi ft);$

pulse: $\mathrm{pul}(V_{\min}, V_{\max}, f, t) = \begin{cases} V_{\min}, & 0 \leqslant t < \dfrac{1}{2f}, \\[2mm] V_{\max}, & \dfrac{1}{2f} \leqslant t < \dfrac{1}{f}; \end{cases}$

triangle: $\mathrm{tri}(V_{\min}, V_{\max}, f, t) = \begin{cases} 2f(V_{\max} - V_{\min})t + V_{\min}, & 0 \leqslant t < \dfrac{1}{2f}, \\[2mm] 2f(V_{\max} - V_{\min})\left(\dfrac{1}{f} - t\right) + V_{\min}, & \dfrac{1}{2f} \leqslant t < \dfrac{1}{f}; \end{cases}$

pwm: $\mathrm{pwm}(V_{\min}, V_{\max}, f, p, t) = \begin{cases} V_{\max}, & 0 \leqslant t < \dfrac{p}{f}, \\[2mm] V_{\min}, & \dfrac{p}{f} \leqslant t < \dfrac{1}{f}. \end{cases}$

The entire periodic signal $x_1(t)$ can be defined using relational techniques along with judicious use of the INT function. Combined with other deterministic signals and appropriate additive noise $w(t)$, the simulation input $x(t) = x_1(t) + w(t)$ can be used directly in the simulation at hand.

6.2   NOMENCLATURE

It is not the intent of this text to espouse a particular methodology or notation. On the other hand, good graphical nomenclature can enhance the modeling process and understanding. Figure 6.7 gives a list of common symbols and their interpretation in modeling specifications. Each will be described in this and succeeding sections of this chapter.

Arithmetic operations allow for the addition, subtraction, multiplication or division of signals. Terms "+" add or subtract, and factors "*" multiply or divide. Operators can be used in very generic ways. For instance, the box identifier refers to a mathematical description defined by the user. Examples are shown in Figure 6.8. These can be nonlinear algebraic operators (Figure 6.8a), linear differential equations where care is taken to specify initial conditions (Figure 6.8b), or difference equations, which also require initial conditions (Figure 6.8c). Many other possibilities exist, such as the low-pass filter (LPF) described in Example 6.2.

The integral operator, which is represented by the $\mathbf{D}^{-1}$ block of Figure 6.7, is special since it occurs so often. Its output is simply the integral of the input, which means that it acts like an accumulator:

$$y(t) = y(t_0) + \int_{t_0}^{t} x(t)\,dt. \tag{6.4}$$

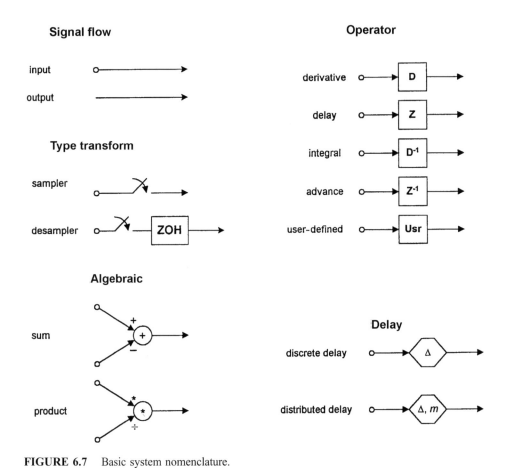

FIGURE 6.7   Basic system nomenclature.

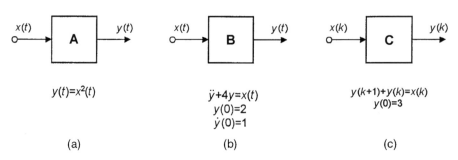

(a)   (b)   (c)

FIGURE 6.8   User-specified component blocks.

Generally, this process cannot be accomplished formally, since the input is likely either empirically defined or the result of a numerical procedure. Thus, it is better to view it by the equivalent differential system,

$$\dot{y}(t) = x(t),$$
$$y(t_0) = y_0.$$

(6.5)

Equation (6.5) can be approximated using the numerical methods described earlier. For instance, the Eulerian update equation is simply $y^{(new)} = y + hx(t)$, from which it is evident that the differential Equation (6.5) behaves much like an accumulator with initial value y_0.

The integral operator tends to smooth the input. Whether by Euler, Taylor, or Runge–Kutta, the solution to either Equation (6.4) or (6.5) is less choppy because of the operation. In contrast to the integral, the derivative is much more volatile, since it involves dividing by the very small number h. So, if at all possible, derivatives should be avoided in simulations. If this is impossible, the low-pass filter of (LPF) introduced in Example 6.2 helps considerably. This is illustrated in the following example.

EXAMPLE 6.3

Consider a signal with two components, an ideal signal $x_1(t) = \cos\frac{1}{2}t$ and an additive white noise component $w(t)$, so that the input is $x(t) = x_1(t) + w(t)$.
(a) Numerically compute the integral of $x(t)$. Compare against the expected ideal results.
(b) Numerically compute the derivative of $x(t)$. Compare against the expected ideal results.
(c) Note the effects of applying a low-pass filter to the output of the derivative output.

Solution
If there were no input noise and the granularity were zero, the integral output would be $z_{int}(t) = 2\sin\frac{1}{2}t$ and the derivative output would be $z_{der}(t) = -\frac{1}{2}\sin\frac{1}{2}t$. Of course, in a computer simulation, the granularity cannot be zero, so there is inevitably an error. Also, if there is no noise, this error will be relatively small, as was shown in Chapter 2. However, with the introduction of noise, the results deteriorate.

(a) The introduction of noise will make the situation worse. Even so, the integral is an accumulation operation, and the erratic noise behavior tends to cancel out over the long haul. For instance, consider the effect of uniformly distributed white noise with unit variance. Since, by definition, white noise has mean $\mu = 0$, we consider the random variate generated by $2 * RND - 1$, which is uniform on $[-1, 1]$. Recalling that uniform random variates on $[-1, 1]$ have variance $\sigma^2 = \frac{1}{3}$ (see Appendix B), the white noise called for is generated by the formula

$$w(k) = \sqrt{3}(2 * RND - 1).$$

In this example, let us consider noise with variance $\sigma^2 = 0.1$. Thus, on discretizing the input signal using $t = hk$ prior to adding the white noise, the input signal is

$$x(k) = \cos \tfrac{1}{2}hk + \sqrt{0.3}(2 * \text{RND} - 1).$$

Using Eulerian integration, the basic update equation for the integral is $y(k + 1) = y(k) + hx(k)$. Assuming an initial condition $y(0) = 0$, the simulation algorithm proceeds as follows:

```
y=0
for k=1 to n
        t=hk
        x₁=cos (½t)
        w=√0.3 (2*RND-1)
        x=x+w
        y=y+hx
    next k
```

The results of this integration are shown in Figure 6.9. The integral $y(t)$ is smoother than the input $x(t)$. Also notice that, despite the noise, the integration of $x(t)$ closely follows the ideal integration formula $z_{\text{int}}(t) = 2\sin\tfrac{1}{2}t$. Thus, we see that the integration operator has a natural smoothing effect that tends to neutralize the effect of input noise on the output.

(b) The basic difference formulas from Chapter 4 estimate the numerical derivative by $y(k) = [x(k) - x(k - 1)]/h$. Since the initial ideal function value is $x(0) = \cos(0) = 1$, the simulation algorithm proceeds as shown below. Note the use of a temporary variable x_2 to keep track of the previous x-value. This is required since both the new and past x-values are needed in the update equation.

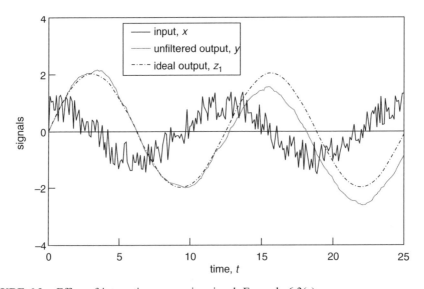

FIGURE 6.9   Effect of integration on a noisy signal, Example 6.3(a).

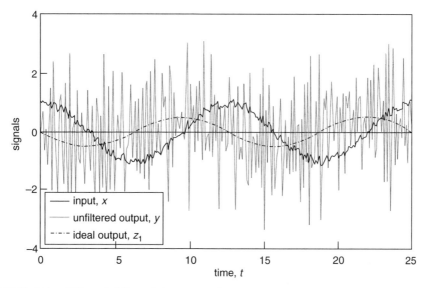

FIGURE 6.10   Effect of differentiation on a noisy signal, Example 6.3(b).

```
t=0
x=1
print t, x, 0
for k=1 to n
        x₂=x
        x₁=cos(½t)
        w=√0.03(2*RND-1)
        x=x₁+w
        y=(x-x₂)/h
        t=t+h
        print t, x, y
    next k
```

The above simulation assumes a white noise variance of $\sigma^2 = 0.01$, and the results are graphed in Figure 6.10. Notice that even though the noise variance is less than that used for the integration example and is reduced from $\sigma^2 = 0.1$ to $\sigma^2 = 0.01$, the fluctuation in the derivative is far more severe than that of the integral shown in Figure 6.9. The simulation block diagram outlining the above

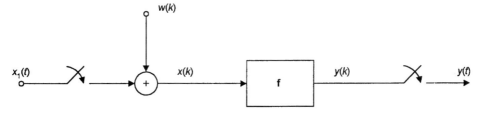

FIGURE 6.11   Block diagram for unfiltered recovery of a noisy differentiated signal.

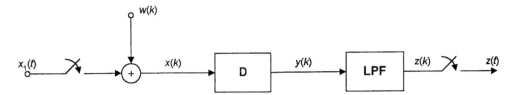

FIGURE 6.12   Block diagram for filtered recovery of a noisy differential signal.

steps is given in Figure 6.11, where block $\mathbf{f}$ is $\mathbf{D}^{-1}$ for the integration and $\mathbf{D}$ for differentiation.

(c) In order to recover the ideal derivative output, the choppy behavior of the derivative can be reduced by introducing the elementary low-pass filter described in Example 6.2. Thus, the block diagram is adjusted to that given in Figure 6.12. The derivative estimator, after being processed by an LPF using $\tau = 0.8$, is given in Figure 6.13. Even though the choppiness is still in evidence, the amplitude is far better than the unfiltered output, and the signal is close to that of the ideal The corresponding program is given in Listing 6.2.   ○

Example 6.3 showed the debilitating results of simply differentiating in a simulation. However, filtering helps by smoothing and averaging the signal. Still, the differentiated signal exhibits a phase-shift characteristic that is noisy. This can be helped either by using a centered-difference formula for the derivative such as those derived in Chapter 4 or by using a different filter. If a centered-difference equation is used, the derivative is estimated by not only past values but future values as well. This is fine if the independent variable is spatial rather than temporal, if functional values are known for all times a priori, or if the signal processing is done off-line rather than in real time.

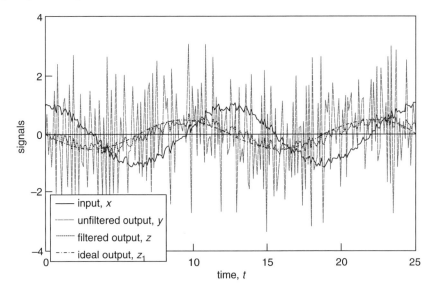

FIGURE 6.13   Effect of filtering a differentiating signal, Example 6.3(c).

```
t=0
x=1
z=0
print t, x, 0, y
for k=1 to n
        x₂=x
        x₁=cos(½t)
        w=√0.03(2*RND-1)
        x=x₁+w
        y=(x-x₂)/h
        z=z+h(y-z)/τ
        t=t+h
        print t, x, y, z
next k
```

LISTING 6.2   Filtered recovery of noisy differentiated signal simulation.

Other types of filters are also available, but their study is beyond the scope of this text. Clearly, filtering is a powerful method, but a word of warning regarding its use is in order. Filtering is only warranted in modeling engineered systems. Nature doesn't do filtering, so only if we are modeling a physical filter can a filter be part of the model. For instance, if an action results from the positiveness or negativeness of a signal, filtering is called for to pre-smooth the signal so that erratic spurious behavior is eliminated. This is often done in communications, timing, and control systems very effectively. Performance is enhanced by increasing the sampling rate significantly, and filtering to infer the signal's underlying intended behavior rather than its actual value.

On the other hand, filtering can also be effectively used in simulation outputs. Systems working with either real data or simulated real data will have a significant amount of noise incorporated into their signals. This means that the system outputs will also appear somewhat random and full of high-frequency fluctuation. By eliminating this high-frequency activity, the analysis of the simulation results is much more straightforward. After the reduction of the raw data into low-frequency effects, standard statistical time series analysis techniques are in order. Again, the rule is that simulations only filter at the outputs. Having understood this, let us give one further example to demonstrate the power of a low-pass filter on noisy output data.

EXAMPLE 6.4

Consider the system of Example 6.3 shown in Figure 6.12. Again the results will be numerically differentiated and passed through a low-pass filter. However, this time, let us create a non-white-noise source for $w(k)$.

(a) Use the autoregressive noise with uniformly distributed random shock of variance $\sigma^2 = 0.04$ described by

$$w(k) = \tfrac{4}{5}w(k-1) + \sqrt{0.12}(2 * \text{RND} - 1). \tag{6.6}$$

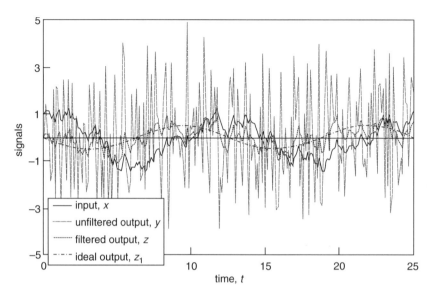

FIGURE 6.14   Effect of a low-pass filter on a differential autoregressive additive noise signal model, Example 6.4.

(b) Use the marginally stable ARIMA model with uniformly distributed random shock of variance $\sigma^2 = 0.01$ described by

$$w(k) = w(k-1) + \sqrt{0.03}(2\,\text{RND} - 1) \tag{6.7}$$

Solution

In case (a), this is an autoregressive (AR) process, and there is a pole at $z = -0.8$. Since this is within the unit circle, the process is stationary. The simulation proceeds exactly as in Listing 6.2, except that Equation (6.6) replaces line 8 and introduces an initializing statement prior to the loop. The results of the simulation are given in Figure 6.14. Notice once again that the filter helps considerably. However, with the highly correlated nature of the AR noise model, the ideal signal is less obvious than in Example 6.3. Evidently, the LPF is no panacea.

In case (b), there is a pole at $z = -1$, and the noise signal is therefore only marginally stationary. The results of this, shown in Figure 6.15, demonstrate that $w(k)$ tends to drift away from its mean at $w = 0$. This should be expected, since a non-stationary process will never appear to settle down, even over the long haul. However, the remarkable thing is that even though the input diverges from its ideal, the filtered signal follows the output ideal. In this sense, the low-pass filter seems to give us something for nothing – garbage in, along with the proper filter, can produce good results! Never underestimate the power of a filter – it not only smooths the signal, but introduces a significant measure of stability as well. ○

The delay and advance operators $\mathbf{Z}^{-1}$ and $\mathbf{Z}$ are defined only for discrete signals. The operator $\mathbf{Z}^{-1}$ acts as a memory element where the signal is stored for one time step. It

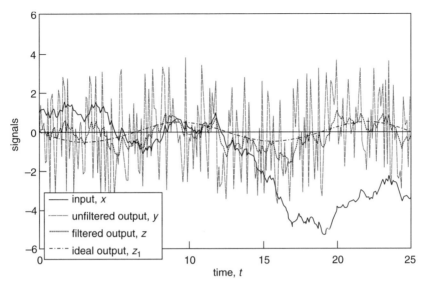

FIGURE 6.15   Efffect of a low-pass filter on a differentiated ARIMA additive noise signal model, Example 6.4.

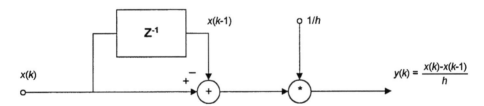

FIGURE 6.16   Estimating differentiation using difference equations.

follows from this that $\mathbf{Z}^{-n}$ is the application of n consecutive time delays with the signal being stored for n time steps. Thus, mathematically, the $\mathbf{Z}$ operator describes difference equations. From a modeling perspective, it is usually applied only after sampling of a continuous signal has taken place.

As was shown in the discussion on derivatives and integrals, signals that are sampled at the granularity level can perform integration and differentiation. Sampled signals, in the form of difference equations, can estimate integrals and derivatives as well. Figures 6.16 and 6.17 illustrate how each of these operations can be executed using the $\mathbf{Z}^{-1}$ operator.

6.3   DISCRETE DELAYS

The other special-purpose operators outlined in Figure 6.7 are delays. These are comparable to $\mathbf{Z}$ and $\mathbf{Z}^{-1}$, but are defined only in the continuous-time domain. Delays come in two primary varieties: discrete and distributed. The discrete delay is actually quite

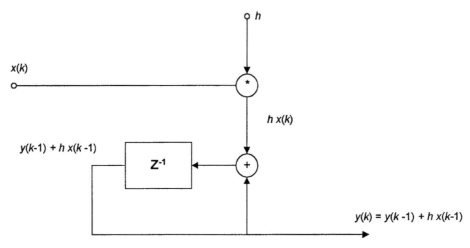

FIGURE 6.17   Estimating integration using difference equations.

simple. Discrete delays wait Δ time units to forward the data on to the output. Mathematically, this can be converted to a difference equation, which can be written in either continuous or discrete time as follows:

$$y(t) = x(t - \Delta),$$

$$y(k) = x\left(k - \frac{\Delta}{\delta}\right), \tag{6.8}$$

where $1/\delta$ is the sampling frequency. The two Equations (6.8) are equivalent, and their relationship is important. Often a problem will be stated using continuous time t, but in performing a simulation, the discrete time domain k is required.

EXAMPLE 6.5

Consider the system defined in Figure 6.18. Assuming a sampling frequency of two samples per second, generate the output for discrete delays (a) $\Delta = 0$ seconds and (b) $\Delta = 2$ seconds.

In the case of $\Delta = 2$ seconds, generate the response to an input of $x(k) = 1 + \sin \frac{1}{10} k$. Characterize this response and determine the *basin of attraction* for this nonlinear system.

Solution
(a) Since $\Delta = 0$, there is no delay. It follows that the nonlinear operator **A** is algebraic rather than dynamic. In this case, the output is square of the input–output difference:

$$y(t) = [x(t) - y(t)]^2. \tag{6.9}$$

Expansion gives $y^2(t) - [1 + 2x(t)]y(t) + x^2(t) = 0$. This quadratic equation has two roots, but we retain only the solution with a negative radical, since when

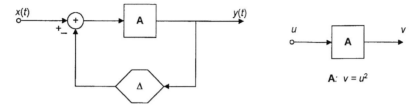

FIGURE 6.18   Nonlinear feedback system of Example 6.5.

$x(t) = 0$, one should expect the output $y(t) = 0$ from Equation (6.9):

$$y(t) = x(t) + \tfrac{1}{2} - \sqrt{x(t) + \tfrac{1}{4}}. \tag{6.10}$$

Equation (6.10) shows that there is a simple algebraic relationship between the input and output. The results are independent of both the sampling frequency δ^{-1} and the granularity h.

    (b) If $\Delta = 2$ seconds, the defining equation becomes dynamic, since the current output $y(t)$ depends on past outputs $y(t-2)$. Using continuous time t,

$$y(t) = [x(t) - y(t-2)]^2. \tag{6.11}$$

Since the sampling rate is two samples per second, $k = 2t$, and the discrete counterpart is

$$y(k) = [x(k) - y(k-4)]^2. \tag{6.12}$$

Notice that Equation (6.12) is a fourth-order nonlinear difference equation. Needless to say, an explicit solution is non-trivial, if not impossible, even if the input is known, deterministic, and given by an explicit formula. Also, Equations (6.11) and (6.12) are algebraically different.

    Although an explicit solution of Equation (6.12) is unrealistic, the output sequence can still be generated recursively. Since Equation (6.12) is of fourth order, four initial values ($k = 0, 1, 2, 3$) must be known a priori. If this were a linear system, the initial values would affect only the transient behavior of the output response, and the steady-state output would be virtually identical. However, since this system is nonlinear, there is the likelihood that different initial conditions will result in different steady-state behavior. Listing 6.3 is a program that produces the output response to the given input so that we can test this model.

    Executing this simulation gives different results for different initial conditions and different input functions $x(k)$. For instance, if the initial conditions are $y(0) = 1.877$, $y(1) = 2.014$, $y(2) = 1.417$ and $y(3) = 0.571$, the output is as given in Figure 6.19, assuming an input signal of $x(k) = 1 + \sin \tfrac{1}{10}k$. In fact, the

```
for k=0 to 3
        x(k)=1+sin(1/10 k)
        y(k)="given"
        print k, x(k), y(k)
next k
for k=4 to n
        y(k)=[x(k)-y(k-4)]²
        print k, x(k), y(k)
next k
```

LISTING 6.3   Simulation for Example 6.5.

same basic output results from all initial conditions satisfying the following criteria:

$$-0.486 \leqslant y(0) \leqslant 1.877,$$
$$-0.461 \leqslant y(1) \leqslant 2.014,$$
$$-0.348 \leqslant y(2) \leqslant 1.417,$$
$$-0.375 \leqslant y(3 \leqslant 0.571.$$

(6.13)

The response for any initial conditions outside the region specified by the inequalities (6.13) will be unbounded. The region defined by (6.13) is called the *basin of attraction*, since all initial conditions in this region exhibit the same steady-state behavior. Thus, the choice of initial conditions (sometimes called the *initial loading*) is extremely important in nonlinear systems.

Similarly, the system is extremely sensitive to the input function. Simply changing the input to $x(k) = 1 + \cos\frac{1}{10}k$ or $x(k) = 1.5 + \sin\frac{1}{10}k$ will create totally different regions of attraction and thus very different resulting outputs as well.   ○

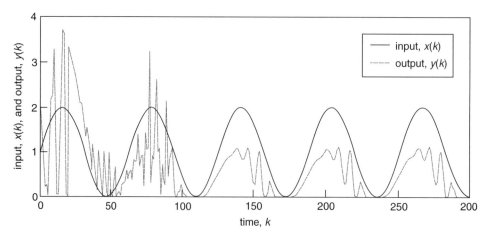

FIGURE 6.19   Simulation results, Example 6.5.

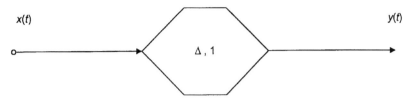

FIGURE 6.20   The single-stage distributed delay.

6.4   DISTRIBUTED DELAYS

Discrete delays occur instantaneously, so that their effects occur exactly Δ time units later. In contrast, distributed delays occur gradually over time. We begin with the simplest case, the *single-stage distributed delay* shown in Figure 6.20. The number "1" indicates there is only one stage of delay Δ.

By definition, the rate of change in the output of a distributed delay is proportional to the difference between input and output. The proportionality constant is $1/\Delta$. Mathematically,

$$\frac{d}{dx}y(t) = \frac{1}{\Delta}[x(t) - y(t)]. \tag{6.14}$$

If an explicit formula is given for the input $x(t)$, this first-order linear equation can be solved explicitly. On the other hand, if there is no formula, table look-ups and/or the noise modeling techniques discussed earlier for digitized or sampled data must be employed. In either case, the first-order differential equation requires a single initial condition, the initial state $y(t_0)$.

EXAMPLE 6.6

Consider a single-stage distributed delay with constant delay Δ. Assuming a zero initial state at $t = 0$ and a unit-step input, derive an explicit solution for the output $y(t)$.

Solution
Mathematically, the problem description reduces to

$$\frac{dy}{dt} = \frac{1}{\Delta}(1 - y), \quad t \geqslant 0,$$

$$y(0) = 0.$$

Equivalently, this can be written as

$$\dot{y} + \frac{1}{\Delta}y = \frac{1}{\Delta},$$

which has an integrating factor of $e^{t/\Delta}$. Therefore, multiplying both sides of this equation by $e^{t/\Delta}$ makes the left-hand side an elementary derivative:

$$e^{t/\Delta}\dot{y} + e^{t/\Delta}\frac{1}{\Delta}y = \frac{1}{\Delta}e^{t/\Delta},$$

$$\frac{d}{dt}[e^{t/\Delta}y(t)] = \frac{1}{\Delta}e^{t/\Delta}.$$

Integrating,

$$e^{t/\Delta}y(t) = e^{t/\Delta}|_0^t$$

$$= e^{t/\Delta} - 1,$$

and the solution is $y(t) = 1 - e^{-t/\Delta}$, for $t \geq 0$. While it is reassuring to achieve explicit results such as this, remember that input data such as the unit step assumed for this example are rare indeed.

One interpretation of the single-stage distributed delay is that of the simple *RC* electrical network of Figure 6.3. If a constant source voltage $x(t) = 1$ is applied, charge builds up on the capacitor. The output $y(t)$, which is the voltage across the capacitor, asymptotically approaches unity at a rate consistent with the time constant $\Delta = RC$.   ○

A *multiple-stage delay* is simply a *cascaded* sequence of single-stage delays, each with a proportional portion of delay time Δ. Specifically, an m-stage distributed delay with delay Δ will reduce to m single-stage distributed delays, each with delay Δ/m, as shown in Figure 6.21. Thus, an m-stage distributed delay reduces to m first-order differential equations with m initial conditions. Each delay block is specified by its state $r_i(t)$ for stages $i = 1, 2, \ldots, m$. Each state is also the output signal $r_i(t)$ of stage i and the input to stage $i + 1$, as shown in Figure 6.21. The last stage $r_m(t)$ is the output $y(t)$ of the distributed delay. Thus, each equation has an initial condition, and is defined as follows:

$$\dot{r}_1(t) = \frac{m}{\Delta}[x(t) - r_1(t)], \qquad\qquad r_1(t_0) = r_{10},$$

$$\dot{r}_2(t) = \frac{m}{\Delta}[r_1(t) - r_2(t)], \qquad\qquad r_2(t_0) = r_{20},$$

$$\vdots$$

$$\dot{y}(t) = \dot{r}_m(t) = \frac{m}{\Delta}[r_{m-1}(t) - r_m(t)], \qquad r_m(t_0) = r_{m0},$$

$$(6.15)$$

where r_{i0} are initial conditions at time $t = t_0$ for $i = 1, 2, \ldots, m$.

It is apparent from Equations (6.15) that the output $y(t)$ depends not only on the parameters Δ and m, but also on the initial conditions $r_i(t_0) = r_{i0}$, $i = 1, 2, \ldots, m$. This vector of initial conditions is called the *initial loading* of the distributed delay, and must be specified at the outset.

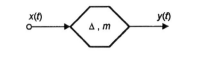

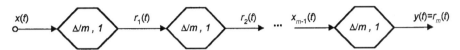

FIGURE 6.21　Equivalent multistage distributed delays.

EXAMPLE 6.7

Describe the unit-step response to the distributed delay shown in Figure 6.22. Assume zero initial loading at time $t = 0$.

Solution

It can be shown that an *m*-stage distributed delay with time delay Δ, unit-step input, and zero initial load has output

$$y(t) = 1 - e^{-mt/\Delta} \sum_{i=0}^{m-1} \left(\frac{m}{\Delta}\right)^i \frac{t^i}{i!}. \tag{6.16}$$

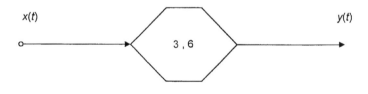

input:　　　　　　　　　　$x(t)=1$

initial loading:　　　　　　$r_1(0)=0$
　　　　　　　　　　　　　$r_2(0)=0$
　　　　　　　　　　　　　$r_3(0)=0$
　　　　　　　　　　　　　$r_4(0)=0$
　　　　　　　　　　　　　$r_5(0)=0$
　　　　　　　　　　　　　$y(0)=r_6(0)=0$

FIGURE 6.22　Multistage distributed delay for Example 6.7.

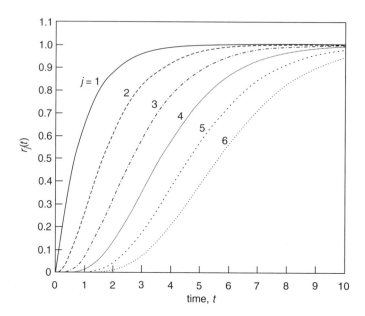

FIGURE 6.23   The staging of Equation (6.17), assuming a unit-step input with a zero initial loading.

Substituting $m = 1$ reduces Equation (6.16) to the single-stage solution discussed in Example 6.6. Notice once again the exponential decay characteristic of charged capacitors.

It can also be shown that the output $r_j(t)$ at stage j of an m-stage delay is given by

$$r_j(t) = 1 - e^{-mt/\Delta} \sum_{i=0}^{j-1} \left(\frac{m}{\Delta}\right)^i \frac{t^i}{i!}.$$ (6.17)

Equation (6.17) is actually a generalization of Equation (6.16) with $j = m$, recalling that $y(t) = r_m(t)$.

Equation (6.17) is graphed for $j = 1, 2, \dots, 6$ in Figure 6.23. It shows that each signal propagates through successive stages; in stage 1 as an exponential decay, then succeeding states in a progression of sigmoid functions, each approaching unity asymptotically. This is sometimes called the *staging effect*.

Other than the simplest of systems with idealistic input, such as the unit-step response considered in Example 6.7, Equations (6.15) cannot be solved analytically. In general, their solution will require numerical assistance. For instance, using the Euler integration

method, the difference equations are

$$r_1(k+1) = r_1(k) + \frac{hm}{\Delta}[x(k) - r_1(k)],$$

$$r_2(k+1) = r_2(k) + \frac{hm}{\Delta}[r_1(k) - r_2(k)], \qquad (6.18)$$

$$\vdots$$

$$y(k+1) = r_m(k+1) = r_m(k) + \frac{hm}{\Delta}[r_{m-1}(k) - r_m(k)],$$

with initial conditions

$$r_1(0) = r_{10},$$
$$r_2(0) = r_{20},$$

$$\vdots$$

$$r_m(0) = r_{m0},$$

where $r_{10}, r_{20}, \ldots, r_{m0}$ are constants. These can be implemented algorithmically, as demonstrated in the following example.

EXAMPLE 6.8

Produce the output of the distributed delay system shown in Figure 6.24 using input $x(t) = \sin\frac{1}{2}t$ and a "back-end load" of 2 units. Integrate using a step size $h = 0.1$ for $0 \leqslant t \leqslant 6$. By *back-end load* we mean that the entire initial 2-unit load is in the first stage. Similarly, a *front-end load* would imply that the entire non-zero load was in the last stage.

FIGURE 6.24 Defining system for Example 6.8.

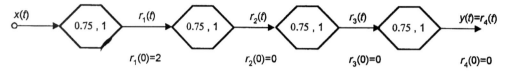

FIGURE 6.25   System of single-stage distributed delays equivalent to that shown in Figure 6.24.

```
t=0
y=0
r₃=0
r₂=0
r₁=2
x=sin(½t)
print t, x, r₁, r₂, r₃, y
for k=1 to n
        t=t+h
        y=y+⁴⁄₃h(r₃-y)
        r₃=r₃+⁴⁄₃h(r₂-r₃)
        r₂=r₂+⁴⁄₃h(r₁-r₂)
        r₁=r₁+⁴⁄₃h(x-r₁)
        x=sin(½t)
        print t, x, r₁, r₂, r₃, y
next k
```

LISTING 6.4   Computer simulation of Example 6.8.

Solution

The distributed delay system described above is equivalent to that shown in Figure 6.25 using only first-order distributed delays. A program segment solving this system using Euler is given in Listing 6.4. Notice in the listing that care has been taken to write the Euler update equations "backwards", that is, by following the system blocks against the system flow. This practice removes the need for intermediate or subscripted variables, and is a good principle to follow in general. The results of this simulation are shown in Figure 6.26, where the input and output as well as the results at intermediate stages are presented. Clearly, the results of the initial load are to gradually "push" the initial load 2 through the system, then settle down to a distorted sinusoid. This is because the input is a sinusoid, and since the distributed delay is a linear system, the output will always tend to follow the input signal.   ○

The initial loading of a delay is important during the transient phase of system simulation. Since in a good many simulations this is the critical aspect, care must be taken in the initial prescription. Often, the total load of a distributed delay is immediately obvious, but the actual distribution is not. For instance, each stage might share the load equally, in which case the initial loading distribution is said to be *uniform*. The distribution

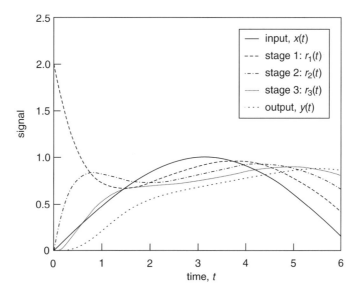

FIGURE 6.26   Simulation results, Example 6.8.

is *geometric* when the initial ratio between the loading of adjacent stages is constant. Similarly, the distribution is *arithmetic* when the initial difference between the loading of adjacent stages is constant.

For instance, suppose there are $m = 4$ stages in a distributed delay that has an initial load of 30 units. If the initial loading is distributed as $r_1(0) = 7.5$, $r_2(0) = 7.5$, $r_3(0) = 7.5$, $y(0) = r_4(0) = 7.5$, the loading is uniform. If the loading is $r_1(0) = 3$, $r_2(0) = 6$, $r_3(0) = 9$, $y(0) = r_4(0) = 12$, it is arithmetic. On the other hand, if the loading is $r_1(0) = 2$, $r_2(0) = 4$, $r_3(0) = 8$, $y(0) = 16$, the loading is geometric. A front-end load gives $r_1(0) = 0$, $r_2(0) = 0$, $r_3(0) = 0$, $y(0) = r_4(0) = 30$ while a back-end load gives $r_1(0) = 30$, $r_2(0) = 0$, $r_3(0) = 0$, $y(0) = r_4(0) = 0$. Clearly other possibilities exist as well.

Suppose that the initial loading is L units and that it is distributed across m stages as indicated below. Each of the protocols described above is characterized in Table 6.1. While the back-end, front-end, and uniform protocols are straightforward, the arithmetic and geometric loading regimens require a bit more explanation. The sums of arithmetic and geometric progressions are well known, and are indicated in the "total" row. For the arithmetic loading, it remains to find the loading a and difference d such that there are a total of L units. Thus, specifying either a or d leaves the other to be determined algebraically. For instance, assuming that a is specified,

$$d = \frac{2(L - ma)}{m(m - 1)}.$$

Similarly, for a geometric loading, and assuming that the ratio r is specified,

$$a = \frac{L(1 - r)}{1 - r^m}.$$

TABLE 6.1   Common Initial Loading Protocols for Distributed Delay

Stage	Back-end	Front-end	Uniform	Arithmetic	Geometric
1	L	0	L/m	a	a
2	0	0	L/m	$a+d$	ar
3	0	0	L/m	$a+2d$	ar^2
$\vdots$	$\vdots$	$\vdots$	$\vdots$	$\vdots$	$\vdots$
m	0	L	L/m	$a+(m-1)d$	ar^{m-1}
Total	L	L	L	$L = ma + \frac{1}{2}dm(m-1)$	$L = a\dfrac{1-r^m}{1-r}$

Thus, there is more freedom in the geometric and arithmetic loading protocols, but their application is straightforward, nonetheless.

The integration step size or *granularity* is a simulation parameter, because it depends totally on the simulation and not on the system model. Even so, its choice is critical for achieving accurate results. It can be shown that for linear systems, there is a theoretical requirement that h must be chosen so that $h < 2\Delta/m$ in order to ensure numerical stability. As a rule of thumb, we usually make an even more restrictive choice:

$$h < \frac{\Delta}{4m}. \qquad (6.19)$$

If there is more than one distributed delay, the step size is taken as the minimum h satisfying the requirement (6.17). Thus, when in doubt, the granularity is taken to be the most restrictive integration step size. Regardless of the step size chosen, it is almost always necessary to make extensive simulation studies to ensure sufficient accuracy.

6.5   SYSTEM INTEGRATION

It is useful to model more than just physically engineered systems. Naturally occurring systems exist in the life, social, and physical sciences as well. To demonstrate the use of the concepts defined so far, we will illustrate with an elementary biological system.

To begin, consider a closed (i.e., no exogenous input) system in which there are three types of population: *adults*, which are individuals able to reproduce, *young*, which are developed but not able to reproduce and *immature* individuals that are not yet developed enough to function. It is required to find the populations of adults $a(t)$ and youths $y(t)$ and the total $T(t)$ as functions of time.

In this contrived system, we assume a two-month distributed delay from the initial egg stage to functioning (but unable to reproduce) young. The delay is spread over three stages of equal length (eggs, larvae, and pupae). When the young finally do emerge, 60% die or are eaten. For those youths who survive, there are two stages (childhood and adolescence) to be negotiated before reaching reproductive maturity as adults. Once they have matured, the adults accumulate in a pool, and 20% die. Of those that live, eggs are produced at the rate of five per adult.

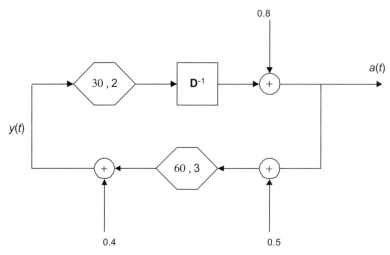

FIGURE 6.27   Simulation diagram of simplified population model – a first cut.

Certainly, it is not difficult to argue the over-simplification of this system. However, assuming its accuracy, Figure 6.27 models the above paragraph. For this problem, let us further assume that there are initially 6 mature adults, 6 youths (children and adolescents), and 6 pre-youth (eggs, larvea and pupae) uniformly distributed over each stage. It is our goal to perform a simulation to determine the number of adults and youths and the total population as functions of time. In performing the simulation, let us assume a granularity $h = 5$ days.

EXAMPLE 6.9

Let us begin by assuming the delays are discrete rather than distributed. To do this, first create a dynamical mathematical description using Figure 6.27. Next, deduce the initial conditions. Finally, generate the output sequence for $a(t)$, $y(t)$, and $T(t)$ for each $h = 5$ days.

Solution
Figure 6.27 is redrawn assuming discrete delays and including intermediate signals as Figure 6.28. Since the delays are 30 and 60 days in continuous time, the basic dynamics are described by the following equations:

$$y(t) = \frac{2}{5}c_4(t),$$
$$c_4(t) = c_3(t - 60),$$
$$c_3(t) = \frac{1}{2}a(t),$$
$$a(t) = \frac{4}{5}c_2(t),$$
$$c_2(t) = c_2(t - h) + hc_1(t - h),$$
$$c_1(t) = y(t - 30),$$
$$T(t) = a(t) + y(t).$$

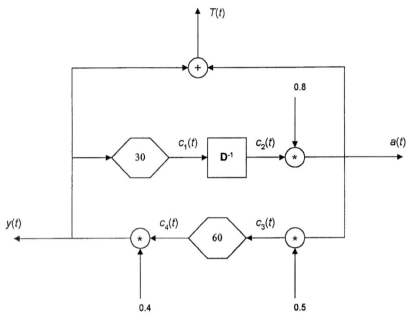

FIGURE 6.28   Simulation diagram of simplified population model, assuming discrete delays.

From the continuous-time equations, it is necessary to convert to the discrete domain. Since $h = 5$ days, the delays of 30 and 60 days are specified as 6 and 12 step-size units respectively. Notice how the integrator is handled to accumulate the adult population in that $h = 5$ in the fifth equation below. Thus, the difference equations are

$$
\begin{aligned}
y(k) &= \tfrac{2}{5} c_4(k), \\
c_4(k) &= c_3(k - 12), \\
c_3(k) &= \tfrac{1}{2} a(k), \\
a(k) &= \tfrac{4}{5} c_2(k), \\
c_2(k) &= c_2(k - 1) + 5c_1(k - 1), \\
c_1(k) &= y(k - 6), \\
T(k) &= a(k) + y(k).
\end{aligned} \tag{6.20}
$$

It is straightforward to eliminate the auxiliary variables c_1, c_2, c_3, and c_4 and reduce Equations (6.20) to the following two dynamic and one algebraic equations:

$$
\begin{aligned}
y(k) &= 2a(k - 12), \\
a(k) &= a(k - 1) + 8y(k - 13), \\
T(k) &= a(k) + y(k).
\end{aligned} \tag{6.21}
$$

Notice that the largest difference in the arguments is 13 and that the dynamic equations involve both state variables a and y. Thus, the model described by Figure 6.27 is actually a 13th-order system of coupled difference equations. Since this is an autonomous system (there are no exogenous inputs), and it is linear, an analytical solution can, in principle, be found if the initial conditions are given.

However, interpreting the problem statement regarding initial conditions in terms of Equation (6.21) is not easy. This is because the original problem statement describes the population distribution of the intermediate variables defined by the delays rather than the output variables y, a, and T described by Equation (6.21). As with all initial-value problems involving difference equations, initial conditions are incorporated by buffering up a pre-sequence involving non-positive times before $k = 1$. Recall from Chapter 2 that this is in contrast to differential equations, where derivative values are specified only at continuous time $t = 0$.

It is useful to visualize the variables in the form of a spreadsheet for non-positive times $k = 0, -1, -2, \ldots, -12$ as shown in Table 6.2. Notice that the dynamic equations (those specifying c_1, c_2, and c_4) are all defined in terms of previous input signals. Thus, the input signals (c_3, y, and a) must be pre-loaded so as to define the required buffering. It follows that in establishing initial conditions for differencing systems, one should pre-load non-state (input) signals for negative times $k < 0$ and initiate the state variables only at time $k = 0$. Since the output of the immature pre-youth delay is specified by the variable c_4 in terms of c_3, and the state of that delay is uniform over the 60-day period, all 12 buffer positions of c_3 must be 0.5. This is shown in column 3 of Table 6.2. Similarly, c_1 requires 6 buffer positions for variable y, as shown in column 5.

Since the accumulator is defined in terms of the variable c_1 and itself, it is initialized in column 6 at time zero. The value 7.5 was derived from the specification that 6 adults exist at the outset ($0.8 \times 7.5 = 6$). Other than columns 3 and 5, no others require pre-loading. However, at the initial time $k = 0$, the state

TABLE 6.2   State Pre-Loading Assuming Discrete Delays, Example 6.9

k	t	c_3	c_4	y	c_1	c_2	a	T
−12	−60	0.5						
−11	−55	0.5						
−10	−50	0.5						
−9	−45	0.5						
−8	−40	0.5						
−7	−35	0.5						
−6	−30	0.5		1.0				
−5	−25	0.5		1.0				
−4	−20	0.5		1.0				
−3	−15	0.5		1.0				
−2	−10	0.5		1.0				
−1	−5	0.5		1.0				
0	0	3.0	0.5	0.4	0.16	7.5	6.0	6.4

TABLE 6.3   The First Few Simulation Results for Population Models Using Discrete Delays, Example 6.9

k	t	c_3	c_4	y	c_1	c_2	a	T
0	0	3.00	0.5	0.4	0.16	7.5	6.0	6.4
1	5	3.32	0.5	0.4	0.16	8.3	6.6	7.0
2	10	3.64	0.5	0.4	0.16	9.1	7.3	7.7
3	15	3.96	0.5	0.4	0.16	9.9	7.9	8.3
4	20	4.28	0.5	0.4	0.16	10.7	8.6	9.0
5	25	4.60	0.5	0.4	0.16	11.5	9.2	9.6

variables (c_1, c_2, and c_4) are defined in terms of the pre-loaded variables, and the output variables (y, a, and T) are defined algebraically in terms of the state variables. Thus, the pre-loading and initializing conditions are shown in Table 6.2.

In fact, the spreadsheet idea is useful for more than establishing pre-loading and initializing conditions. While a traditional procedural program, such as those shown throughout the majority of this text, can easily be written, a spreadsheet program can do just as well. The results of such a spreadsheet for $k \geqslant 0$ are shown in Table 6.3. Each cell is defined by the recursive equations (6.20) and the pre-loading shown in Table 6.2. Of course, Tables 6.2 and 6.3 are usually combined and a graph created. The graphical results of this simulation are shown in Figure 6.29.

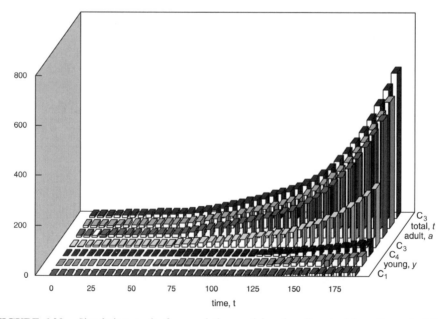

FIGURE 6.29   Simulation results for population models using discrete delays, Example 6.9.

EXAMPLE 6.10

Reconsider the system described in Example 6.9, but this time assume distributed delays. Write a program to simulate the operation of the system described.

Solution
Figure 6.27 is redrawn decomposing each of the multiple-stage distributed delays into single stages as shown in Figure 6.30. Intermediate signals are included.

Taking our cue from Example 6.8, we traverse the loop "backward" so as to eliminate the need for subscripted variables in the simulation. The following sets of differential and algebraic equations then describe this system:

$$\begin{aligned}
\dot{q}(t) &= r_5(t), \\
\dot{r}_5(t) &= \tfrac{1}{15}[r_4(t) - r_5(t)], \\
\dot{r}_4(t) &= \tfrac{1}{15}[y(t) - r_4(t)], \\
\dot{r}_3(t) &= \tfrac{1}{20}[r_2(t) - r_3(t)], \\
\dot{r}_2(t) &= \tfrac{1}{20}[r_1(t) - r_2(t)], \\
\dot{r}_1(t) &= \tfrac{1}{20}[p(t) - r_1(t)];
\end{aligned} \tag{6.22}$$

$$\begin{aligned}
a(t) &= \tfrac{4}{5}q(t), \\
y(t) &= \tfrac{2}{5}r_3(t), \\
\alpha(t) &= \tfrac{1}{5}a(t), \\
T(t) &= r_4(t) + r_5(t) + a(t).
\end{aligned} \tag{6.23}$$

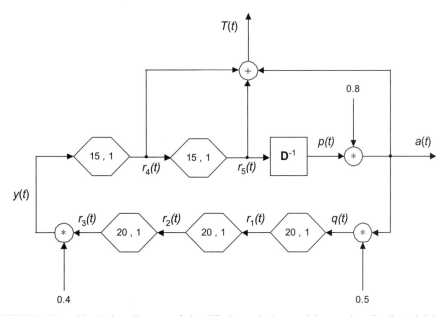

FIGURE 6.30   Simulation diagram of simplified population model assuming distributed delays.

Equations (6.22) and (6.23) can be reduced to the following set of six dynamic and one algebraic equations that fully describe the system:

$$\dot{a}(t) = \tfrac{4}{5} r_5(t),$$
$$\dot{r}_5(t) = \tfrac{1}{15}[r_4(t) - r_5(t)],$$
$$\dot{r}_4(t) = \tfrac{1}{15}[y(t) - r_4(t)],$$
$$\dot{y}(t) = \tfrac{1}{30} r_2(t) - \tfrac{1}{12} y(t), \qquad\qquad (6.24)$$
$$\dot{r}_2(t) = \tfrac{1}{20}[r_1(t) - r_2(t)],$$
$$\dot{r}_1(t) = \tfrac{1}{4} a(t) - \tfrac{1}{20} r_1(t),$$
$$T(t) = r_4(t) + r_5(t) + a(t).$$

Notice that the order of the system is therefore reduced from 13 in Example 6.9 to 6 here. Unlike the case with discrete delays, applying initial conditions to distributed delays is much more straightforward. There are a number of possibilities, but let us use variables $a, y, r_1, r_2, r_4,$ and r_5 to define the system state. The adults must begin with a population of 6, so $z(0) = 6$ is easy. Similarly, there must be 6 children and adolescents, so r_4 and r_5 must split the difference. However, the immature must also total 6. Distributing them evenly gives 2 to y and 5 to each of r_2 and r_1, since their numbers are reduced by 0.4 on entry into the "youth" world. Of course, there might also be another interpretation of this problem specification too. Are the 6 pre-youths 6 organisms yet to be eaten or are they 15 organisms but effectively 6 as specified? These questions are the essence of systems analysis and the basis for the detail necessary for rigorous simulation studies. Thus, we choose the following initial conditions:

$$a(0) = 6,$$
$$r_5(0) = 3,$$
$$r_4(0) = 3,$$
$$y(0) = 2, \qquad\qquad (6.25)$$
$$r_2(0) = 5,$$
$$r_1(0) = 5.$$

The differential equations (6.24) can be approximated by the Euler technique for a suitable step size. Using the rule of thumb given by Equation (6.19), we initially choose

$$h = \min\left(\frac{1}{4 \cdot 2}, \frac{2}{4 \cdot 3}\right) = \tfrac{1}{8}.$$

This choice guarantees numerical stability, but it does not guarantee accurate results. A quick study reveals that an integration step of $h = 0.01$ is much more reasonable.

A program segment performing the indicated integration is shown in Listing 6.5. Note that coding the dynamic variables against the system flow avoids the use of subscripted variables. The results are given in Figure 6.31.   ○

```
t=0
a=6
r₅=3
r₄=3
y=2
r₂=5
r₁=5
print t, r₄+r₅, a, r₄+r₅+a
for k=1 to n
        t=t+h
        a=⁴⁄₅q
        q=q+hr₅
        r₅=r₅+¹⁄₁₅h(r₄-r₅)
        r₄=r₄+¹⁄₁₅h(r₃-r₄)
        y=²⁄₅r₃
        r₃=r₃+¹⁄₂₀h(r₂-r₃)
        r₂=r₂+¹⁄₂₀h(r₁-r₂)
        r₁=r₁+h(¼a-¹⁄₂₀r₁)
        print t, r₄+r₅, a, r₄+r₅+a
next k
```

LISTING 6.5   Simulation of population model assuming distributed delays per Example 6.10.

6.6   LINEAR SYSTEMS

It is possible to generalize the concept of a delay. An mth-order distributed delay is a special case of an mth-order linear differential equation, and a discrete delay is a special case of a linear difference equation. Linear systems are ubiquitous, especially in engineering, where engineers strive to design systems as such. Historically, there has been a concerted effort to model systems linearly, if this is at all possible. Even when a system is decidedly nonlinear, the preferred method of attack has been to linearize it over a region and simply consider several special cases to achieve a reasonable approximation to reality. The computer, of course, has changed all this. It is no longer necessary to simply "overlook" difficulties. Rather, the potential for intensive numerical work nowadays has lead to a revolutionary understanding of system phenomena (e.g., chaos and catastrophe theory). Even so, many systems are inherently linear or are close to linear. In either case, they can be modeled linearly anyway.

Recall that a linear system is one in which the *superposition principle* holds. Superposition states that $y(t) = f(x_1, x_2) = f(x_1, 0) + f(0, x_2)$, for system inputs x_1 and x_2 and output y, as shown in Figure 5.20. That is to say, the output of a system can be computed by finding the response to each of the individual inputs, then simply summing the total.

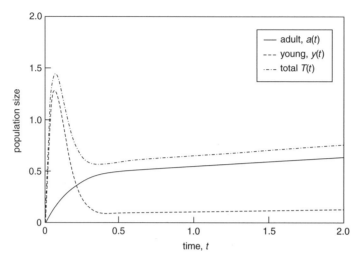

FIGURE 6.31 Simulation results for the simplified population model, Example 6.10.

It is well known that linear differential equations satisfy the superposition principle. In fact, that is why we call them linear. Differential equations are sometimes written using the "Big-**D**" notation, where $\mathbf{D} = d/dt$ is the derivative operator (usually with respect to time). This is convenient, since the derivative is itself linear, amd **D** behaves algebraically very much like a variable. For example, some of the properties of **D** are as follows:

$$\mathbf{D}[x(t)] = \frac{d}{dt}x(t),$$

$$\mathbf{D}^{-1}[x(t)] = \frac{1}{\mathbf{D}}x[t] = x(0) + \int_{t_0}^{t} x(t)\, dt,$$

$$\mathbf{D}[x(t) + y(t)] = \mathbf{D}[x(t)] + \mathbf{D}[y(t)],$$

$$\mathbf{D}[\alpha x(t)] = \alpha \mathbf{D}[x(t)],$$

$$\mathbf{D}^{-1}[\mathbf{D}[x(t)]] = \mathbf{D}[\mathbf{D}^{-1}[x(t)]] = x(t),$$

$$\mathbf{D}^{n}[x(t)] = \frac{d^n}{dt^n}x(t).$$

(6.26)

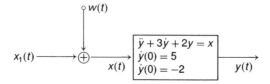

FIGURE 6.32 System block diagram for Example 6.11.

These properties lead to the very important concept of a *transfer function*. By definition, the transfer function of a (sub)system is defined as the ratio of output to input. If the system is linear, the transfer function itself is independent of the input and output. This is best described by an example.

EXAMPLE 6.11

Consider a system block defined by the block diagram in figure 6.32 where $x(t)$ is the input and $y(t)$ is the output. Find the transfer function and numerically solve the system subject to the initial conditions $y(0) = 5$ and $\dot{y}(0) = -2$. Also assume the input to be a ramp function $x_1(t) = \frac{1}{2}t$ superimposed on Gaussian white noise of variance $\sigma^2 = \frac{1}{4}$.

Solution
The input signal is evidently an additive noise model with $x_1(t) = \frac{1}{2}t$ and $w(t) = \frac{1}{2}\sqrt{-2\ln(\text{RND})}\cos(2\pi\,\text{RND})$. Therefore,

$$x(t) = \frac{1}{2}t + \frac{1}{2}\sqrt{-2\ln(\text{RND})}\cos(2\pi\,\text{RND})$$

The system is also linear, so it can be described using the "Big-**D**" notation described above. Its transfer function follows immediately:

$$x(t) = \mathbf{D}^2 y + 3\mathbf{D}y + 2y$$
$$= (\mathbf{D}^2 + 3\mathbf{D} + 2)y(t)$$
$$= (\mathbf{D} + 1)(\mathbf{D} + 2)y(t).$$

Thus, the transfer function is

$$\frac{y(t)}{x(t)} = \frac{1}{\mathbf{D}^2 + 3\mathbf{D} + 2}$$
$$= \frac{1}{(\mathbf{D} + 1)(\mathbf{D} + 2)}$$
$$= \frac{1}{\mathbf{D} + 1} - \frac{1}{\mathbf{D} + 2}.$$

Diagrammatically, this system can be represented in any of the ways shown in Figure 6.33, where it should be noted that the block description (transfer function) is independent of both x and y. Since the transfer function can be factored, the single block can be *decomposed* into two blocks, each representing a first-order differential equation as opposed to the original second-order equation. Decomposition can be handled either as two blocks in *cascade* (serially, one after the other) for products, or as two blocks in *parallel* for sums or differences. These options are shown in Figures 6.33(b) and 6.33(c), respectively. Each diagram gives a different, but equivalent approach to the system solution.

Method 1
Let us first solve this system without factoring the transfer function. To do this, first define an intermediate variable $r(t) = \dot{y}(t)$. Then,

$$\dot{r}(t) + 3r(t) + 2y(t) = x(t),$$
$$\dot{y}(t) = r(t).$$

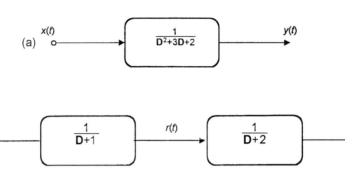

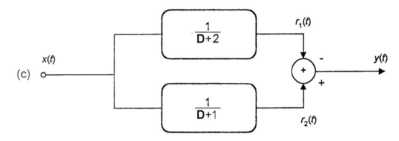

FIGURE 6.33   Equivalent decompositions of the transfer function of Example 6.11.

Discretizing via Euler,

$$r(k + 1) = r(k) + h[x(k) - 3r(k) - 2y(k)],$$
$$y(k + 1) = y(k) + hr(k).$$

In an actual implementation, note that neither equation can be updated without affecting the other. Therefore, the use of an array or intermediate variables will be necessary.

The initial conditions follow immediately from the definition of variable $r(t)$: $r(0) = \dot{y}(0) = -2$ and $y(0) = 5$.

Method 2
By factoring the transfer function first and defining the intermediate variable $r(t)$ as shown in Figure 6.33(b),

$$\dot{r}(t) + r(t) = x(t),$$
$$\dot{y}(t) + 2y(t) = r(t).$$

Since the intermediate variable $r(t)$ is defined differently, its initial value is too:

$$r(0) = \dot{y}(0) + 2y(0) = 5 + 2(-2) = 1$$

```
t=0
x=½t+½√(-2 * ln(RND)) cos(2π*RND)
r=1
y=5
print t, x, r, y
for k=1 to n
        t=t+h
        x=½t+√(-2 * ln(RND))cos(2π*RND)
        y=y+h(x-r)
        r=r+h(x-r)
        print t, x, r, y
next k
```

LISTING 6.6   Simulation using cascade decomposition.

The update equations follow from Euler:

$$r(k + 1) = r(k) + h[x(k) - r(k)],$$
$$y(k + 1) = y(k) + h[r(k) - 2y(k)].$$

Here we note that the "back-to-front" rule for updating the difference equations works as before. After establishing suitable initial conditions, the system solution is straightforward; a partial listing is shown in Listing 6.6.

The simulation results are shown in Figure 6.34. Notice how the output is smooth, even though the system is driven by a very noisy input. Also, as is always

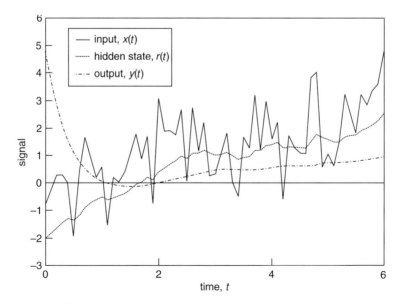

FIGURE 6.34   Simulation results using cascade decomposition, Example 6.11.

the case for linear systems, the output tries to follow the input. Even though they never catch it, both the output and the hidden state signals tend toward $x(t)$.

Method 3

Linear systems described by differential equations will always have rational polynomial transfer functions. If the denominator can be factored, the cascade approach of method 2 is most appealing, but, once it has been factored, it can also be reduced to the sum of lower-degree rational polynomials by partial fractions or simple inspection. This leads to the block diagram shown in Figure 6.33(c), illustrating a parallel decomposition.

Using the intermediate variables $r_1(t)$ and $r_2(t)$ as defined in Figure 6.33(c), it follows immediately that

$$\dot{r}_1(t) + r_1(t) = x(t),$$
$$\dot{r}_2(t) + 2r_2(t) = x(t),$$
$$y(t) = r_1(t) + r_2(t).$$

Using the Euler procedure,

$$r_1(k+1) = r_2(k) + h - [x(k) - r_1(k)],$$
$$r_2(k+1) = r_2(k) + h[x(k) - 2r_2(k)], \qquad (6.27)$$
$$y(k) = r_1(k) + r_2(k).$$

Equations (6.27) clearly show the advantage of this decomposition method to be that the equations are totally decoupled. That is, $r_1(k)$ and $r_2(k)$ can be solved independently of each other, then simply added or subtracted to produce output $y(k)$.   ○

EXAMPLE 6.12

Consider the system defined in Figure 6.35 assuming the initial state is $y(0) = 5$. Further assume the same input signal as Exmaple 6.11. Solve this system numerically using Euler's technique.

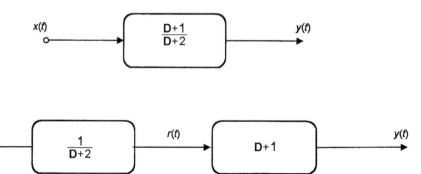

FIGURE 6.35   Functional decomposition, Example 6.12.

Solution
We begin by noting that the transfer function can be decomposed into two blocks in cascade with intermediate variable $r(t)$ as shown in Figure 6.33(b). Noting that $r(t)$ is both the output of the first block and the input to the second block, it follows that

$$\dot{r}(t) + 2r(t) = x(t), \tag{6.28a}$$

$$\dot{r}(t) + r(t) = y(t) \tag{6.28b}$$

Since Equation (6.28a) possesses derivative of the output but not the input, it is a normal differential equation. However, Equation (6.28b) has derivatives of the input and not the output; thus making it in reality an *integral equation*. Therefore, we use a backward-difference formula (Chapter 4) for the derivative:

$$\dot{r}(t) \approx \frac{r(k) - r(k-1)}{h}.$$

This leads to the following difference equations:

$$r(k+1) = r(k) + h[x(k) - 2r(k)],$$

$$y(k+1) = \frac{r(k+1) - r(k)}{h} + r(k),$$

where the first is Euler's equation and the second follows from the backward-difference formula. After substituting the expression for $r(k+1)$ into that for $y(k+1)$,

$$r(k+1) = r(k) + h[x(k) - 2r(k)],$$
$$y(k+1) = x(k) - r(k).$$

In this form, updates are easily performed, since the left-hand sides are the new values (at time $k+1$) of variables on the right-hand sides, which are defined at the current time k. Note that only one initial condition is required, since there is only one equation defined by the transfer function. An appropriate program segment is given in Listing 6.7.

The results of this simulation are shown in Figure 6.36. Here we note that, unlike Example 6.11, the state variable $r(t)$ is just as noisy as the input $x(t)$. Still the output $y(t)$ tends to follow the input. This is because the transfer function is a quotient of two first-order operators. Thus, they approximately cancel out. In fact, all that is left is a residual transient effect. In other words, there is in effect only one state variable, $y(t)$, rather than two as illustrated in Example 6.11.   O

Even though the transfer function approach to system modeling and simulation was illustrated only by example, it is very general. There is even an algebra of blocks by which the rules of transfer functions can be stated graphically, yet at the same time retain their rigor. They are also similar to the **Z**-operators defined for discrete variables as described in Chapter 5.

```
t=0
x=½t+½√‾‾‾‾‾‾‾‾‾‾‾‾‾-2 * ln(RND)cos(2π*RND)
r=-2
y=5
print t, x, r, y
for k=1 to n
        t=t+h
        x=½t+½√‾‾‾‾‾‾‾‾‾‾‾‾‾-2 * ln(RND)*cos(2π*RND)
        y=x-r
        r=r+h(x-2r)
        print t, x, r, y
next k
```

LISTING 6.7   Simulation for linear system of Example 6.12.

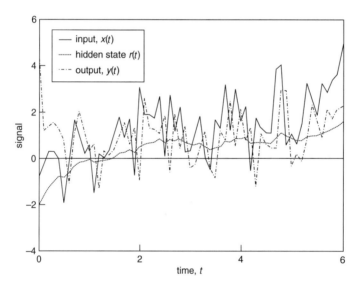

FIGURE 6.36   Simulation results, Example 6.12.

6.7   MOTION CONTROL MODELS

The design of control systems constitutes a major application area of engineering modeling endeavors. Of these, *motion control* is of major importance. At its center is the common DC servomotor, which is an industry workhorse used in feedback control systems. The DC motor is a transducer in the sense that it changes electrical energy (in the form of a current) into mechanical energy (in the form of a motion). A motor system is illustrated in Figure 6.37.

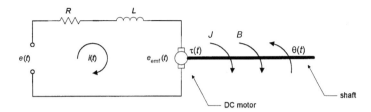

FIGURE 6.37   Cartoon of simplified motion control system with DC drive.

Figure 6.37 shows a driving voltage $e(t)$, which creates a current $i(t)$ through a series circuit consisting of a resistance R and the inductance L of the motor windings. Motion of the motor creates a torque $\tau(t)$ that is resisted by the inertial mass (characterized by the moment of inertia J) and frictional force (characterized by the friction coefficient B). The motion results in angular velocity $\omega = \dot{\theta}$, where θ is the displacement angle. This scenario can be modeled mathematically by the following equations:

$$\tau(t) = K_1 i(t), \tag{6.29a}$$

$$e_{\text{emf}}(t) = K_2 \omega(t), \tag{6.29b}$$

$$e(t) = Ri(t) + L\frac{di}{dt} + e_{\text{emf}}(t), \tag{6.29c}$$

$$\tau(t) = J\frac{d\omega}{dt} + B\omega(t), \tag{6.29d}$$

$$\omega(t) = \frac{d\theta}{dt}. \tag{6.29e}$$

The constants K_1 and K_2 are also characteristics of the specific motor design, and are referred to as the *torque* and *back-emf* constants, respectively. The back-emf $e_{\text{emf}}(t)$ is a "frictional" voltage created by the motor, and acts proportionally to the speed, but in a negative sense. It follows from Equations (6.29a) and (6.29b) that a DC motor's speed is proportional to the voltage, and the torque is proportional to the current. Thus, the voltage and current are two state variables used to control motion systems.

Bearing in mind which quantities are constants (K_1, K_2, R, L, J, and B) and which are variables (τ, i, e_{emf}, ω, e, and θ), the transfer functions resulting from each equation are straightforward. These are as follows:

$$\frac{\tau(t)}{i(t)} = K_1,$$

$$\frac{\omega(t)}{e_{\text{emf}}(t)} = K_2,$$

$$\frac{i(t)}{e(t) - e_{\text{emf}}(t)} = \frac{1}{R + LD}, \tag{6.30}$$

$$\frac{\omega(t)}{\tau(t)} = \frac{1}{B + JD},$$

$$\frac{\theta(t)}{\omega(t)} = \frac{1}{D}.$$

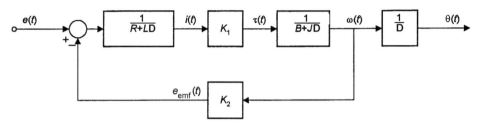

FIGURE 6.38 Block diagram of simplified motion control system with DC drive.

By putting the blocks together, the complete system block diagram is formed. This is shown in Figure 6.38, from which we can build a simulation. It is especially important to note that Equations (6.29b) and (6.29c) together imply negative feedback. Without this, the system would not be stable, and control would be impossible.

It is clear from Figure 6.38 that the parameters K_1 and K_2 are design parameters by which the system can be tuned so as to optimize performance. This can be done using either the feedforward or the feedback loop.

Each of Equations (6.29) is then converted to the appropriate difference equation using Euler. Working in the traditional back-to-front fashion,

$$\theta(k+1) = \theta(k) + h\omega(k),$$

$$\omega(k+1) = \omega(k) + \frac{h}{J}[\tau(t) - B\omega(k)],$$

$$i(k+1) = i(k) + \frac{h}{L}[e(k) - e_{emf}(k) - Ri(t)],$$

$$\tau(k) = K_1 i(k),$$

$$e_{emf}(k) = K_2\omega(k).$$

In order to completely specify the system, initial conditions must be defined for each state variable. These are variables that output from the *integral blocks*. (Loosely speaking, an integral block is one in which the transfer function has a **D** in the denominator.) Thus, $\theta(0)$, $\omega(0)$, and $I(0)$ must be given a priori.

EXAMPLE 6.13

Simulate the motion of a DC motor for the following system parameters:

$$K_1 = 1000,$$
$$K_2 = 5,$$
$$J = 3,$$
$$L = 10,$$
$$B = 30,$$
$$R = 100.$$

Use a step size of 0.05 and zero initial conditions. Print each of the state variables $\theta(t)$, $\omega(t)$, and $i(t)$ at each iteration, and graph. Let the ideal reference input be a pulse train of ± 5 volts that alternates each 3 seconds with a 50% duty cycle. Assume additive Gaussian white noise with variance $\sigma_w^2 = \frac{1}{4}$.

Solution

Using relational notation for the deterministic input, the ideal signal has formula

$$e_1(t) = 5[(t - 3 * \text{INT}(\tfrac{1}{3}t) < 1.5) - (t - \text{INT}(\tfrac{1}{3}t) < 1.5)].$$

It follows from this formula that the noise is given by $w(t) = \frac{1}{2}\sqrt{-2\ln(\text{RND})}\cos(2\pi\,\text{RND})$, and the resulting signal is $e(t) = e_1(t) + w(t)$. The complete listing, assuming Eulerian integration, is given in Listing 6.8. The results are shown in Figure 6.39. The position $\theta(t)$ steadily increases at a rate of approximately 0.62 radians per second after a short initial transient stage (after 1 second), before being driven down by the inverted pulse. Clearly, the current $i(t)$ leads the speed $\omega(t)$, and both the speed and the current exhibit system ringing and overshoot throughout the transient stage before settling down to reasonable steady-state values as well.

It should be noted that the parameter values specified in this example were not arbitrary. In this case, the system is stable, since the bounded input of $e(t) = 5$ results in a bounded output of $\omega(t)$. However, a change in almost any parameter will likely result in unbounded behavior. The systems engineer's task is to design the system so that it will remain bounded and optimize performance (i.e., minimize overshoot, etc.). The most likely method by which to accomplish this is to vary the gains K_1 and K_2. ○

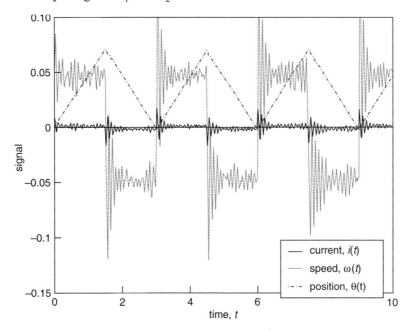

FIGURE 6.39   Response to the pulsed input signal of motion control, Example 6.13.

```
- system parameters
        K₁=1000
        K₂=5
        J=3
        L=10
        B=30
        R=100
- modeling parameters
        t₀=0
        tₙ=10
        h=0.01
        n=(tₙ-t₀)/h
        t=0
- ideal input function
        fne(t)=5[(t-3*INT(⅓t)>1.5)-(t-3*INT(⅓t)<1.5)]
- simulation
        i=0
        ω=0
        θ=0
        τ=K₁i
        e_emf=K₂ω
        print t, i, ω, θ
        for k=1 to n
                t=t+h
                w=½√(-2 * ln(RND)) cos(2π*RND)
                e=fne(t)+w
                i=i+ h/L (e-e_emf-Ri)
                τ=K₁i
                ω=ω+ h/J (τ-Bω)
                e_emf=K₂ω
                θ=θ+hω
                print t, i, ω, θ
        next k
```

LISTING 6.8   Simplified motion control system with DC servo feedback.

6.8   NUMERICAL EXPERIMENTATION

Systems methodology closely resembles the scientific method in that it is impossible to simply deduce the right answer from first principles. Both of these procedures require an insightful researcher to form a *hypothesis* (model) from which *experiments* (simulations) can produce critical decision points where either the model is rejected or its acceptance is strengthened. It is impossible to prove a model – only to disprove it or demonstrate its limitations. Acceptance of a model is just like acceptance of a scientific fact, in that it is "true" only within the scope imposed by the initial limitations of creating the model.

Regardless of the truth or falsity of a model, the critical phase is always creating it. This requires insight on the part of the researcher, and is not automatic.

Simulations are another matter. Once a model has been validated and accepted, simulation studies should be rather straightforward. However, this topic can be a textbook in itself. Let us just agree that there are far too many results published with false claims due to inappropriate simulation experimentation or analysis of the experiments. A good simulation study will inevitably address each of the following issues in detail:

1. **Parameter Determination**   Virtually all models have a set of parameters that need to be identified, subject to specification, experimental determination, or system performance requirements. This procedure, sometimes called *system identification*, requires systematic studies in optimization and sensitivity analysis. For instance, it makes very little sense to endorse claims that work for one parameter value but do not for a small perturbation of that parameter. How sensitive the system is with respect to any parameter is always important.

2. **Input Modeling**   Just as systems need to be modeled, so do the input signals that will be used to make claims using those models. Unless statistical studies are undertaken to characterize the input signal noise characteristics and random processes involved, there is little hope that the output will be of any value. Even though the model is valid, inappropriate inputs will render any results useless. This is especially true for nonlinear systems, where sensitivity to inputs is the norm.

3. **Initial Conditions**   There is a tendency among many researchers to either approach initial conditions carelessly or simply be arbitrary in their application. This is usually not a problem if the system is linear or at least behaves linearly, since performance in the steady state is independent of the initial conditions for linear systems. Let us just admit that many systems are nonlinear and that transient phases are also important. For that matter, even unstable linear systems need to be modeled, and they will be just as sensitive to initial conditions as nonlinear systems.

Once a system has been modeled and the simulations are in hand, the final step is to analyze the results. Not surprisingly, these analyses are anything but cut and dried. Even though what appear to be very definitive results are produced under what many consider the most conservative of conditions, there is falsity here too! Usually, this is due to either missing or inadequate sensitivity studies. Researchers continue to make claims for sets of parameter values, input signals, and initial conditions that are simply those that demonstrate a previously held bias rather than a universal principle. Beware of any study where insufficient attention has been paid to sensitivity!

The techniques introduced in this chapter are very powerful, and it is tempting to get carried away in their application. Especially in the social and life sciences, where deductive models are less the norm, the system integration approach is very tempting. However, as outlined above, a great deal of care needs to be used to justify the models, their simulations, and their results.

BIBLIOGRAPHY

Antsaklis, P. J. and A. N. Michel, *Linear Systems*. McGraw-Hill, 1997.
Bussel, H., *Modeling and Simulation*. A. K. Peters, 1994.

Bhonsle, S. R. and K. J. Weinmann, *Mathematical Modeling for Design of Machine Components*. Prentice-Hall, 1999.

D'Azzo, J. J. and C. H. Houpis, *Linear Control System Analysis and Design – Conventional and Modern*, 3rd edn. McGraw-Hill, 1988.

DeCarlo, R. A. *Linear Systems – A State Variable Approach with Numerical Implementation*. Prentice-Hall, 1989.

Dorf, R. and R. Bishop, *Modern Control Systems*, 9th edn. Prentice-Hall, 1995.

Fishwick, P., *Simulation Model Design and Execution: Building Discrete Worlds*. Prentice-Hall, 1995.

Gernshenfeld, N. A., *The Nature of Mathematical Modeling*. Cambridge University Press, 1999.

Johnson, D. E., J. R. Johnson, and J. L. Hilburn, *Electric Circuit Analysis*, 3rd edn. Prentice-Hall, 1992.

Kalmus, H., *Regulation and Control in Living Systems*, Wiley, 1966.

Khalil, H. K., *Nonlinear Systems*. MacMillan, 1992.

Kuo, B. C., *Automatic Control Systems*, 7th edn. Prentice-Hall, 1991.

Ljung, L., *System Identification – Theory for the User*, 2nd edn. Prentice-Hall, 1999.

Mesarovic, M. C., *Systems Theory and Biology*. Springer-Verlag, 1968.

Neff, H. P. *Continuous and Discrete Linear Systems*. Harper and Row, 1984.

Press, W. H., B. P. Flannery, S. A. Teukolsky, and W. T. Vettering, *Numerical Recipes: The Art of Scientific Computing*. Cambridge University Press, 1986.

Rosen, R., *Dynamical System Theory in Biology*. Wiley-Interscience, 1970.

Thompson, J. R., *Simulation: A Modeler's Approach*. Wiley-Interscience, 2000.

Vemuri, V., *Modeling of Complex Systems*. Academic Press, 1978.

───── EXERCISES

6.1 Use relational notation to specify each of the following signals.

(a) $\sin(0, 6, 0.5, t)$:

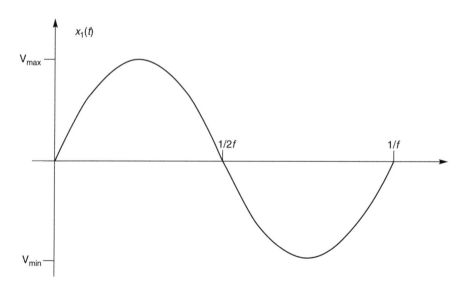

(b) pul(0, 5, 0.25, t):

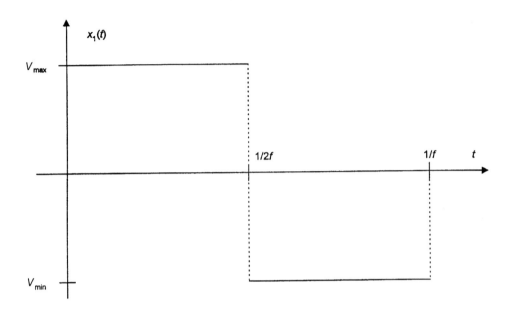

(c) tri(0, 5, 0.25, t):

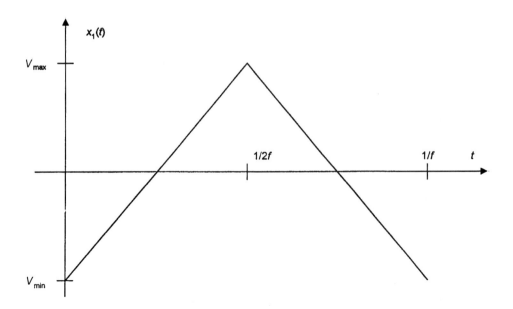

(d) pwm(0, 6, 0.125, 0.25, *t*):

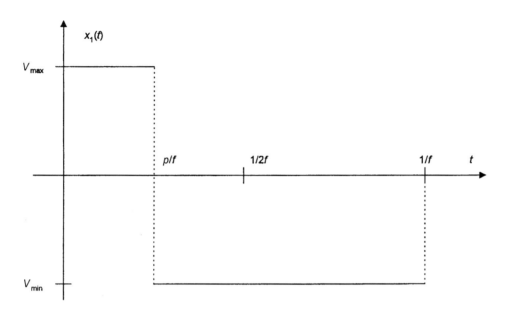

6.2 Using relational notation, generalize each of the signals in Exercise 6.1 to possess arbitrary maxima, minima, frequencies, and proportions as defined in the text:

    (i) $\sin(V_{min}, V_{max}, f, t)$;

    (ii) $\text{pul}(V_{min}, V_{max}, f, t)$;

    (iii) $\text{tri}(V_{min}, V_{max}, f, t)$;

    (iv) $\text{pwm}(V_{min}, V_{max}, f, p, t)$.

Verify correct working by plotting each of the following signals over the interval [0, 20]:

    (a) $x_1(t) = \text{pul}(-2, 3, 0.33, t)$;

    (b) $x_1(t) = \text{tri}(1, 3, 0.25, t)$;

    (c) $x_1(t) = \text{pwm}(-5, 5, 0.25, 0.25, t)$;

    (d) $x_1(t) = \sin(0, 6, 0.5, t) + 2\text{tri}(1, 3, 0.25, t)$;

    (e) $x_1(t) = \text{pul}(0, 3, 0.25, t) + \text{pul}(-2, 0, 0.2, t)$.

6.3 Consider an *m*-stage distributed delay with zero initial loading. Assuming a unit-step input, show that:

    (a) for output $y(t)$, $y(t) = 1 - e^{-mt/\Delta} \sum_{i=0}^{m-1} \left(\frac{m}{\Delta}\right)^i \frac{t^i}{i!}$;

    (b) for each stage *j* in the delay, $r_j(t) = 1 - e^{-mt/\Delta} \sum_{i=0}^{j-1} \left(\frac{m}{\Delta}\right)^i \frac{t^i}{i!}$.

6.4 Consider a system with a five-stage distributed delay of length $\Delta = 10$ seconds and initial loading of 31 units. The input has both a deterministic and noise component defined as follows:

Input signal: deterministic: $x_1(t) = \text{pul}(-5, 5, 1, t)$;
                    additive noise: Gaussian, $\sigma_w^2 = 1$.
Output signal: $y(t)$.
System: dynamics: see block diagram;
                    initial load: 31 units.

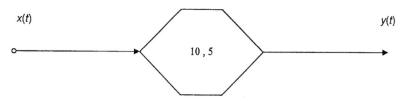

(a) Write the set of differential equation describing the model.
(b) Write a program to simulate the system behavior over [0,50].
(c) Assuming geometric initial loading, perform the simulation.
(d) Repeat part (c), assuming uniform, front-end, and back-end loads.
(e) Compare the results for each simulation in parts (c) and (d). In particular, how do the steady-state and transient behavior differ with each loading?

6.5 Consider the ideal triangular signal $x_1(t) = \text{tri}(0, 10, 1)$ superimposed with additive white noise $w(t)$. The noise is uniformly distributed on $[-3, 3]$ and the actual signal is $x(t) = x_1(t) + w(t)$, as illustrated by the following block diagram:

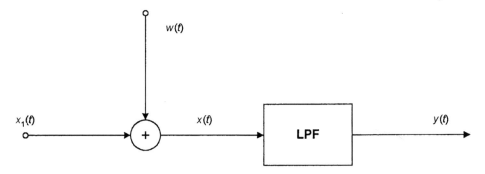

The signal is passed through an *RC* low-pass filter with adjustable parameter τ to produce an output $y(t)$. The problem is to find a good value for τ. To do this, create a simulation and print the ideal input $x_1(t)$ and output $y(t)$ as functions of time. Superimpose the graphs of the results for $\tau = 50, 5, 0.5$, and 0.05 so as to compare the output against the ideal input.

6.6 This problem is a continuation of Exercise 6.5 in which we wish to find an optimal filter. To do this, define a performance measure as the average error between the ideal input and filtered output as follows:

$$\text{PM} = \frac{1}{1+n} \sum_{k=0}^{n} |x_1(t) - y(t)|.$$

Clearly it is desirable to choose the value of τ so as to make PM as small as possible, since this means that the most noise will have been filtered out.

(a) Rewrite the program created in Exercise 6.5 as a subroutine that receives the parameter τ and sends PM back to the calling program in return.

(b) Write a calling program that accepts as input a value of τ, computes PM by calling the subroutine of part (a), and prints the ordered pair (τ, PM).

(c) By numerous executions of part (b), create a set of points $(\tau_1, \text{PM}_1), (\tau_2, \text{PM}_2), \ldots, (\tau_m, \text{PM}_m)$, and graph them. From the graph, estimate the minimum point, and thereby find an estimator for the optimal filter parameter τ_{opt}.

6.7 Simulate the following system over the time interval $0 \leqslant t \leqslant 12$:

Input signal:	deterministic:	$x_1(t) = \text{pul}(-2, 5, 2, t)$;
	additive noise:	white Gaussian, $\sigma_w^2 = 3$.
Output signal:	$y(t)$.	
System:	dynamics:	$\ddot{y}(t) + 4\dot{y}(t) + 8y(t) = x(t)$;
	initial state:	$y(0) = 0, \dot{y}(0) = 0$.

6.8 Simulate the following system over the time interval $0 \leqslant t \leqslant 12$:

Input signal:	deterministic:	$x_1(t) = \sin(0, 4, 3, t)$;
	additive noise:	uniform on $[-1, 1]$;
	sampling frequency:	$f = 2$ samples per second.
Output signal:	$y(t)$.	
System:	dynamics:	see block diagram;
	initial state:	$y(0) = 0$;
	initial loading:	uniform load of 5 in top,
		uniform load of 4 in bottom.

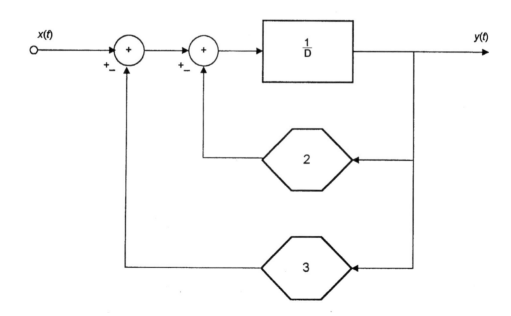

6.9 Simulate the following system over the time interval $0 \leqslant t \leqslant 10$:

Input signal:	deterministic:	$x(t) = \sin(0, 6, 3, t) + \frac{1}{2}t$;
	additive noise:	Gaussian, $\sigma_w^2 = 3$;
	sampling frequency:	$f = 2$ samples per second;
	deterministic:	$y_1(t) = \text{pul}(-5, 5, 1)$;
	additive noise:	Gaussian, $\sigma_w^2 = 1$;
	sampling frequency:	$f = 2$ samples per second.
Output signal:	$z(t)$.	
System:	dynamics:	see block diagram;
	initial state:	$y(0) = 0$;
	initial loading:	uniform load of 10 in left, geometric load of 26 in right

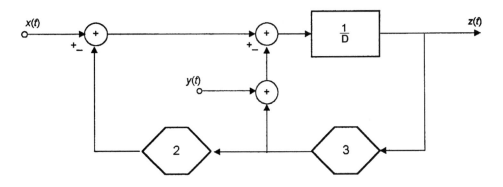

6.10 Consider the simple open-loop system with a single distributed delay as shown below. Simulate this system over the time interval $0 \leqslant t \leqslant 10$, clearly showing input, output, and (internal) staging effects. Use the following specifications:

Input signal:	deterministic:	$x_1(t) = \text{pwm}(-2, 3, 2, 0.25, t)$;
	additive noise:	white Gaussian, $\sigma_w^2 = 0.25$.
Output signal:	$y(t)$.	
System:	dynamics:	see block diagram;
	initial load:	geometric distribution of 80 units.

6.11 Consider the same system as that in Exercise 6.10, but this time assume a different and sampled input signal as follows:

Input signal:	deterministic:	$x_1(t) = \sin(-2, 4, 3, t) + \text{pwm}(-1, 1, 2, -.25, t)$;
	additive noise:	white Gaussian, $\sigma_w^2 = 0.25$;
	sampling frequency:	8 samples per second.
Output signal:	$y(t)$.	
System:	dynamics:	see block diagram for Exercise 6.10;
	initial load:	geometric distribution of 80 units.

6.12 Consider the open-loop system with two distributed delays in parallel shown below. Simulate this system over the time interval $0 \leqslant t \leqslant 5$, clearly showing the input, output, and (internal) staging effects. Use the following specifications:

Input signal:	deterministic:	$x(t) = \sin(-2, 2, 1, t) + \frac{1}{4}t$.
Output signal:	$y(t)$.	
System:	dynamics:	see block diagram;
	initial loading:	front-end load of 12 in top,
		back-end load of 12 in bottom.

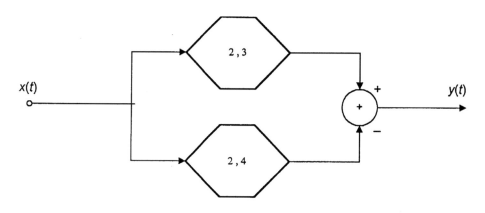

6.13 Repeat Exercise 6.12 using the following feedback system:

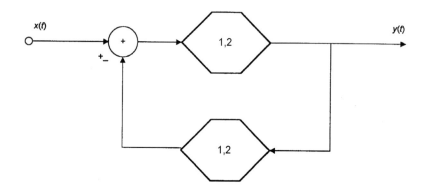

6.14 Repeat Exercise 6.12 using the following open-loop cascade system:

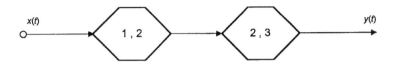

6.15 Consider the system with both distributed and discrete delays shown below. Simulate this system over the time interval $0 \leqslant t \leqslant 10$, clearly showing the input, output, and (internal) staging effects. Use the following specifications:

Input signal:	deterministic:	$x_1(t) = \text{tri}(0, 8, 6)$;
	additive noise:	autoregressive, $x_2(t) = x_2(t-2) + \frac{1}{2}w(t)$,
		$w(t)$ is white uniform on $[-1, 1]$.
Output signal:	$y_1(t), y_2(t)$.	
System:	dynamics:	see block diagram;
	initial state:	zero loading: $r_i(0) = 0$;
	special functions:	A_1: output $= |\text{input}|$,
		A_2: output $= (\text{input})^2$.

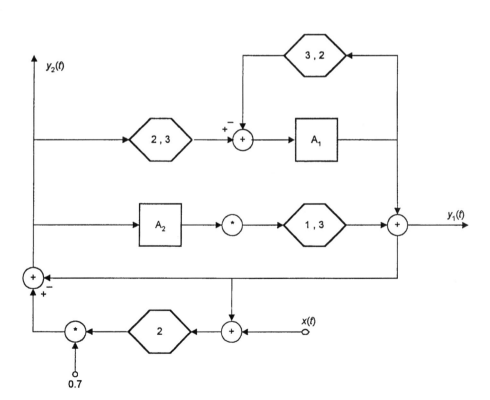

6.16 Consider the following linear system with inputs $x(t)$ and $y(t)$ and output $z(t)$. Simulate the system's behavior over the time interval $[0,15]$ using an appropriate integration step size h.

Input signal:	deterministic:	$x_{11}(t) = \sin\frac{1}{2}t$;
	additive noise:	ARMA, $x_{12}(t) = x_{12}(t-2) + \frac{1}{2}w_1(t) + \frac{1}{4}w_1(t-1)$,
		$w_1(t)$ is white uniform on $[-1, 1]$;
	deterministic:	$x_{21}(t) = \cos\frac{1}{2}t$;
	additive noise:	ARMA, $x_{22}(t) = x_{22}(t-1) + \frac{1}{2}w_2(t) + \frac{1}{3}w_2(t-1)$,
		$w_2(t)$ is white uniform on $[-1, 1]$.
Output signal:	$y_1(t), y_2(t)$.	
System:	dynamics:	see block diagram;
	initial state:	zero loading: $r_i(0) = 0$;
	special functions:	A_1: output $= |input|$,
		A_2: output $= (input)^2$.

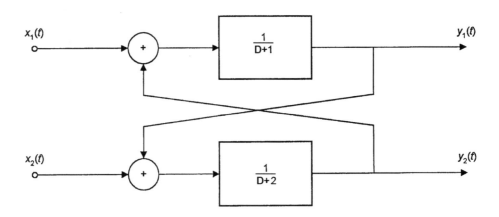

6.17 Consider the following system, with input $x(t)$ as defined below and output $y(t)$ over the time interval $[0, 20]$.

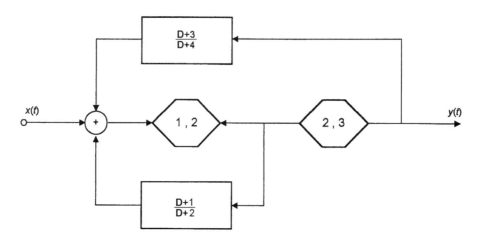

(a) One cycle of a periodic signal $x_1(t)$ is defined as follows by $x_0(t)$:

$$x_0(t) = \begin{cases} 2t, & 0 \leqslant t < 1, \\ 2, & 1 \leqslant t < 2, \\ 6 - 2t, & 2 \leqslant t < 4, \\ -2, & 4 \leqslant t < 5, \\ 2t - 12, & 5 \leqslant t < 6. \end{cases}$$

Write a function describing the ideal deterministic x_1 signal defined for all real t.

(b) Formulate an additive noise signal x_2 as the following ARMA process with Gaussian white noise $w(t)$:

$$x_2(t) = x_2(t - \tfrac{1}{4}) - \tfrac{1}{2}x_2(t - \tfrac{1}{2}) + w(t) + \tfrac{1}{2}w(t - \tfrac{1}{4}).$$

Give an explicit formula for the superimposed signal $x(t) = x_1(t) + x_2(t)$.

(c) Write the difference equations, assuming an integration step size $h = 0.25$ and a uniform initial loading of both cells.

(d) Simulate the system behavior using an appropriate integration technique over $[0, 20]$.

6.18 Consider the simplified motion control system described in the block diagram below. It is desired to simulate the working of this system for a number of different reference inputs, different amplifier definitions, and different parameter values. In particular, use the following parameter and gain values: $J = 3$, $B = 20$, $K_I = 80\,000$, $L = 600$, $R = 1.15$, $K_P = 15.5$, $K_L = 50\,000$, and $K_S = 1$.

(a) Implement this system using a "perfect" amplifier with gain $K_A = 20$. That is, the amplifier "characteristic" is defined by $f(x) = 20x$. Generate the system response for a step input $r(t) = 10$ superimposed with white noise that is uniform on $[-0.5, 0.5]$. Plot the input, position, and speed outputs over time for $0 \leqslant t \leqslant 50$.

(b) Repeat part (a) using the sinusoidal input

$$r(t) = 10(1 - \cos \tfrac{1}{5}t).$$

(c) Using the results of part (b), find the lag or lead of the output signal compared with the input.

(d) Now change the amplifier to an "ideal" *operational amplifier* with a characteristic defined by

$$f(x) = \begin{cases} -10, & x < -0.2, \\ 50x, & -0.2 \leqslant x \leqslant 0.2, \\ 10, & x > 0.2. \end{cases}$$

Repeat parts (a)–(c).

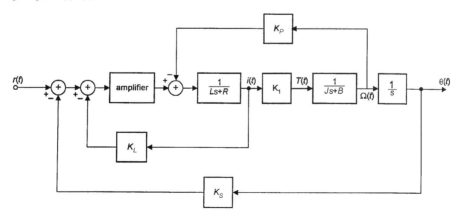

6.19 Consider the following block diagram of a simplified motion control system.

   (a) From the block diagram, draw a working system diagram with which to simulate the system. Assume linear electrical devices and a "perfect" amplifier.

   (b) Write a computer program with which to simulate the system behavior for any set of physical parameters and gains.

   (c) Find a set of parameters for which the system is stable using a step input: $\Omega_r(t) = \mathrm{pul}(-5, 5, 3)$ for $t \geqslant 0$. Start your simulation from zero initial conditions.

   (d) Validate your results by plotting the input and output over $[0,15]$ seconds.

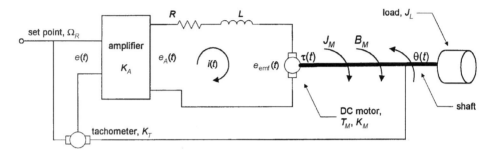

6.20 Model a simplified wildlife system where an animal population interacts with a food source that supplies its nutrition. Simulate the system dynamics.

The live animal birth rate, BR (animals per year), is

$$BR(t) = f(\text{ANPA}) * \text{SMF}(t),$$

where SMF is the population of surviving mature females, ANPA is the available nutrition per animal, and $f(\text{ANPA})$ is the function f shown in the following figure:

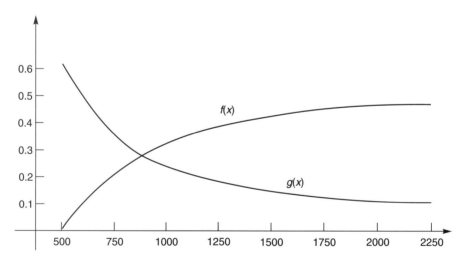

Births are equally divided between males and females, and the average maturation time for females (the time between birth and entrance into the mature population) is 3 years. Model this maturation process with a distributed delay with $\delta = 3$ years and $m = 4$ stages. Use the same approach to model maturation of males, except with a maturation time of 2.5 years.

Assume that animals born alive survive until they enter the mature populations. However, death rates for mature males and females are directly proportional to g(ANPA)

$$\text{DRM}(t) \propto g(\text{ANPA}),$$
$$\text{DRF}(t) \propto g(\text{ANPA}),$$
$$\text{SMM} = [1 - g(\text{ANPA})] * \text{PMM}$$
$$\text{SMF} = [1 - g(\text{ANPA})] * \text{PMF}$$

where DRM is the death reate for males, DRF is the death rate for females, PMM is the population of mature males, SMM is the population of surviving mature males, SMF is the population of surviving mature females and g(ANPA) is the function g shown in the figure above.

The functions f and g should be approximated by using either the table look-up procedure or a modeling function. Whichever you choose, be very explicit in showing your development of this step.

The variable ANPA is the annual nutrition available (AN) divided by the total population (TP), where mature animals count as 1.0 and maturing animals as 0.7 each. Assume initially that AN is 20 million pounds per year. Also assume that the initial population sizes are

$$\begin{aligned}
\text{SMF} &= 15\,000 \quad &\text{(population of surviving mature females),}\\
\text{PIF} &= 6\,000 \quad &\text{(population of immature females),}\\
\text{SMM} &= 5\,000 \quad &\text{(population of surviving mature males),}\\
\text{PIM} &= 4\,000 \quad &\text{(population of imature males).}
\end{aligned}$$

Initialize the delays so as to produce the required numbers of animals at $t = 0$ and satisfy one of the protocols discussed in the text.

As in the case of many natural systems of this kind, the food supply is a function of the population pressure. Thus,

$$\frac{d}{dt} \text{AN}(t) = -7.5 \times 10^{-6}[\text{TP}(t) - 20\,000]\,\text{AN}(t).$$

(a) Draw an explicit system block diagram describing the system.
(b) From the diagram, write a computer program to simulate this model. Print the time t, AN, ANPA, PMF, PIM, PMM, PIM, and TP at 6-month intervals for 20 years of system behavior.
(c) Simulate 10 years of system behavior. Graph the variables PMF, PIM, PMM, and PIF as functions of time.
(d) Determine a value for the step size h that will ensure that population levels from the simulation are within 5% of those of the "true" solution of the mathematical model for $0 \leqslant t \leqslant 10$ years. Show that your results satisfy this 5% requirement.

CHAPTER 7

Exogenous Signals and Events

In general, systems comprise modules that are connected by signals, which are either exogenous or endogenous. Exogenous signals come from or go to the environment through entry/exit points called ports. On the other hand, endogenous signals are entirely contained within the system and just connect the system modules. For instance, in Figure 7.1, the module X has two inputs and one output. The signal (a) is an exogenous input, (d) is an endogenous input, and (b) is an endogenous output with respect to X. On the other hand, the exogenous signals (a) and (c) form the input and output for the system S.

Endogenous signals are either feedforward signals or feedback signals. Feedforward signals are those signals that progress from the input toward the output (left to right in Figure 7.1), thus causing no module to be visited more than once. On the other hand, feedback signals form loops in that they revisit modules or move from the output side toward the system input. In Figure 7.1, the signal (b) is feedforward and the signal (d) is feedback. It will be noticed that the signals (b) and (d) form a loop.

Endogenous inputs are sometimes further subdivided into *overt signals* and *disturbances*. Loosely speaking, overt inputs are those signals that are desired, while disturbances are nuisances that simply need to be dealt with. For instance, in flying an airplane, the throttle controls a pilot's wish to go faster or slower, and is therefore an overt input to the aircraft system. On the other hand, a good stiff headwind cannot be ignored, even

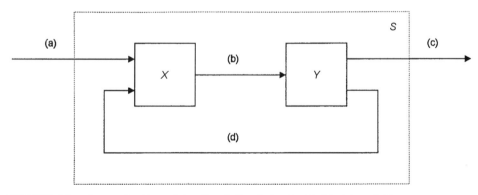

FIGURE 7.1   System S with modules X and Y and signals (a), (b), (c), and (d).

though it is not "wanted" and therefore constitutes a disturbance. Even so, both the throttle position and the headwind must be accounted for as inputs into a precision aircraft system. Since disturbances are by definition not under control, they are usually not given by explicit formulas. Rather, they are often taken as random processes and characterized by their statistical properties.

7.1 DISTURBANCE SIGNALS

The models considered in the previous chapter were largely deterministic. If the state was known at a given time, the model could predict exactly what the state would be at any succeeding time. In general, this is not the case – more often than not, future states can only be inferred in a statistical sense. Such models are called *probabilistic*, in that they provide inexact, probably erroneous, results for a specific instance but excellent results for an entire ensemble. Probabilistic models often arise when an autonomous system receives exogenous input in an apparently random form called a *disturbance*.

Strictly speaking, both the "input" and "disturbance" in the system diagram shown in Figure 7.2 are inputs. However, usually we think of an input as regular and predictable, and perhaps exogenous. It could be a deterministic signal or it could even be a random process, but, regardless, we think of it as under our control. In contrast, the disturbance is thought of as outside of our purview and somewhat of a shock to the system.

As an example, consider a motor driving a constant load. A constant input voltage signal (called a *set-point*) directs the motor to turn at a fixed rate so that the load also rotates at a constant speed. However, if the load is suddenly increased at some random time, the system will suddenly slow down because the input signal does not change to compensate. This extra load then acts as a disturbance. Engineers use feedback to solve this problem. By measuring the output speed with a tachometer and feeding back the voltage corresponding to the actual speed, the system can automatically self-adjust, regardless of a disturbance. This is shown in Figure 7.3. Note that by comparing the set-point against the feedback speed, there is only a system input if an error occurs. As long as the actual speed matches the desired speed, the system status quo is maintained.

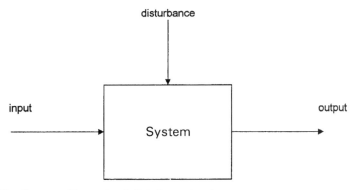

FIGURE 7.2 System with overt and disturbance inputs.

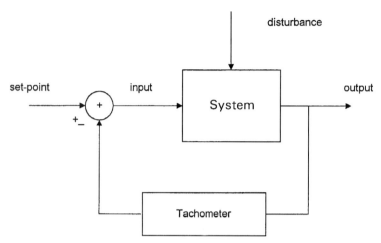

FIGURE 7.3   Closed-loop motion control system.

EXAMPLE 7.1

Consider the open-loop system shown in Figure 7.4 with set-point $x(k) = 12$, transfer function $\frac{1}{2}$, and a deterministic disturbance $d(k) = 2 \sin \frac{1}{10} \pi k$. Analyze the system and implement a feedback control so as to reduce the disturbance effects.

Solution
It is clear from the simulation diagram in Figure 7.4 that without the disturbance, the output will be $z(k) = 6$. Combining this with the indicated disturbance gives an additional ± 2 units of low-frequency "wiggle", producing an output $z(k) = 6 + 2 \sin \frac{1}{10} \pi k$, which will be confined to the interval [4,8]. In this example, we think of the disturbance as being somewhat of an additive error, so it is desirable to minimize the wiggle as much as possible. Taking our cue from the discussion above, let us introduce feedback as illustrated in Figure 7.5, where we have incorporated a feedback with gain C into the input signal. The idea is to find C so that the disturbance $d(k)$ is reduced.

First it is necessary to find an expression for the output $z(k)$. This is straightforward if we remember to work from the output back to the input as

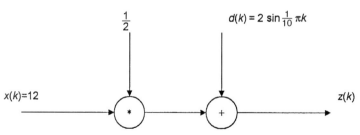

FIGURE 7.4   Open-loop system, Example 7.1.

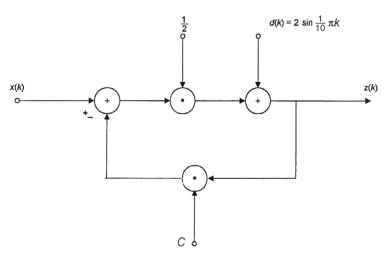

FIGURE 7.5   Feedback compensation, Example 7.1.

described in Chapter 6:

$$z(k) = d(k) + \tfrac{1}{2}[12 - Cz(k)].$$

Solving for $z(k)$ and substituting the expression for $d(k)$,

$$z(k) = \frac{4}{C+2}(3 + \sin \tfrac{1}{10} k\pi). \tag{7.1}$$

Close inspection of Equation (7.1) shows that making $C = 0$ results in an output $z(k) = 6 + 2 \sin \tfrac{1}{10} \pi k$, just as deduced for the case of no feedback. However, on increasing C, the effect of $d(k)$ is reduced, and there will be a less pronounced wiggle. Unfortunately, the desired set-point of $z(k) = 6$ is also reduced, and the mean output signal is $z(k) = 12/(2 + C)$. The resulting residual error is

$$e(k) = 6 - \frac{12}{2+C} = \frac{6C}{2+C},$$

which approaches $e = \lim_{c \to \infty} 6C/(2 + C) = 6$ as C becomes large.

So, there is evidently a trade-off as the feedback gain C becomes large. It is impossible to maintain the desired set-point and reduce the additive noise disturbance at the same time. This is illustrated in Figure 7.6, where the system response is graphed for different gains. It will be noticed that as C becomes larger, the mean signal becomes further from the set-point of $x(k) = 6$ and the amplitude of the wiggle becomes smaller.   ○

The word "disturbance" tends to imply the idea of noise and therefore undesirability. This should be distinguished from many exogenous signals that are overt actions from external sources such as switches, message arrivals, termination of interrupt service routines, and the like. Exogenous signals come in several varieties, depending on exactly what is random, the magnitude of the signal, or the time at which exceptional events occur. If the time is regular, the signal is called *synchronous* or clock-driven. On the other hand, if the

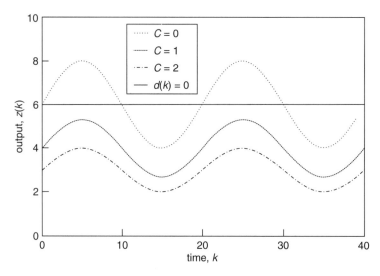

FIGURE 7.6   System response for different feedback gains, Example 7.1.

time of an event's occurrence is irregular and cannot be predicted, the signal is called *asynchronous* or event-driven. Similarly, the magnitude can be either *deterministic* or *random*. This leads to the following special cases of disturbance signals:

1. *Synchronous, deterministic:*   This case is essentially a deterministic signal that can be explicitly combined with the input. The resulting signal is then handled just like input signals in Chapter 6. It is called a *time-driven deterministic model*.

2. *Synchronous, random:*   In this case, the time is regular and clock-like. However, the magnitude is random and based on probabilistic models. For instance, the disturbance $d(k) = 2k + \text{RND}$ is random, but its mean $\bar{d}(k) = 2k + \frac{1}{2}$ is not random and varies with time, as does its standard deviation. This is called a *time-driven probabilistic model*.

3. *Asynchronous, deterministic:*   In this case, the signal magnitude is often unity; that is, the disturbance is either present (1) or absent (0). For instance, a machine might be either busy or idle, but when it becomes busy, it is random and subject to the arbitrariness of external forces. This is called an *event-driven model*.

4. *Asynchronous, random:*   This is the most random of all cases. Both the magnitude and occurrence times are random. Models such as this are rare, and lack a consensus name. While this is easy to simulate, there is no existing theory with which to handle it.

7.2 ___ STATE MACHINES

Models are common in which the overt input is one of a finite set X, the disturbance signal is one of a finite set D, and there are a finite number of states Y. If both of the exogenous input sets are finite and the number of states is finite, it follows that the output set Z is also finite. Specifically, if $X = \{x_i\}$, $D = \{d_i\}$, and $Y = \{y_i\}$, then $Z = \{x_i\} = \{(x_i, d_i, z_i)\}$ is the set of ordered triples of X, D, and Y, as illustrated in Figure 7.7. Notice that the current output is a function of the current endogenous and exogenous inputs as well as the current state. For any specific time k, the next state is a function solely of the inputs and current state. Similarly, the future state is a function of the current inputs and state. Thus, for suitably defined vector functions f and g,

$$y(k + 1) = f(x(k), d(k), y(k)),$$
$$z(k) = g(x(k), d(k), y(k)). \tag{7.2}$$

These relationships are shown graphically as the block diagram of Figure 7.7. Such systems are called *finite state machines* (*FSM*). Here we consider the case of finite state machines that are *independent of time*. All this is best illustrated by an example.

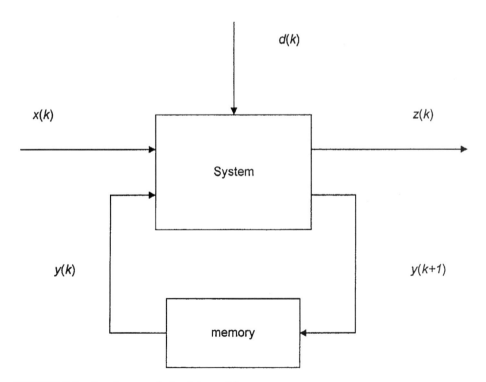

FIGURE 7.7   Synchronous finite state machine.

EXAMPLE 7.2

Consider a digital sequencer that runs sequentially through the alphabet A, B, C, D to the beat of a regular clock. Whether this device sequences up or down or simply stays in place depends on an exogenous input $d(k)$, which we assume to be independent of time. There is no overt input, and the output is simply whatever the next state will be. Specifically, assuming the initial state is $y(0) = A$:

- If the input signal is $d(k) = 1$, the sequencer just stays in place: A, A, A, ...
- If the input signal is $d(k) = 2$, the state sequence is listed in ascending order: A, B, C, D, A, B, C, D, ...
- If the input signal is $d(k) = 3$, the state sequence is listed in descending order: A, D, C, B, A, D, C, B, ...

Solution
Since the state in this example can only change on a clock beat, it is a synchronous system. Also, there are obviously a finite set of inputs and states, which are enumerated as follows:

$$X = \emptyset,$$
$$D = \{1, 2, 3\},$$
$$Y = \{A, B, C, D\}.$$

Since the endogenous set X is empty, the output Z is just the set of ordered pairs of D and Y as listed in the following so-called *state table*. The table body is the next state, and is a function of the left column (current input) and top row (current state).

	$y(k) = A$	$y(k) = B$	$y(k) = C$	$y(k) = D$
$d(k) = 1$	A	B	C	D
$d(k) = 2$	B	C	D	A
$d(k) = 3$	D	A	B	C

A device that is equivalent to the state table is the *state diagram*. A state diagram shows the same information in a graphical format in which vertical lines represent each of the finite number of states, and horizontal arrows show allowable transitions from one state at time k to the next state at time $k + 1$. Numbers placed directly above the arrows show the input required that actually causes the transition indicated. Figure 7.8 shows the state diagram for this example.

Notice that if the input is $d = 1$, which is denoted in the state diagram as (1), the FSM diagram takes every state back to itself. This implies that an input of (1) causes our machine to *idle* or show no change. An input of (2) causes the next state to go from A to B, from B to C, C to D, and D to A, exhibiting a sequential wrap-around effect. Similarly, an input of (3) causes the machine to go in reverse. Since the system is assumed to be time-independent, there is but one diagram that

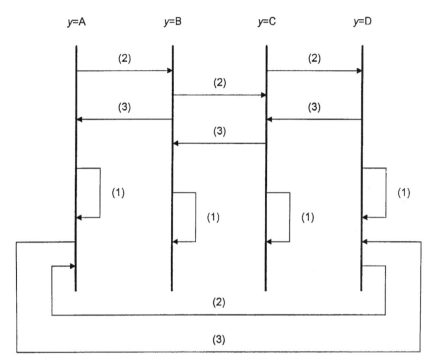

FIGURE 7.8   State diagram, Example 7.2.

works for all time k. If the time-independence assumption were not in effect, there would need to be a different diagram, or equivalently a different table, for each value of k.   ○

There is no reason for state changes to be confined to regular times. If these events occur at irregular times, thereby causing the inter-event time to be variable or random, the system is asynchronous. In this case, the system block diagram is given by Figure 7.9. Notice the obvious relationship between Figures 7.7 and 7.9, where time "k" is replaced by "current" and time "$k + 1$" is replaced by "future". This is because the exact time is known in the synchronous case, but it is unknown a priori in the asynchronous case. Similarly, the update equations analogous to Equations (7.2) are

$$y_{\text{future}} = f(x_{\text{current}}, d_{\text{current}}, y_{\text{current}}),$$
$$z_{\text{currrent}} = g(x_{\text{current}}, d_{\text{current}}, y_{\text{current}}). \tag{7.3}$$

A practical example of an asynchronous finite state machine is the ordinary telephone. Exogenous inputs include events such as lifting the receiver, dialing a number, receiving a phone call, and getting a busy signal. These events are asynchronous in that they are initiated by persons external to the telephone system itself and they are not dependent on a clock. What happens as a function of the input depends on what state the system is in. For instance, the response to lifting the receiver (input) depends on whether the phone is in a ringing or not ringing state – one will send a message and the other a dial tone.

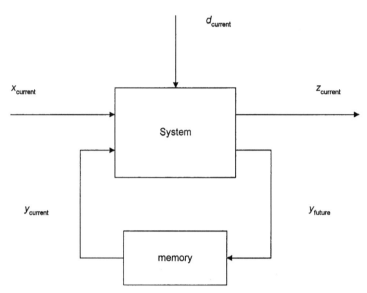

FIGURE 7.9   Asynchronous finite state machine.

There are two problems regarding finite state machines that are of concern to us. First, it is necessary to understand the nature of an implementation. Once implementation is understood, the simulation follows immediately. Regardless of whether we are implementing or simulating, the essence of finite state machines is summarized in *FSM programming*. FSM programming is essentially finding the values of the two functions y_{future} and $z_{current}$ indicated in Figure 7.9 and given explicitly by Equation (7.2). In a realistic situation, it is only necessary to find the functions f and g of Equation (7.2). Since the inputs, states, and outputs are all finite sets, this can be done by explicitly listing every possible combination rather than finding a single formula. For instance, the FSM of Example 7.2 calculate the functions f and g using inputs d and y with outputs y and z. Since there are no elements in the set X, 3 elements in the set D, and 4 elements in the set Y, there are must be $1 \times 3 \times 4 = 12$ possible combinations for the arguments of y_{future} and $z_{current}$. These functions are calculated using the subroutine given in Listing 7.1, which we denote by $\Phi(d, y, z)$. Actually, this code is quite straightforward. It simply takes every input combination of the current d- and y-values, produces the future y- and current z-values, and returns to the function evaluation routine.

Now consider the special case of a synchronous system. In an implementation, the system keeps its current state memorized and provides a clock. From the clock, the system samples the current d-value and places the updated state y and current output z. This is executed by the following code:

```
[1]     get d
        call Φ(d, y, z)
        put y, z
        goto [1]
```

In a simulation, the simulating computer must keep the clock. This requires an initial time t_0, a sampling interval δ, and an initial state y_0. Since the input d is unknown, we

```
subroutine Φ(d, y, z)
      case y
      if y=A
            if d=1 then y₁=A ; z₁=A
            if d=2 then y₁=B ; z₁=B
            if d=3 then y₁=D ; z₁=D
      if y=B
            if d=2 then y₁=B ; z₁=B
            if d=2 then y₁=C ; z₁=C
            if d=3 then y₁=A ; z₁=A
      if y=C
            if d=2 then y₁=C ; z₁=C
            if d=2 then y₁=D ; z₁=D
            if d=3 then y₁=B ; z₁=B
      if y=D
            if d=2 then y₁=D ; z₁=D
            if d=2 then y₁=A ; z₁=A
            if d=3 then y₁=C ; z₁=C
      end case
            y=y₁,
            z=z₁
      return
```

LISTING 7.1   The subroutine $\Phi(d, y, z)$ sending d and y and returning y and z as the future state and output.

make a random guess by taking each possible d-value as equally likely. This is accomplished by using the INT (integer) and RND functions: $d = \text{INT}(1 + 3 * \text{RND})$. A simulation producing n epochs is produced by the code given in Listing 7.2.

Execution of this code results in a sequence of successive states encountered assuming random exogenous input. The results of a typical simulation run are shown in Figure 7.10, where it will be noticed that each state encountered by the system is held statically over a 0.2-second time interval, after which time a new state is found. It is easy to compute an estimate for the probability that the system is in any particular state by simply counting the number of times that A, B, C, and D are visited on a run. If this is done, the percentages are very close to 25% for each state. If statistics on the relative input probabilities are known,

```
k=0
t=t₀
y=y₀
print t, y
for k=1 to n
      t=t+δ
      d=INT(1+3*RND)
      call Φ(d, y, z)
      print t, y, z
next k
```

LISTING 7.2   Synchronous simulation of Example 7.2.

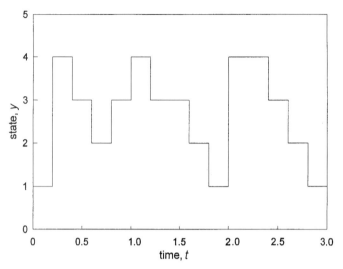

FIGURE 7.10   FSM of Example 7.2 with synchronous exogenous input.

this program can be adjusted accordingly. For instance, the following code produces probabilities $Pr\{D = 1\} = 0.1$, $Pr\{D = 2\} = 0.3$, and $Pr\{D = 3\} = 0.6$ for d:

```
r=10*RND
if r<1 then d=1
if 1 ⩽ r<4 then d=2
if r ⩾ .4 then d=3
```

Application of this code also results in each state being visited 25% of the time. Interestingly enough, no matter what the relative probabilities are for the inputs, the proportion of times each state is visited are always equal – in this case, 25% of the time. This will be shown to conform to theory later in the text.

There is yet another possibility. Rather than being regular or clock-driven, the exogenous signal can also be asynchronous. In this case, the implementation simply just waits for an external event, and the system determines the next state as required when the event occurs. This is usually handled in software by an interrupt. In fact, the computer can be busy doing whatever its required tasks are, but when the interrupt occurs, a special routine called a *interrupt service routine* is executed before control is returned to the main program. Since interrupts are initiated in hardware, there are no program statements comparable to calls or subroutines. Setting up an interrupt amounts to initiating interrupt vectors that reference another part of memory where the interrupt routine actually occurs in the processing unit. In the case of Example 7.2, this is especially simple in that after updating the state, a new state is put out:

```
call Φ(d, y, z)
put y, z
```

From a simulation point of view, the time of the event must be random. Often this is done by assuming an exponential inter-event time, since exponential inter-arrivals are non-negative and less likely to allow a lengthy wait (although this is not impossible). This is

```
          t=t₀
          y=y₀
          print t, y
    [1]   t=t-μ*ln(RND)
          d=INT(1+3*RND)
          call Φ(d, y, z)
          print t, y, z
          if t<tₘₐₓ then goto [1]
```

LISTING 7.3   Simulation of an asynchronous execution of Example 7.2.

called *scheduling* the event. Under the exponentially distributed inter-event time assumption, scheduling is accomplished by updating time as $t^{(new)} = t - \mu \ln(\text{RND})$. A suitable module simulating this model is given in Listing 7.3. Of course, other distributions are also possible. Applying the exponential assumption to Example 7.2 gives the graph shown in Figure 7.11, where the effect of exponentially distributed inter-arrival time ($\mu = 1$) is clear. Compare these results against those of the synchronous graph of Figure 7.10.

7.3   PETRI NETS

Probably the best known synchronous finite state machine is the digital computer. Since a computer has a finite memory of n cells and each memory element is binary in nature, there are 2^n different memory configurations or states. Most computers are also sequential machines, where only one change can occur at a time. These two characteristics – finite memory and sequential transitions between these states – make traditional computers finite state machines. In practice, this is obvious, because there can be only one

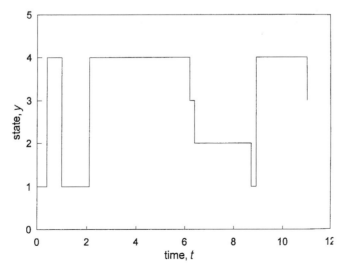

FIGURE 7.11   FSM of Example 7.2 with asynchronous exogenous signal.

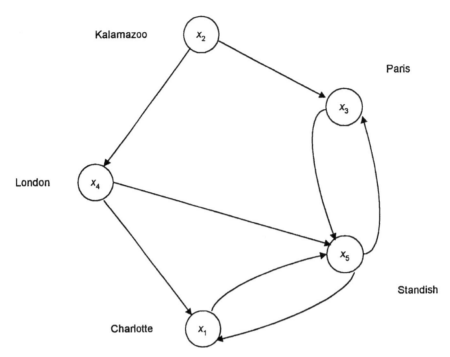

FIGURE 7.12   Network of message centers.

program operating at a time, and within the program steps are handled sequentially one after the other. This structure, called the *von Neumann architecture*, has been a mainstay of computing machines since their early days. However, there are other alternatives and strategies available too. Multiprocessing enables machines to operate on several processes in parallel and at the same time. Thus there exists the need for modeling concurrency. These more general structures are called *Petri nets*.

Consider a network of message centers such as the one shown in Figure 7.12. Legal communication links are indicated by directed arcs joining the nodes. Each center, represented in Figure 7.12 as a circular node for $i = 1, 2, 3, 4, 5$, has a queue of messages of length x_i waiting to be sent to another city. x_i is called the *local state*, since it represents the state of the particular node in question. In contrast, the *system state* is given as the vector of local states, since they must all be known in order to characterize the total system, $\mathbf{x} = [x_1, x_2, x_3, x_4, x_5]$. This is summarized in Table 7.1. For instance, the vector $\mathbf{x} = [2, 0, 1, 3, 0]$ indicates that Charlotte has two messages waiting to be transmitted, Paris has one, and London has three. Kalamazoo and Standish have no messages in queue.

It is desirable to simulate the system's distribution of messages over time, and here is how we propose to do it. Using Figure 7.12, we construct Table 7.2, which lists all possible transitions assuming an initial state $\mathbf{x} = [2, 0, 1, 3, 0]$. For instance, the transition from Kalamazoo to London is impossible, since there are no messages in Kalamazoo to start with, even though Figure 7.12 implies that such a transition is legal. Of the four possible transitions, only one can actually be executed, so we choose one at random. This puts the system in a new state, from which a new Table 7.2 can be constructed showing all the

TABLE 7.1   Message Center System States

City Center	Node	Local state
Charlotte	1	x_1
Kalamazoo	2	x_2
Paris	3	x_3
London	4	x_4
Standish	5	x_5

TABLE 7.2   All Possible Next States
Assuming an Initial State $\mathbf{x} = [2, 0, 1, 3, 0]$

Transition	$\mathbf{x}$ (future)
Charlotte→Standish	[1,0,1,3,1]
Kalamazoo→Paris	Impossible
Kalamazoo→London	Impossible
Paris→Standish	[2,0,0,3,1]
London→Standish	[2,0,1,2,1]
London→Charlotte	[3,0,1,2,0]
Standish→Charlotte	Impossible
Standish→Paris	Impossible
Standish→Kalamazoo	Impossible

new future states. This process is repeated over and over, creating one instance of the ensemble of all possible *trajectories* of the FSM state over time. Notice that in so doing, we require that only one transition takes place at any time and that we do not know just when this time is chronologically speaking. Thus, we have only a statistical study at best. Such a system is called a *Petri net*.

Petri net nomenclature calls the message center nodes described above *places*, the messages *tokens*, and the number of messages in each queue the *local state*. The set of possible legal transitions is called the *action set*, and the randomly selected transition is an *action*. These definitions and their symbols are shown in Figure 7.13.

Symbolically, legal transitions are shown "blocked" by a heavy line crossing each arc. Each transition is also given a unique numerical label. The transition is included in the action set $\mathbf{S}$ only if all inputs x_i to the transition are positive. This is illustrated graphically in Figure 7.14. The action set is the set of all transactions satisfying this rule.

One element is selected at random from the action set $\mathbf{S}$. Assuming this to be transition k, all input local states to k are then decremented and all output local states are incremented. This is illustrated graphically in Figure 7.15. This update process creates a new system state, and the entire process is repeated until the end of the simulation.

It follows that Petri net simulations take the form of a three-step process embedded in a loop. After printing the initial conditions, the loop is repeated through the time horizon. Within the loop, legal transitions are added to the action set, the action is randomly selected from the action set, and the nodes local to the selected action transition are updated. Algorithmically, it takes the form of Listing 7.4.

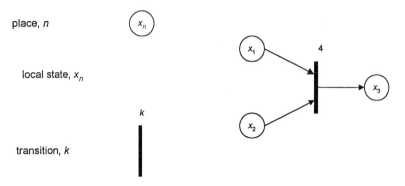

FIGURE 7.13   Standard Petri net nomenclature.

It will be recalled from the discussion of finite state machines that state diagrams show the system state and transitions between the states. However, an FSM shows no explicit demarcation of exactly what state is currently occupied. In the more general context of multiprocessing, there can be a number of processes taking place at the same time. Thus, the current state, as well as transitions between those states, must be explicitly defined for each process as well. Petri net places can be thought of as *system conditions* such as "the valve is open", "the buffer is full", and "the frequency is larger than 60 hertz". If a process occupies a place (i.e., the local state is positive), it is represented by a token. As such, places and their tokens characterize the total system condition, which is called the *system state*. Every Petri net place diagram is followed by a transition symbol, except for

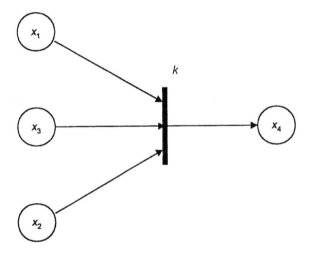

if $x_1 > 0$, $x_2 > 0$, $x_3 > 0$, then $S_{new} = S_{old} \cup \{k\}$

FIGURE 7.14   Updating the action set.

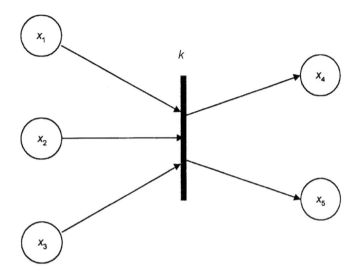

if k is an element of **S**, then $x_1=x_1-1$
$x_2=x_2-1$
$x_3=x_3-1$
$x_4=x_4+1$
$x_5=x_5+1$

FIGURE 7.15   Updating selected nodes.

the terminal place. The transition can be thought of as a *system action* such as "opening a value", "moving data into a buffer", and "increasing the frequency". Each action moves tokens from one place into another.

Tokens represent abstractions of processes; there is one token for each process in the system. The distribution of tokens at any time characterizes the system state at that time. In general, if there are n places and m tokens, there are

$$\binom{n}{m} = \frac{n!}{m!(n-m)!}$$

```
Set initial place, token distribution
k=0
print k, state
for k=1 to n
        Establish all legal transitions; place in action set
        Randomly chose one transition from the action set
        Update only those nodes local to this transition
        print k, state
next k
```

LISTING 7.4   Generic simulation of a Petri net.

different configurations of the system. For the case of a finite state machine, there is only one token, which can occupy any of the n places in the Petri net. The resulting n different configurations constitute the n different states of the system.

In the more general context of multiple tokens, the system state is characterized by the distribution of tokens throughout the network. Events occur when system conditions change or, in the Petri net sense, a token moves from one place to another. This can happen when sufficient tokens are available in the places prefacing a transition so as to make a transition possible. If that transition takes place, a token from each of the prefacing places is deleted and a new token is added to the places immediately following the transition. Thus, it appears that an enabled transition allows tokens to "pass through" to the successor place.

As a rule, Petri nets are not used so much for timing simulations. Usually, they are created to ensure that all places are well defined and that transitions behave as desired. Making the assumptions that events occur asynchronously and non-simultaneously, it is only necessary to establish all possible event orderings rather than simulate an actual instance. This can be done by establishing a synchronous timing sequence in which random transitions are chosen from those that are eligible. By selecting a random transition out of all those eligible, it is assumed that every possible occurrence and combination will be realized. This is important, since a successful simulation can verify that no system hang-ups such as deadlock (to be discussed later) exist, that every system state is accessible, that no abnormal endings will occur, and that the statistical distribution of states is as desired. Thus, even without temporal information, there are meaningful results.

EXAMPLE 7.3

Consider the Petri net shown in Figure 7.16, in which there are initially 8 tokens in place 1 and 5 tokens in place 2, with all other places empty. Simulate the transient and long-term behavior using the elementary Petri net model described above.

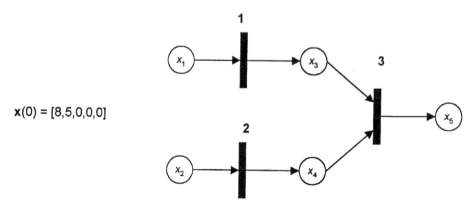

FIGURE 7.16   Example 7.3.

Solution

There are three stages to a Petri net simulation. First, all legal transitions must be determined and marked. From these, one must be selected at random and executed. In this particular example, there are three transitions. Transitions 1 and 2 can be executed in parallel, but since we do not know which one begins, one must still be chosen as the first. Transition 3 cannot take place until places x_3 and x_4 have tokens. Finally, transition 3 can move the token to place x_5, but in so doing note that places x_1 and x_2 will eventually be depleted. The following algorithm segments do what is outlined above. (Although the algorithm is stated in terms of the action set **S**, in practice it is more likely that **S** will be a string and that standard string functions are used to perform the actual operations.)

1. **Determine and mark all legal transitions**   This is done by creating a special string **S** (the action set) which lists all legal transitions. **S** is assembled by cycling through each transition and adding the transition to **S** if it is found to be legal. An empty string indicates there are no legal transitions available to any token.

```
S=∅
if x₁>0 then S=S∪{1}
if x₂>0 then S=S∪{2}
if x₃>0 and x₄>0 then S∪{3}
```

2. **Choose one of the legal transitions randomly**   This can be done using the function RAND(**S**), which returns a random element of the action set **S** unless **S** is empty, in which case it returns a zero:

$$\mathrm{RAND}(\mathbf{S}) = \begin{cases} 0, & \mathbf{S}=\emptyset, \\ y \in \mathbf{S}, & \mathbf{S} \neq \emptyset. \end{cases}$$

Thus, it will be noted that RAND acts on a finite set similarly to the way in which RND acts on the continuous unit interval [0,1).

An implementation of RAND is given in Listing 7.5. This version considers **S** to be a string of characters. The operator MID(**S**, i) returns a single character from the ith position of **S**. The operator VAL converts this character to a number, and LEN gives the number of characters in **S**. Thus,

```
function RAND(S)
        r=RND
        for i=1 to LEN(S)
                if r<i/LEN(S) then
                        y=VAL(MID(S,i))
                        goto [1]
                end if
        next i
        y=0
[1]     RAND=y
        return
```

LISTING 7.5   The function RAND(S) returns a random character from the string S.

```
n=20
k=0
read   x₁, x₂, x₃, x₄, x₅
print k, x₁, x₂, x₃, x₄, x₅
for k=1 to n
        S=Ø
        if x₁>0 then S=S∪{1}
        if x₂>0 then S=S∪{2}
        if x₃>0 and x₄>0 then S=S∪{3}

        y=RAND(S)

        if y=0 then "deadlock"
        if y=1 then x₁=x₁-1 : x₃=x₃+1
        if y=2 then x₂=x₂-1 : x₄=x₄+1
        if y=3 then x₃=x₃-1 : x₄=x₄-1 : x₅=x₅+1
        print k, x₁, x₂, x₃, x₄, x₅
next k
```

LISTING 7.6   Petri net simulation, Example 7.3.

$y = \text{RAND}(S)$ provides the required random transition. If there is no candidate transitions (that is, if **S** is empty), then $y = 0$ is returned. This indicates a *deadlock* condition, and no further transitions can take place.

3. **Update the places local to y**   Once the random transition has been selected, [1] performs the actual token changes. The results are printed and the cycle repeats.

```
[1]   if y=1 then x₁=x₁-1 : x₃=x₃+1
      if y=2 then x₂=x₂-1 : x₄=x₄+1
      if y=3 then x₃=x₃-1 : x₄=x₄-1 : x₅=x₅+1
```

A complete listing of this simulation is shown as Listing 7.6. A typical run for $n = 6$ executions is given in the following table:

k	x_1	x_2	x_3	x_4	x_5
0	8	5	0	0	0
1	8	4	0	1	0
2	7	4	1	1	0
3	7	4	0	0	1
4	7	3	0	1	1
5	7	2	0	2	1
6	7	1	0	3	1

○

Petri nets have several standard constructs that are especially useful in practice. As in combinational logic, AND and OR are fundamental. These are shown in Figures 7.17 and 7.18, where places x_1 and x_2 act as inputs to place x_3. In Figure 7.17, it is clear that

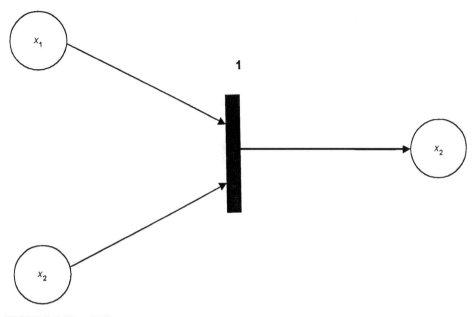

FIGURE 7.17   AND gate.

transition 1 will be enabled only when places x_1 and x_2 are both occupied. Assuming 1 is a marked transition, x_3 will be activated as follows:

```
if x₁>0 and x₂>0 then y=1                enables transition 1
S={1}                                    activation set
if y=1 then x₁=x₁-1 : x₂=x₂-1 : x₃=x₃+1  activates place 3
```

Similarly, Figure 7.18 illustrates a logical inclusive OR in that x_3 will be activated only when transition 1 is enabled or transition 2 is enabled. This happens when either x_1 or x_2 or both are occupied. Assuming 1 is a marked transition, x_3 will be activated as follows:

```
if x₁>0 then y=1                 enables transition 1
if x₂>0 then y=2                 enables transition 2
S={1, 2}                         activation set
if y=1 then x₁=x₁-1 : x₃=x₃+1    activates place 3
if y=2 then x₂=x₂-1 : x₃=x₃+1    activates place 3
```

Of all sequential logic devices, memory is of fundamental importance. For instance, if a contact switch is pressed, the electrical current must continue to flow, even after the switch retracts. Thus, a memory element can be activated and remain activated for either a set period of time or until it is deactivated. The principle is the same in both cases, as illustrated in Figure 7.19, where we assume an initial state of $x(0) = [1, n, 0]$. Note that place x_1 acts as a switch, since if transition 1 is not enabled, none of the n tokens in place x_2 can be moved to place x_3. However, once transition 1 is enabled, it allows a continuous flow of n tokens from place x_2 to x_3, since x_3 will be occupied and this feeds back to transition 2 even after transition 1 is no longer enabled. Clearly, this will continue only as long as there remain some of the original n tokens in place x_2. This is realized as follows:

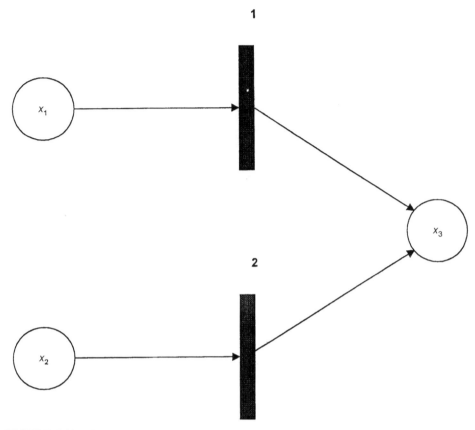

FIGURE 7.18   OR gate.

```
if  x₁>0  then  y=1                        enables  transition  1
S={1}
if  y=1  then  x₁=x₁-1  :  x₃=x₃+1         activates  place  3
if  x₂>0  and  x₃>0  then  y=2             enables  transition  2
S={2}
if  y=2  then  x₂=x₂-1                     activates  place  3
```

Notice that once place x_3 is in the "on" state, activation both increments and decrements x_3 because of the feedback. It follows that x_3 will remain "on" only until all n tokens in place 2 are expended, at which time the local state $x_3 = 0$ turns "off". These principles are best illustrated with an example. Thus, memory is maintained for exactly n cycles.

EXAMPLE 7.4

Consider an automobile system in which a key is used to initiate a starter sequence, followed by the car idling and the vehicle moving. Model this sequence using the basic Petri net constructs described above.

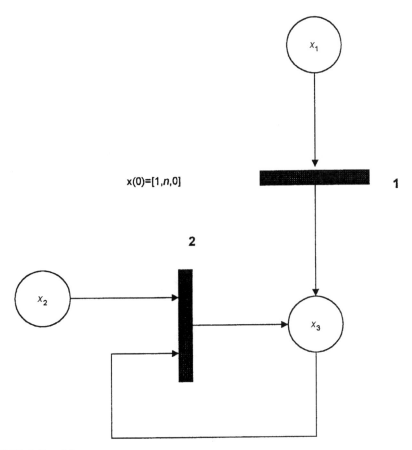

x(0)=[1,*n*,0]

FIGURE 7.19   Memory.

Solution
We begin by assembling a list of places and transitions. Remembering that places
are local states while transitions are actions taking us from one state to another, a
simplified list is as follows:

- Places (local states):   1: key in ignition
                            2: gasoline in tank
                            3: battery charged
                            4: gasoline in carburetor
                            5: distributor powered
                            6: engine running
                            7: car moving
                            8: starter engaged
- Transitions (actions):   1: move gas to carburetor
                            2: move gas to cylinder
                            3: move key to start position

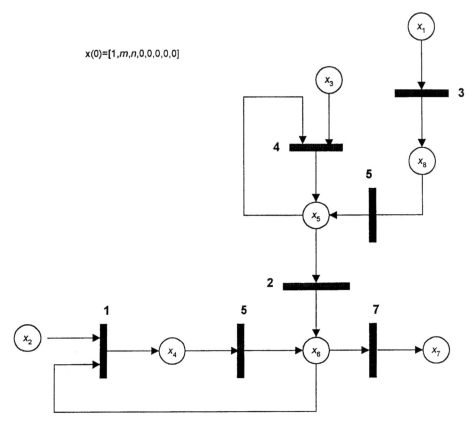

x(0)=[1,*m*,*n*,0,0,0,0,0]

FIGURE 7.20   Automobile start-up model, Example 7.4.

4:   move current to distributor
5.   close contacts
6:   create spark in cylinder
7:   depress accelerator

Initially, there must be one key in the ignition ($x_1 = 1$), some gasoline in the tank ($x_2 = m > 0$), and charge in the battery ($x_3 = n > 0$), where m and n are given a priori. All other initial local states are zero. Thus, the initial state vector is $\mathbf{x}(0) = [1, m, n, 0, 0, 0, 0, 0]$. Using these lists, a suitable Petri net model is shown in Figure 7.20. Notice the use of two memory elements to maintain electrical current from the battery and gasoline power over a period of time after initial start-up. In a more realistic model, the fuel tank and charge on the battery would be shut off using feedback sensors rather than the timed operations shown here. This would require a slightly different memory element.   ○

The Petri net transitions can be generalized further. For instance, a number of modelers use non-zero thresholds to establish transition criteria. Also, by "coloring" tokens, one can establish a myriad of different transition rules, depending which kind of token is being considered. Even so, the basic structure and transition principles are always the same: pass a token along only if a transition is enabled and the activation takes place.

7.4   ANALYSIS OF PETRI NETS

Petri nets are a very general tool to describe generic discrete models. Usually, they are not used to describe timing behavior, but rather action sequences. Therefore, one of their common applications is to verify the correctness of programs and protocols. This is not a trivial task, since in either a concurrent or distributed environment, it is important to know that bad things do not happen and that good things eventually will happen. Neither of these goals is obvious, since there are a number of semi-independent procedures being executed simultaneously, and cooperation cannot be taken for granted.

A distributed model is built around a number of independent *processes*, each of which is vying for the use of a finite number of resources. For instance, a computer network might have ten computers but only two printers available. If several users wish to print their work at the same time, competition for the printer resource is inevitable. If the printer also requires the use of an additional disk drive for creating a print spooler, multiple resources are required, and if different computers tie each up, then no single user can ever have sufficient resources at his disposal to complete his task. This circumstance, called a *deadlock*, is common, so establishing a protocol to avoid it is a good thing. Since Petri nets can model processes, system simulation amounts to testing a proposed protocol so as to verify that deadlocks do not occur.

A process can be thought of as an independent, almost self-contained, sequential program. As a part of a distributed system, there can be several processes, or even several instances of the same process, acting at once. In practice, such concurrent processes need to communicate with each other. For instance, an application program must request the use of a community printer or disk drive. As an abstraction, a distributed system has independent processes (each of which has both places and transitions) and resources (each of which has only a place). The processes will either compete or coordinate their use of the resource so as to establish the system function. The process coordination, defined by the state sequence, is called the *system protocol*. In this section, we consider several classical distributed systems, beginning with the so-called *producer/consumer system with buffer*.

EXAMPLE 7.5

Consider a producer/consumer system with an infinite buffer. This system has two processes: the producer process, which creates an item, and the consumer process, which disposes of the item. The buffer serves as a temporary holding area for the item as an interface between the producer and consumer. Even a small finite buffer will tend to smooth out the action of the entire system, since, without it, items that are not consumed will stop the producer. On the other hand, if an infinite buffer is in place, there will be no need to curtail actions of the producer at any time. Model this system using a Petri net, then simulate and analyze the results.

Solution

The producer can take one of two possible actions (transitions): *producing an item* and *appending an item* to the buffer queue. Associated with these transitions are two places: *ready to produce an item* and *ready to append an item* to the buffer. The buffer is a shared resource, and therefore has only one state, *buffer has item*, and that is characterized by the number of items stored.

On the other hand, the consumer's actions include *remove an item* from the buffer or *consume an item*. Each has an associated place: *ready to remove an item* and *ready to consume an item*. Each of these transitions and places, along with an English language statement of the producer/consumer protocol, is summarized in Figure 7.21. An equivalent Petri net interpretation is shown in Figure 7.22. The state vector, which is simply a 5-tuple of all system places, is given by

Places (conditions)

Place name	Local state
ready to produce item	x_1
ready to append item	x_2
ready to remove item	x_3
ready to consume item	x_4
buffer has item	x_5

Transitions (actions)

Action name	Label
produce item	1
append item	2
remove item	3
consume item	4

Protocol
- To produce an item:
    produce item
    append item to buffer

- To consume an item:
    remove item from buffer
    consume item

FIGURE 7.21   Characterization of producer/consumer system with infinite buffer.

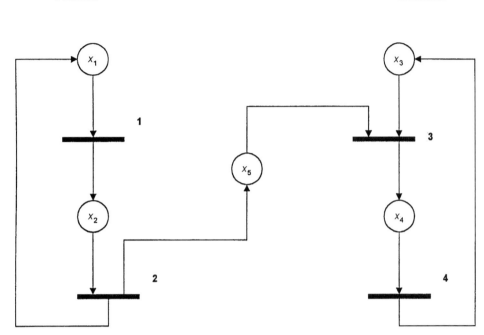

FIGURE 7.22   Producer/consumer system with infinite buffer.

$\mathbf{x} = [x_1, x_2, x_3, x_4, x_5]$. The action y is a randomly selected member of the action set $\mathbf{S} = \{1, 2, 3, 4\}$. Place variable names and labels are defined in Figure 7.21.

In testing this model, it is useful to begin with single tokens in the "ready to produce" and "ready to remove" places, with all other places being empty. Thus, the initial state is $\mathbf{x}(0) = [1, 0, 1, 0, 0]$. Using the same procedures as outlined in the previous section, Listing 7.7 shows a simulation for the producer/consumer system with an infinite buffer as outlined above. The results of a simulation for $n = 100$ events are summarized in Table 7.3. Even though the sequence of states encountered is non-deterministic, there are clearly only certain successor states that are permitted. Also, as one would expect, allowing for the infinite buffer in this example means that there are an infinite number of legal states and no end to a legal state sequence. Table 7.3 exhibits this, since place 5 (buffer has item) apparently grows without bound. It follows from this that the entire producer/consumer system described here is unbounded. It also seems that this system is "safe" in that it never comes to an untimely end: there are always legal transitions to be made and no deadlock occurs. Clearly, the simulation conveys a wealth of information.   ○

Of course, infinite buffers cannot exist in reality. A more realistic finite buffer can be modeled by simply adding another place named *buffer not empty* (x_6) in parallel to the

```
read n
read x₁, x₂, x₃, x₄, x₅
k=0
print k, x₁, x₂, x₃, x₄, x₅
for k=1 to n
        S=∅
        if x₁>0 then S=S∪{1}
        if x₂>0 then S=S∪{2}
        if x₅>0 and x₃>0 then S=S∪{3}
        if x₄>0 then S=S∪{4}

        Y=RAND(S)

        if y=0 then print "deadlock!" : stop
        if y=1 then x₁=x₁-1: x₂=x₂+1
        if y=2 then x₂=x₂-1: x₁=x₁+1: x₅=x₅+1
        if y=3 then x₅=x₅-1: x₃=x₃-1: x₄=x₄+1
        if y=4 then x₄=x₄-1: x₃=x₃+1

        print k, x₁, x₂, x₃, x₄, x₅
next k
```

LISTING 7.7   Simulation of producer/consumer system with infinite buffer, Example 7.5.

TABLE 7.3   Results of Producer/Consumer
Simulations, Example 7.5; $x(0) = [1, 0, 1, 0, 0]$

k	x_1	x_2	x_3	x_4	x_5
0	1	0	1	0	0
1	0	1	1	0	0
2	1	0	1	0	1
3	1	0	0	1	0
4	0	1	0	1	0
5	1	0	0	1	1
6	1	0	1	0	1
7	0	1	1	0	1
8	0	1	0	1	0
9	0	1	1	0	0
10	1	0	1	0	1
20	0	1	0	1	0
30	0	1	0	1	1
40	0	1	0	1	0
50	0	1	0	1	1
60	1	0	1	0	1
70	0	1	0	1	5
80	1	0	1	0	6
90	1	0	1	0	3
100	1	0	1	0	6

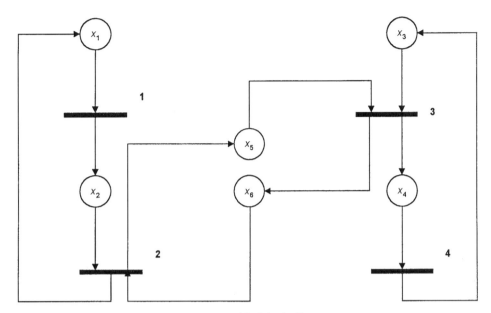

FIGURE 7.23   Producer/consumer system with finite buffer.

buffer item place, as shown in Figure 7.23. By initiating x_6 with m tokens, a maximum of m items can be stored at any given time. This gives rise to the state vector $\mathbf{x} = [x_1, x_2, x_3, x_4, x_5, x_6]$, where the first two components x_1 and x_2 are associated with the producer, components x_3 and x_4 with the consumer and the last two components x_5 and x_6 associated with the finite buffer control. As in the infinite case, x_1, x_2, x_3 and x_4 are each bits: 0 or 1. Since the buffer is finite, x_5 and x_6 are bounded. It follows that since there are a finite number of legal states, the system is also bounded.

In the case that $m = 1$, each element in state vector $\mathbf{x}$ is 0 or 1, and therefore the state can be represented as a bit-vector. Creating a simulation similar to that of Table 7.3, but this time with the six local states, shows that there are only eight possible system states: 00 10 11, 00 11 10, 01 10 01, 01 11 00, 10 00 11, 10 01 10, 11 00 01 and 11 01 00. Aside from the obvious symmetry, the most interesting thing to note is the transitions between the states. A simulation of $n = 200$ transitions is made, producing the statistics shown in Table 7.4, in which is shown the proportion of time the system is in each state, and Table 7.5, where the proportion for each transition is estimated. The associated state diagram is shown in Figure 7.24. It is evident from this diagram that every state has at least one entry and one exit. Further, it is possible to trace a path from every state to any other state, thus making every state accessible. This is a very important property. Assuming stationarity, the transition probabilities can be used to estimate the system performance.

The state diagram in Figure 7.24 follows immediately from Table 7.5. For instance, from the state 00 11 01 there are apparently two possible transitions: to the state 00 11 10 and to the state 00 00 01. Since the relative transition probabilities are 11% and 8%, respectively, there is a $11/(11 + 8) \approx 58\%$ probability for the transition 00 11 01→ 00 11 10. Similarly, there is approximately a 42% chance that the transition

TABLE 7.4   Producer/Consumer System
with Finite Buffers: Percentage of Time in
Each State

State	Number	Percentage
00 11 10	38	19.0
00 10 11	34	17.0
01 01 10	21	10.5
01 00 11	12	6.0
10 11 00	10	5.0
10 10 01	29	14.5
11 01 00	16	8.0
11 00 01	40	20.0

00 11 01→00 11 01 will occur. The remaining transition probabilities are computed in like manner. Figure 7.24 also provides a means by which to simulate Petri net dynamics. This is best illustrated by the following example.

EXAMPLE 7.6

Consider the produce/consumer system with finite buffer described by the Petri net of Figure 7.23. Suppose that events take place asynchronously on average one event per unit time and with an exponential inter-event time. Simulate the behavior of this system dynamically over the interval [0, 25].

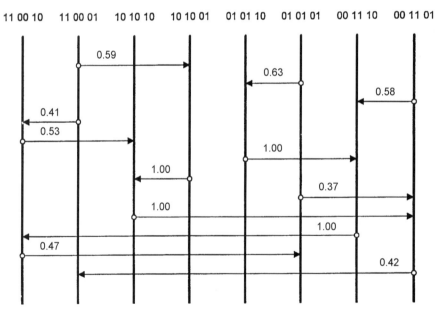

FIGURE 7.24   State diagram of the FSM associated with the producer/consumer system with finite buffer.

TABLE 7.5   Producer/Consumer System with Finite
Buffers: Absolute Percentage of Each Transition

Transition			
Past state	Present state	Number	Percentage
00 11 01	00 11 10	22	11.0
00 11 01	11 00 01	16	8.0
00 11 10	11 00 10	34	17.0
01 01 01	00 11 01	7	3.5
01 01 01	01 01 10	12	6.0
01 01 10	00 11 10	12	6.0
10 10 01	10 10 10	10	5.0
10 10 10	00 11 01	31	15.5
11 00 01	10 10 01	10	5.0
11 00 01	11 00 10	7	3.5
11 00 10	01 01 01	19	9.5
11 00 10	10 10 10	20	10.0

Solution

It is easiest if we make a simple state assignment with a single variable y for the
eight legal states as follows:

y	State
0	Deadlock
1	11 00 10
2	11 00 01
3	10 10 10
4	10 10 01
5	01 01 10
6	01 01 01
7	00 11 10
8	00 11 01

It follows that an asynchronous procedure such as that of Listing 7.3 will perform
the task at hand. Listing 7.8 gives the details where the function evaluation is
handled using a case statement. Results of this using $\mu = 1$ and $t_{max} = 25$ are
given in Figure 7.25. If the only print statement used was the one at the bottom of
the loop, a time and state (as interpreted by the above table) would be given for
each event. However, by using two print statements per loop – one before the
computation of the new state and one after – both starting and ending points are
printed. This creates the graph of Figure 7.25.

Even though Petri nets are most often used only as a modeling tools in which
timing is unimportant, it is also possible to perform dynamic simulations with
them. Thus, parallel processes can be effectively simulated and the models tested
using sequential algorithms and a von Neumann machine.   ○

```
           t_max=25
           μ=1
           t=0
           y=1
           print t, y
[1]        t=t-μ*ln(RND)
           print t, y
           case y
              if y=1
                     if RND<0.47 then y=6
                     else y=3
              if y=2
                     if RND<0.59 then y=4
                     else y=1
              if y=3 then y=8
              if y=4 then y=3
              if y=5 then y=7
              if y=6
                     if RND<0.63 then y=5
                     else y=8
              if y=7 then y=1
              if y=8
                     if RND<0.58 then y=7
                     else y=2
           end case
           print t, y
           if t<t_max then goto [1]
```

LISTING 7.8   Program to simulate the system dynamics of the producer/consumer system, Example 7.6.

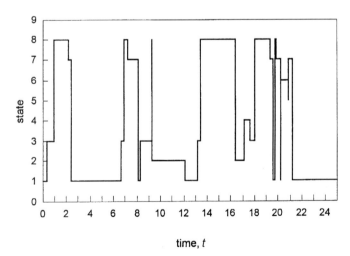

FIGURE 7.25   Simulation of the system dynamics of the producer/consumer system of Example 7.6.

The producer/consumer system illustrates how a Petri net can be used to model cooperating processes and how best to use a resource (the buffer) to smooth out system performance. Another application is where processes compete for the use of shared resources. Here we describe a competition model where each process has the same procedure. In particular, we consider a system called the *dining philosophers problem*, where m philosophers, all seated in a circle, like to do nothing more than eat and think. Between each philosopher is a fork. Initially, all philosophers begin thinking, which requires no external assistance. But, eventually, they each want to eat. Eating requires the use of two forks: a left fork and a right fork. Since the philosophers are seated in a circle, if one philosopher is eating, his colleagues on either side of him must continue to think, even though they are getting hungry. We do not allow the philosophers to negotiate or communicate in any way. But, they still choose to stop eating occasionally to think again. What we need to is find a protocol that allows this to happen ad infinitum; that is, that does not allow the system to become deadlocked.

EXAMPLE 7.7

Consider the dining philosophers problem described above for the case of $m = 3$ philosophers with the following protocol:

- Each philosopher begins in a thinking state.
- To eat, the philosopher must:
    (1) pick up the left fork;
    (2) pick up the right fork (and immediately commence eating).
- To think, the philosopher must:
    (1) set the left fork down;
    (2) set the right fork down (and immediately commence thinking).

Model this problem using Petri nets with three identical processes (one for each professor) and three resources (the forks). Use simulation to test for system safety. That is, is deadlock a problem or not?

Solution
Following the protocol defined above, we must have the following local states (places) associated with each philosopher:

- philosopher thinking (has no fork).
- philosopher eating (has both left and right forks).
- philosopher has only left fork.
- philosopher has only right fork.

The forks act as resources, each of which is also associated with a place:

- fork available at philosopher's left

Since there are three philosophers and three forks, there are $3 \times 4 + 3 = 15$ places. These are enumerated along with their variable names below.

Place (conditions)	Philosopher 1	Philosopher 2	Philosopher 3
Philosopher thinking	x_1	x_6	x_{11}
Philosopher has left fork	x_2	x_7	x_{12}
Philosopher eating	x_3	x_8	x_{13}
Philosopher has right fork	x_4	x_9	x_{14}
Fork available at philosopher's left	x_5	x_{10}	x_{15}

This is a distributed system, in that each philosopher knows only his own local state. There is no philosopher who knows what the entire system state is. However, collectively, the system is defined by the following state vector:

$$\mathbf{x} = [x_1, x_2, x_3, x_4, x_5, \quad x_6, x_7, x_8, x_9, x_{10}, \quad x_{11}, x_{12}, x_{13}, x_{14}, x_{15}].$$

Since each local state is binary, we can treat x as a bit vector. For instance, if

$$\mathbf{x} = [10001\ 10001\ 10001],$$

each philosopher is busy thinking and all forks are available. Since this is a process-recourse model, all transitions (actions) are of the type where a philosopher, is acting on a fork. For each philosopher, they are as follows:

Transition (action)	Philosopher 1	Philosopher 2	Philosopher 3
Philosopher picking up left fork	1	5	9
Philosopher picking up right fork	2	6	A
Philosopher putting down left fork	3	7	B
Philosopher putting down right fork	4	8	C

Figure 7.26 shows the graphical relationship between each of the 15 places and the 12 transitions.   ○

It is clear that the dining philosophers in the preceding example cannot all eat at the same time. For example, if philosopher 2 is dining, neither philosopher 1 or 3 can eat, since neither can get the necessary two forks. Of course, if the three all cooperate, it is possible that they take turns eating. This being the case, as they randomly try to occasionally pick up a fork, the hungry thinkers could conceivably go on forever. The problem is that they cannot communicate with one another. This can lead to trouble if each happens to have a left fork in hand. Since the only legal next step is to pick up a right fork, there can be no further progress. This situation is called *deadlock*.

One of the reasons for doing a simulation is to determine whether or not deadlock can occur. Since the actual transitions are randomly chosen from a set of legal transitions, eventually every possible occurrence will occur.

EXAMPLE 7.8

Write and perform a simulation for the dining philosophers problem. From this simulation, determine whether or not deadlock can occur.

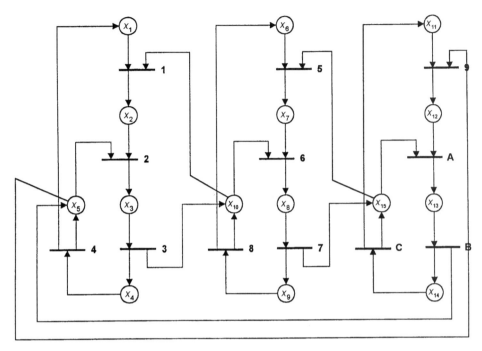

FIGURE 7.26   Petri net model of the $m = 3$ dining philosophers problem.

Solution

Recalling our Petri net simulations of previous examples, we note that there are three distinctive parts. The first section checks places prefacing each transition to see if that transition is legal or not. If it is legal, it is added to the action set **S** of all legal transitions. The second section randomly selects exactly one transition from **S**. The third section executes the chosen transition by updating predecessor places (by decrementing them) and successor places (by incrementing them).

In the dining philosophers problem, there are three analogous code sequences by which each philosopher checks for legal transitions. Each philosopher has four transitions, which are numbered. In the case of philosopher 1, the numbers are 1, 2, 3, and 4; the code proceeds as follows:

```
S=∅
if x₁>0 and x₅>0 then S=S∪{1}
if x₂>0 and x₁₅>0 then S=S∪{2}
if x₃>0 then S=S∪{3}
if x₄>0 then S=S∪{4}
```

Since there are 12 legal transitions, philosopher 2 uses codes 5, 6, 7, and 8, while philosopher 3 uses 9, A, B, and C (the characters A, B, and C being used instead of integers 10, 11, and 12).

The number of legal transitions can vary from one time to the next, so the length of string S can be anywhere from 0 if there are no legal transitions (i.e., deadlock occurs!) to 12, in which case all transitions are permissible. The following code determines the number of transitions and picks a random one to actually execute. If there is deadlock, the default $y = 0$ is used.

```
        r=RND
        for i=1 to LEN(S)
                if r<i/LEN(S) then y=MID(S,i)) : goto [1]
        next i
        y=0
[1]
```

Once the transition to be executed has been determined, the predecessor places are decremented and the successor places are incremented as described in the previous section. For instance, if transition 3 (philosopher 1 putting down left fork) is activated, the execution is as follows:

```
        if y=3 then
                x₃=x₃-1
                x₂=x₂+1
                x₅=x₅+1
        end if
```

To accomplish this entire simulation, the three sections (compile legal list, select transition, execute transition) are placed inside a loop of $k = 1$ to n cycles and the state is printed at the end of each cycle.

Let us assume that initially each philosopher begins thinking and that every fork is available. That is, the initial state vector is given by $x(0) =$ [10001 10001 10001]. A typical sequence is shown in Table 7.6. Note that the group goes about happily thinking and eating for awhile, but eventually their luck runs out when they are all thinking and have a left fork in hand. Thus, the system deadlocks when the state vector becomes $x(11) = $ [01000 01000 01000].   ○

TABLE 7.6   One Instance of a Dining Philosophers Simulation Culminating in Deadlock

k	Philosopher 1	Philosopher 2	Philosopher 3
0	10001	10001	10001
1	10001	10001	11000
2	10001	01000	01000
3	10000	00100	01000
4	10000	10000	01000
5	10001	10001	11000
6	10001	10000	10100
7	01000	10000	10100
8	01000	01000	10011
9	01000	01000	10001
10	01000	01000	11000
11	01000	01000	01000
	Deadlock!		

7.5   SYSTEM ENCAPSULATION

Since Petri nets are so general, they can be used to model a great number of systems. However, unless the system is small, the detail can be overwhelming. Therefore, it is natural to establish a set of building blocks that can be used to construct larger systems. From a simulation point of view, this allows us to proceed using a subroutine library for large-scale distributed modeling.

Recall that the basic distributed system is modeled by a set of quasi-independent (often identical) processes that have sparse communication with each other or with mutually shared resources. This allows us to model each process as a sequence of steps, each of which receives and passes information in an inter-process or intra-process fashion. Thus, we consider a process to be a list of modules receiving information from a predecessor module and sending local information to a successor module. Some modules also send and receive remote messages to other processes.

As an example, consider the three processes in communication with one another shown in Figure 7.27. Process $P(1)$ is composed of a sequence of three modules: M_1, M_2, and M_3. Module M_1 has an exogenous input and an endogenous output with its successor M_2. Since these two inputs are local to process $P(1)$, they are called *local signals*. Module M_2 also has local input (from M_1) and output (to M_3). However, it also has remote communication links set up with module M_5 of process $P(2)$. In general, we define the following kinds of inter- and intra-process communication. Similarly, M_3 has local input and output as well as remote output to module M_3 of process $P(2)$.

Communication	Interpretation
Local input	Intra-process message from predecessor module
Local output	Intra-process message to successor module
Remote input	Inter-process message to another process or resource
Remote output	Inter-process message from another process or resource

Modules, like the systems they make up, are composed of places, transitions, and arcs. The set of places within the module defines the local state of the module. By convention, input arcs (messages from another module) always enter a module at a place. Outputs (messages going to another module) always leave the module from transitions. These conventions ensure that any module can communicate directly to any other module since transitions separate places in all Petri nets.

There are two basic types of communication: synchronous and asynchronous. Combined with input–output combinations, there are four possibilities, characterized as follows:

- Passthrough: Local input–output only. There is no remote communication.
- Asynchronous send: Local input–output and a remote send.
- Asynchronous receive: Local input–output and a remote receive.

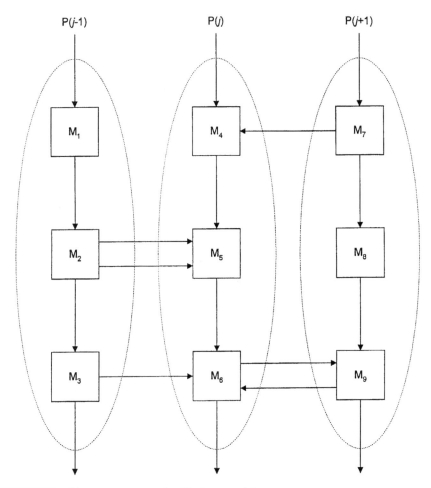

FIGURE 7.27   Three processes, each with three modules.

- Synchronous send: Local input–output and remote input–output. The local output does not proceed until the remote message is received.

These basic communication modules, along with their Petri net realizations, are shown in Figure 7.28. Convention has it that vertically placed signals are intra-process signals and horizontally placed signals are inter-process signals. For instance, in the synchronous send module, x_1 is a local input and z_1 is a local output, while x_2 is a remote input and z_2 is a remote output.

These communication modules are very elementary building blocks, and they can be extended in some obvious ways. The local input and outputs must be single links, since there can be only one predecessor and one successor. However, the remote links can be any number. For instance, the internal workings of a 2-port asynchronous send module are given in Figure 7.29. It should be noticed that this realization satisfies all module

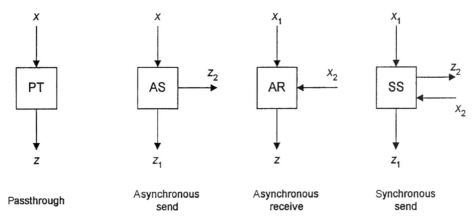

FIGURE 7.28   Fundamental communication modules.

requirements in that the input is a place and the outputs are all transitions. It is asynchronous in the sense that when a local input signal is received, notifications is sent immediately to the remote processes as well as the successor module (in the form of local output); there is no waiting to synchronize the local output signal with some other event.

The synchronous send is different from the asynchronous send in that the local output signal of a synchronous send is not immediately forwarded to the successor modules.

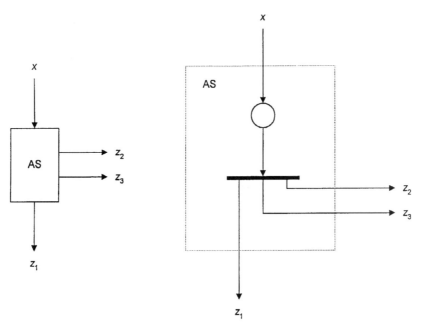

FIGURE 7.29   Asynchronous send communication module.

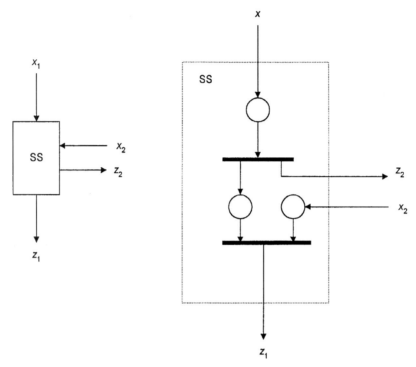

FIGURE 7.30   Synchronous send communication module.

When a local input arrives, remote notification is immediate, but the local output is only forwarded to the successor module when a remote acknowledgment is also received. The Petri net internals show this, as illustrated in Figure 7.30, which is the 1-port synchronous send module. Notice that there can be no local output z_1 until the remote input x_2 is also received.

Petri nets model systems, but it takes computer programs to simulate them. As might be expected in a simulation program, a main or driving program describes the overall system behavior, and subroutines perform the detail internals within each module. Actually, two subroutines are required to simulate modules, since the action set of legal transitions must be global. Here we call them module MA and module MB, where M identifies the module type (PT for passthrough, AS for asynchronous send, AR for asynchronous receive, and SS for synchronous send). Subroutine A forms the local legal transitions within a module. Combining all locally legal transitions into one large string allows us to randomly choose exactly one to actually execute, which is done in subroutine B. Parameters to be passed include the local state variables, output ports, an offset, the legal transition string, and the transition executable, each of which is defined below.

- local state vector: all places within module;
- output port vector: signals going from transitions to output ports;
- offset: number preceding the first transition label within the module;

- legal transition string: the action set of legal transitions within the module;
- transition executable: the label of the action to be executed.

For a communication module M, subroutine MA requires an input of the local places, along with the offset. If the module's transitions are consecutively numbered, the offset is the number of the first transition. This device allows the same subroutine to be used for all modules of the same type, since they behave in the same way, even though they have different transition numbers. The subroutine produces an output of all locally legal transitions in the form of a string. Thus, it takes on the following form:

$$\text{subroutine MA ([local state], } \textit{offset}, \mathbf{S}),$$

where $\mathbf{S}$ is the string of legal transitions.

Subroutine MB requires an input of all local places, along with the output ports and the number of transitions to be executed. If the executable transitions exist within the particular module, this subroutine will update the state variables and output signals, then pass the results back to the calling program. Thus, it takes on the following form:

$$\text{subroutine MB ([local state], [output ports], } \textit{offset}, y),$$

where y is the executable transition number. These concepts are best illustrated by an example.

EXAMPLE 7.9

Write subroutines simulating the behavior of the synchronous send modules shown in Figure 7.30. Specifically, this requires SSA and SSB.

Solution
Noting the synchronous send module in Figure 7.30, there are two transitions (numbered 1 and 2), three places (two are input places and one is hidden), and two output ports. For simplicity, we use the following names:

- local states: x_1 (top place),
  x_2 (lower right place),
  x_3 (lower left place);
- transitions: 1 (top transition),
  2 (bottom transition);
- output ports: z_1 (lower output signal),
  z_2 (right output signal).

Thus, the subroutine can be called using SSA($x_1, x_2, x_3, 0, \mathbf{S}$). Note that the same subroutine can be used for another synchronous send module SSB with transitions labeled 13 and 14 by using SSA($x_1, x_2, x_3, 12, \mathbf{S}$). The actual code producing the string $\mathbf{S}$ is given in Listing 7.9.

If the random selection of the executable transition is y, the subroutine SSB will actually update the affected places and produce outputs z_1 and z_2, as also shown in Listing 7.9.   ○

```
subroutine SSA(x₁, x₂, x₃, offset, S)
    S=∅
    if x₁>0 then S=S∪{1+offset}
    if x₃>0 and x₂>0 then S=S∪{2+offset}
return
```

```
subroutine SSB(x₁, x₂, x₃, z₁ ,z₂, offset, y)
    z₁=0
    z₂=0
    if y=1+offset then
            x₁=x₁-1
            x₃=x₃+1
            z₂=1
    end if
    if y=2+offset then
            x₂=x₂-1
            x₃=x₃-1
            z₁=1
    end if
return
```

LISTING 7.9   Subroutines SSA and SSB for the synchronous send communication module.

Simulations of module-based Petri nets assume a fictitious overseer in the form of a main program. Such an overseer cannot really exist, since in a real distributed system, there is no one in charge. In a simulation, however, this is permitted, since the overseer acts the part of nature and calls subroutines (which simulate the modules) as per the protocol established by the model. All processes act in an manner determined a priori and without any global communication. After initializing all places, the simulation cycles through a number of transitions while printing the state vector each time through. Without the cycle, all functions of the MA subroutine are performed, a transition is selected, and all functions of the MB subroutine are performed. Again, this is best shown by an example.

EXAMPLE 7.10

Consider a model characterized by two synchronous send modules as shown in Figure 7.31. Simulate the results assuming different initial states. Determine whether or not deadlock is a potential problem.

Solution
According to the modeling diagram of Figure 7.31, there are two external places and transitions in addition to the six places and four transitions inside the two SS (synchronous send) modules. These can be numbered in an arbitrary fashion, and we choose the one listed here.

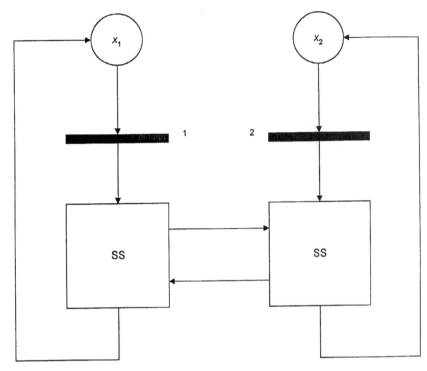

FIGURE 7.31   Example 7.10.

	Left SS module	Right SS module
Top place	x_3	x_6
Lower right place	x_5	x_8
Lower left place	x_4	x_7
Top transition	3	5
Bottom transition	4	6

It follows from these definitions that the main program given in Listing 7.10, in conjunction with the previously defined subroutines, will provide $n + 1$ event cycles of the simulation. Simulations for two different runs are given in Tables 7.7(a) and (b), where it will be noted that deadlock eventually becomes a problem in either case. ○

The ability to model a system as a set of submodules, each with a rudimentary communication structure of local inputs and outputs, is called *encapsulation*. Schematically this amounts to spatially partitioning a system into subsystems. Algorithmically, it is done by temporally partitioning tasks into subroutines. Clearly, encapsulation is powerful because it enables modeling on a series of *levels*. High levels can be described using very

```
k=0
read n
read x₁,  x₂, x₃, x₄, x₅, x₆, x₇, x₈
print k, x₁, x₂, x₃, x₄, x₅, x₆, x₇, x₈
for k=1 to n
        S=∅
        if x₁>0 then S=S∪{1}
        if x₂>0 then S=S∪{2}
        call SSA(x₃, x₅, x₄, 2, S1)
        call SSA(x₆, x₈, x₇, 5, S2)
        S=S∪S1∪S2

        y=RAND(S)

[1]     if y=1 then x₁=x₁-1: x₃=x₃+1
        if y=2 then x₂=x₂-1: x₆=x₆+1
        call SSB(x₃, x₅, x₄, x₁, x₈, 2, y)
        call SSB(x₆, x₈, x₇, x₂, x₅, 5, y)

        if y=0 then print "deadlock!": stop
        print k, x₁, x₂, x₃, x₄, x₅, x₆, x₇, x₈
next k
```

LISTING 7.10   Simulation of the system defined in Figure 7.31 using the synchronous send module subroutines SSA and SSB, Example 7.10.

TABLE 7.7   Two Simulation Results, Example 7.10

(a)

k	x_1	x_2	x_3	x_4	x_5	x_6	x_7	x_8
0	1	1	0	1	0	0	0	1
1	0	0	0	1	0	1	0	0
2	0	0	0	1	1	0	1	0
3	1	0	0	0	0	0	1	0
4	0	0	1	0	0	0	1	0
							Deadlock!	

(a)

k	x_1	x_2	x_3	x_4	x_5	x_6	x_7	x_8
0	1	1	0	0	0	0	0	0
1	0	0	0	0	0	1	0	0
2	0	0	0	0	1	0	1	0
							Deadlock!	

general behavioral descriptions, and low levels are full of detail. In between these extremes, there lie a number of intermediate levels as well.

The power of encapsulation lies in its understandability. Rather than having to look at the lowest level, one can model at a level more suitable to the problem at hand, then proceed to subdivide. As an illustration, let us reconsider the dining philosophers problem. But, this time, we associate each fork with a philosopher, thereby requiring a hungry adjacent philosopher to ask for the use of the fork from his peer. This different view does not change the problem, but it does permit a solution using just three modules – one for each philosopher (along with his associated fork). Now each philosopher has two higher-level local states – thinking and eating. The associated higher-level transitions are communication modules that take care of requesting and receiving tokens, so that the forks can move to where they are needed. Of course, these "higher level" transitions have both places and transitions contained within them. Following the philosopher's goals, the procedure goes as follows:

- Think:
    get left fork;
    get right fork.
- Eat:
    release left fork;
    release right fork.
- Repeat.

The system shown in Figure 7.32 implements this process along with detailed views of the required communication modules.

EXAMPLE 7.11

Find the linkages required for the $m = 3$ dining philosophers problem. Generalize this to a generic number of m philosophers.

Solution
In order for a fork request to be granted, the last philosopher to use the fork must first have returned it. Thus, we see that at the first requesting stage (request left fork), a request is made to the left-most philosopher. However, there can be no further request until this synchronous send module is in receipt of a release fork from the other side. It follows that all request left fork modules must link with their left neighbor and that all request right fork modules must link with their right neighbor. Also, when forks are released, an asynchronous send to the requesting modules is sufficient for the request to proceed. The links are clearly shown in Figure 7.33. Generalizing this to any integral number $m > 3$ should be clear.   ○

The dining philosophers problem is classic, because it is an elementary example of a process. In particular, note that it exhibits the following properties:

1. All philosophers (processes) are the same. Even though they each have "free will" in when they wish to eat and think, they are all composed of the same modules, communication attributes, and behaviors.

2. The model is *extensible*. That is, even though we explicitly handled the case for $m = 3$, the general case can be modeled just as easily.

3. The processes do not need to be synchronized. In fact, philosophers can come and go, and there is no need even to maintain a constant number of philosophers in the system over time.

Philosopher Module

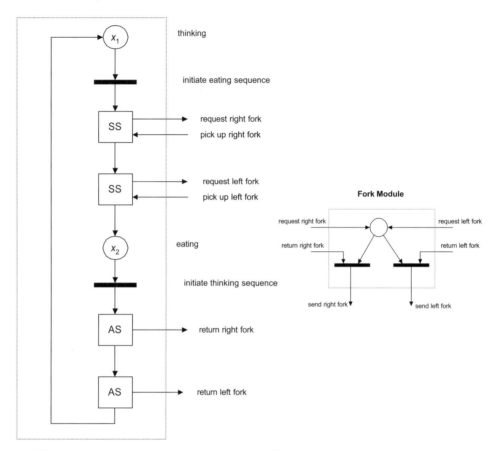

FIGURE 7.32   High-level Petri net view of one philosopher's view in the dining philosophers problem.

4.   There are interesting problems. In particular, deadlock can occur. Let's face it, we wouldn't define the dining philosophers problem at all unless there was trouble to be dealt with!

At the same time, there are solutions. If we were to create a authorizing agent to which all philosophers had to "apply" for eating status, we could probably avoid deadlock. Unfortunately, this is a high price to pay. Such an entity would have to be a *global agent* in that it would need to know the total system state when applications were received. In real problems, such centralized control is extremely expensive and thus undesirable.

Processes connote temporal models in that they are, like traditional computer software, a series of sequential steps. The chronological times at which these steps occur are

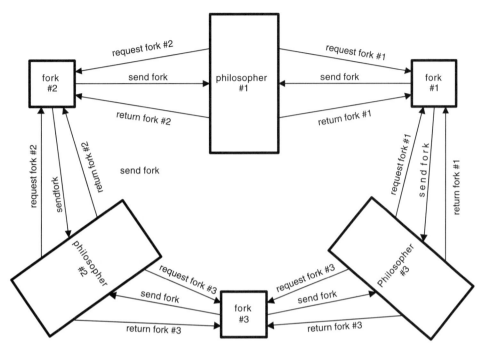

FIGURE 7.33   Modular high-level Petri net view of the dining philosophers problem, Example 7.11.

not as important as the order in which they occur. For this reason, processes are used to model digital communication systems, computer networks, and multitasking operating systems. In each of these systems, there are entities that create many process instances and there is no centralized control mechanism in place. Progress is made strictly by local control using inter-process communication. For example, a communication network has many nodes that send and receive messages. While each message is probably unique ("free will"), they all share the same structure (headers, address bits, footers, etc.). Thus, as nodes flood the system with many process instances, each instance can practice inter-process communication with its fellows as required. This is especially true for multitasking computer operating systems, where processes correspond to tasks.

On the other hand, many practical computer systems are anything but traditional. Real-time operating systems (RTOS) impose chronological time requirements on the multi-tasking process oriented systems described above. This amounts to placing synchronous demands on an asynchronous system, so an inevitable structure conflict follows. Indeed, this is the fundamental problem facing real-time system designers today. Solutions to this problem are anything but elegant, and are the topic of much current research.

For the modeler, multitasking systems and process modeling present similar problems. The reader has probably noted already that Petri nets are tools that do not directly lend themselves to chronological time analysis. However, it should be noted that a Petri net model can be rephrased as a finite state machine (recall the producer/consumer finite buffer problem). This FSM can then be simulated either synchronously or asynchronously as outlined at the beginning of this chapter. Unlike previous chapters of this text, there are no formulas and no state-versus-time graphs, just algorithms. At the same time, such analysis is sorely needed. This is also a rich source of research material in simulation and modeling.

BIBLIOGRAPHY

Attiya, H. and J. Welch, *Introduction to Computer System Performance Evaluation*. McGraw-Hill, 1998.

Fishwick, P., *Simulation Model Design and Execution: Building Discrete Worlds*. Prentice-Hall, 1995.

Hammersley, J. M. and D. C. Handscomb, *Monte Carlo Methods*. Methuen, 1964.

Knuth, D. E., *The Art of Computer Programming*, Vol. 1: *Fundamental Algorithms*. Addison-Wesley, 1997.

MIL 3, Inc., *Modeling and Simulating Computer Networks: A Hands on Approach Using OPNET*. Prentice-Hall, 1999.

Pinedo, M., *Scheduling: Theory, Algorithms and Systems*. Prentice-Hall, 1994.

Roberts, N., D. Andersen, R. Deal, M. Garet, and W. Shaffer, *Introduction to Computer Simulation*. Addison-Wesley, 1983.

Sape, M. (ed.), *Distributed Systems*, 2nd edn. Addison-Wesley, 1993.

Shatz, S. M., *Development of Distributed Software*. MacMillan, 1993.

Shoup, T. E., *Applied Numerical Methods for the Micro-Computer*. Prentice-Hall, 1984.

EXERCISES

7.1 Consider the following discrete feedback system in which $x(k)$ is the periodic sequence 4, 4, 4, 0, 0, 4, 4, 4, 0, 0, ... and $d(k)$ is the periodic sequence 0, 1, 2, 0, 1, 2, ... :

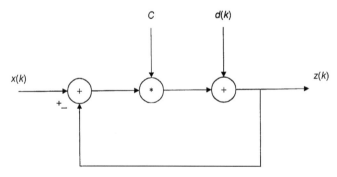

(a) Create a simulation for this system for a generic C-value.
(b) Execute and graph the simulations for $C = 1, \frac{1}{2}, \frac{1}{3}$, and $\frac{1}{4}$.
(c) Execute and graph the simulations for $C = 1, 2, 3$, and 4.

(d) What is the effect of increasing and decreasing C?

7.2 Consider a finite state machine with four states: 1, 2, 4, and 8. If the input $x(k)$ is 0, it counts up (1, 2, 4, 8, 1, 2, 4, 8, ...) and if $x(k)$ is 1, it counts down (1, 8, 4, 2, 1, 8, 4, 2, ...). The system is complicated in that a disturbance causes the system to jump over the next number. In particular, if $d(k) = 2$ and $x(k) = 0$, the state sequence is 1, 4, 1, 4, 1, 4, ..., and if $d(k) = 3$ and $x(k) = 0$, the sequence is 1, 8, 4, 2, 1, 8, 4, 2, 1, ... If $d(k) = 1$, the state sequence is normal. The behavior is analogous but backward for the case where $x(k) = 1$.

    (a) Create the defining state table and diagrams for this FSM.

    (b) Assuming that the input signal is regular and that it takes $\frac{1}{3}$ second between each input, simulate the output over the time horizon [0,10].

    (c) Assuming an asynchronous event-driven system with an average $\frac{1}{3}$ second between events and that the inter-event time is exponentially distributed, simulate the output over the time horizon [0,10].

7.3 Consider the following sampled feedback system with overt input $x(t)$ and disturbance $d(t)$. Assume that the signals are sampled every $\delta = 0.2$ second and that $x(t) = 5 + 2\cos\frac{1}{8}t$ and $d(t) = \frac{1}{4}\sin 3t + w(t)$, where $w(t)$ is Gaussian white noise with variance $\sigma_w^2 = 0.25$.

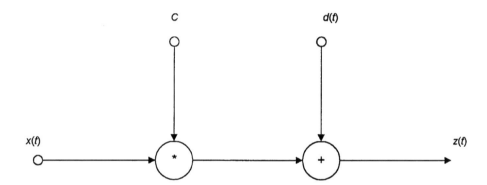

    (a) Assume for now that $C = 1$. Find and graph the output $z(t)$.

    (b) Now allow a generic C and put a feedback loop from the output to in front of the C so that C represents the forward-loop gain. Write a formula for the output z as a function time.

    (c) Defining the error signal as $e(t) = x(t) - z(t)$, find a formula for $e(t)$. What is the mean error?

    (d) Determine C so that the effect of $d(t)$ is minimized.

    (e) Determine C so that the error is less than 1 unit and the effect of the disturbance $d(t)$ is minimized.

7.4 Construct a Petri net that acts like an infinitely held switch; that is, it is initially "off", but turns "on" and stays "on" for the rest of the simulation. Verify the correct operation of this system.

7.5 Implement a simulation of the dining philosophers problem. Verify its proper working, and test for deadlock.

7.6 Implement the single-stage buffer producer/consumer problem described in the text. Verify the correctness of Tables 7.4 and 7.5.

7.7 Consider the finite state machine described in Example 7.2 and the simulations summarized in Figures 7.10 and 7.11. There are two sets of interpretations available directly, each graph involving the states, transitions, and time.

(i) **The time the system is in a given state**   This can be measured empirically by finding the time difference $t_k - t_{k-1}$ of the time the system is in each state, and tabulating. The proportion of time for each state is a useful statistical measure.

(ii) **The proportionate number of transitions from one state to another**   If it is known that there are no transitions from a state to itself, this can be found by counting the number of transitions from each state to any other state. From this, an empirical transition table can be created.

(a) Implement simulations to create graphs similar to Figures 7.10 and 7.11 over the time horizon [0,100]. Automate your program so as to (i) add up the amount of time in each state and (ii) count the number of transitions from one state to another.

(b) By execution of the programs of part (b), create the tables described above and draw the implied state diagram. Note that this will not be the same as Figure 7.8, since no transitions from a state to itself are allowed. Even so, the simulations will be identical!

7.8 Consider a system with two overt inputs x_1 and x_2, which can take on Boolean values 0 and 1. There is also one output z that is initially 0 but changes to 1 when x_1 is toggled (that is, it changes from 0 to 1 or from 1 to 0). Output z changes back to zero when x_2 is toggled.

(a) Construct a Petri net description of this system. Define the state vector carefully.

(b) Create a simulation for this system. Notice that either the inputs will have to be entered through a main program that acts whimsically like nature or you will need to implement a keyboard interrupt or polling mechanism by which to enter x_1 and x_2.

(c) Verify the correct working of this simulation.

7.9 Consider the following state transition table defining an FSM for only input $d(k)$ and output $z(k) = y(k)$:

	$y(k) = 1$	$y(k) = 2$	$y(k) = 4$
$d(k) = 1$	2	4	1
$d(k) = 2$	4	2	1
$d(k) = 3$	2	1	4

(a) Create the equivalent state diagram.

(b) Assuming a synchronous clock that beats four times per second, simulate the output over the time horizon [0,5].

(c) Assuming an asynchronous event-driven system that has an average of four events each second and is exponentially distributed, simulate the output over the time horizon [0,5].

(d) Notice that the mean state of an FSM is in is the total area of the state-versus-time graph divided by the time horizon. Compute this for each graph produced in part (c).

7.10 Change the dining philosophers problem to that of a finger-food buffet. Here philosophers have no private forks but instead use one of two serving forks (kept at the center of the table) to put hors d'oeuvres on their plates. Thus, now two philosophers can eat simultaneously.

(a) Will there be a deadlock?

(b) Draw an appropriate Petri net model for this problem, assuming $m = 3$ philosophers. Be sure to define all places and transitions.

(c) Implement a simulation of this model.

(d) Using the results of part (c), make a statistical analysis of each state and legal transitions between each state.

7.11 Use the basis communication modules to define a generic philosopher in the buffet problem defined in Exercise 7.10.

(a) From this single philosopher, create a more general Petri net model like that shown in Figure 7.33 for $m = 4$ philosophers from which a simulation can be made.

(b) Simulate the performance of this model, and verify its correctness.

(c) Using the results of part (b), make a statistical analysis of each state and legal transitions between each state.

(d) Extend this same model for general m. Write a program to accept m and simulate the performance of this model.

7.12 Write pairs MA and MB of subroutines for each of the basic communication modules defined in Section 7.5:

- passthrough: $\quad$ $PTA(x, \textit{offset}, \mathbf{S})$,
  $PTB(x, z, \textit{offset}, y)$;
- asynchronous send: $\quad$ $ASA(x, \textit{offset}, \mathbf{S})$,
  $ASB(x, z_1, z_2, z_3, \textit{offset}, y)$;
- asynchronous receive: $\quad$ $ARA(x_1, x_2, z, \textit{offset}, \mathbf{S})$,
  $ARB(x_1, x_2, z, \textit{offset}, y)$;
- synchronous send: $\quad$ $SSA(x_1, x_2, x_3, \textit{offset}, \mathbf{S})$,
  $SSB(x_1, x_2, x_3, z_1, z_2, \textit{offset}, y)$.

7.13 Implement a simulation of the producer/consumer problem with a finite buffer described in Example 7.5. Using a long-run simulation, estimate the percentage of time that the buffer is in each state.

(a) Do this for $n = 1000, 2000, 5000$, and $10\,000$ steps.

(b) Are the proportions of time constant for each case in part (a) or do they vary? Use this as a measure of whether or not a measurement is dependable.

(c) Estimate an empirical result by finding a reasonable formula for the proportion of time the buffer is in state m, for $m = 0, 1, 2, \ldots$.

7.14 Consider the single-stage buffer producer/consumer problem described in the text. Using the transition diagram shown in Figure 7.24 and the numerical values discovered in Figures 7.4 and 7.5 to create an FSM asynchronous simulation. Assume an average of four events per second and an exponentially distributed inter-event time.

(a) Draw an appropriate transition diagram.

(b) Create the simulation.

(c) Simulate a single instance of the model on the time horizon $[0,10]$.

(d) Using the results of a model generated on $[0,1000]$, estimate the proportion of time the system spends in each state. Note that this should be automated in your program!

(e) Verify the consistency of the results produced in part (d) with those given in Figures 7.4 and 7.5.

7.15 Consider the Petri net defined by the following figure.

(a) Simulate this using the initial condition given.

(b) Is this system bounded? If it is, list all possible states.

(c) Assuming that the answer to part (b) is "yes", find the proportion of time the system is any of the legal states.

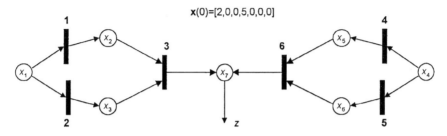

x(0)=[2,0,0,5,0,0,0]

7.16 Consider the Petri net defined by the following figure:

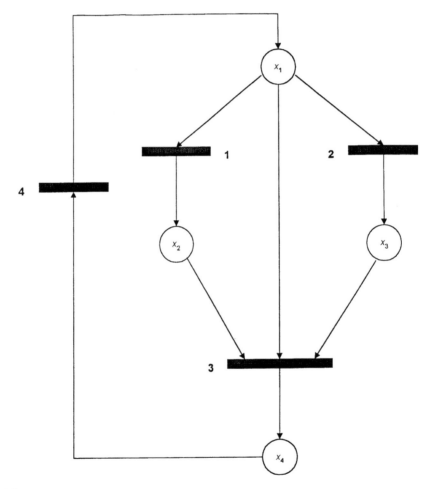

Part 1

(a) Simulate this using the initial condition $x(0) = [0, 1, 1, 0]$ and updating the nodes as described in the text.

(b) Is this system bounded? If it is, list all possible states.

(c) Assuming that the answer to part (b) is "yes", find the proportion of time the system is any of the legal states.

Part 2

(d) Simulate this using the initial condition $\mathbf{x}(0) = [0, 1, 1, 0]$ and updating the nodes by defining the increment and decrement steps as follows:

$$\begin{aligned} \text{increment:} \quad & x = 1, \\ \text{decrement:} \quad & x = 0. \end{aligned}$$

(e) Is this system bounded? If it is, list all possible states.

(f) Assuming that the answer to part (e) is "yes", find the proportion of time the system is any of the legal states.

(g) Create the associated FSM.

7.17 "Fix" the dining philosophers problem by instituting a waiter/maitre d' who takes reservations for meals as each of the philosophers become hungry. This should eliminate deadlock, since if any one philosopher is eating, the competitors will be blocked until she is finished.

(a) Draw the appropriate Petri net for $m = 3$ philosophers.

(b) Perform a simulation, thus showing that deadlock is no longer a problem.

(c) Verify the correctness of your program.

7.18 Extend Exercise 7.17 to the case of a generic m by:

(a) Creating a generic philosophers model by using the communication modules describe in the text.

(b) Illustrating how the general m philosophers would be configured in a Petri net diagram.

(c) Performing a simulation for $m = 5$ philosophers and noting that there is still no problem with deadlock.

(d) Do a statistical analysis showing:

    (i) all possible states, along with their interpretation;

    (ii) the proportion of time the system is in each state;

    (iii) the list of all legal transitions between the states.

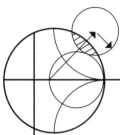

Markov Processes

8.1   PROBABILISTIC SYSTEMS

Time is traditionally partitioned into one of three sets: the past, present, and future. Presumably, we know where we are in the here and now, and hopefully we can remember the past. The idea is to somehow anticipate or predict the future. In this chapter we will do just that. Toward this end, we consider that discrete time k represents the present, $k - 1, k - 2, \ldots$ denotes the past, and $k + 1$ the immediate future. A system responds to a sequence of inputs by producing a sequence of system states, which creates a sequence of outputs.

	Past	Present	Future
Time	$k - 1, k - 2, \ldots$	k	$k + 1, k + 2, \ldots$
Input	$x(k - 1), x(k - 2), \ldots$	$x(k)$	$x(k + 1), x(k + 2), \ldots$
State	$y(k - 1), y(k - 2), \ldots$	$y(k)$	$y(k + 1), y(k + 2), \ldots$
Output	$z(k - 1), z(k - 2), \ldots$	$z(k)$	$z(k + 1), z(k + 2), \ldots$

In general, the future state will depend on the historical input signals and sequence of states. If the future is deterministic, the future is uniquely predictable given the historical record, and $y(k + 1) = f[x(0), x(1), \ldots, x(k); y(0), y(1), \ldots, y(k)]$, for a suitable function f.

The problem is that the future is often not deterministic and there is little certainty. In this case, it is useful to make probabilistic statements, since we only know likelihoods of event occurrences. For instance, consider models implied by statements such as the following:

1. The stock market was high yesterday and is even higher today. Therefore, it should be higher yet tomorrow.
2. If a telephone system has three calls currently on hold, can we expect that there will be at most two calls on hold one minute from now?
3. Today is the first day of the rest of your life.

In model (1), let $y(k)$ be a random variable representing the value of a certain stock on day k. According to the statement, the value $y(k + 1)$ is heavily influenced by $y(k)$ and $y(k - 1)$. A typical probabilistic statement might be $\Pr[y(k + 1) = 8700 \mid y(k) = 8600$ and $y(k - 1) = 8500]$, where the vertical line is read "given" and separates the conjecture (left of "|") and the hypothesis (right of "|"). In other words, assuming that today's stock value is 8600 and that yesterday's is 8500, we infer a prediction regarding the likelihood of tomorrow's value being 8700. Although it is not the case in this example, it is possible that the future state $y(k + 1)$ might also be influenced by another variable, such as the unemployment rate $x(k)$.

In the second model, we note that the past is apparently irrelevant. Interpreting the state to be the number of phone calls queued up, there can be $y(k) \in \{0, 1, 2, \ldots\}$ different states. A probabilistic statement regarding the future state is thus based only on the present state. Specifically, assuming that $y(k)$ represents the number of calls on hold at time k, $\Pr[y(k + 1) \leqslant 2 \mid y(k) = 3]$ mathematically describes the English language statement made in (2) above. Model (3) is a more casual statement espousing the same model. It implies that, regardless of how badly life has treated you in the past, tomorrow's happenings are solely a result of your current actions. Oh if that were only true!

In general, a probabilistic model makes a conjecture regarding a future state based on the present and historical state and inputs. For input x and state y, the most general probabilistic model is $\Pr[f(y(k + 1)) \mid g(x(k), x(k - 1), x(k - 2), \ldots), h(y(k), y(k - 1), y(k - 2), \ldots)]$, where f, g, and h are suitable defined functions. However, in this chapter, we consider the simplest of these probabilistic models. Specifically, we make the following assumptions:

P1. The system has a finite number of possible states $j = 1, 2, \ldots, n$.

P2. The future state of the system is dependent only on the current state. It is autonomous (there is no input) and is independent of time and the historical record.

P3. Denote by $p_j(k)$ the probability that the system will be in state j. Since the current state can be any of the n different states, by Bayes' theorem it follows that $p_{j+1}(k)$ is linearly dependent on the current state probabilities and that

$$p_j(k + 1) = p_{1j}\, p_1(k) + p_{2j}\, p_2(k) + \ldots + p_{nj}\, p_n(k),$$

where

$$p_{1j} = \Pr[X(k + 1) = j \mid X(k) = 1],$$
$$p_{2j} = \Pr[X(k + 1) = j \mid X(k) = 2],$$
$$\vdots$$
$$p_{nj} = \Pr[X(k + 1) = j \mid X(k) = n].$$

These are called the *transition probabilities*. Notice the notational implication that p_{ij} is the probability that the system goes from the current state i to the future state j in the next time increment. Also, it is independent of time.

Such systems are called *Markovian* and the probabilistic models characterizing them are called *Markov chains*.

Since Markov chains are finite, it is possible to totally characterize the chain's state by a vector function of time. Assuming n states, the probability state vector $\mathbf{p}(k)$ has n components, each of which is the probability of being in that state. The jth component of this vector gives the probability that the system is in state $p_j(k)$ at time k:

$$\mathbf{p}(k) = [p_1(k), \; p_2(k), \ldots, \; p_n(k)].$$

Because Markov chains are linear, the set of transition probabilities p_{ij} from the current state to any future state can be displayed as the following so-called state *transition matrix* $\mathbf{P} = [p_{ij}]$:

$$\mathbf{P} = \begin{bmatrix} p_{11} & p_{12} & \cdots & p_{1n} \\ p_{21} & p_{22} & \cdots & p_{2n} \\ \vdots & \vdots & \ddots & \vdots \\ p_{n1} & p_{n2} & \cdots & p_{nn} \end{bmatrix}.$$

The fact that p_{ij} is time-independent means that $\mathbf{P}$ is constant and ensures that the stationarity requirement **P3** is maintained.

The possible states for a Markovian system are mutually exclusive and exhaustive. Therefore, the probability that the system is in state j at time $k + 1$ is given by the *total probability* of the state the system was in at time k. Specifically, the probability that the new state will be j is the following linear combination of all possible previous states and times k:

$$\text{state } j: \quad p_j(k + 1) = p_1(k)p_{1j} + p_2(k)p_{2j} + \ldots + p_n(k)p_{nj}.$$

Since this is true for every state $j = 1, \; 2, \ldots, \; n$, there is actually a whole set of n relationships:

$$\text{state 1:} \quad p_1(k + 1) = p_1(k)p_{11} + p_2(k)p_{21} + \ldots + p_n(k)p_{n1},$$
$$\text{state 2:} \quad p_2(k + 1) = p_1(k)p_{12} + p_2(k)p_{22} + \ldots + p_n(k)p_{n2},$$
$$\vdots \qquad\qquad\qquad\qquad \vdots$$
$$\text{state } n: \quad p_n(k + 1) = p_1(k)p_{1n} + p_2(k)p_{2n} + \ldots + p_n(k)p_{nn}.$$

This is more conveniently stated using vector and matrix notation: $\mathbf{p}(k + 1) = \mathbf{p}(k)\mathbf{P}$.

8.2 DISCRETE-TIME MARKOV PROCESSES

Since a Markov chain has n finite states, a convenient graphical way by which to view a Markov process is a state diagram. Each of the states $1, 2, \ldots, n$ is represented by a vertical line. Directed arcs joining vertical lines represent the transition probabilities. For instance, the diagram drawn in Figure 8.1 shows $n = 4$ states with transition probabilities

$p_{11} = \frac{3}{4}$ (the probability the state will remain unchanged),
$p_{13} = \frac{1}{4}$,

$p_{23} = 1$ (it is certain that the state will change 2 to 3),
$p_{31} = \frac{1}{3}$,

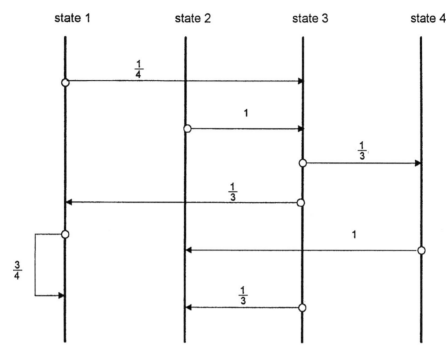

FIGURE 8.1   State diagram for $n = 4$ states.

$p_{32} = \frac{1}{3}$,

$p_{34} = \frac{1}{3}$,

$p_{42} = 1$ (this is also a certainty).

All other transition probabilities are zero, since there are no other arcs connecting states in the diagram. It should be noted that a necessary consequence of this construction is that the sum of all transition probability arcs *leaving* each state must be unity. This is because, for any given state, some transition must happen and the sum of all possible probabilities is one. Mathematically, $\sum_{j=1}^{n} p_{ij} = 1$ for each state $i = 1, 2, \ldots, n$. In other words, summing over the second index yields unity. Summing on i has a much different interpretation.

The transition matrix **P** can best be thought of as an $n \times n$ matrix in which each row corresponds to the current state and each column corresponds to the future state. For example, the probabilistic system defined by the state diagram of Figure 8.1, has the probability transition matrix

$$P = \begin{bmatrix} \frac{3}{4} & 0 & \frac{1}{4} & 0 \\ 0 & 0 & 1 & 0 \\ \frac{1}{3} & \frac{1}{3} & 0 & \frac{1}{3} \\ 0 & 1 & 0 & 0 \end{bmatrix}.$$

The entry $p_{32} = \frac{1}{3}$ of the transition matrix is the probability that a transition from state 3 to state 2 will occur in a single time instant. Since the transition probabilities sum to unity on the second (column) index, each row in the transition matrix **P** always totals one.

Graphically, this is equivalent to noting that the transition probabilities *leaving* each state must also sum to unity.

EXAMPLE 8.1

A game of chance has probability p of winning and probability $q = 1 - p$ of losing. If a gambler wins, the house pays him a dollar; if he loses, he pays the house the same amount. Suppose the gambler, who initially has 2 dollars, decides to play until he is either broke or doubles his money. Represent the night's activity as a Markov chain.

Solution

In general, state labels can be quite arbitrary. There is no reason why they must begin at 1 or even be consecutive integers for that matter. So, for convenience, let us define the state to be the number of dollars our gambler currently has – in this case, the only possibilities are 0, 1, 2, 3, or 4. Thus, he begins in state 2, and the game is over once he attains state 4 (doubles the initial 2 dollars) or state 0 (he goes broke). The state transition diagram is shown in Figure 8.2.

This diagram clearly shows that once our gambler is in state 0 or state 4, his state remains fixed. From here he either remains impoverished or rich. In this case, we say there are two *absorbing barriers*, since once there, he never leaves. For the intermediate states, the likelihood of advancing is p and that of regressing

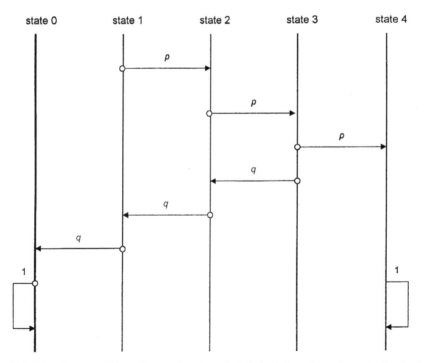

FIGURE 8.2   State transition diagram for Example 8.1, including the option of going broke.

is $q = 1 - p$. The corresponding state transition matrix is

$$
\mathbf{P} = \begin{bmatrix}
1 & 0 & 0 & 0 & 0 \\
q & 0 & p & 0 & 0 \\
0 & q & 0 & p & 0 \\
0 & 0 & q & 0 & p \\
0 & 0 & 0 & 0 & 1
\end{bmatrix}.
$$

As an alternative, suppose the casino has a liberal policy that never allows patrons to go broke. In other words, upon a client losing his last dollar, the bank gives the client another one. This time, the zero state is said to be a *reflecting barrier*, since once in state (0) our gambler is certain to go to state (1) in the next round. The corresponding state diagram is shown in Figure 8.3, and the adjusted state transition matrix is

$$
\mathbf{P} = \begin{bmatrix}
0 & 1 & 0 & 0 & 0 \\
q & 0 & p & 0 & 0 \\
0 & q & 0 & p & 0 \\
0 & 0 & q & 0 & p \\
0 & 0 & 0 & 0 & 1
\end{bmatrix}.
$$

Finally, suppose that our friend decides "the sky's the limit". He will gamble forever, unless he goes broke. In this case, we assume that the bank will never run out of money. There is now a countably infinite number of states, and this is

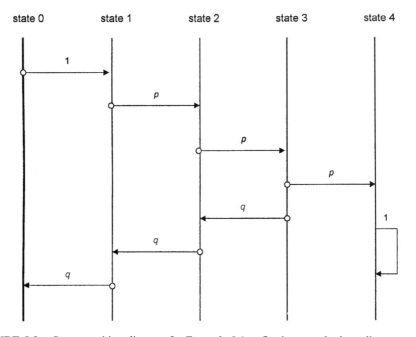

FIGURE 8.3 State transition diagram for Example 8.1, reflecting a no-broke policy.

therefore not a Markov chain, since it requires infinitely many rows and columns in the transition matrix. Even so, the system can be represented as an infinite set of partial difference equations. The transition probabilities are

$$
p_{ij} = \begin{cases} 1, & i = 0,\ j = 0, \\ q, & i > 0,\ j = i - 1, \\ p, & i > 0,\ j = i + 1, \\ 0, & \text{otherwise.} \end{cases} \qquad \bigcirc
$$

Markov probabilistic models seek to find the state vector $\mathbf{p}(k)$ for any time k given that the transition matrix P is known. Using vectors and matrices, we know that $\mathbf{p}(k + 1) = \mathbf{p}(k)\mathbf{P}$. But, this is a recursive relationship in that $\mathbf{p}(k)$ is defined in terms of itself. For instance, $\mathbf{p}(4)$ can only be found if $\mathbf{p}(3)$ is known, and $\mathbf{p}(3)$ can only be found if $\mathbf{p}(2)$ is known, and so forth, forming a chain of intermediate probability state vectors. Thus, we call this the *Markov chain problem*. What is really needed is an explicit formula for $\mathbf{p}(k)$ where the intermediate states do not have to be computed along the way.

Typically, the Markov chain problem is stated as an initial-value problem. The initial state is given for time $k = 0$, and it is desired to find the probability of each state for generic time k. That is, the problem is to find $\mathbf{p}(k)$ for

$$
\mathbf{p}(0) \text{ given,}
$$
$$
\mathbf{p}(k + 1) = \mathbf{p}(k)\mathbf{P}. \tag{8.2}
$$

Recursively, this problem is not difficult. Setting $k = 0$, $\mathbf{p}(1) = \mathbf{p}(0)\mathbf{P}$, and letting $k = 1$, $\mathbf{p}(2) = \mathbf{p}(1)\mathbf{P} = [\mathbf{p}(0)\mathbf{P}]\mathbf{P} = \mathbf{p}(0)\mathbf{P}^2$. Continuing in this fashion for $k = 2$, $\mathbf{p}(3) = \mathbf{p}(0)\mathbf{P}^3$. This can continue on as long as needed, and, the problem reduces to that of raising the matrix $\mathbf{P}$ to an arbitrary power k:

$$
\mathbf{p}(k) = \mathbf{p}(0)\mathbf{P}^k. \tag{8.3}
$$

EXAMPLE 8.2

Consider a three-state Markov probabilistic system that begins in state 2 at time $k = 0$. Find explicit formulas for the matrix $\mathbf{P}^k$ and state vector $\mathbf{p}(k)$, assuming that the states are 1, 2 and 3, and

$$
\mathbf{P} = \begin{bmatrix} 1 & 0 & 0 \\ \frac{1}{2} & \frac{1}{2} & 0 \\ 0 & \frac{1}{2} & \frac{1}{2} \end{bmatrix}.
$$

Solution

We begin by computing powers of **P**. This is straightforward by repeatedly multiplying **P** as follows:

$$\mathbf{P}^2 = \begin{bmatrix} 1 & 0 & 0 \\ \frac{1}{2} & \frac{1}{2} & 0 \\ 0 & \frac{1}{2} & \frac{1}{2} \end{bmatrix} \begin{bmatrix} 1 & 0 & 0 \\ \frac{1}{2} & \frac{1}{2} & 0 \\ 0 & \frac{1}{2} & \frac{1}{2} \end{bmatrix} = \begin{bmatrix} 1 & 0 & 0 \\ \frac{3}{4} & \frac{1}{4} & 0 \\ \frac{1}{4} & \frac{1}{2} & \frac{1}{4} \end{bmatrix},$$

$$\mathbf{P}^3 = \mathbf{P}^2\mathbf{P} = \begin{bmatrix} 1 & 0 & 0 \\ \frac{7}{8} & \frac{1}{8} & 0 \\ \frac{1}{2} & \frac{3}{8} & \frac{1}{8} \end{bmatrix},$$

$$\mathbf{P}^4 = \mathbf{P}^3\mathbf{P} = \begin{bmatrix} 1 & 0 & 0 \\ \frac{15}{16} & \frac{1}{16} & 0 \\ \frac{11}{16} & \frac{1}{4} & \frac{1}{16} \end{bmatrix}.$$

By noting the pattern, $\mathbf{P}^k$ is evident, and we conclude that

$$\mathbf{P}^k = \begin{bmatrix} 1 & 0 & 0 \\ \left(1 - \frac{1}{2^k}\right) & \frac{1}{2^k} & 0 \\ \left(1 - \frac{k+1}{2^k}\right) & \frac{k}{2^k} & \frac{1}{2^k} \end{bmatrix}. \tag{8.4}$$

Since the problem statement gives that the system is initially in state 2, the initial state vector is $\mathbf{p}(0) = [0, 1, 0]$. It follows from Equation (8.3) that the state vector for generic time k is given by

$$\mathbf{p}(k) = [0, 1, 0] \begin{bmatrix} 1 & 0 & 0 \\ \left(1 - \frac{1}{2^k}\right) & \frac{1}{2^k} & 0 \\ \left(1 - \frac{k+1}{2^k}\right) & \frac{k}{2^k} & \frac{1}{2^k} \end{bmatrix} = \left[\left(1 - \frac{1}{2^k}\right), \frac{1}{2^k}, 0\right]. \tag{8.5}$$

Since many problems deal only with long-term behavior, it is often sufficient to deal with the steady-state solutions to Markov systems. By definition, these solutions are given by

$$\mathbf{P}_{ss} = \lim_{k \to \infty} \mathbf{P}^k,$$
$$\boldsymbol{\pi} = \lim_{k \to \infty} \mathbf{p}(k). \tag{8.6}$$

It can be shown that Equation (8.2) always has a steady-state solution and the solution is independent of the initial conditions. By applying the formulas (8.6) to

$\mathbf{P}^k$ and $\mathbf{p}(k)$ given by Equations (8.4) and (8.5), the steady-state solutions follow immediately:

$$\mathbf{P}_{ss} = \begin{bmatrix} 1 & 0 & 0 \\ 1 & 0 & 0 \\ 1 & 0 & 0 \end{bmatrix},$$

$$\boldsymbol{\pi} = [1, \ 0, \ 0]. \qquad \qquad \bigcirc$$

The procedure outlined in Example 8.2 is straightforward, except that one cannot always expect to detect the pattern to determine the matrix $\mathbf{P}^k$. However, the system described by Equations (8.2) is a linear first-order initial-value problem, and the solution of such a system is well understood mathematically. The general solution to Equation (8.2) is beyond the scope of this section, but it is a standard topic in linear systems texts. Even so, there is one special case that is of particular interest, namely, that of the binary system shown in Figure 8.4. This system has only states 0 and 1. Recalling that each row must sum to unity, the transition matrix has only two degrees of freedom. Therefore, if α and β are two problem-specific parameters such that $0 \leqslant \alpha \leqslant 1$ and $0 \leqslant \beta \leqslant 1$, the state transition matrix is

$$\mathbf{P} = \begin{bmatrix} 1 - \alpha & \alpha \\ \beta & 1 - \beta \end{bmatrix}. \qquad (8.7)$$

It can be shown that an explicit formula for $\mathbf{P}^k$ is given by

$$\mathbf{P}^k = \frac{1}{\alpha + \beta} \begin{bmatrix} \beta & \alpha \\ \beta & \alpha \end{bmatrix} + \frac{(1 - \alpha - \beta)^k}{\alpha + \beta} \begin{bmatrix} \alpha & -\alpha \\ -\beta & \beta \end{bmatrix}. \qquad (8.8)$$

Assuming that the initial probability state vector $\mathbf{p}(0)$ is known, Equation (8.8) leads directly to the generic state vector $\mathbf{p}(k)$ using Equation (8.3).

Since the probabilities α and β must be between 0 and 1, the sum must be between 0 and 2. Thus, the factor $1 - \alpha - \beta$ is between -1 and $+1$. Accordingly,

$$\lim_{k \to \infty} (1 - \alpha - \beta)^k = 0, \qquad 0 < \alpha + \beta < 2.$$

Therefore, the steady-state solution to the binary Markov system is

$$\mathbf{P}_{ss} = \lim_{k \to \infty} \mathbf{P}^k = \frac{1}{\alpha + \beta} \begin{bmatrix} \beta & \alpha \\ \beta & \alpha \end{bmatrix}. \qquad (8.9)$$

Assuming that the steady-state state vector is $\boldsymbol{\pi} = [\pi_0, \ \pi_1]$, it can be shown on multiplying $\mathbf{P}_{ss}$ by $\boldsymbol{\pi}$ that

$$\pi_0 = \frac{\beta}{\alpha + \beta},$$

$$\pi_1 = \frac{\alpha}{\alpha + \beta}. \qquad (8.10)$$

EXAMPLE 8.3

A carrier-sense, multiple-access/collision detect (CSMA/CD) is a broadcast communication protocol for local area networks. In a CSMA/CD network, a

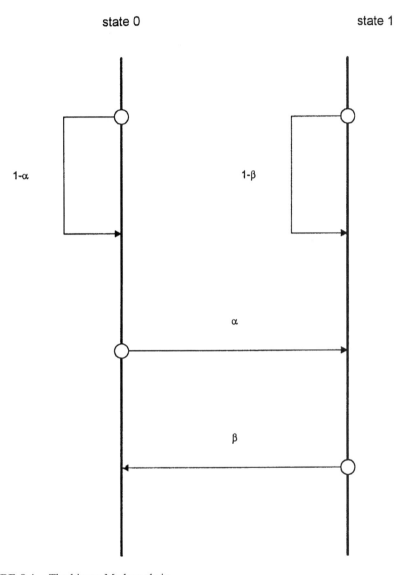

FIGURE 8.4   The binary Markov chain.

node wishing to transmit first checks for traffic. If traffic is sensed, no message is sent (since this would cause an erroneous message), and the medium is tested again at the next clock cycle. On the other hand, if no traffic is observed, the message is sent, beginning at the next clock cycle. While this works most of the time, it is possible that a *clear* medium on one clock cycle will be *busy* during the next, resulting in a collision. This causes a garbled message, and both message initiators must retransmit after waiting a random period of time.

Suppose that it is known that the probability a clear medium at time k will also be clear at time $k + 1$ is 0.9. Further, the probability a busy medium at time k will still be busy at time $k + 1$ is 0.8. Analyze the traffic load of this system.

Solution

It is evident from the description that the system has two states: *clear* and *busy*. Let us agree that if $y(k) = 0$, the medium is clear, and that if $y(k) = 1$, the medium is busy. Then, by restating the problem using random variables,

$$p_{00} = \Pr[Y(k + 1) = 0 \mid Y(k) = 0] = 0.9,$$
$$p_{11} = \Pr[Y(k + 1) = 1 \mid Y(k) = 1] = 0.8.$$

Using the nomenclature of Equation (8.7) with $\alpha = 0.1$ and $\beta = 0.2$, the probability transition matrix is

$$\mathbf{P} = \begin{bmatrix} 0.9 & 0.1 \\ 0.2 & 0.8 \end{bmatrix}.$$

Equation (8.8) gives the explicit formula for the kth transition matrix, which, as we have learned, is just the kth power of $\mathbf{P}$:

$$\mathbf{P}^k = \begin{bmatrix} \frac{2}{3} & \frac{1}{3} \\ \frac{2}{3} & \frac{1}{3} \end{bmatrix} + \left(\tfrac{7}{10}\right)^k \begin{bmatrix} \frac{1}{3} & -\frac{1}{3} \\ -\frac{2}{3} & \frac{2}{3} \end{bmatrix}$$

$$= \begin{bmatrix} \frac{1}{3}[2 + (0.7)^k] & \frac{1}{3}[1 - (0.7)^k] \\ \frac{2}{3}[1 - (0.7)^k] & \frac{1}{3}[1 + 2(0.7)^k] \end{bmatrix}. \tag{8.11}$$

However, such a local area network operates in steady-state mode virtually all the time. Therefore, we let $k \to \infty$, and observe that the second term goes to zero, resulting in

$$\mathbf{P}_{ss} = \frac{1}{3} \begin{bmatrix} 2 & 1 \\ 2 & 1 \end{bmatrix},$$
$$\boldsymbol{\pi} = \begin{bmatrix} \frac{2}{3}, & \frac{1}{3} \end{bmatrix}.$$

It follows that it is reasonable to expect the system to remain clear from one cycle to the next two-thirds of the time, but to turn busy one-third of the time.   ○

These results show that of the two terms in Equation (8.8), the first term, which is independent of k, is the steady-state transition matrix and the second is only applicable to the transient phase. In practice the transient phase is often of little interest and the steady state alone is sufficient. In this case, it is not necessary to have an explicit formula for $\mathbf{P}^k$; the probability transition matrix $\mathbf{P}$ alone is sufficient. It is then preferable to have an alternative technique to find $\boldsymbol{\pi}$ so that $\mathbf{P}^k$ and $\mathbf{p}(k)$ don't have to be found first. Toward this end, notice that, at steady state, $\mathbf{p}(k + 1)$, $\mathbf{p}(k)$, and $\boldsymbol{\pi}$ are interchangeable, and, by Equation (8.1), $\boldsymbol{\pi} = \boldsymbol{\pi}\mathbf{P}$. The application of this important formula is best illustrated with an example.

EXAMPLE 8.4

Suppose that $\mathbf{P}$ is defined as in Example (8.3). Since there are two states, $\boldsymbol{\pi} = [\pi_0, \ \pi_1]$. We look for only the steady-state solution, so $\boldsymbol{\pi} = \boldsymbol{\pi}\mathbf{P}$, and, therefore,

$$[\pi_0, \ \pi_1] = [\pi_0, \ \pi_1]\begin{bmatrix} 0.9 & 0.1 \\ 0.2 & 0.8 \end{bmatrix}.$$

Equating coefficients, there are two equations and two unknowns:

$$\pi_0 = \tfrac{9}{10}\pi_0 + \tfrac{2}{10}\pi_1,$$
$$\pi_1 = \tfrac{1}{10}\pi_0 + \tfrac{8}{10}\pi_1.$$

Unfortunately, these two equations are redundant, since $\pi_0 = 2\pi_1$. However, recalling that $\pi_0 + \pi_1 = 1$ and solving simultaneously, it follows that $\pi_0 = \tfrac{2}{3}$ and $\pi_1 = \tfrac{1}{3}$. Therefore, $\boldsymbol{\pi} = \left[\tfrac{2}{3}, \tfrac{1}{3}\right]$ as was determined in Example 8.3.   ○

Markov chains can be applied to a number of examples where the immediate future state of a system can be made to depend linearly only on the present state: $\mathbf{p}(k+1) = \mathbf{p}(k)\mathbf{P}$. This implies that $\mathbf{p}(k) = \mathbf{p}(k-1)\mathbf{P}$, which leads immediately to $\mathbf{p}(k+1) = \mathbf{p}(k)\mathbf{P} = \mathbf{p}(k-1)\mathbf{P}^2$. Accordingly, it follows that every future state can be predicted from as remote a history as desired. However, the converse is not true. For instance, the probabilistic system defined by $\mathbf{p}(k+1) = \tfrac{1}{2}\mathbf{p}(k) + \tfrac{2}{3}\mathbf{p}(k-1)$ is apparently not a Markov chain, since the future depends on both the present and past states. At the same time, the Markov requirement is not obvious, since $\mathbf{p}(k)$ could be a binary vector in which the system could be restated in Markov form.

EXAMPLE 8.5

Consider the random process with two states 1 and 2 defined by

$$\mathbf{p}(k+1) = \mathbf{p}(k)\begin{bmatrix} \tfrac{1}{2} & \tfrac{1}{2} \\ \tfrac{1}{3} & \tfrac{2}{3} \end{bmatrix}. \tag{8.12}$$

This is a Markov process, since the future state is predicted on the basis of the present state, the elements of the matrix $\mathbf{P}$ are between 0 and 1, and the sum of each row is unity. It follows that

$$p_1(k+1) = \tfrac{1}{2}p_1(k) + \tfrac{1}{3}p_2(k), \tag{8.13a}$$
$$p_2(k+1) = \tfrac{1}{2}p_1(k) + \tfrac{2}{3}p_2(k). \tag{8.13b}$$

Doubling Equation (8.13a) and subtracting it from (8.13b) eliminates $p_2(k)$ and results in $p_2(k+1) = 2p_1(k+1) - \tfrac{1}{2}p_1(k)$. Replacing k by $k-1$ in this equation gives $p_2(k) = 2p_1(k) - \tfrac{1}{2}p_1(k-1)$. Combining this with Equation (8.13a) gives the following second-order difference equation for the isolated variable p_2:

$$p_1(k+1) = \tfrac{7}{6}p_1(k) - \tfrac{1}{6}p_1(k-1). \tag{8.14}$$

Even though Equation (8.14) represents the future state as a combination of both present and past states, it is possible to express it as a two-state vector by which the future is expressed as a function of only the present, as is clear from Equation (8.13) in the original statement of the problem.

It can be shown that in general the second-order difference equation $p_1(k+1) = Ap_1(k) + Bp_1(k-1)$ represents a two-state Markov chain if and only if $A + B = 1$. From the above result, we note that this statement is verified. On the other hand, just what those states are is another matter. For instance, if $p_2(k) = p_1(k-1)$, it follows that

$$\mathbf{p}(k+1) = \mathbf{p}(k) \begin{bmatrix} \frac{7}{3} & 1 \\ -\frac{1}{6} & 0 \end{bmatrix}.$$

This is clearly not a Markov model, since one of the elements of the matrix $\mathbf{P}$ is negative and the sum of each row is not unity. So, even though we know that an equivalent Markov model exists, this is not it!   ○

8.3   RANDOM WALKS

In Section 3.8 we discussed the actions of a random walker who would either advance with probability p or regress with probability $g = 1 - p$. We also found that the walker's actions on a discrete random walk in discrete time are easily simulated. Even so, it is desirable to have a theory with which to predict these experimental results. It turns out that these results can be predicted using Markov chains. In fact, random walks present a particularly useful application of Markov chains, since, from the geometry, future destinations depend only on the present location of a random walker. Further, the destination of the walker can only be adjacent to the present location. This is true, regardless of the dimension of the space in question. As shown earlier, this results in a sparse probability transition matrix $\mathbf{P}$ that is banded.

EXAMPLE 8.6

Consider the domain shown below, in which a walker traverses a path of four nodes 1, 2, 3, 4. We allow walks in which half of the time he remains in place and half the time he travels to an adjacent node. Which direction he travels, left or right, is equally likely. Create a simulation with which we must describe the walker's travels.

Solution
Since random walks are probabilistic in nature, the description must be statistical. They can be simulated by performing a series of experiments or, more formally speaking, instances of the underlying processes. In this case, the logical statistical choice is the proportion of time the walker is in each state, y. Since simulations require long-term averages to be computed, they are analogous to steady-state

probability vectors $\boldsymbol{\pi}$. Accordingly, the initial state is not important, since such systems are linear, guaranteed to be stable and to always converge to the same proportions, regardless of where the system begins.

1. Beginning at node 1, the walker can either stay at $y = 1$ or go to $y = 2$, each with probability $\frac{1}{2}$. From a simulation point of view, this can be stated as

    ```
    r=INT(1+2*RND)
    if y=1 then
            if r=1 then
                    y=1
            else
                    y=2
    ```

    At node 2, the walker can go back to $y = 1$ or forward to $y = 3$, each with probability $\frac{1}{4}$ The probability that he remains at $y = 2$ is $\frac{1}{2}$. Thus,

    ```
    r=INT(1+4*RND)
    if y=2 then
            if r=1 then
                    y=1
            else
                    if r=4 then
                            y=3
                    else
                            y=2
    ```

    There is an obvious symmetry between boundary nodes $y = 1$ and $y = 4$ and interior nodes $y = 2$ and $y = 3$. Putting this all together and including accumulators by which to calculate the proportion of time he spends in each state gives Listing 8.1. Since we were not told just where the walker began, we arbitrarily let him start at node $y = 1$. Recall that in the steady state, the initial state is irrelevant so long as time k becomes large enough.

    The algorithm also includes four accumulators T_1, T_2, T_3, and T_4, which count the number of times the walker visits each node. Execution of the algorithm for a value of $n = 1000$ in Listing 8.1 gives ratios leading us to believe that

    $$\pi_1 \approx \frac{T_1}{n} \approx \frac{1}{6}, \qquad \pi_2 \approx \frac{T_2}{n} \approx \frac{1}{3}, \qquad \pi_3 \approx \frac{T_3}{n} \approx \frac{1}{3}, \qquad \pi_4 \approx \frac{T_4}{n} \approx \frac{1}{6}.$$

2. This result can also be obtained analytically. We begin by considering each of the walk's nodes as a different state. From the description, the state diagram is as shown in Figure 8.5. It follows from this that the probability transition matrix is given by

    $$\mathbf{P} = \begin{bmatrix} \frac{1}{2} & \frac{1}{2} & 0 & 0 \\ \frac{1}{4} & \frac{1}{2} & \frac{1}{4} & 0 \\ 0 & \frac{1}{4} & \frac{1}{2} & \frac{1}{4} \\ 0 & 0 & \frac{1}{2} & \frac{1}{2} \end{bmatrix}. \tag{8.15}$$

```
T₁=0
T₂=0
T₃=0
T₄=0
y=1
for k=1 to n
        if y=1 then
                r=INT(1+2*RND)
                if r=1 then y₁=1
                if r=2 then y₁=2
        end if
        if y=2 then
                r=INT(1+4*RND)
                if r=1 then y₁=1
                if r=2 then y₁=4
                if r=3 then y₁=4
                if r=4 then y₁=3
        end if
        if y=3 then
                r=INT(1+4*RND)
                if r=1 then y₁=3
                if r=2 then y₁=1
                if r=3 then y₁=3
                if r=4 then y₁=4
        end if
        if y=4 then
                r=INT(1+2*RND)
                if r=1 then y₁=3
                if r=2 then y₁=4
        end if
        y=y₁
        print k, y
        if y=1 then T₁=T₁+1
        if y=2 then T₂=T₂+1
        if y=3 then T₃=T₃+1
        if y=4 then T₄=T₄+1
next k
print T₁/n, T₂/n, T₃/n, T₄/n
```

LISTING 8.1   Estimating the components of the steady-state state vector $\boldsymbol{\pi}$, Example 8.6.

Using the technique illustrated in Example 8.4 to solve for the steady-state probability state vector gives the following four equations:

$$\pi_1 = \tfrac{1}{2}\pi_1 + \tfrac{1}{4}\pi_2,$$
$$\pi_2 = \tfrac{1}{2}\pi_1 + \tfrac{1}{2}\pi_2 + \tfrac{1}{4}\pi_3,$$
$$\pi_3 = \tfrac{1}{4}\pi_2 + \tfrac{1}{2}\pi_3 + \tfrac{1}{2}\pi_4,$$
$$1 = \pi_1 + \pi_2 + \pi_3 + \pi_4.$$

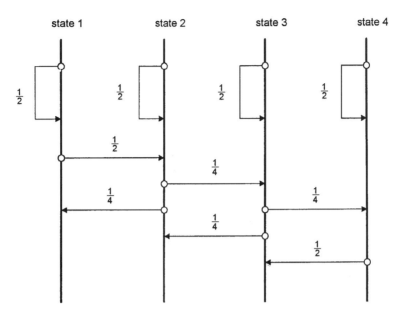

state 1   state 2   state 3   state 4

FIGURE 8.5   State diagram for a four-node random walk with reflecting borders, Example 8.6.

The first three come from applying the steady-state relationship $\boldsymbol{\pi} = \boldsymbol{\pi}\mathbf{P}$ to the state transition matrix (8.15). Since the rank of $\mathbf{P}$ is 3 (in general, the rank of $\mathbf{P}$ is $n - 1$), we ignore the (arbitrarily chosen) fourth equation $\pi_4 = \frac{1}{2}\pi_3 + \frac{1}{2}\pi_4$. In its place, we make use of the fact that all steady-state probabilities sum to unity. Solving simultaneously gives the unique solution

$$\pi_1 = \tfrac{1}{6}, \qquad \pi_2 = \tfrac{1}{3}, \qquad \pi_3 = \tfrac{1}{3}, \qquad \pi_4 = \tfrac{1}{6}.$$

Again we note the walk's symmetry.

3.   Owing to the symmetry of the random walk, there is a very handy simplification to this problem. Rather than thinking of each step in the walk denoting a different state, we should think of there being two boundary nodes (1 and 4) and two interior nodes (2 and 3). Each pair of nodes behave in the same way. Formally, we define these new states as follows:

state A: boundary nodes (1 and 4);

state B: interior nodes (2 and 3).

From state A, transitions to both A and B are equally likely. From state B, the probability that the walker moves to state A is $\frac{1}{4}$. On the other hand, the probability of transition to an interior node (state B) is $\frac{3}{4}$. This gives the state diagram shown in Figure 8.6, along with the probability transition matrix

$$\mathbf{P} = \begin{bmatrix} \frac{1}{2} & \frac{1}{2} \\ \frac{1}{4} & \frac{3}{4} \end{bmatrix}.$$

state A   state B

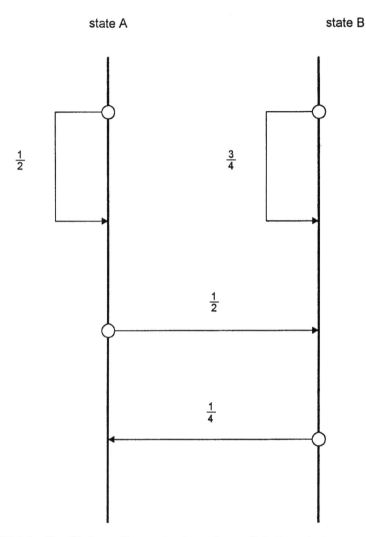

$\frac{1}{2}$ $\frac{3}{4}$

$\frac{1}{2}$

$\frac{1}{4}$

FIGURE 8.6   Simplified state diagram for the random walk in Example 8.6.

Solving this system for the steady-state vector $\boldsymbol{\pi} = [\pi_A, \pi_B]$ results in $\pi_A = \frac{1}{3}$ and $\pi_B = \frac{2}{3}$. Since each of the two boundary nodes is equally likely, the probability of each node is $\pi_1 = \pi_4 = \frac{1}{2}\pi_A = \frac{1}{6}$. Similarly, the probability of each of the two interior nodes is $\pi_2 = \pi_3 = \frac{1}{2}\pi_B = \frac{1}{3}$, which confirms each of our previous results. ○

The important lesson from this example is that states are not always geometrical or physical objects. Nor is the state representation necessarily unique. Even so, whether by simulation or by exhaustive redundant states or with reduced states, all results are the same. This is true for random walks of more than one dimension as well.

EXAMPLE 8.7

Consider the 3×3 two-dimensional grid with nine nodes shown in Figure 8.7. A random walker can go in any direction either horizontally or vertically with equal probability, but cannot remain in place. For instance, if the walker is at the upper left node 1, the next step can be to either node 2 or node 4, each with probability $\frac{1}{2}$. Describe this probabilistic system.

Solution

1. Taking a cue from Example 8.6, we notice that there are essentially three different types of nodes: corners (4), edges (4), and an interior (1). Noting that the walker can only step horizontally or vertically, we formally define these classes as follows:

   class A: corner (bordered by 2 edges);

   class B: edges (bordered by 2 corners and 1 interior);

   class C: interior (bordered by 4 edges).

   This leads to the 3-state state diagram of Figure 8.8, in which transitions can take place only with adjacent nodes. From the state diagram, the probability transition matrix is

$$\mathbf{P} = \begin{bmatrix} 0 & 1 & 0 \\ \frac{2}{3} & 0 & \frac{1}{3} \\ 0 & 1 & 0 \end{bmatrix}.$$

   Finding the three steady-state probabilities $\boldsymbol{\pi} = [\pi_A, \ \pi_B, \ \pi_C]$ from $\mathbf{P}$ gives $\pi_A = \frac{1}{3}$, $\pi_B = \frac{1}{2}$, and $\pi_C = \frac{1}{6}$. It follows from this that the probability that any of the nodes 1, 3, 7, or 9 are visited is $\pi_1 = \pi_3 = \pi_7 = \pi_9 = \frac{1}{4}\pi_A = \frac{1}{12}$. Similarly, for nodes 2, 4, 6, or 8, the probability is $\pi_2 = \pi_4 = \pi_6 = \pi_8 = \frac{1}{4}\pi_B = \frac{1}{8}$. Since there is only one interior node (5), its probability is $\pi_5 = \frac{1}{6}$.

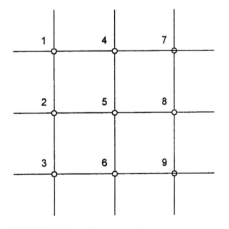
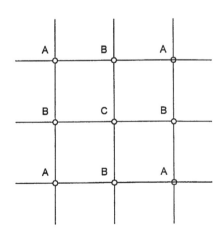

FIGURE 8.7   Example 8.7.

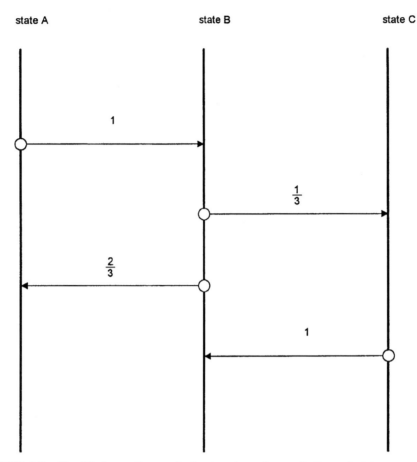

FIGURE 8.8 Simplified state diagram for 3×3 node random walk, Example 8.7.

2. If a simulation is desired, each of the nodes must be handled individually. Since each corner node has the same behavior, they can be treated as a class, node 1:

```
if x=1 then
        r=INT(1+2*RND)
        if r=1 then y₁=2
        if r=2 then y₁=4
    end if
```

Treating node 2 as a typical edge node,

```
if x=2 then
        r=INT(1+3*RND)
        if r=1 then y₁=1
        if r=2 then y₁=3
        if r=3 then y₁=5
    end if
```

For the interior node 5,

```
if x=5 then
        r=INT(1+4*RND)
        if r=1 then y₁=2
        if r=2 then y₁=4
        if r=3 then y₁=6
        if r=4 then y₁=8
end if
```

The results of such a simulation verify the theoretical results derived using the Markov chain model.   ○

8.4   POISSON PROCESSES

Up to this point, we have only considered Markov chains in discrete time. At each successive k-value, a new state is chosen at random, leading to one particular instance of the random process. In order to achieve a statistically meaningful result, it is necessary to either simulate sufficiently many instances to obtain meaningful proportions or proceed with the theoretical Markov results. This is possible because any system describable as a finite state machine can be defined uniquely by a state diagram. Assuming a regular synchronous system, a state diagram can in turn be modeled as a discrete Markov probabilistic system and solved using the Markov chain techniques described above.

It is also possible to model asynchronous event-driven models in continuous time. For instance, consider an old-fashioned telephone switchboard system, with the state being the number of callers requesting service. Each incoming phone call would be an event that appears to occur spontaneously in continuous time from an outside source. An event's occurrence increments the system state by one, but the exact time at which it occurs is unknown. Even if the calls occur asynchronously, the event times can still be determined statistically by characterizing their inter-event intervals. If the inter-event times are exponentially distributed, the process is said to be *Poisson*, and models of this type are continuous Markovian.

Suppose that $N(t)$ is a *random variable* that represents the number of events that occur during the continuous time interval $[0, t]$. Further, let $n = 0, 1, 2, \ldots$ represent the number of events in the infinite sample space. Specifically, we define $P_n(t) = \Pr[N(t) = n]$ to be the probability that exactly n events occur over the time interval $[0, t]$. This probability may be characterized by the following three rather general so-called *Poisson postulates*:

P1. The number of events occurring over any single time interval is independent of the number of events that have occurred over any other non-overlapping time interval. That is, the numbers of events occurring over times $[t_1, t_2]$ and $[t_2, t_3]$ are statistically independent. Mathematically stated, $P_{m+n}(t_3 - t_1) = P_m(t_2 - t_1)P_n(t_3 - t_2)$.

P2. The probability that a *single* event occurs over a *short* time interval $[t, t+h]$ is approximately proportional to h. More precisely, $P_1(h) = \lambda h + o(h^2)$, where the notation $o(h^2)$ (read "small oh of h square") denotes any function having the properties

$$\lim_{h \to 0} \frac{o(h^2)}{h^2} \neq 0,$$

$$\lim_{h \to 0} \frac{o(h^2)}{h} = 0.$$

That is, the Taylor series representation of $o(h^2)$ comprises terms of degree 2 and higher, but no constant or linear terms in h.

P3. The probability that *more than one* event occurs during a short time interval $[t, t + h]$ is essentially nil. Stated mathematically, $P_n(h) = o(h^2)$, $n > 1$.

Any random variable that satisfies these three postulates is said to be a *Poisson random variable* and $P_n(t)$ is called a *Poisson process*.

Although these postulates might appear to be too vague and general to be of any use, they lead to very powerful mathematical results. By first considering only *small* time intervals h in postulates P2, and P3, we can infer a general formula for $P_n(h)$ for all non-negative integers n:

$$P_n(h) = \begin{cases} 1 - \lambda h + o_1(h^2), & n = 0, \\ \lambda h + o_2(h^2), & n = 1, \\ o_3(h^2), & n > 1, \end{cases} \tag{8.16}$$

where h is *small* in the sense that $o_1(h^2)$, $o_2(h^2)$, and $o_3(h^2)$ can be ignored since they are *infinitesimally* small. Equation (8.16) follows since P_n is a probability mass function and thus must sum to unity: $\sum_{n=0}^{\infty} P_n(h) = 1$ over each point of the *sample space* $n = 0, 1, 2, \ldots$. However, Equation (8.16) is useful only for "small" h, where we can ignore second-order effects and $o(h^2)/h \to 0$. Therefore, let us proceed to extend this result to $P_n(t)$ for any time t, be it large or small.

Consider a Poisson process defined on the interval $[0, \ t + h]$ as shown in Figure 8.9. We begin by finding the probability of a non-event, $P_0(t + h)$. The subintervals $[0, t]$ and $[t, t + h]$ are disjoint, and the only way for zero events to occur on $[0, t + h]$ is for

0 events to occur on $[0, t]$ and 0 events to occur on $[t, t + h]$.

The mathematical consequence of this is that

$$P_0(t + h) = P_0(t)P_0(h). \tag{8.17}$$

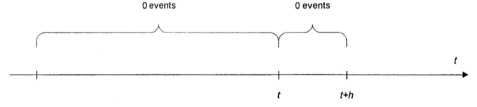

FIGURE 8.9   Poisson non-events $P_0(t + h)$ over the time interval $[0, t + h]$.

From combining Equations (8.16) and (8.17), it follows that $P_0(t+h) = P_0(t)[1 - \lambda h + o(h^2)]$. Rearranging algebraically,

$$\frac{P_0(t+h) - P_0(t)}{h} = -\lambda P_0(t) + \frac{o(h^2)}{h} P_0(t).$$

By taking the limit of both sides and recalling the definition of the derivative along with the definition of $o(h^2)$,

$$\lim_{h \to 0} \frac{P_0(t+h) - P_0(t)}{h} = \lim_{h \to 0}\left[-\lambda + \frac{o(h^2)}{h} \right] P_0(t),$$

$$\frac{dP_0}{dt} = -\lambda,$$

$$(8.18)$$

since λ is a constant and $P_0(t)$ is independent of h. Thus, the Poisson non-event probability is given by an elementary differential equation. Combined with the initial condition $P_0(0) = 1$ (i.e., it is a *certainty* that no events will occur at time $t = 0$), the solution is

$$P_0(t) = e^{-\lambda t}. \tag{8.19}$$

Unlike Equation (8.16), which is true only for small time intervals h, Equation (8.19) is valid for any time $t \geq 0$. Intuitively, it states that to receive no calls is quite likely for $t \approx 0$ (probability close to 1); however, the probability tapers off exponentially with larger time t. It seems that peace and quiet might work in the short term – but not for long! Certainly, this equation makes intuitive sense, since one should expect ever smaller probabilities of no incoming phone calls as time increases.

A similar argument can be made for $n > 0$. For instance, in order that there be $n = 1$ event over the interval $[0, \, t + h]$, exactly one of the following mutually exclusive instances happens:

1 event occurs on $[0, t]$ *and* 0 events occur on $[t, t + h]$

or

0 events occur on $[0, t]$ *and* 1 event occurs on $[t, t + h]$.

Again, since the intervals do not overlap, these are independent events. However, by noting the key word "*or*", this time there are two terms that contribute to the net probability, since the cases are mutually exclusive:

$$P_1(t+h) = P_1(t)P_0(h) + P_0(t)P_1(h),$$

$$P_1(t+h) = P_1(t)[1 - \lambda h + o_1(h^2)] + P_0(t)[\lambda h + o_2(h^2)],$$

$$\frac{P_1(t+h) - P_1(t)}{h} = \left[-\lambda + \frac{o_1(h^2)}{h} \right] P_1(t) + \left[\lambda + \frac{o_2(h^2)}{h} \right] P_0(t). \tag{8.20}$$

Upon taking limits, Equation (8.20) reduces to the following differential equation:

$$\frac{d}{dt} P_1(t) = -\lambda P_1(t) + \lambda P_0(t). \tag{8.21}$$

Equations (8.19) and (8.21) combine to give

$$\frac{d}{dt} P_1(t) = -\lambda P_1(t) + \lambda e^{-\lambda t}. \tag{8.22}$$

Thus, the Poisson process once again reduces to solving a first-order linear differential equation. Taking a cue from the case $n = 0$, we multiply each side of Equation (8.22) by the integration factor $e^{-\lambda t}$ so that the left-hand side of the equation now becomes the derivative of a product. From this, it is just a matter of integrating and solving for $P_1(t)$ subject to the initial condition $P_1(0) = 0$ (since it is impossible to receive a call at the instant time $t = 0$):

$$e^{\lambda t}\frac{d}{dt}P_1(t) + \lambda e^{\lambda t}P_1(t) = \lambda,$$

$$\frac{d}{dt}[e^{\lambda t}P_1(t)] = \lambda,$$

$$P_1(t) = e^{-\lambda t}\lambda t. \tag{8.23}$$

Equations (8.19) and (8.23) provide explicit formulas for the probability of $n = 0$ and $n = 1$ events during the time interval $[0, t]$. By continuing in a similar fashion, this procedure can produce a formula for generic n. In particular, it can be shown that, in general,

$$\frac{d}{dt}P_n(t) = -\lambda P_n(t) + \lambda P_{n-1}(t). \tag{8.24}$$

Equation (8.24) is the most general first-order differential equation describing a Poisson process. The accompanying initial condition is

$$P_n(0) = \begin{cases} 1, & n = 0, \\ 0, & n > 0. \end{cases} \tag{8.25}$$

The system given by Equations (8.24) subject to Equation (8.25) can be solved recursively. Once $P_{n-1}(t)$ is known, Equation (8.24) reduces to a non-homogenous linear first-order differential equation that can be solved by elementary means. The general solution follows by induction:

$$P_n(t) = e^{-\lambda t}\frac{(\lambda t)^n}{n!}, \qquad n = 0,\ 1,\ 2,\dots. \tag{8.26}$$

It should be remembered that Equation (8.26) follows directly from the very general Poisson postulates P1, P2, and P3. It is remarkable that these apparently innocuous properties imply the explicit probability that exactly n events occur over a time interval $[0, t]$.

It is sometimes convenient to denote the product λt by μ. Since this removes explicit time dependence, an analogous *Poisson random variable* P_n is defined by

$$P_n = e^{-\mu}\frac{\mu^n}{n!}, \qquad n = 0,\ 1,\ 2,\dots,$$

where $\mu = \lambda t$ and $P_n = P_n(t)$. With this substitution, λ is the mean event *occurrence rate* and μ is the *mean number of events* during time t.

Since Equation (8.26) gives the probability that exactly n events occur, it represents how much probability "mass" exists at each of the $n = 0,\ 1,\ 2,\dots$ points of the sample space. As such, Equation (8.26) is called the *probability mass function* for the random variable $N(t)$. Since it is a function of time $t \geqslant 0$, it is time-dependent. Even so, it should be noted that the time depends only on the interval length and not on absolute time. Thus, $P_n(t)$ is a *memoryless* process. That is, regardless of the current state, the probability of new entries remains unchanged.

Properties of the Poisson Process

While the explicit probabilities given by $P_n(t)$ are useful, summary information such as the *mean* and *variance* is often sufficient. The mean of this process is straightforward. Using the *expectation operator E*, the mean, is computed as follows:

$$
\begin{aligned}
E[N] &\equiv \sum_{n=0}^{\infty} n P_n \\
&= \sum_{n=0}^{\infty} n e^{-\mu} \frac{\mu^{n-1}}{n!} \\
&= e^{-\mu} \mu \sum_{n=1}^{\infty} \frac{\mu^{n-1}}{(n-1)!} \\
&= \mu e^{-\mu} \sum_{n=0}^{\infty} \frac{\mu^n}{n!} \\
&= \mu e^{-\mu}(e^{\mu}) \\
&= \mu.
\end{aligned}
\tag{8.27}
$$

Thus, while λ represents the event occurrence rate (the number of events per unit time), μ represents the mean *number* of events that occur during time t. Similarly it can be shown that the second moment for a Poisson process is given by $E[N^2] = \mu(\mu + 1)$. Therefore, the variance is given by

$$
\begin{aligned}
\sigma^2 &= E[\{N - E[N]\}^2] \\
&= E[N^2] - E^2[N] \\
&= \mu(\mu + 1) - \mu^2 \\
&= \mu.
\end{aligned}
\tag{8.28}
$$

Interestingly, both the mean and variance of a Poisson distribution are μ. This property is unique to Poisson distributions.

EXAMPLE 8.8

Consider the specific Poisson distribution characterized by $\lambda = 0.9$ events per unit time. Graph and interpret the Poisson probability mass and distribution functions.

The Poisson mass function $P_n(t)$ is a function of two variables: n and t. Therefore, a three-dimensional technique is required to sketch their graphs. Since there is an explicit formula for the mass function, it is probably easiest to simply use a spreadsheet for this example. Using rows for n-values and columns for t-

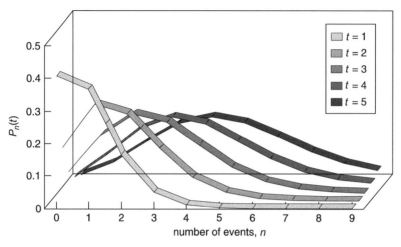

FIGURE 8.10   Family of Poisson density functions, Example 8.8.

values, the following tabulates values of Equation (8.26) assuming an event rate of $\lambda = 0.9$:

n	$t = 1$	$t = 2$	$t = 3$
0	0.407	0.165	0.067
1	0.366	0.298	0.181
2	0.165	0.268	0.245
3	0.049	0.161	0.220
4	0.011	0.072	0.149

By extending this table to $t = 1, 2, 3, 4, 5$ and $n = 0, 1, 2, \ldots, 9$ it creates a family of three-dimensional line graphs. A family of five such graphs are shown in Figure 8.10. It should be noticed that the case of $t = 1$, which is equivalent to a mean of $\mu = 0.9$, resembles a discrete exponential function. For larger t-values, the function is similar to a discrete gamma distribution.

Since the distribution function is simply the cumulative probability function, it can also be found using a spreadsheet by simply summing successive rows. The results are shown using a family of "three-dimensional area graphs" in Figure 8.11. As with all distribution functions, they are each monotonically increasing, and asymptotically approach unity as n increases.   ○

EXAMPLE 8.9

Suppose that telephone calls arrive randomly throughout the day at an office, at an average rate of two calls per minute. Assuming this to be a Poisson process:
(a) How many calls can be expected between 2:00 and 2:05?
(b) What is the probability that no calls will be received between 2:00 and 2:01?

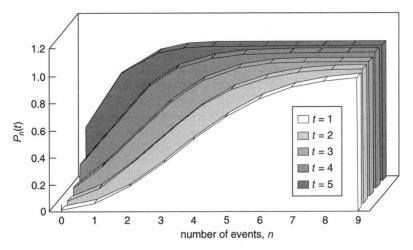

FIGURE 8.11   Family of Poisson distribution functions, Example 8.8.

(c) What is the probability that more than one call will be received between 2:03 and 2:04?

Solution

Since we are assuming this to be a Poisson process, exactly when we start counting is not important. The only thing that matters is the length of the time interval in question.

(a) The event rate is $\lambda = 2$ calls per minute. Since the time interval is $t = 5$ minutes, the mean number of calls is $\mu = 2 \times 5 = 10$ calls.

(b) Since $t = 1$ minute, the required probability of exactly $n = 0$ calls is $P_0(1) = e^{-2}(2^0/0!) \approx 0.1353$.

(c) The key phrase is "more than one" event. Since each of the $n = 2, 3, 4, \ldots$ events are mutually exclusive, the total probability is

$$\Pr[N > 1] = \sum_{n=2}^{\infty} P_n(1)$$

$$= e^{-2} \sum_{n=2}^{\infty} \frac{2^n}{n!}.$$

However, rather than computing or approximating the infinite series, it is easier to just calculate the complementary event and subtract from unity:

$$\Pr[N > 1] = 1 - P_0(1) - P_1(1)$$

$$= 1 - e^{-2} - 2e^{-2}$$

$$\approx 0.5950. \qquad \bigcirc$$

8.5 THE EXPONENTIAL DISTRIBUTION

The Poisson distribution is defined for a discrete random variable $N(t)$. It is closely related to another, but continuous, random variable called the *exponential random variable*. Rather than looking at the number of events in a fixed time interval, the exponential distribution considers the time T between each event, called the *inter-event time*. Since the process is Poisson and there is no memory, the time between two successive events is the same as the time until the first event t. That is, $\Pr[t_1 < T < t_1 + t] = \Pr[T < t]$, for all t_1. Since there are no events that occur on the interval $[0, t]$,

$$\Pr[T > t] = \Pr[N = 0]$$
$$= P_0(t)$$
$$= e^{-\lambda t}.$$

But this is, by definition, the complement of the distribution function $F_T(t)$, and, therefore, $F_T(t) = \Pr[T < t] = 1 - e^{-\lambda t}$. Since the continuous probability density function $f_T(t)$ is the derivative of the distribution function,

$$f_T(t) = \frac{dF_T}{dt}$$
$$= \lambda e^{-\lambda t}. \tag{8.29}$$

The relationship between the Poisson and exponential distributions is especially important in that they are duals of one another. The Poisson distribution measures the *number of events* that occur over the time interval $[0, t]$, and the exponential distribution measures the *time between events*. In addition, there is a complementary relationship between the statistics of the two distributions as well. For instance, calculation of the mean is straightforward:

$$E[T] = \int_{-\infty}^{\infty} t f_T(t)\, dt$$
$$= \int_{0}^{\infty} t \lambda e^{-\lambda t}\, dt \tag{8.30}$$
$$= \frac{1}{\lambda}.$$

Since λ is the number of events per unit time, the mean of T represents the time per event. More specifically, it is the average *inter-event* time.

Similarly, the second moment can be shown to be $E[T^2] = 2/\lambda^2$, and the variance becomes

$$\sigma^2 = E[T^2] - E^2[T]$$
$$= \frac{2}{\lambda^2} - \left(\frac{1}{\lambda}\right)^2 \tag{8.31}$$
$$= \frac{1}{\lambda^2}.$$

TABLE 8.1   Comparison of the Poisson and Exponential Distributions

	Poisson distribution	Exponential distribution
Measurement	Number of events n during time t	Time t between each event
Type	Discrete	Continuous
Mean	$\mu = \lambda t$	$\mu = 1/\lambda$
Variance	$\sigma^2 = \lambda t$	$\sigma^2 = 1/\lambda^2$

Therefore, the standard deviation $\sigma = 1/\lambda$. These statistics, along with their Poisson statements are summarized in Table 8.1.

EXAMPLE 8.10

The service times for six bank customers are 30, 50, 120, 20, 80, and 30 seconds. Assuming that these times are representative of an underlying exponential distribution:

(a) Determine the parameter λ.
(b) What is the probability that a customer will take more than 90 seconds?
(c) If 500 customers are serviced in one day, how many can be expected to take more than 90 seconds?

Solution
It can be shown that the best *unbiased estimator* $\hat{\lambda}$ of the exponential distribution is the reciprocal of the average experimental time. This should not come as a surprise, since we have already shown that $\mu = 1/\lambda$ is also the theoretical mean of the distribution. This can be proven using maximum-likelihood estimators as discussed in Appendix A.

(a) The average of the service times is

$$\hat{\mu} = \tfrac{1}{6}(30 + 50 + 120 + 20 + 80 + 30)$$
$$= 55 \text{ seconds per customer.}$$

Thus, the estimator for λ is

$$\hat{\lambda} = \frac{1}{\hat{\mu}}$$
$$\approx 0.0182 \text{ customers per second.}$$

(b) Using this estimate, the density function follows immediately,

$f_T(t) = \frac{1}{55}e^{-t/55}$. This, in turn, leads to the required probability:

$$\Pr[T > 90] = \int_{90}^{\infty} f_T(t)\, dt$$

$$= \int_{90}^{\infty} \frac{1}{55}e^{-t/55}\, dt$$

$$= e^{-90/55} \approx 0.1947.$$

(c) The estimated number of customers receiving service taking 90 seconds or greater is $(500)(0.1947) \approx 97$ customers of the 500 total.   ○

The Poisson process takes the system's view. It characterizes the state of the system by giving the probability a number of events occur. On the other hand, the exponential distribution characterizes the event's view. It gives the expected time between event occurances. These two views are equivalent. Another interpretation of the Poisson process is that of customers entering a service queue. The system "sees" the customers as arriving for service according to the Poisson postulates. Equivalently, the inter-event times are exponentially distributed. It is also reasonable to believe that customers depart a service queue exponentially in that the service time is random but is more likely to be shorter than longer. (However, arguing from past experience, one can certainly think of supermarket counter-examples!)

Poisson models are similar to a telephone system, where calls constitute events and length of call constitutes service time. In this way, it is possible to think of the traffic problem as a queuing system. Messages arrive according to the Poisson postulates at a rate of λ messages per second. These messages are serviced, and therefore depart the system, after μ seconds per message. This is illustrated in Figure 8.12, which can be characterized by noting the number of events in the system – either queued or being serviced – as a function of time. This very important problem, which is considered in detail in Chapter 9, was treated initially by A. K. Erlang. Erlang was the Danish electrical engineering genius who pioneered the basis of traffic problems for American Telephone and Telegraph (ATT).

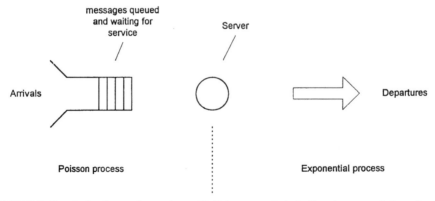

FIGURE 8.12   A simple queuing system with Poisson events (rate λ) and exponential service time (length $\tau = 1/\mu$).

His work characterizes the foundations of queuing systems still used so heavily in operations research today.

8.6 SIMULATING A POISSON PROCESS

It is not always possible to find an explicit formula for probabilistic models. Often, the mechanism seems clear and fairly well understood, but an analytical formula cannot be found by which to answer interesting questions. For this reason, purely mathematical models tend to ask questions that lend themselves to approximations and can be easily answered. In such cases, we often pose the question mathematically, but answer it using numerical simulation techniques.

A defining characteristic of Poisson processes is that events have exponential inter-event times. Using this fact, we can create an instance of a Poisson process by generating a sequence of n exponentially distributed random variates to represent the times at which each of the n events occur. This can be done by repeated application of the formula $t^{(\text{new})} = t - \mu \ln(\text{RND})$, where RND is uniformly distributed on [0,1]. The following algorithm can be used to generate a single instance of a Poisson process for $j = 0, 1, \ldots, n$ event times $t(j)$:

```
t(0)=0
for j=1 to n
        t(j)=t(j-1)-μ*ln(RND)
next j
```

Since each event in a Poisson process corresponds to a new "arrival", the system state is j and the time at which it expresses itself is $t(j)$. The results for a typical run using $\mu = 3$ are shown in Figure 8.13.

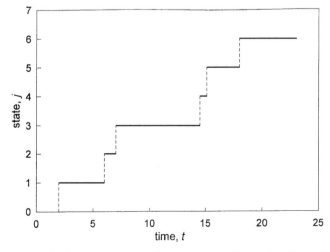

FIGURE 8.13 A single instance of a Poisson process with $\mu = 3$ and $m = 7$ arrivals.

```
for i=1 to m
      t(i,0)=0
      for j=1 to n
            t(i,j)=t(i,j-1)-μ*ln(RND)
      next j
next i

for j=0 to n
      for i=1 to m
            print t(i,j),
      next i
      print
next j
```

LISTING 8.2   Generating an ensemble of *m* Poisson processes of *n* events and printing them in columns.

In reality, Figure 8.13 only shows a single instance of the entire family of possible Poisson processes. This family, usually called an *ensemble*, exists only as an abstraction, since it is impossible to list every possible instance. However, we can still get a good idea of what it would look like by computing *m* instances of the *n* event times $t(i,j)$ for $i = 1, \ldots, m$ and $j = 0, 1, \ldots, n$. This can be accomplished by using the algorithm shown in Listing 8.2.

In executing Listing 8.2, it is good to actually visualize the output beforehand. In this case, we see each instance printed row-wise, with each column as a separate instance of the process. This is shown for 10 event times for each of 5 instances in Table 8.2. Note that in order to reproduce this, the use of indices *i* (the instance number) and *j* (the event number) is critical. In particular, note that in the generation section, *i* is the outside index, while in the print section, *j* is the outside index.

A good methodology in doing simulations of this type is to generate the simulation per se using a standard high-level language and put the results to a simple ASCI file. This file can then be exported to a standard spreadsheet or special-purpose graphics software to

TABLE 8.2   Poisson Random Process: 5 Instances of 10 Events in a Spreadsheet Format

j	$i = 1$	$i = 2$	$i = 3$	$i = 4$	$i = 5$	Statistical average	First difference
0	0.00	0.00	0.00	0.00	0.00	0	
1	1.47	0.29	0.73	1.81	0.13	0.89	0.89
2	4.08	1.30	7.93	4.58	8.03	5.18	4.30
3	4.96	11.70	8.59	7.68	17.97	10.18	5.00
4	11.01	14.43	11.12	11.88	25.08	14.70	4.52
5	15.59	16.42	15.54	15.24	25.36	17.63	2.93
6	18.67	23.08	16.53	16.66	27.00	20.39	2.76
7	23.47	34.03	16.55	18.94	28.18	24.23	3.84
8	26.50	35.73	17.65	20.76	31.64	26.46	2.22
9	28.40	36.17	19.16	21.16	37.52	28.48	2.02
10	28.52	36.30	24.92	21.79	39.74	30.25	1.77
							Mean = 3.03

statistically analyze and sketch your conclusions. It is amazing what good quality comes from using just a simple spreadsheet such as Lotus, Quattro Pro, or even Excel. In the case shown here, we observe the first six columns as the event number (state) j and the event time $t(j)$ for each of the $i = 1, 2, 3, 4, 5$ instances. The seventh column averages the event times for each state. This is actually called a "statistical average", since the event time is fixed and the resulting statistic is computed over all instances.

The other useful average that could be used is the "time average", and would be an average of the rows by first fixing each column. The significance of the statistical average is not obvious in the instance shown here unless the average is *differenced*. The difference is shown in the last column, and gives the average time that the system is in each state. Not surprisingly, we note here that the differences are approximately the same, regardless of the time (fundamental to the Poisson postulates). In fact, the overall average of 3.03 corresponds to the mean $\mu = 3$ used in the original simulation.

Differencing a time series is a technique that is very useful in statistics. From a theoretical point of view, it is necessary to difference any marginally stationary random process, since, strictly speaking, it is not stationary. Therefore, it will never achieve a final (though random) value. From a practical point of view, differencing the statistical average column in Table 8.2 is necessary because the exponential distribution describes inter-event times and not the times of the events per se. Thus, the differenced column (inter-event times) is expected to be stationary, where the statistical average column gives the cumulative inter-event times.

8.7 CONTINUOUS-TIME MARKOV PROCESSES

Discrete-time Markov processes are excellent tools for modeling synchronous finite state machines. Knowledge regarding the statistical properties of the input translates directly to transition probabilities, and this leads to a well-defined state diagram, from which an explicit mathematical realization follows. This is true for both synchronous systems where events occur on a regular basis and asynchronous systems where inter-event times must be scheduled. Asynchronous finite state machines are different only in that the times at which event occur are no longer deterministic. Still, if we understand the inter-event-time statistics, we can model and simulate these systems too.

Discrete-time Markov chains lead to difference equations. Not surprisingly, continuous-time models lead to differential equations. The transformation of continuous to discrete time is effected by the usual $t = hk + t_0$ relationship. The analogous functions for each case are as follows:

Discrete time	Continuous time
k	t
$p(k)$	$p(t)$
$p(k+1)$	$p(t+h)$

Since, for discrete time, the future probability of being in generic state i is $p_i(k+1) = p_1(k)p_{1i} + p_2(k)p_{2i} + \ldots + p_n(k)p_{ni}$. In continuous time,

$$p_i(t+h) = \sum_k p_k(t)p_{ki}(h)$$

$$= p_i(t)p_{ii}(h) + \sum_{k\neq i} p_k(t)p_{ki}(h).$$

Algebraically rearranging,

$$\frac{p_i(t+h) - p_i(t)}{h} = p_i(t)\frac{p_{ii}(h) - 1}{h} + \sum_{k\neq i} p_k(t)\frac{p_{ki}(h)}{h}$$

$$= p_i(t)\frac{p_{ii}(h) - 1}{h} + \sum_{k\neq i} \frac{p_k(t)p_{ki}(h)}{h}.$$

Upon taking limits,

$$\frac{d}{dt}p_i(t) = \lim_{h\to 0}\frac{p_i(t+h) - p_i(t)}{h}$$

$$= p_i(t)\lim_{h\to 0}\frac{p_{ii}(h) - 1}{h} + \sum_{k\neq i} p_k(t)\lim_{h\to 0}\frac{p_{ki}(h)}{h} \tag{8.32}$$

$$= p_i(t)\gamma_{ii} + \sum_{k\neq i} p_k(t)\gamma_{ki}$$

$$= \sum_k p_k(t)\gamma_{ki},$$

where

$$\gamma_{ij} = \begin{cases} \lim_{h\to 0}\dfrac{p_{ij}(h)}{h}, & i\neq j, \\ \lim_{h\to 0}\dfrac{p_{ii}(h) - 1}{h}, & i = j. \end{cases} \tag{8.33}$$

The ratios $p_{ij}(h)/h$ are *transition probability rates*, which, by the Poisson postulates P2 and P3, are well defined, and the limit exists. They differ from transition probabilities in that they are probabilities per unit time. On the other hand, the transition probabilities discussed in conjunction with discrete-time Markov models are just probabilities per se. The ratios γ_{ij} are called *transition intensities* if $i\neq j$ and *passage intensities* if $i=j$; the constant matrix given by $\mathbf{\Gamma} = [\gamma_{ij}]$ is called the *intensity matrix*.

Equation (8.32) can be written succinctly using vector and matrix notation as

$$\frac{d}{dt}\mathbf{p}(t) = \mathbf{p}(t)\mathbf{\Gamma},$$

where $\mathbf{p}(t) = [p_1(t), p_2(t), \ldots, p_n(t)]$ is the probability state vector. For continuous systems, this differential equation is fundamental, and the basic problem is usually posed as an initial-value problem in matrix form:

$$\mathbf{p}(0) \text{ given,}$$

$$\frac{d}{dt}\mathbf{p}(t) = \mathbf{p}(t)\mathbf{\Gamma}. \tag{8.34}$$

TABLE 8.3   Comparison between Discrete- and Continuous-Time Poisson Processes

	Discrete-time Poisson process	Continuous-time Poisson process
Transition matrix	Probability matrix $\mathbf{P} = [\,p_{ij}]$	Intensity matrix $\mathbf{\Gamma} = [\gamma_{ij}]$
Nodal relationship	$\sum_j p_{ij} = 1$	$\sum_j \gamma_{ij} = 0$
Recursive relation	$\mathbf{p}(k+1) = \mathbf{p}(k)\mathbf{P}$	$\dfrac{d}{dt}\mathbf{p}(t) = \mathbf{p}(t)\mathbf{\Gamma}$
Explicit relation	$\mathbf{p}(k) = \mathbf{p}(0)\mathbf{P}^k$	$\mathbf{p}(t) = \mathbf{p}(0)e^{\mathbf{\Gamma}t}$
Steady state	$\boldsymbol{\pi} = \boldsymbol{\pi}\mathbf{P}$	$\boldsymbol{\pi}\mathbf{\Gamma} = 0$

The solution to Equation (8.34) is relatively straightforward, but it should be remembered that matrix differential equations have unique idiosyncrasies that need special attention. Even though application of elementary calculus produces $\mathbf{p}(t) = \mathbf{p}(0)e^{\mathbf{\Gamma}t}$, this begs the issue in that it requires that e be raised to a matrix power. While this problem is in principle well understood, it is in general non-trivial and beyond the scope of this text.

It should come as no surprise that there are many analogies between discrete- and continuous-time systems. Table 8.3 presents a short list of corresponding properties that illustrate this closeness while at the same time being useful in solving specific problems.

In particular, it should be recalled that the so-called *nodal relationship* for a discrete Poisson system is that the probability transition matrix $\mathbf{P}$ is of rank $n-1$ and that its rows each sum to one. The corresponding relationship for the intensity transition matrix is that it too has rank $n-1$, but each row sums to zero. The reason for this follows from Equation (8.33), where it will be noted that the passage intensities are the probability minus one per unit time. Therefore, the nodal relationship $\sum_j \gamma_{ij} = 0$ follows immediately.

EXAMPLE 8.11

Consider a single-server queuing system that allows a single customer to wait (if needed). Events occur according to a Poisson process at the average rate of 1 customer each 3 minutes and are served with an exponential service time of mean 15 seconds. Analyze the steady-state character of the system.

Solution
In this problem, events correspond to customer arrivals and states correspond to the number of customers in the system. The system apparently has three possible states:

- zero customers: state 0;
- one customer (being serviced): state 1;
- two customers (one being serviced and the other waiting): state 2.

From the problem description, the probability rate of an arrival is $p_{12} = \frac{1}{3}$ and $p_{23} = \frac{1}{3}$ customers per minute. Since a customer is serviced every 15 seconds on the average, the probability departure rate is $\frac{1}{4}$ customer per minute. Thus, $p_{32} = \frac{1}{4}$ and $p_{21} = \frac{1}{4}$. There is also a possibility of retention (no net gain or loss of customers in the system), which may be found by noting that the probabilities

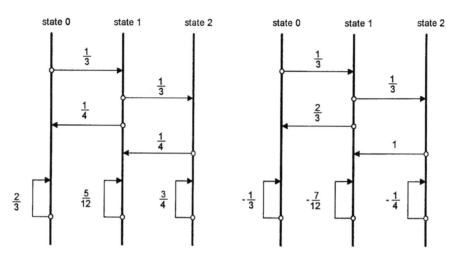

FIGURE 8.14   Transition probability and transition intensity diagrams, Example 8.11.

leaving each node sum to unity. An appropriate state transition diagram and corresponding transition intensity diagram are shown in Figure 8.14. It will be noted that the two diagrams are identical except for the transition from a state to itself, in which case the two differ by unity. While exiting transitions from the probability diagram sum to unity, the transitions exiting the intensity diagram sum to zero.

The transition intensities, which label the directed arcs between various states, can be read directly from the *transition intensity diagram* (Figure 8.14). Since the passage intensities are represented by arcs from a mode to itself minus unity (Equation (8.33)), the intensity matrix is as follows:

$$\Gamma = \begin{bmatrix} -\frac{1}{3} & \frac{1}{3} & 0 \\ \frac{1}{4} & -\frac{7}{12} & \frac{1}{3} \\ 0 & \frac{1}{4} & -\frac{1}{4} \end{bmatrix}.$$

At steady state, the state vector levels off and become constant. Therefore in a continuous system, the *derivatives must vanish* and $\pi\Gamma = 0$. Since in this example there are three states, the steady-state state vector is $\pi = [\pi_0, \pi_1, \pi_2]$ and

$$[\pi_0, \pi_1, \pi_2] \begin{bmatrix} -\frac{1}{3} & \frac{1}{3} & 0 \\ \frac{1}{4} & -\frac{7}{12} & \frac{1}{3} \\ 0 & \frac{1}{4} & -\frac{1}{4} \end{bmatrix} = 0.$$

This matrix equation is actually three equations with two unknowns. However, the rank of the matrix Γ is 2 (in general, it is always $n - 1$), and we must again incorporate the help of the *normalizing condition*, $\pi_0 + \pi_1 + \pi_2 = 1$. This leads to the following set of four equations, any two of the first three along with the last of which can be used to solve for the unique state vector $\boldsymbol{\pi}$:

$$-\tfrac{1}{3}\pi_0 + \tfrac{1}{4}\pi_1 = 0,$$

$$\tfrac{1}{3}\pi_0 - \tfrac{7}{12}\pi_1 + \tfrac{1}{4}\pi_2 = 0,$$

$$\tfrac{1}{3}\pi_1 - \tfrac{1}{4}\pi_2 = 0,$$

$$\pi_0 + \pi_1 + \pi_2 = 1.$$

The solution is $\boldsymbol{\pi} = [\tfrac{9}{37}, \tfrac{12}{37}, \tfrac{16}{37}]$.   ○

It is important to understand both the power and the limitations of this procedure. Assuming exponentially distributed inter-arrival times for the customers, this solution gives us long-term predictive results. It turns out that these results are generally rather robust. This tells us that even though the arrival discipline might not quite be exponential, the simulation results will still be close to that predicted by the Markov theory. For this reason, along with the fact that other theoretical distributions do not lead to such elegant closed-form solutions, practitioners often use continuous Markov models, even in cases where they do not apply!

As in the case of discrete time, the binary (two-state) system is of fundamental interest. In order to find $\mathbf{p}(t)$ from Equation (8.34), it is first necessary to evaluate $e^{\Gamma t}$. Since the rows of the intensity matrix must each sum to zero, there are two degrees of freedom. Therefore, the most general intensity matrix Γ is given by

$$\Gamma = \begin{bmatrix} -\alpha & \alpha \\ \beta & -\beta \end{bmatrix}. \tag{8.35}$$

It can be shown that

$$e^{\Gamma t} = \frac{1}{\alpha + \beta}\begin{bmatrix} \beta & \alpha \\ \beta & \alpha \end{bmatrix} + \frac{e^{-(\alpha+\beta)t}}{\alpha + \beta}\begin{bmatrix} \alpha & -\alpha \\ -\beta & \beta \end{bmatrix}. \tag{8.36}$$

As with discrete-time Markov chains, there are two terms: a steady-state term and a transient term. These, together with initial system state $\mathbf{p}(0) = [p_0(0), p_1(0)]$, form the general solution.

EXAMPLE 8.12

Consider a telephone system where calls are received only when the system is not busy. In other words, calls cannot be queued, and incoming calls are blocked when the system is busy. Suppose it is found that the probability of a call being made when the system is *busy* is $\tfrac{3}{4}$. Also, the probability that a call is made when the system is *clear* (not busy) is $\tfrac{1}{3}$. Assuming exponentially distributed inter-event times for the phone calls, describe both the transient and steady-state behavior of this system.

Solution
Let $y(t)$ be the system state at time t. The system is clearly binary in that the line is either busy or it is clear, so we arbitrarily assign states as follows: $y(t) = 0$ if the

line is busy and $y(t) = 1$ if the line is clear. From the problem description, it follows that the probability that the next state is clear is

$$\Pr[Y(t+h) = 1 \mid Y(t) = 1] = \tfrac{3}{4},$$
$$\Pr[Y(t+h) = 1 \mid Y(t) = 0] = \tfrac{1}{3}.$$

From these, the probability the next state is busy follows immediately. Thus, the transition probabilities are $p_{00}(t) = \tfrac{2}{3}$, $p_{01}(t) = \tfrac{1}{3}$, $p_{10}(t) = \tfrac{1}{4}$, and $p_{11}(t) = \tfrac{3}{4}$. From these, the intensities follow:

$$\gamma_{00} = \tfrac{2}{3} - 1 = -\tfrac{1}{3},$$
$$\gamma_{01} = \tfrac{1}{3},$$
$$\gamma_{10} = \tfrac{1}{4},$$
$$\gamma_{11} = \tfrac{3}{4} - 1 = -\tfrac{1}{4},$$

and

$$\Gamma = \begin{bmatrix} -\tfrac{1}{3} & \tfrac{1}{3} \\ \tfrac{1}{4} & -\tfrac{1}{4} \end{bmatrix}. \tag{8.37}$$

Figure 8.15 shows the corresponding state transition and intensity diagrams. Equations (8.35) and (8.36) with $\alpha = \tfrac{1}{3}$ and $\beta = \tfrac{1}{4}$ lead to

$$e^{\Gamma t} = \begin{bmatrix} \tfrac{3}{7} + \tfrac{4}{7}e^{-7t/12} & \tfrac{4}{7} - \tfrac{4}{7}e^{-7t/12} \\ \tfrac{3}{7} - \tfrac{3}{7}e^{-7t/12} & \tfrac{4}{7} + \tfrac{3}{7}e^{-7t/12} \end{bmatrix}.$$

Transition probability diagram   **Transition intensity diagram**

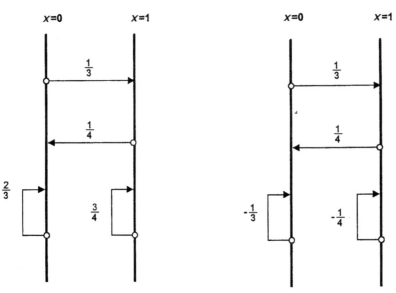

FIGURE 8.15   State transition and intensity diagrams, Example 8.12.

Assuming generic initial conditions of $\mathbf{p}(0) = [\,p_0(0), p_1(0)]$,

$$\mathbf{p}(t) = \mathbf{p}(0)e^{\Gamma t}$$

$$= [\,p_0(0), p_1(0)]\begin{bmatrix} \frac{3}{7} + \frac{4}{7}e^{-7t/12} & \frac{4}{7} - \frac{4}{7}e^{-7t/12} \\ \frac{3}{7} - \frac{3}{7}e^{-7t/12} & \frac{4}{7} + \frac{3}{7}e^{-7t/12} \end{bmatrix}$$

$$= [\tfrac{3}{7} + \{p_0(0) - \tfrac{3}{7}\}e^{-7t/12}, \; \tfrac{4}{7} + \{p_1(0) - \tfrac{4}{7}\}e^{-7t/12}].$$

Component-wise,

$$p_0(t) = \tfrac{3}{7} + \{p_0(0) - \tfrac{3}{7}\}e^{-7t/12}$$
$$p_1(t) = \tfrac{4}{7} + \{p_1(0) - \tfrac{4}{7}\}e^{-7t/12}. \tag{8.38}$$

Thus, at steady state,

$$\pi_0 = \lim_{t\to\infty} p_0(t) = \tfrac{3}{7},$$
$$\pi_1 = \lim_{t\to\infty} p_1(t) = \tfrac{4}{7}.$$

Or, written in vector form,

$$\boldsymbol{\pi} = \lim_{t\to\infty} \mathbf{p}(t) = \left[\tfrac{3}{7}, \; \tfrac{4}{7}\right].$$

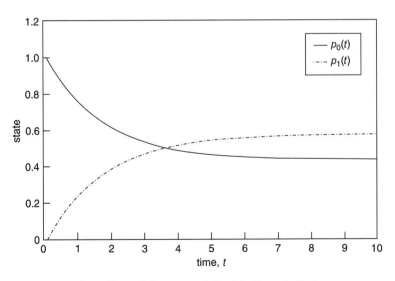

FIGURE 8.16　Transient effects of the system defined in Example 8.12.

Again notice that the steady-state solution is independent of the initial state of the system. This is true for all linear systems. Equations (8.38) are graphed in Figure 8.16 using an initial *clear* state: $\mathbf{p}(0) = [1,0]$.   ○

The special case where state transitions only occur between a state and an adjacent state are especially important. Such systems are called *birth–death* systems, since, by considering the state to be the number of individuals in a population, the state goes up or down as individuals are born (increment) or die (decrement). As with Petri nets and Markov chains generally, we assume that only one event occurs at any event time. For any state k, arcs exiting that state fall into one of three categories:

- *Birth processes*: the arc goes from the current state to a higher state.
- *Death processes*: the arc goes from the current state to a lower state.
- *Retention processes*: the arc goes from the current state to the same state.

The retention arcs can be inferred since the sum of all arcs (i.e., transition rates) leaving state k is zero. By ignoring the retention arcs, one obtains a *simplified intensity diagram*, which is the same as the unsimplified diagram except that no arc existing a state can return directly to itself. Two equivalent intensity diagrams are shown in Figure 8.17. Note that the only difference is that the simplified diagram has no transitions reflecting back to itself.

The simplified diagram is especially useful for steady-state problems, because in the steady state, the following conservation principle applies:

Transition intensity diagram   **Simplified intensity diagram**

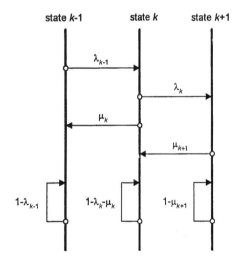
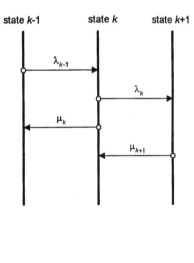

FIGURE 8.17   Simplified and unsimplified intensity diagrams.

- *At every node, the sum of the inputs to the node equals the sum of the outputs from that node.*

Application of this principle follows immediately from the simplified intensity diagram, and creates the so-called *balance equations* of the steady-state system. For instance, at node k in Figure 8.17, the balance equations are

$$\lambda_{k-1}\pi_{k-1} + \mu_{k+1}\pi_{k+1} = (\lambda_k + \mu_k)\pi_k, \tag{8.39}$$

for birth rates λ_i, deaths μ_i, and steady-state transition probabilities π_i. This can be shown formally using the unsimplified diagram of Figure 8.17 as well. Even so, in the steady state they are equivalent.

EXAMPLE 8.13

Reconsider the system described in Example 8.11. Find the steady-state state vector using balance equations.

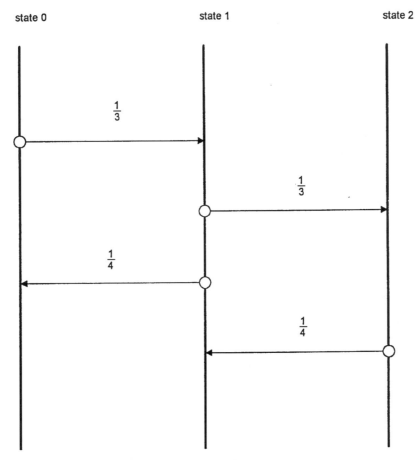

FIGURE 8.18   Simplified intensity diagram, Example 8.13.

Solution

The equivalent simplified transition rate diagram is illustrated in Figure 8.18. Accordingly, the balance equations should be written for each node:

$$\text{state } 0: \quad \tfrac{1}{4}\pi_1 = \tfrac{1}{3}\pi_0,$$
$$\text{state } 1: \quad \tfrac{1}{3}\pi_0 + \tfrac{1}{4}\pi_2 = \left(\tfrac{1}{3} + \tfrac{1}{4}\right)\pi_1,$$
$$\text{state } 2: \quad \tfrac{1}{3}\pi_1 = \tfrac{1}{4}\pi_2.$$

As in the case of discrete Markov chains, the three equations are redundant. As has been done in earlier examples, the normalizing equation $\pi_0 + \pi_1 + \pi_2 = 1$ is also required. It follows that the solution is $\boldsymbol{\pi} = [\tfrac{9}{37}, \tfrac{12}{37}, \tfrac{16}{37}]$, as was found previously in Example 8.11. ○

A word of warning is in order. Poisson, Markov and balance equation techniques are very powerful, and it is possible to become careless. The formulas derived in this section really only apply in cases where the Poisson postulates hold, or, equivalently, where exponential inter-event times can reasonably be assumed. If studies show a different inter-event protocol, the theory presented here is of little value, and simulation is required. The literature is replete with Markov techniques being applied to non-Markov distributions. Simply put, their results are of little value. When in doubt it, is necessary to do an empirical study and apply simulation techniques instead of direct easy to apply, but invalid, formulas.

BIBLIOGRAPHY

Banks, J., J. S. Carson, and B. L. Nelson, *Discrete-Event System Simulation*, 2nd edn. Prentice-Hall, 1996.

Buzacott, J. A. and J. G. Shanthikumar, *Stochastic Models of Manufacturing Systems*. Prentice-Hall, 1993.

Fishwick, P., *Simulation Model Design and Execution: Building Discrete Worlds*. Prentice-Hall, 1995.

Freeman, H., *Discrete Time Systems*. Polytechnic Press, 1965.

Hammersley, J. M. and D. C. Handscomb, *Monte Carolo Methods*. Methuen, 1964.

Hillier, F. and G. Lieberman, *Introduction to Operations Research Guide*, 6th edn. McGraw-Hill, 1995.

Kleinrock, L., *Queuing Systems*. Wiley, 1976.

Knuth, D. E., *The Art of Computer Programming*, Vol. 1: *Fundamental Algorithms*, 3rd edn. Addison-Wesley, 1997.

Law, A. and D. Kelton, *Simulation, Modeling and Analysis*. McGraw-Hill, 1991.

MIL 3, Inc., *Modeling and Simulating Computer Networks: A Hands on Approach Using OPNET*. Prentice-Hall, 1999.

Nelson, B., *Stochastic Modeling, Analysis and Simulation*. McGraw-Hill, 1995.

Pinedo, M., *Scheduling: Theory, Algorithms and Systems*. Prentice-Hall, 1994.

Shoup, T. E., *Applied Numerical Methods for the Micro-Computer*. Prentice-Hall, 1984.

Thompson, R. R., *Simulation: A Modeler's Approach*. Wiley-Interscience, 2000.

White, J. A., J. W. Schmidt, and G. K. Bennett, *Analysis of Queuing Systems*. Academic Press, 1975.

Wolff, R. W., *Stochastic Modeling and the Theory of Queues*. Prentice-Hall, 1989.

_____ EXERCISES

8.1 Consider an urn that initially contains one red and one black ball. A single ball is drawn at random. If the ball is black, it is replaced with a red ball and put back. On the other hand, if it is red, the ball is not replaced. The experiment ends when the urn is empty.
(a) Define and list the set of possible states.
(b) Draw an appropriate state transition diagram.
(c) Find the probability transition matrix.
(d) Find the steady-state vector $\boldsymbol{\pi}$.

8.2 The probability transition matrix $\mathbf{P}$ is given by

$$\mathbf{P} = \begin{bmatrix} \frac{1}{2} & \frac{1}{3} & \frac{1}{6} \\ 0 & \frac{1}{2} & \frac{1}{2} \\ \frac{1}{4} & \frac{1}{4} & \frac{1}{2} \end{bmatrix}.$$

(a) Draw an appropriate steady-state diagram.
(b) Find the steady-state transition matrix $\mathbf{P}_{ss}$.

8.3 Consider a telephone line as a single-server queuing system that allows a maximum of two customers to wait while another one is being served. Assume that calls arrive as a Poisson process at the rate of 1 call every 10 minutes and that each call lasts an average of 8 minutes. Analyze the steady-state character of the system; i.e., find $\boldsymbol{\pi}$.

8.4 Modems networked to a mainframe computer system have a limited capacity. The probability that a user dials into the network when a modem connection is available is $\frac{2}{3}$ and the probability that a call is received when all lines are busy is $\frac{1}{2}$.
(a) Describe the transient behavior of the system; i.e., find the vector $\mathbf{p}(t)$.
(b) Describe the steady-state behavior of the system; i.e., find the vector $\boldsymbol{\pi}$.

8.5 Consider a system described by the following *state transition diagram*:

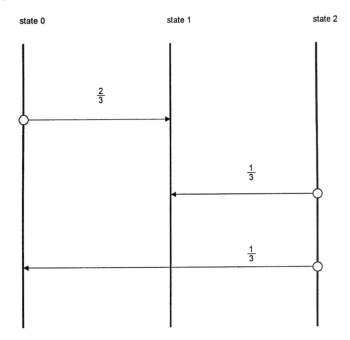

(a) Find the probability transition matrix **P**.

(b) Show that

$$\mathbf{P}^k = \begin{bmatrix} \dfrac{1}{3^k} & 1 - \dfrac{1}{3k} & 0 \\ 0 & 1 & 0 \\ \dfrac{k}{3^k} & 1 - \dfrac{k+1}{3^k} & \dfrac{1}{3^k} \end{bmatrix}.$$

(c) Find the steady-state transition matrix $\mathbf{P}_{ss}$ and state vector $\boldsymbol{\pi}$.

8.6 Verify that if $\mathbf{P} = \begin{bmatrix} 1-\alpha & \alpha \\ \beta & 1-\beta \end{bmatrix}$, then

$$\mathbf{P}^k = \frac{1}{\alpha+\beta}\begin{bmatrix} \beta & \alpha \\ \beta & \alpha \end{bmatrix} + \frac{(1-\alpha-\beta)^k}{\alpha+\beta}\begin{bmatrix} \alpha & -\alpha \\ -\beta & \beta \end{bmatrix}.$$

8.7 It has been found that a certain car wash receives, on the average, 20 cars per hour between the hours of 7 a.m. and 10 a.m. The average drops to 8 cars per hour for the rest of the business day, which finishes at 6 p.m.

(a) What is the expected number of cars washed in one complete business day?

(b) What is the probability that no more than 2 cars will enter the car wash between 9:00 and 9:30 a.m.?

(c) What the probability that more than 3 cars will arrive between 4:00 and 5:00 p.m.?

(d) Find the probability of getting exactly 2 cars in the interval between 9:55 and 10:10 a.m.

8.8 A local area network of 8 computers share a single printer. It has been found that, on the average, each computer sends one print request every 10 minutes. Also, it takes an average of 45 seconds to complete a print job.

(a) Find the expected number of print jobs in an 8-hour period.

(b) Determine the probability that not more than 3 print jobs will be submitted in 1 hour's time.

(c) Calculate the probability of processing exactly 10 print jobs by the printer in 1 hour's time.

(d) Find the expected time of completion of exactly 10 print jobs.

8.9 (a) Show from first principles (i.e., the Poisson postulates) and the fact that $P_1(t) = e^{-\lambda t}\lambda t$ that

$$\frac{d}{dt}P_2(t) = -\lambda P_2(t) + \lambda P_1(t).$$

(b) Show that the solution of this differential equation is given by

$$P_2(t) = \tfrac{1}{2}\lambda^2 t^2 e^{-\lambda t}.$$

8.10 Let $P_n(t)$ be a Poisson process.

(a) Show from first principles that

$$\frac{d}{dt}P_n(t) = -\lambda P_n(t) + \lambda P_{n-1}(t),$$

$$P_n(0) = \begin{cases} 0, & n = 0, \\ 1, & n > 0. \end{cases}$$

(b) Prove, using mathematical induction, that

$$P_n(t) = e^{-\lambda t}\frac{(\lambda t)^n}{n!}, \qquad n = 0,\ 1,\ 2,\ \ldots\ .$$

8.11 A computer network receives an average of 200 connect requests in 1 hour's time. It has also been found that the response to a connect request varies and is exponentially distributed. For example, a sample of 8 consecutive connect requests and their corresponding response times are given as follows:

Request number	Response time (seconds)
1	20
2	25
3	23
4	27
5	27
6	23
7	21
8	31

(a) Determine the average response time for the sample; thus calculate the mean number of requests per unit time.

(b) Find the probability of getting a response at least 32 seconds after the request is made.

(c) Calculate the expected number of requests for which the response time is more than 32 seconds in 1 hour's time.

8.12 Show that for a Poisson random variable, the second moment is given by $E[N^2] = \mu(\mu + 1)$, where $\mu = E[N]$.

8.13 Show that the second moment of an exponential random variable is $E[T^2] = 1/\lambda^2$. Generalize to the kth moment: $E[T^k] = k!/\lambda^k$.

8.14 Let $N(t)$ be a Poisson process with *variable rate* $\lambda(t) = 3t^2$.
(a) Calculate the expected number of events of the process in $(0,1]$.
(b) Calculate the expected number of events of the process in $(1,2]$.
(c) Let T be the time of occurrence of the first event of the process. Show that $\Pr[T > t] = e^{-t^3}$.
(d) Find the probability density function for the random variable T.

8.15 Verify that if $\Gamma = \begin{bmatrix} -\alpha & \alpha \\ \beta & -\beta \end{bmatrix}$, then

$$ e^{\Gamma k} = \frac{1}{\alpha + \beta}\begin{bmatrix} \beta & \alpha \\ \beta & \alpha \end{bmatrix} + \frac{e^{-(\alpha+\beta)t}}{\alpha + \beta}\begin{bmatrix} \alpha & -\alpha \\ -\beta & \beta \end{bmatrix}. $$

8.16 Suppose that the probability that a functioning machine becomes out of order on a given day is α. Conversely, the probability that the machine, when broken, will be fixed by the next day is β. Clearly, a small value for α is desirable, as is a large value for β.
(a) Draw the state diagram for this system.
(b) Using $\alpha = 0.1$ and $\beta = 0.85$, describe the transient and steady-state behavior of this system. Assuming the machine is initially out of order, find equations for the probabilities of it being up or down.
(c) Simulate 100 trials of an experiment to confirm your results, being careful to focus on the transient phase. Compare the experimental average against your calculations.

8.17 Consider the two-dimensional random walk described in Example 8.7.
(a) Create a simulation of this walk.
(b) Verify the correctness of the theoretically described steady-state state vector π described in Example 8.7.

8.18 Consider a system described by the following *state transition rate diagram*:

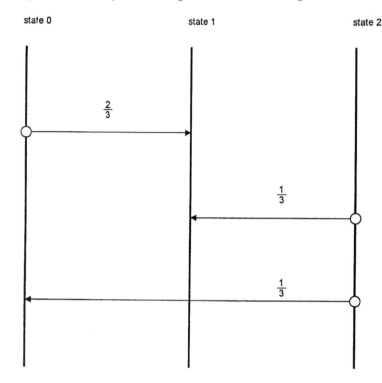

(a) Find the transition rate matrix Γ.

(b) Show that

$$
e^{\Gamma t} = \begin{bmatrix}
e^{-2t/3} & 0 & \frac{1}{3}te^{-2t/3} \\
1 - e^{-2t/3} & 1 & 1 - e^{-2t/3} + te^{-2t/3} \\
0 & 0 & e^{-2t/3}
\end{bmatrix}.
$$

(c) Find the steady-state state vector $\boldsymbol{\pi}$.

8.19 Consider the machine described above in Exercise 8.16. Suppose that $\alpha = 0.2$ and $\beta = 0.75$, and that you come to work on Monday and see the machine working.

(a) What is the probability that the machine will be working on Thursday?

(b) What is the probability that the machine will not break through at least Thursday?

8.20 Suppose that a football player can be in exactly one of three states on a given Sunday:

$0 = $ disabled: not available for play;
$1 = $ marginally healthy: just barely limping along;
$2 = $ healthy: can play regularly.

Assume that, when he is available to play, the probability that he will incur an injury is α. If he is indeed injured, the probability that it is serious (he becomes disabled) is β and the probability that it is only temporary (he is now marginally healthy) is $1 - \beta$. If his health is marginal, assume that the probability is γ that he will be disabled next week and there is no chance whatsoever that he will be healthy. Also, if he is disabled today, the probability is δ that he will be able to play next week, but there is only a probability ϵ that he will be fully recovered.

(a) Show that the one-week probability transition matrix is

$$\mathbf{P} = \begin{bmatrix} 1-\alpha & \alpha(1-\beta) & \alpha\beta \\ 0 & 1-\gamma & \gamma \\ \delta\epsilon & \epsilon(1-\delta) & 1-\epsilon \end{bmatrix}.$$

(b) Draw the state diagram.

(c) Starting with a perfectly healthy football player, derive a steady-state prediction of his health for the coming football season. Assume $\alpha = 0.1$, $\beta = 0.4$, $\gamma = 0.6$, $\delta = 0.5$, and $\epsilon = 0.7$.

8.21 Consider a four-state random walk with reflecting barriers where the probability of advance is $\frac{1}{4}$ (and the probability of regress is $\frac{3}{4}$).

(a) Sketch an appropriate transition diagram for this problem. Write the difference equations defining transition probabilities for the system and determine the (theoretical) steady-state state vector $\boldsymbol{\pi}$.

(b) Simulate a single random walk of approximately 25 time steps, starting in the state $\mathbf{p}(0) = [1, 0, 0, 0]$. Graph these results on a state (y-axis) versus time (x-axis) graph. Estimate the steady-state vector using this graph to compute the time average $\langle \mathbf{p}(k) \rangle$. Repeat this for the other possible original states. What conclusion can you draw?

(c) For a single initial state of your choosing, simulate a number of random walks. Estimate the statistical average $E[\boldsymbol{\pi}]$ for several different time slices. What conclusion can you draw?

(d) Compare the results of parts (a)–(c). Comment on the *Ergodic Hypothesis*; that is, does $\boldsymbol{\pi} = \langle \mathbf{p}(k) \rangle = E[\boldsymbol{\pi}]$?

8.22 Consider a two-dimensional random walk over the nine-point square region defined by the points $(1,1)$, $(1,2)$, $(1,3)$, $(2,1)$, ..., $(3,3)$ in the x–y plane. For any given point, travel is possible to only points directly above, below, left, or right within the region. It is not possible to remain at any point. Also, the probability of travel to any legal point is equally likely.

Part 1

(a) Start by defining the state as the point that your walker is at. For instance, call the point $(1,1)$ state 1, the point $(3,3)$ state 9, and so forth. Sketch an appropriate transition diagram for this problem and find the probability transition matrix $\mathbf{P}$.

(b) Do a simulation to determine the probability of visiting any particular point. Run this simulation at least 1000 times to estimate the probability.

(c) By looking at the geometry, observe that there are certain states that can be combined. Now redefine your states (so as to greatly reduce the number) that can also solve the problem.

(d) You now have three ways to solve this problem, and they should all give the same result.

    (i) Solve part (a) by raising $\mathbf{P}$ to a large even power. MatLab is really good here, but Lotus can do it too. As a last resort, writing a program is also rather straightforward. Also raise $\mathbf{P}$ to a large odd power. What is the difference. How should you interpret these results?

    (ii) Do the simulation.

    (iii) Solve the reduced state system that you prescribed. It should give you consistent results in the steady state.

    (iv) Compare all these results in a table and graphically. Explain.

Part 2

(a) Now do the same thing for a 16-element 4-by-4 array by writing the 16-state transition diagram and the 16-by-16 state transition matrix $\mathbf{P}$. Be thankful that you don't need to manipulate it in any way!

(b) Now combine appropriate states and create a theoretical steady-state solution. From this, find the probability of visitation for each point.

(c) Do a simulation, and compare your results of parts (a) and (b).

Part 3

(a) Generalize your results to a 25-element 5-by-5 array. Do not simulate, but rather find the probabilities directly.

(b) Generalize your results to an n^2-element n-by-n array.

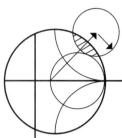

Event-Driven Models

Time-driven models characterize systems in which signals march along in a synchronous fashion in regular time. In contrast, event-driven models behave asynchronously and usually have very simple signals that occur at irregular, often random, time intervals. Most often, the signal is binary: a one (1) if an event occurs and a zero (0) if it doesn't. What makes these models interesting is just when events happen rather than exactly what event occurs.

There are many phenomena that resemble event-driven models, especially those involving digital computer systems. For instance, a digital communication system is often characterized by random, but statistically distributed, message arrivals into a computer network. In characterizing system performance, the message itself is not so important as the arrival rates and message length. If these are known, the network load and reliability can be modelled with precision. In certain cases, a purely analytical result can be found. However in general, computers are required to perform simulations.

In the digital communications example, the message is either in the system (1) or it isn't (0). The defining characteristics are statistical: when each message arrives, the length of the message and how long it remains in the network are all random variables. Typically, the event arrivals are at times $t_0, t_1, t_2, \ldots$, where $t_0 = 0$ is a reference, and are queued up in a first-in/first-out (FIFO) queue, so that the system induces a wait before each message can be processed. The processing time or service times are also usually statistically distributed in that their lengths are random, but their statistics are known. To be precise, the statistics of event times t_k are non-stationary random processes, but their inter-event times $t_{k-1} - t_k$ are stationary. Thus, a simple random variable suffices to describe their steady-state behavior.

9.1   SIMULATION DIAGRAMS

Event-driven models have simulation diagrams, just like time-driven models, but with different basic structures (see Figure 9.1). The first structure is an *event sequencer*, which accepts a sequence of events that *arrive* at times $t_1, t_2, t_3, \ldots$. For convenience, we associate a time $t_0 = 0$ to be the *null event* as a reference, and assume the sequence to be

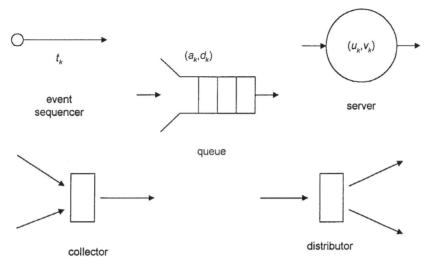

FIGURE 9.1   Event-driven system simulation structures.

finite, with $k = n$ occurring at time t_n. The notation for the sequencer is $T = [t_0, t_1, t_2, \ldots, t_n]$, for monotonic times t_k. T determines the system state, because if T is known, it is possible to determine exactly how many events have occurred for any time t. Specifically, T is in state k at time t if $t_k < t < t_{k+1}$. While the arrival time t_k is monotonic, the inter-arrival time $t_{k+1} - t_k$ is much tamer. The sequence $\{t_k\}$ is often stochastic, and individual differences $t_{k+1} - t_k$ vary, but the long-run average is constant. If

$$\lambda \lim_{n \to \infty} \frac{1}{n} \sum_{k=1}^{n} (t_{k+1} - t_k)$$

is constant, the inter-arrival time is stationary and λ is the mean inter-arrival time.

Sequenced events are often called *customers*, since they resemble customers arriving for service at a bank waiting to be served. This is a powerful analogy, and simulation diagrams have structures that match. Upon arrival, event occurrences enter a *queue*, in which the first customer in is also the first one to be serviced. Thus, we call this a first-in/first-out or FIFO structure. The representation of a queue requires a vector of ordered pairs $Q = [(a_1, d_1), (a_2, d_2), \ldots, (a_n, d_n)]$ in which a_k is the arrival time and d_k is the departure time of event k. For simplicity, this is sometimes abbreviated $Q = [(a_k, d_k)]$, where the event number k is understood to range from 1 to n. It follows that if $T = [a_k]$ is the arrival sequence and $Q = [(a_k, d_k)]$ is a queue, then d_k is monotonic, since a_k must be monotonic and queues are FIFO structures.

A third important structure is the *server*, which can be thought of as a device that actually performs work or an operation on the message or customer in question. Servers can only service a single customer at a time, during which the server is said to be "busy"; otherwise the server is "idle" or "available". For instance, in a digital communications system, messages arrive and queue up waiting for use of a file server. If the server is busy, it is said to be in state (1), but when it is available it goes to state (0), at which time it performs a file transfer on the message. The server is represented as a vector of ordered pairs, in which the on–off times for each event are given: $S = [(u_1, v_1), (u_2, v_2), \ldots, (u_n, v_n)]$ or $S = [(u_k, v_k)]$. At time t, the server is in state (1) if $u_k < t < v_k$ for some k and is in state (0) if $v_k < t < u_{k+1}$ for some k. Since, in general, service times are variable and often stochastic, the mean service time μ is defined as the average on-time:

$$\mu = \lim_{n \to \infty} \frac{1}{n} \sum_{k=1}^{n} (v_k - u_k).$$

Interfaces between event sequencers, queues, and servers are also required. These include *distributors*, which usually disperse assignments to different servers, and *collectors*, which reassemble queues after being acted on by the servers. These are also shown in Figure 9.1. The *distributor* takes the output of a single queue and routes it to one or more servers or other queues. There are many kinds of distributors, each of which is governed by a protocol. For instance, the queue might send an event to a random server, it might send it to the server with least time left for completion of its current job (the "first available" protocol), or it might simply send the event to server number one. Selecting the best distribution protocol for a given situation is an important problem. The *collector* combines the event times of two or more queues to form a single queue. One of the most common collectors is called a *merge*. In a merge, queue Q and queue R are collected to form queue T so that the elements of T are the union of Q and R are sorted into ascending order.

EXAMPLE 9.1

Consider a so-called M/M/1 queuing system in which data packets of random lengths arrive at random times at a single communications server. The server simply forwards the packets along the communications network. Notice the similarity between this system and the drive-through service window at a fast food store, where data packets are replaced by customers and message length is replaced by order size. Model the system, and graph illustrative results for mean event rate $\lambda = 3$ messages per minute and $\mu = 0.5$ minutes per message average message time.

Solution
From the description, this system requires a single queue and a single server, along with the definition of a distributor and collector as shown in Figure 9.2. Since the instruction "model the system" is rather vague, this needs further specification as well.

A reasonable assumption for the incoming messages is that their inter-arrival times are exponentially distributed with mean λ. This means that short intervals between messages are much more likely than long ones, but even very long ones are not out of the question. Such an arrival protocol is characterized by the

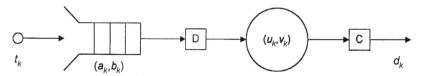

FIGURE 9.2   M/M/1 queue simulation diagram, Example 9.1.

Poisson postulates of Section 8.4. Since there are an average $\lambda = 3$ messages arriving each minute, the mean time between message arrivals is $\tau = \frac{1}{\lambda} = \frac{1}{3}$ minutes. Thus, recalling the means by which to generate an exponential random variate (Section 3.3) a sequence of arrival times is generated by $t_k = t_{k-1} - \frac{1}{3}\ln(\text{RND})$. Thus, the code by which to form an input event sequencer is given by

```
t₀=0
for k=1 to n
        tₖ=tₖ₋₁-1/λln(RND)
        aₖ=tₖ
    next k
```

The event sequencer immediately puts each message into a queue, where the arrival times a_k are given by the event sequencer, but the departure times b_k need to be determined by the distributor joining with the server defined by service times u_k, v_k. The output should also be an event sequencer, as is usually the case in event-driven modeling. In this case, a queue of finishing times for each message is reasonable. From this, statistical analysis is straightforward.

Another reasonable assumption is that the service times are also exponentially distributed with mean time μ and that the message either waits to be forwarded if the server is busy or is immediately served if the server is idle. Thus, the code for the distributor D to execute this protocol is

```
for k=1 to n
        if tₖ>vₖ₋₁ then
                    bₖ=tₖ
                    uₖ=tₖ
                    vₖ=tₖ-μ*ln(RND)
        else
                    bₖ=vₖ₋₁
                    uₖ=vₖ₋₁
                    vₖ=vₖ₋₁-μ*ln(RND)
        end if
    next k
```

Since there is nothing to do once the server releases the data packet, the collector C, whose output is d_k, is especially simple:

```
for k=1 to n
        dₖ=vₖ
    next k
```

Before actually running a simulation, it is useful to first decide on what statistics are most appropriate. Generally, this calls for random processes that exhibit stationarity, since these are more easily characterized, analyzed, and understood by lay people than time-varying statistics. For this simple model, there are likely three such candidates. Even though the arrival times of each message are clearly monotonic, the inter-arrival times should be stationary. But we already knew this from our assumption concerning how arrivals were generated. Thus, we ignore this obvious choice. However, the mean waiting time – that is, the time a message waits in line after arriving in the queue until service is initiated – should also be stationary. Also, the time for service once in the server should be obvious from our assumptions, but let's use it anyway. Finally, the total time a message is in the system, both waiting in the queue and being processed and delivered is of interest. By this we mean the time from entry into the queue to the time of exit from the server. Here is a summary:

- *System:*

input queue:	$[t_k]$;
queue:	$[(a_k, b_k)]$;
server:	$[(u_k, v_k)]$;
output queue:	$[d_k]$.

- *Summary measures:*

waiting time in queue:	$Q_k = b_k - a_k$;
service time:	$S_k = v_k - u_k$;
total time in system:	$T_k = d_k - t_k$.

There is still one more important item. In order to graph the state k as a function of time t, it should be noted that the simulation generates the event times t_k as a function of each event number k, rather than event number as a function of time, as we would prefer. However, by simply reversing their roles, the proper event number-versus-time graph is in place. Thus we use t_k, u_k, v_k, etc. as horizontal "x" values and k as the vertical "y" value. By connecting the points as in the following code, we produce the input queue arrival times for each arrival:

```
set (t₀,k)
for k=1 to n
        line (t_{k-1},k-1)-(t_{k-1},k)
        line (t_{k-1},k)-(t_k,k)
next k
```

The resulting graphs are shown in Figures 9.3–9.5.

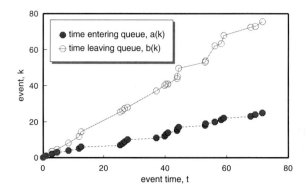

FIGURE 9.3 (a)   Arrival and departure times from queue, Example 9.1.

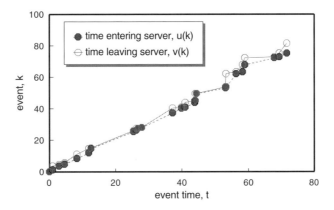

FIGURE 9.3 (b)   Arrival and departure times from server, Example 9.1.

Figures 9.3, 9.4 and 9.5 display different views of this system. Notice how the state changes only as new arrivals enter the queue and that there are small differences in the system variables as messages queue up and are sent. At the same time, there is roughly linear growth in both message arrivals and departures.

Figure 9.3 (a) displays the times at which customers enter the queue (a_k) and leave the queue (b_k). The relative slope of the two curves show that messages are submitted for transmission at roughly one rate and leave for processing at a different one. The ever-widening gap between the two curves show that in this simulation late-messages pay the price of longer waiting times in the queue. This is in contrast to Figure 9.3(b) which compares the processing (service) times – server entry time (u_k) and sever exit time (v_k) – of each message. Clearly the rates of departure and arrival are roughly the same, thus indicating that once the server begins to process the message, there is no longer a late penalty.

Figures 9.4(a), (b) and (c) show a different veiw. Instead of giving the system status (how many messages there are thoughout the system) at each time, Figures 9.4 show the critical event times for a given message – in other words, the historical record for each message. That these views are equivalent leads to an impressive duality principle that holds for all queuing systems. Even-so, Figure 9.4(a) shows that messages that come later also wait in the queue longer. Figure 9.3(b) indicates that, within randomness constraints of statistics, the processing times for each message are roughly the same. Figure 9.4(c) summarizes the system by giving the total time each event remains in the system.

A long-term look at the summary measures is even more revealing. In Figure 9.5, we see that the jump was really just a "burst". In fact, the summary variables tend to show a lot of *bursty* behavior. This is in keeping with out experience, if not with our intuition. We have all been in a supermarket when, for no apparent reason, our time in line was extraordinarily long. No, there wasn't necessarily a conspiracy; it just sort of happened that way, even though the arrival rates are statistically constant. Thus, the performance measures are close to being stationary, as expected earlier.

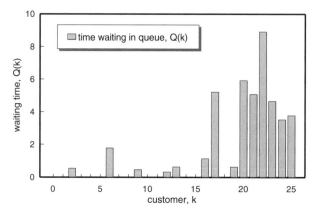

FIGURE 9.4 (a)   Time each customer waits in queue, Example 9.1.

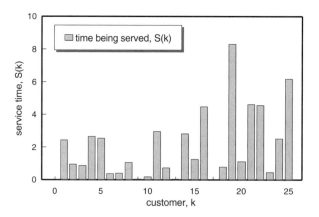

FIGURE 9.4 (b)   Time each customer spends being served, Example 9.1.

Code producing the required event sequence for the input and output queues using the "first available" protocol for the server and graphing the results is shown in Listing 9.1. By importing the results into a spreadsheet, analysis is straightforward using the many built-in tools available. Table 9.1 shows the results after a typical simulation for $\lambda = 3$ and $\mu = 2$.   ○

The reason that the model of Example 9.1 is called an M/M/1 queue is because the input has exponentially distributed or "Markovian" inter-arrival times, the service time is also exponentially distributed, and there is a single queue. Certainly we can imagine more

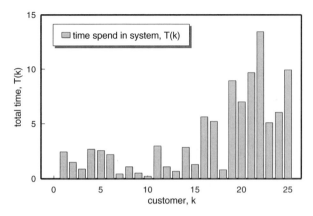

FIGURE 9.4 (c)   Time each customer spends in system, Example 9.1.

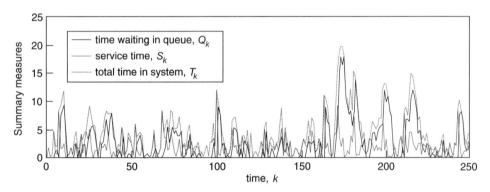

FIGURE 9.5   Long-term summary measures for the M/M/1 queue, Example 9.1.

creative queuing mechanisms than just the M/M/1. To start with, many more statistical distributions are just as easily implemented. Also, there could be several input queues and several servers. As an example, consider the case of an M/M/2 queuing model.

EXAMPLE 9.2

Simulate an M/M/2 queuing model.

Such a system is like that of many commercial airport check-in counters. There is one queue where customers wait until one of (in this case) 2 servers are ready to process their ticket.

Solution

An M/M/2 queue also has a single input, but this time there are two servers. Despite the fact that there are two servers, there is only one output of satisfied customers. Again we will use the "first available server" protocol for the

```
μ = 1/λ
t(0)=0
for k=1 to n
        t(k)=t(k-1)-μ*ln(RND)
        a(k)=t(k)
next k
for k=1 to n
        if t(k)>v(k-1) then
                    b(k)=t(k)
                    u(k)=t(k)
                    v(k)=t(k)-μ*ln(RND)
        else
                    b(k)=v(k-1)
                    u(k)=v(k-1)
                    v(k)=v(k-1)-μ*ln(RND)
        end if
next k
for k=1 to n
        d(k)=v(k)
next k

for k=0 to n
    print t(k), a(k), b(k), u(k), v(k), d(k)
next k

set (t(0, 0)
set (d(0), 0)
for k=1 to n
    line (t(k-1), k-1)-(t(k),k-1)
    line (t(k), k-1)-(t(k), k)
    line (d(k-1), k-1)-(d(k), k-1)
    line (d(k), k-1)-(d(k), k)
next k
```

TABLE 9.1 Summary Measure Simulation
Results for the M/M/1 queue, Example 9.1

k	Q_k	S_k	T_k
0	0.00	0.00	0.00
1	0.00	6.19	6.19
2	4.30	1.76	6.06
3	4.43	0.30	4.72
4	1.01	0.47	4.48
5	0.00	1.97	1.97
Average	2.03	1.89	3.92

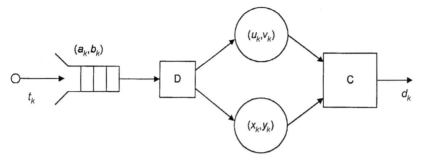

FIGURE 9.6   M/M/2 queue simulation diagram, Example 9.2.

distribution filter. A suitable simulation diagram is shown in Figure 9.6. It follows that the system state is completely defined by the following entities:

input sequencer: $[t_k]$;
queue: $[(a_k, b_k)]$;
first server: $[(u_k, v_k)]$;
second server: $[(x_k, y_k)]$;
output sequencer: $[d_k]$.

The primary difference in the implementation of an M/M/2 queue is in the distributor protocol and collector merge. Whereas the M/M/1 queue's distributor was based on the relationship between only two entities, t_k and v_{k-1}, the M/M/2 queue's distributor is based on the relative magnitudes of t_k v_{k-1} and y_{k-1}. Since there are three entities, there are $3! = 6$ defining relationships for the distributor. It should be clear that the "first available" protocol is characterized as follows:

$$\text{if } t_k < v_{k-1} < y_{k-1} \text{ then}$$
$$b_k = v_{k-1}$$
$$u_k = v_{k-1}$$
$$v_k = v_{k-1} - \mu \ln(\text{RND})$$
$$S_2 \text{ unchanged}$$
$$\text{if } t_k < y_{k-1} < v_{k-1} \text{ then}$$
$$b_k = y_{k-1}$$
$$x_k = y_{k-1}$$
$$y_k = y_{k-1} - \ln(\text{RND})$$
$$S_1 \text{ unchanged}$$
$$\text{if } v_{k-1} < t_k < y_{k-1} \text{ then}$$
$$b_k = a_k$$
$$u_k = a_k$$
$$v_k = t_k - \mu \ln(\text{RND})$$
$$S_2 \text{ unchanged}$$

TABLE 9.2 M/M/2 Queue Simulation, Example 9.2

k	t_k	a_k	b_k	u_k	v_k	x_k	y_k	d_k	$\overline{d}_k$
0	0.00	0.00	0.00	0.00	0.00	0.00	0.00	0.00	0.00
1	1.51	1.51	1.51	1.51	5.06	0.00	0.00	5.06	2.83
2	1.90	1.90	1.90	1.51	5.06	1.90	2.83	2.83	4.54
3	2.82	2.82	2.83	1.51	5.06	2.83	4.54	4.54	5.06
4	2.97	2.97	4.54	1.51	5.06	4.54	5.67	5.67	5.67
5	4.60	4.60	5.06	5.06	7.93	4.54	5.67	7.93	7.93
6	4.66	4.66	5.67	5.06	7.93	5.67	21.99	21.99	10.25
7	6.47	6.47	7.93	7.93	10.25	5.67	21.99	10.25	10.31
8	9.49	9.49	10.25	10.25	10.31	5.67	21.99	10.31	19.76
9	9.90	9.90	10.31	10.31	19.76	5.67	21.99	19.76	21.99
10	10.46	10.46	19.76	19.76	24.57	5.67	21.99	24.57	24.46
11	11.58	11.58	21.99	19.76	24.57	21.99	24.46	24.46	24.57
12	13.08	13.08	24.46	19.76	24.57	24.46	26.33	26.33	26.33
13	13.40	13.40	24.57	24.57	27.72	24.46	26.33	27.72	27.72
14	14.35	14.35	26.33	24.57	27.72	26.33	28.04	28.04	28.04
15	15.37	15.37	27.72	27.72	29.37	26.33	28.04	29.37	29.37
16	15.78	15.78	28.04	27.72	29.37	28.04	32.16	32.16	31.28
17	16.24	16.24	31.28	31.28	33.79	28.04	32.16	33.79	33.79

if $v_{k-1} < y_{k-1} < t_k$ then
$$b_k = a_k$$
$$u_k = a_k$$
$$v_k = t_k - \mu \ln(\text{RND})$$
$$S_2 \text{ unchanged}$$
if $y_{k-1} < t_k < v_{k-1}$ then
$$b_k = a_k$$
$$x_k = a_k$$
$$y_k = t_k - \mu \ln(\text{RND})$$
$$S_1 \text{ unchanged}$$
if $y_{k-1} < v_{k-1} < t_k$ then
$$b_k = a_k$$
$$x_k = a_k$$
$$y_k = t_k - \mu \ln(\text{RND})$$
$$S_1 \text{ unchanged.}$$

Similarly, the collector merges the output sequencer on the basis of whichever variable changed in the servers:

$$\text{if } t_k < v_{k-1} < y_{k-1} \text{ then } d_k = v_k;$$
$$\text{if } t_k < y_{k-1} < v_{k-1} \text{ then } d_k = y_k;$$
$$\text{if } v_{k-1} < t_k < y_{k-1} \text{ then } d_k = v_k;$$
$$\text{if } v_{k-1} < y_{k-1} < t_k \text{ then } d_k = v_k;$$
$$\text{if } y_{k-1} < t_k < v_{k-1} \text{ then } d_k = y_k;$$
$$\text{if } y_{k-1} < v_{k-1} < t_k \text{ then } d_k = y_k.$$

Results of a simulation giving the input sequencer t_k, the queue $[a_k, b_k]$, the two servers $[u_k, v_k]$ and $[x_k, y_k]$, and the output sequencer d_k are compiled in Table 9.2. From this, interesting statistics such as the waiting time follow immediately.

However, notice that in the M/M/2 queue, the departing customers do not necessarily exit in the same order as they arrived. Even so, the M/M/2 queue departures are monotonic with time. Table 9.2 also has a column labeled $\bar{d}_k$, which is column d_k sorted into ascending order. This is especially useful, because graphing t_k and $\bar{d}_k$ versus time t gives the true state of the queuing system over time, where k is the arrival and departure number rather than the customer number.

○

9.2   QUEUING THEORY

There is one special event-driven model that is especially pervasive. This structure is the *common queue*, in which customers line up waiting for service and wait again while the service is actually being performed. These queues, of which the M/M/1 and M/M/2 queues are examples, can always be simulated. However, if the arrival time and service time statistics are well understood (each being Markovian), there are also closed-form solutions to this problem. This theory is called *queuing theory*, and forms the basis of modern-day operations research.

Queuing theory describes the behavior of waiting lines. Whenever the demand for a service exceeds the capacity to render that service, queues naturally form. This is true in banks or supermarkets where multiple checkout clerks are involved, a telephone network where callers are blocked by a "busy" signal, or a computer operating system limiting processes from utilizing various resources such as disks or printers. The essential components of a queuing system are event occurrences (customers or messages), which need to be serviced by a server. If events happen too quickly, the server cannot handle them, a queue builds up, and the customers must wait. After waiting in line, more time is required for the service to take place. The customer then exits the system.

The event occurrence rate is characterized by the *arrival rate* λ. If λ is relatively small, the queue will ebb and flow with time, but will not grow over the long run, creating a stable system. On the other hand, if λ is too large, the queue will grow without bound and the system will go unstable. In these cases, design adjustments must be made to deal with instability, such as either adding servers (a cost) or to *block* new arrivals, leading to lack of service. While the terminology smacks of commerce (customers, servers, etc.), applications are typically technological. In fact, the field was largely developed by A. K. Erlang of Bell Laboratories in the early 20th century for use in communication networks.

The classical queuing system problem is illustrated in Figure 9.7, where the words "customer arrival" are used instead of "event". Similarly, "arrival rate" is synonymous with "event occurrence rate". Arrival events occur randomly, and a certain proportion are denied service and blocked. Whether or not blocking occurs, events (messages in a communication system or customer arrivals in a fast food outlet) occur at an effective rate of λ customers per unit time. Customers then queue up, waiting for service from one of the m number of servers. After waiting in the queue for time Q_k, the selected server then services the customer for a (typically random) service time S_k. The service time is characterized by having a mean time of τ time units per customer. Equivalently, there are $\mu = 1/\tau$ customers per unit time serviced by the server.

Taking our cue from the previous section, the time random variables Q_k (the time the kth customer waits in the queue until stepping up for service), S_k (the total time the kth

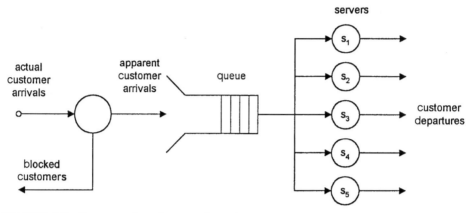

FIGURE 9.7   Representation of a general queuing system.

customer spends being served), and T_k (the total time the kth customer spends in the system), even though they are random and bursty, exhibit rough stationarity. Therefore, we consider the means $\bar{T} = E[T_k]$, $\bar{Q} = E[Q_k]$, and $\bar{S} = E[S_k]$ as useful statistical measures. Clearly, $T_k = Q_k + S_k$. Since these times are random variables, $E[T_k] = E[Q_k] + E[S_k]$. This is more commonly written $\bar{T} = \bar{Q} + \tau$, since $E[S_k] = \tau$.

The summary measures Q_k, S_k, and T_k view the queuing model from the kth customer's point of view. This is well and good, but the system's view of this model is different. Rather than the waiting time for each customer, the system views the number of customers at each time. In particular, these are the analogous summary measures $Q(t)$ (the number of customers waiting in the queue at time t), $S(t)$ (the number of customers being served at time t), and $T(t)$ (the number of customers being served at time t). Clearly, $T(t) = Q(t) + S(t)$. Since these variables are random processes, $\langle T(t) \rangle = \langle Q(t) \rangle + \langle S(t) \rangle$, where $\langle \bullet \rangle$ is the time average of each random process.

Even though it might seem confusing, the standard convention is to define $\bar{N} = \langle T(t) \rangle$ as the average number of customers in the system, $\bar{L} = \langle Q(t) \rangle$ as the average queue length (number of customers waiting in line), and $\rho = \langle S(t) \rangle$ as the average number of customers being served over time. This notation is justified on the basis that each of the measures T, Q, and S is stationary and that statistical averages and time averages are the same. This gives us six statistics of interest for a for a queuing model: three statistical averages and three time averages. Since each is stationary, we use the mean of each measure. These are summarized in Table 9.3.

The classical assumption is that $\lambda(t)$ is constant. While this is often reasonable in the short term, to assume that telephone traffic at 3:00 a.m. is the same as that at 9:00 a.m. is beyond reality! Even so, if the event occurrence rate is constant and the inter-event times are exponentially distributed, an analytical result exists. In the more general case of a variable rate, numerical simulation techniques such as those outlined in the previous section are required. The same is true for service rates $\mu(t)$ and times $\tau(t) = 1/\lambda(t)$. Again, explicit analytical results exist in the case of constant mean service time.

Traditional queuing theory uses the notation $A/B/c/d$ to define the queuing system and its arrival, service disciplines. Here "A" indicates the arrival discipline, "B" indicates

TABLE 9.3 Summary Measures for a Queuing Model

View	Nomenclature	Definition	Interpretation
System	$\bar{N}$	$\langle T(t) \rangle$	Mean number of customers in system
	$\bar{L}$	$\langle Q(t) \rangle$	Mean number of customers waiting in queue
	ρ	$\langle S(t) \rangle$	Mean number of customers being served
Customer	$\bar{T}$	$E[T_k]$	Mean time each customer is in system
	$\bar{Q}$	$E[Q_k]$	Mean time each customer is in queue
	τ	$E[S_k]$	Mean time for each customer to be served

the service discipline, c is the number of servers, and d is the maximum number in the system (if this is omitted, "$d = \infty$" is implied). This text treats only the Markovian (M) arrival and service disciplines. However, other treatments are typically considered, and more general assumptions can always be treated numerically as in the previous section. For example, M/M/2 means that customers arrive in a Markovian manner (Poisson-distributed with exponential inter-arrival times) and are serviced in a Markovian manner (exponentially distributed rates), and there are two servers. Also, M/M/5/5 has Markovian arrivals and service rates, but there are five servers and no queuing exists, since after five customers are in the system, new arrivals are turned away until someone leaves the system.

Queuing systems present two events of interest: customers arrive and customers depart the system. The times of these are important, since every change of state occurs only at one of these times; in between times, the system "sleeps". Therefore, we choose the total number of customers in the system $T(t)$ to represent the system state.

We begin by graphing the arrivals $A(t)$ and departures $D(t)$ at time t. In general, $A(t)$ and $D(t)$ are each monotonically increasing functions and $D(t) \leqslant A(t)$. These are shown in Figure 9.8, where we see both the system and the customer views described in the same graph. For any time t (shown on the horizontal axis), the vertical difference $T(t) = A(t) - D(t)$ gives the number of customers in the system. Similarly, for any customer k (shown on the vertical axis), the horizontal difference $T_k = t_k - t_{k-1}$ gives the length of time that customer k is in the system. This demonstrates that the number of customers in the system, regardless of the view taken, its the area between the arrive and departure curves.

Even without knowing anything regarding the actual distributions of the arrival and service disciplines, there is a remarkable relationship between the quantities $T(t)$ and T_k. It should be clear from Figure 9.8 that the shaded area to the left of time t is equal to the area below customer k. Therefore, there are

$$T(t) = \sum_{i=1}^{A(t)} T_k$$

customers in the system at time t, and

$$E[T_k] = \frac{1}{A(t)} \sum_{i=1}^{A(t)} T_i.$$

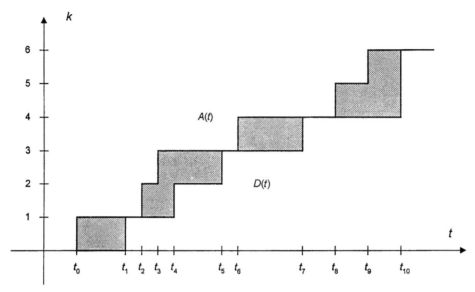

FIGURE 9.8   Graph of arrivals $A(t)$ and departures $D(t)$ as functions of time t.

Assuming a constant arrival rate of λ customers per unit time and $A(t)$ customers at time t,

$$\lambda = \frac{1}{t}\sum_{i=1}^{A(t)} T_i.$$

It follows that

$$\langle T(t)\rangle = \frac{1}{t}\int_0^t [A(t) - D(t)]\, dt$$

$$= \frac{1}{t}\sum_{i=1}^{A(t)} T_i$$

$$= \frac{A(t)}{t}\frac{1}{A(t)}\sum_{i=1}^{A(t)} T_i$$

$$= \lambda E[T_k]. \tag{9.1}$$

This remarkable result is called *Little's formula* (after the famous operations research specialist John D. C. Little). Its significance is that it expresses the duality of the system's view and the customer's view of any queuing system, regardless of the queue's statistical nature. Given the behavior in one view, the other is immediately derivable – changing one directly affects the other. In other words, the average number of customers in a system is directly linked to the time in the system experienced by an individual customer.

Little's formula is very general. In fact, it applies equally well to both the queue and server subsystems. For the queue subsystem, $\langle Q(t)\rangle = \lambda E[Q_k]$ and for the server subsystem, $\langle S(t)\rangle = \lambda E[S_k] = \lambda/\mu = \lambda\tau$. Using the nomenclature introduced earlier, $\bar{N} = \lambda\bar{T}$ and

$\bar{L} = \lambda\bar{Q}$. This states that the number of customers in the system is directly proportional to the time they spend in the system and that the length of the queue is directly proportional to the time customers spend in the queue. The same statement can be made about the number and time spent in the server, and the proportionality constant is the same in all three cases.

9.3   M/M/1 QUEUES

An M/M/1 system has one queue and one server, and can be characterized by the number of customers $T(t)$ in the system at any given time. Using the principles of Chapter 8, it is possible to find differential equation descriptions for $T(t)$, but it is almost always sufficient to consider only the steady-state case. This is because most queuing systems are designed to operate over extended periods of time and the transient phase is short-lived and irrelevant. Assuming a steady-state, T is a stationary random variable representing the state of the queuing system. Therefore, it has a countably infinite number of states $T = 0, 1, 2, \ldots$, and we define

$$\pi_k = \Pr[T(t) = k]$$

to be the probability that there are exactly k customers in the system. The countably infinite vector $\boldsymbol{\pi} = [\pi_0, \pi_1, \pi_2, \ldots]$ gives the probability for all possible states. Our goal is to find $\boldsymbol{\pi}$.

Ignoring the detail of Figure 9.2, the M/M/1 queue can be thought of as shown in Figure 9.9. Event occurrences (customers) arrive at rate of λ customers per unit time, and service takes τ seconds per customer. Assuming the service to be Markovian, an average of $\mu = 1/\tau$ customers are served each unit time, and the system has utilization $\rho = \lambda/\mu = \lambda\tau$. It follows from this that the appropriate simplified transition rate diagram is as given in Figure 9.10. Notice that both the arrival and departure rates are in units of customers per unit time and not time per customer Thus, the probability rate intensities are λ and μ, as indicated.

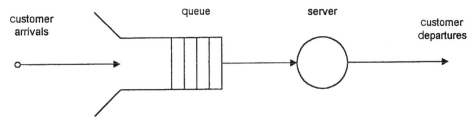

FIGURE 9.9   A graphical representation of the M/M/1 queue.

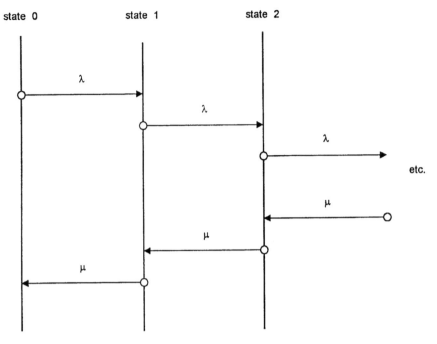

FIGURE 9.10   Simplified state transition diagram for the M/M/1 queue.

The state diagram shows an infinite number of balance equations, one for each state:

$$
\begin{aligned}
k = 0: &\quad \lambda \pi_0 = \mu \pi_1, \\
k = 1: &\quad \lambda \pi_0 + \mu \pi_2 = (\mu + \lambda)\pi_1, \\
k = 2: &\quad \lambda \pi_1 + \mu \pi_3 = (\mu + \lambda)\pi_2, \\
k > 0: &\quad \lambda \pi_{k-1} + \mu \pi_{k+1} = (\mu + \lambda)\pi_k.
\end{aligned}
\tag{9.2}
$$

It follows that each of Equations (9.2) is of the same form except for the first, which defines the left boundary state $k = 0$. Solving the boundary equation ($k = 0$) for π_1 gives

$$
\pi_1 = \frac{\lambda}{\mu}\pi_0 = (\lambda \tau)\pi_0,
$$

since $\tau = 1/\mu$ is the average service time. Combining this into the $k = 1$ state equation produces $\pi_2 = (\lambda/\mu)^2 \pi_0 = (\lambda \tau)^2 \pi_0$. Repeating the process using the $k = 2$ state equation gives $\pi_3 = (\lambda/\mu)^3 \pi_0 = (\lambda \tau)^3 \pi_0$, and the pattern should be obvious. In general,

$$
\pi_k = (\lambda \tau)^k \pi_0.
$$

It remains to find π_0. This is done by using the normalization equation, $\pi_0 + \pi_1 + \pi_2 + \ldots = 1$; in other words, since $\boldsymbol{\pi} = [\pi_0, (\lambda \tau)\pi_0, (\lambda \tau)^2 \pi_0, \ldots]$ is a probability vector, its components must sum to unity. Also, $\pi_k = (\lambda\tau)^k \pi_0$ forms a geometric sequence, so

$$\sum_{k=0}^{\infty} \pi_k = \sum_{k=0}^{\infty} \pi_0 (\lambda\tau)^k$$

$$= \frac{\pi_o}{1 - \lambda\tau} = 1, \qquad 0 \leqslant \lambda\tau < 1.$$

Recalling that $\rho = \lambda\tau$ is called the utilization, $\pi_0 = 1 - \rho$ and

$$\pi_k = (1 - \rho)\rho^k, \qquad k = 0, 1, 2, \ldots \quad . \tag{9.3}$$

Equation (9.3) is important in that it gives the probability that a queuing system is in any specified state k. Since $0 \leqslant \lambda\tau < 1$, it can be seen that π_k is monotonically exponentially decreasing. It should also be noted that π_k forms a geometric probability distribution (see Appendix B). From this, computation of the mean number of customers in the system at any time is straightforward:

$$\bar{N} \equiv \langle T(t) \rangle = \sum_{k=0}^{\infty} k \Pr[T(t) = k]$$

$$= \sum_{k=0}^{\infty} k(1 - \rho)\rho^k$$

$$= \rho(1 - \rho) \sum_{k=0}^{\infty} \frac{d}{d\rho} \rho^k$$

$$= \frac{\rho}{1 - \rho}$$

$$= \frac{\lambda}{\mu - \lambda}. \tag{9.4}$$

Equation (9.4) shows that as the arrival rate λ becomes close to the service time μ, the mean length of the queue becomes large. In the event that $\mu < \lambda$, the system becomes unstable, Equation (9.3) is no longer valid and the queue length goes to infinity. The case where $\mu = \lambda$ (equivalently, $\lambda\tau = 1$) leads to marginal stability.

Using Little's formula, the mean time a customer remains in the system is

$$\bar{T} = \frac{1}{\lambda}\bar{N}$$

$$= \frac{1}{\lambda}\frac{\lambda}{\mu - \lambda}$$

$$= \frac{1}{\mu - \lambda} = \frac{\tau}{1 - \lambda\tau}. \tag{9.5}$$

Since the time customer k spends waiting in the queue is $Q_k = T_k - S_k$,

$$\bar{Q} = \bar{T} - \tau$$

$$= \frac{\tau}{1 - \lambda\tau} - \tau$$

$$= \frac{\lambda}{\mu(\mu - \lambda)} = \frac{\lambda\tau^2}{1 - \lambda\tau}.$$

By Little's formula, $\langle S(t) \rangle = \lambda \bar{S}_k = \lambda\tau$. Also, it follows that the number of customers in the queue at time t is $Q(t) = T(t) - S(t)$, and

$$\bar{L} = \bar{N} - \lambda\tau$$

$$= \frac{\lambda}{\mu - \lambda} - \lambda\tau$$

$$= \frac{\lambda^2}{\mu(\mu - \lambda)} = \frac{\lambda^2\tau^2}{1 - \lambda\tau}.$$

In each of the formulas for the mean number of customers, whether in the queue, in the server, or in total, it is worth noting that the results are independent of both time and event number. This is because of the stationarity properties and steady-state assumptions made at the outset. The same is true in the formulas for the mean time customers remain in the queue, server, or total system.

EXAMPLE 9.3

Consider a data communications system in which messages are transmitted as Poisson processes at the rate of 100 messages per hour. The messages are received at a (single) front-end processor with virtual (essentially unlimited) memory. The messages, which may be assumed to be of exponentially distributed length, are processed by the central processor in an average time of 30 seconds. Describe the behavior of this system.

Solution
Clearly, the problem describes an M/M/1 queue with a message arrival rate of $\lambda = 100$ messages per hour, and a mean service time of $\tau = 30$ messages per second $= 120$ messages per hour, or $100/120 \approx 0.833$ hours to process each message. Thus, $\rho = 100/120 = 5/6$. By Equation (9.2),

$$\pi_k = \tfrac{1}{6}(\tfrac{5}{6})^k, \qquad k = 0, 1, 2, \ldots. \tag{9.6}$$

Enumerating these probabilities in the form of the steady-state state vector,

$$\boldsymbol{\pi} = [\tfrac{1}{6}, \tfrac{5}{36}, \tfrac{25}{216}, \tfrac{125}{1296}, \ldots]$$

$$\approx [0.17, 0.14, 0.12, 0.10, 0.08, 0.07, \ldots].$$

The probability mass function τ_k defined in Equation (9.6) is graphed in Figure 9.11. Note the discrete nature and the "decaying exponential" shape that is characteristic of the geometric distribution. By considering cumulative probabilities, note that approximately one-third of the time, there will be no waiting (i.e.,

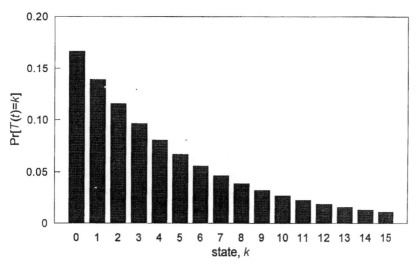

FIGURE 9.11   Probability mass function for the state of the M/M/1 queue, Example 9.3.

there are 0 or 1 messages in the system). However, one-third of the time, there will be a significant wait – there being more than 5 in the system.

Using Equation (9.4), on average there are $\bar{N} = 100/(120 - 100) = 5$ messages in the system at any time. There are also more or less $\lambda/\mu = 100/120 \approx 0.83$ messages being serviced and $\bar{L} = 5 - 5/6 \approx 4.17$ messages in the queue.

By Equation (9.4), the average time in the system (from transmit to receipt) is $\bar{T} = 1/(120 - 100) = 0.05$ hours = 180 seconds. Of this, $\bar{Q} = 100/[120(120 - 100)] = \frac{1}{24}$ hours = 150 seconds is spent waiting in the queue and $\tau = \bar{T} - \bar{Q} = 30$ seconds for processing.

The assumption of infinite queue length is important. For instance, suppose that no more than 10 messages could be accommodated. All others would be blocked. In this case, the probability of blocking is

$$\text{Pr[blocked message]} = \text{Pr}[T(t) > 10]$$

$$= \sum_{k=11}^{\infty} \pi_k$$

$$= \sum_{k=11}^{\infty} \frac{1}{6} \left(\frac{5}{6}\right)^k$$

$$= \frac{b^{11}}{6^{12}} [1 + \frac{5}{6} + \left(\frac{5}{6}\right)^2 + \cdots]$$

$$= \left(\frac{5^{11}}{6^{12}}\right) \frac{1}{1 - \frac{5}{6}} \approx 0.1346. \qquad \bigcirc$$

According to Equation (9.3), the mass function for the random variable $T(t)$ is geometrically distributed, and is a discrete random variable. The related random variable for T_k is different in that it is continuous, since it describes the inter-event time, which is continuous rather than discrete. It must therefore be described by a probability density function rather than a mass function. If there are already k customers in the system, they must wait for $k+1$ services, since the kth customer must also be served before being finished. Since we know that the service times are exponentially distributed, this is the sum of $k+1$ IID exponentially distributed random variables. This is also the defining property of the $(k+1)$-Erlang distribution, which is the Gamma density function with $\alpha = k+1$ and $\beta = \mu$. (Note Appendices B and C, where the formulas for the Gamma and Erlang density functions are given.) However, this is a conditional density function, since we assumed k customers to be in the system at the outset. Therefore, the conditional density function is

$$f_T(t|T_k = k) = \frac{(\mu t)^k}{k!}\mu e^{-\mu t}. \tag{9.7}$$

As we have determined above, the probability that there are exactly k customers in the system is $\Pr[T(t) = k] = (1 - \rho)\rho^k$. By using the fundamental property of $k+1$ mutually exclusive and exhaustive events, the *total probability* is

$$f_T(t) = \sum_{k=0}^{\infty} f_T(t|T_k = k)\,\Pr[N = k]$$

$$= \sum_{k=0}^{\infty} \left[\frac{(\mu t)^k}{k!}\mu e^{-\mu t}\right][(1 - \lambda\tau)(\lambda\tau)^k]$$

$$= \mu(1 - \rho)e^{-\mu t}\sum_{k=0}^{\infty}\frac{(\mu\lambda\tau t)^k}{k!}$$

$$= (\mu - \lambda)e^{-\mu t}\sum_{k=0}^{\infty}\frac{(\lambda t)^k}{k!}$$

$$= (\mu - \lambda)e^{-(\mu-\lambda)t}, \qquad t \geqslant 0. \tag{9.8}$$

Thus, we see that the probability density function of the *time-in-system* is exponential with parameters $\mu - \lambda$. The mean and standard deviations are

$$E[T_k] = \sigma_T = \frac{1}{\mu - \lambda} = \frac{\tau}{1 - \lambda\tau},$$

which is consistent with Equation (9.5).

EXAMPLE 9.4

Recall the scenario of Example 9.3. It will be of interest to know the distribution of T_k probably more than $T(t)$, because the communication system memory requirement is more likely directly related to the time a typical message is in the system, rather than the number of messages in the system. Suppose the computer

system, including both the front end and central processor have combined memory sufficient to handle up to 5 minutes of messages.
(a) What is the probability that an incoming message will be blocked?
(b) If it is known that messages take at least one minute to process, what is the probability that an incoming message will not be blocked?

Solution
Since $\lambda = 100$ messages per minute and $\mu = 120$ messages per minute, the probability density function reduces to

$$f_T(t) = (120 - 100)e^{-(120-100)t}$$
$$= 20e^{-20t}, \quad t \geqslant 0,$$

where t is measured in hours. This is graphed in Figure 9.12, where it will be noted that the critical 5 minutes $= \frac{1}{12} \approx 0.833$ hours is clearly indicated.

(a) As required,

$$\Pr[T_k > 5 \text{ minutes}] = \Pr[T_k > \tfrac{1}{12} \text{ hour}]$$

$$= \int_{1/12}^{\infty} 20e^{-20t}\, dt$$

$$= e^{-5/3} \approx 0.1889.$$

Graphically, this is the shaded area in Figure 9.12. For this computer system, an incoming message will be blocked approximately 19% of the time.
(b) The second problem statement assumes $T_k > 1$ minute and asks the probability of $T_k < 5$ minutes. This is calculated using conditional prob-

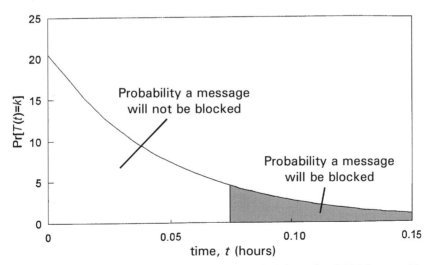

FIGURE 9.12 Probability density function for inter-arrival time of an M/M/1 queue, Example 9.4.

ability as follows:

$$\Pr[T_k < 5 \text{ minutes} | T_k > 1 \text{ minute}] = \Pr[T_k < \tfrac{1}{12} \text{ hours} | T_k > \tfrac{1}{60} \text{ hours}]$$

$$= \frac{\Pr[\tfrac{1}{60} < T_k < \tfrac{1}{12}]}{\Pr[T_k > \tfrac{1}{60}]}$$

$$= \frac{\int_{1/60}^{1/12} e^{-20t}\, dt}{\int_{1/60}^{\infty} e^{-20t}\, dt}$$

$$= 1 - e^{-4/3} \approx 0.7353. \qquad \bigcirc$$

Since so many designed systems are based on FIFO queues, M/M/1 models are ubiquitous. Applications as diverse as operations research, digital communications and multitasking operating systems all use queues and servers as their basis. Table 9.4 summarizes this important model by listing formulas for various probabilities along with the six important summary statistics discussed in the previous section.

TABLE 9.4   Summary of M/M/1 Queues

Description		Symbol	Formula
Probability of:	empty system	π_0	$1 - \rho$
	k customers in system	π_k	$(1-\rho)\rho^k$
	waiting for service		ρ
	blocking		0
Mean number of:	customers in system	$\bar{N}$	$\dfrac{\lambda}{\mu - \lambda}$
	customers in queue	$\bar{L}$	$\dfrac{\lambda^2}{\mu(\mu - \lambda)}$
	customers being served		$\dfrac{\lambda}{\mu}$
Mean time:	in system	$\bar{T}$	$\dfrac{1}{\mu - \lambda}$
	in queue	$\bar{Q}$	$\dfrac{\lambda}{\mu(\mu - \lambda)}$
	being served		$\dfrac{1}{\mu}$
Apparent arrival rate		λ_a	λ

9.4   SIMULATING QUEUING SYSTEMS

Queuing systems form one class of discrete event models. There are but two events: customers arrive and customers depart the system. Therefore, the state of the system is completely determined if one knows the time at which these events occur. The most general technique for simulating these models is to create an event list. This might be more accurately called the event-time list since it lists the times at which each event occurs. Between successive event times, the system lies dormant.

In the case of queuing systems, two event types suffice: *arrivals* and *departures*. As each of the n customers progress through the system, the state is be characterized by two vectors:

$$\mathbf{a} = [a_1, a_2, a_3, \ldots, a_n],$$
$$\mathbf{d} = [d_1, d_2, d_3, \ldots, d_n],$$

the arrival and departure times *of each customer*. To simulate system behavior, one first generates arrivals a_k (typically noting that inter-arrival times are exponentially distributed). For each customer k, the service time (also assumed to be exponential) produces the time of departure d_k.

In general, this produces a monotonically increasing sequence of arrivals, but the sequence of departures is not monotonically increasing, since some events can require inordinate amounts of service time. By sorting the departure sequence d into ascending order, one obtains the sequence of arrival/departure events, characterized by the vector $\mathbf{a}$ and the sorted departures vector $\bar{\mathbf{d}}$. These are not to be confused with $\mathbf{a}$ and $\mathbf{d}$, which give the arrival and departure times of specific customers. The vectors $\mathbf{a}$ and $\bar{\mathbf{d}}$ give the times when arrival and departure events occur. Even so, it is easy to show that the mean time of the kth customer in the system is the same as the mean time between arrival/departure events.

Once the vectors $\mathbf{a}$ and $\bar{\mathbf{d}}$ are known, the arrival and departure functions $A(t)$ and $D(t)$ defined earlier are readily determined. In turn, $\langle T(t) \rangle$ and $E[T_k]$ may be found. Using Little's formula (Equation (9.1)) and Equation (9.3), the mass function of the state vector and density function $f_T(t)$ can be found. From these, the six important summary statics can also be determined. Unless information regarding the subsystem queue or server subsystems are required, nothing more is necessary.

EXAMPLE 9.5

Suppose that the event sequences are as follows:

$$\mathbf{a} = [1.0, 2.5, 4.0, 5.0, 7.0, 9.0],$$
$$\mathbf{d} = [2.0, 4.5, 6.0, 5.5, 8.0, 10.0]. \tag{9.9}$$

Find the functions $A(t)$ and $D(t)$. From these, estimate $\langle T(t) \rangle$, $E[T_k]$, and π_k.

Solution

Evidently, the event vectors are given by

$$\mathbf{a} = [1.0, 2.5, 4.0, 5.0, 7.05, 9.0],$$
$$\bar{\mathbf{d}} = [2.0, 4.5, 5.5, 6.0, 8.0, 10.0]. \qquad (9.10)$$

It is important to note that Equation (9.9) gives the time of the kth message arrival and the time at which that message departs the system. Equation (9.10) gives the time of the kth arrival event, which is the same as the kth message arrival. On the other hand, $\bar{\mathbf{d}}$ in equation (9.10) gives the kth departure event, but this is evidently not the same as the kth message departure, since $\mathbf{d} \neq \bar{\mathbf{d}}$. The graphs of $A(t)$, which is the top curve $\mathbf{a}$ and $D(t)$, which is the lower curve, $\bar{\mathbf{d}}$ follow immediately, and are shown in Figure 9.13.

$\langle T(t) \rangle$ can be thought of as the horizontal average. It is the area of the shaded region averaged over the time interval $0 \leqslant t \leqslant 10$:

$$\langle T(t) \rangle = \frac{1}{t} \sum_{i=1}^{A(t)} T_i$$

$$= \frac{1.0 + 2.0 + 1.5 + 1.0 + 1.0 + 1.0}{10} = 0.75 \text{ customers.}$$

$E[T_k]$ can be thought of as the vertical average. It is the area of shaded region averaged over the number of states $k = 0, 1, \ldots, 6$:

$$E[T_k] = \frac{1.0 + 2.0 + 1.5 + 1.0 + 1.0 + 1.0}{6} = 1.25.$$

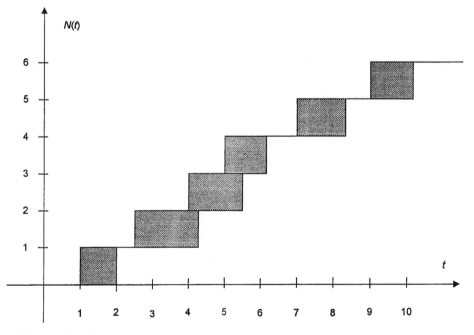

FIGURE 9.13   The arrivals $A(t)$ and departures $D(t)$ as functions of time for Example 9.5.

The mean arrival rate, according to Little's formula, is

$$\lambda = \frac{\langle T(t) \rangle}{E[T_k]} = 0.6$$

The probabilities π_k are approximated by the proportion of time there are k customers in the system. For instance, there are zero customers on the intervals [0.0,1.0], [2.0,2.5], [6.0,7.0], and [8.0,9.0]. Thus,

$$\pi_0 \approx \frac{1.0 + 0.5 + 1.0 + 1.0}{10} = 0.35.$$

Similarly, there is one customer during the intervals [1.0,2.0], [2.5,4.0], [4.5,5.0], [5.5,6.0], [7.0,8.0], and [9.0,10.0], so that

$$\pi_1 \approx \frac{1.0 + 1.5 + 0.5 + 0.5 + 1.0 + 1.0}{10} = 0.55,$$

and two customers during [4.0,4.5] and [5.0,5.5], so that

$$\pi_2 \approx \frac{0.5 + 0.5}{10} = 0.10.$$

Since at no time are there more than two customers in the system, $\pi_{k>2} = 0$. The estimated state probability vector follows immediately as $\hat{\boldsymbol{\pi}} = [0.35, \ 0.55, \ 0.10, \ 0, \ 0, \ldots]$.   ◯

It is sometimes useful to view the explicit drawing shown in Figure 9.13, but usually it is merely a conceptual tool and the probabilities π_k are found directly. For this reason, discrete event simulations come as a two-step process:

- Step 1: Create the event lists a_k and d_k. Sort d_k.
- Step 2: From the event lists, calculate the probabilities π_k (or other statistics of interest).

Step 1 is the simulation, while step 2 is the realization of that simulation into a form to be analyzed. Each of these steps need to be automated to accommodate use of a computer. The procedure shown in Listing 9.2 gives a high-level view of a routine generating event lists. Implementation in a procedural language should take no more than 50 instruction steps.

9.5   FINITE-CAPACITY QUEUES

Since practical systems generally require a finite number of resources, realistic systems often do not lend themselves to infinite queuing models. In fact, the maximum queue length is often a design parameter that directly affects both system cost and

```
        Read input parameters

        Initialize
                Set clock to zero
                Initialize state variables
                Initialize statistical counters
[1]     Find time of next event
                Determine next event type (arrival, departure)
                If event-list is not empty, advance clock
                Else, simulation over

        If event-type=arrival,
                Schedule next arrival
                If server is idle,
                        Make server busy
                        Schedule next departure
                If server is busy,
                        Add customer to queue

        If event-type=departure,
                If queue empty,
                        Set time of departure to infinity
                If queue not empty,
                        Remove customer from system
                        Decrement queue size
                        Schedule departure time
                Advance each customer's queue position

        If simulation not over goto [1]
        Report statistics
        Stop
```

LISTING 9.2   General algorithm for simulation of the M/M/1 queue.

performance. Thus, we study queues whose length is at most n customers in the system at a given time, the so-called *finite-capacity queuing* problem.

Consider an M/M/1/n queuing system in which there are at most n customers in the system and only one server. Recalling that the number of customers in the queue defines the state, there are $n + 1$ possible states for this system. As with the M/M/1 system, customers enter at a mean rate λ and are serviced at a rate $\mu = 1/\tau$. The simplified steady-state transition rate diagram is shown in Figure 9.14. The balance equations are also the same as with the M/M/1 system at node k_0 and the interior, but this time there is also a boundary condition at state n. Therefore, rather than an *initial-value problem*, we now have the following *boundary-value problem*:

$$
\begin{aligned}
k = 0: &\quad \lambda\pi_0 = \mu\pi_1, \\
0 < k < n: &\quad \lambda\pi_{k-1} + \mu\pi_{k+1} = (\mu + \lambda)\pi_k, \\
k = n: &\quad \lambda\pi_{n-1} = \mu\pi_n.
\end{aligned}
\tag{9.11}
$$

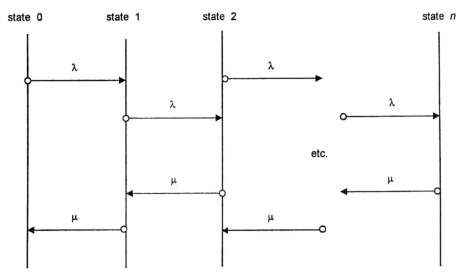

FIGURE 9.14   State transition diagram for a finite-capacity $M/M/1/n$ queue.

The technique for solving this problem is similar to what we have seen before in the $M/M/1$ queue. Beginning with the $k = 0$ node, $\pi_1 = \pi_0\rho$, where $\rho = \lambda/\mu = \lambda\tau$. Proceeding to the general interior node k, $\pi_k = \pi_0\rho^k$ for $k = 0, 1, 2, \ldots n$, as with Equations (9.2). However, this time the probability vector $\boldsymbol{\pi} = [\pi_0, \pi_0\rho, \pi_0\rho^2, \ldots, \pi_0\rho^n]$ is finite and we must find π_0 subject to the finite normalizing equation $\pi_0 + \pi_1 + \pi_2 + \ldots + \pi_n = 1$:

$$\sum_{k=0}^{n} \pi_k = \sum_{k=0}^{n} \pi_0\rho^k$$
$$= \frac{\pi_0(1-\rho)^{n+1}}{1-\rho} = 1.$$

Accordingly,

$$\pi_0 = \frac{1-\rho}{1-\rho^{n+1}}$$

for $\rho \neq 1$. If $\rho = 1$, they each state π_k in the state vector is the same, and since there are $n+1$ states, the net result is

$$\pi_k = \begin{cases} \dfrac{(1-\rho)\rho^k}{1-\rho^{n+1}}, & \rho \neq 1, \\[3mm] \dfrac{1}{n+1}, & \rho = 1, \end{cases} \qquad (9.12)$$

for non-negative integers k. Notice that Equation (9.12) no longer requires that $\mu < \lambda$. Since the number of states is finite and even though the input rate exceeds the output, excess customers are simply turned away and the queue remains bounded. However, if λ

does not exceed μ, the utilization $\rho = \lambda/\mu$ is less than unity, so that $\lim_{n \to \infty} \rho^n = 0$ and we observe that Equation (9.12) reduces to Equation (9.3) for the M/M/1 queue derived in Section 9.3 as a special case.

For finite queuing systems, the key question pertains to the likelihood of *blocking*, which occurs when the queue is full, that is, when there are n customers in the system. Presumably, events that occur when this happens go unserviced. By substituting $k = n$ into Equation (9.12), we obtain the probability that blocking will occur. This important quantity is called the *Erlang A-function* and is denoted by $A(n, \rho)$, which for an M/M/1/n queue is

$$A(n, \rho) = \frac{(1 - \rho)\rho^n}{1 - \rho^{n+1}}, \tag{9.13}$$

where $\rho = \lambda/\mu = \lambda\tau$ is the unblocked utilization.

Since arrivals are blocked, there are two arrival rates. Customers actually arrive at a rate λ but since excess customers are blocked, the system "sees" customers at an apparent rate λ_a. This apparent arrival rate is found by adjusting λ by the probability that there is no blocking, $1 - A(\rho, n)$:

$$\lambda_a = \lambda[1 - A(\rho, n)]$$

$$= \frac{\lambda(1 - \rho^n)}{1 - \rho^{n+1}}. \tag{9.14}$$

EXAMPLE 9.6

An entrepreneur builds a car wash with one service bay and a driveway with room for two customers to wait. It is observed that cars arrive as a Poisson process with a mean rate of 4 cars per hour. Also, it takes an average time of 10 minutes to wash a car. Without using the formulas developed in this section, describe the system behavior.

Solution
From the description there are $\lambda = 4$ cars per hour, $\tau = 10$ minutes per car, and $\mu = 0.1$ cars per minute, or 6 cars washed each hour and a maximum of $n = 2$ vehicles at any time. Presumably, if the lot is full, anyone wishing to wash their car simply drives on and is lost as a customer. The steady-state transition rate diagram is shown in Figure 9.15.

The balance equations for each node are as follows:

$$\begin{aligned} k = 0: && 4\pi_0 &= 6\pi_1, \\ k = 1: && 4\pi_0 + 6\pi_2 &= (4 + 6)\pi_1, \\ k = 2: && 4\pi_1 &= 6\pi_2. \end{aligned} \tag{9.15}$$

There are two boundary and one interior equations. Although there are three equations and three unknowns, it is evident that only two equations are independent (add the first and third of Equations (9.15) to obtain the second).

state 0   state 1   state 2

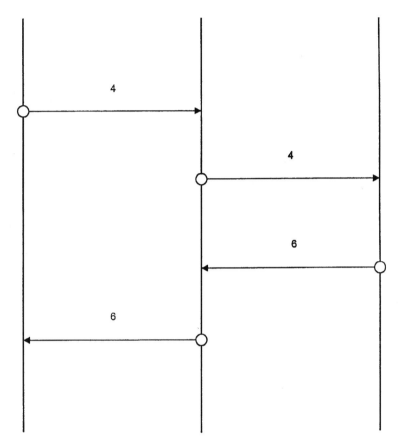

FIGURE 9.15   State transition rate diagram for the M/M/1/2 queue of Example 9.6.

Solving simultaneously, $\pi_1 = \frac{2}{3}\pi_0$ and $\pi_2 = \frac{4}{9}\pi_0$. Using the fact that $\pi_0 + \pi_1 + \pi_2 = 1$, it follows that $\pi_0 = \frac{9}{19}$. Therefore, the steady-state vector is

$$\boldsymbol{\pi} = [\tfrac{9}{19}, \tfrac{6}{19}, \tfrac{4}{19}].$$

The probability of blocking is $A(2, \frac{2}{3}) = \frac{4}{19}$, and the apparent arrival rate is $\lambda_a = (4)(\frac{15}{19}) \approx 3.16$ cars per hour. Also, the expected number of cars in the system is

$$\bar{N} = \sum_{k=0}^{2} k p_k$$

$$= (0)(\tfrac{9}{19}) + (1)(\tfrac{6}{19}) + (2)(\tfrac{4}{19})$$

$$= \tfrac{14}{19} \approx 0.74 \text{ cars.}$$

Since the system "sees" an arrival rate of $\lambda_a = \frac{60}{19}$, Little's formula shows the mean time in the system to be

$$\bar{T} = \frac{1}{\lambda_a}\bar{N}$$

$$= (\tfrac{19}{60})(\tfrac{14}{19}) = \tfrac{7}{30} \text{ hours} = 14 \text{ minutes}.$$

The expected service time was given to be 10 minutes, so the expected waiting time in the queue is $\bar{Q} = 14 - 10 = 4$ minutes. Using the apparent arrival rate in Little's formula gives the expected length of the queue, $\bar{L} = \lambda_a \bar{Q} = (\tfrac{60}{19})(\tfrac{4}{60}) \approx 0.21$ cars. This means that there are an average of $\tfrac{14}{19} - \tfrac{4}{19} = \tfrac{10}{19} \approx 0.21$ cars in the car wash bay. These results should be checked against the formulas derived earlier in this section. Even so, the point is that it is often easier to apply basic principles to specific applications rather than general formulas.   ○

General formulas for the M/M/1/n queue exist for each of the interesting summary statistics $\bar{N}, \bar{T}, \bar{L}$, and $\bar{Q}$ that are comparable to those of the M/M/1 queue. For instance, it can be shown that

$$\bar{N} = \begin{cases} \dfrac{\rho[1 + (n\rho - n - 1)\rho^n]}{(1 - \rho)(1 - \rho^{n+1})}, & \rho \neq 1, \\[2ex] \tfrac{1}{2}n, & \rho = 1, \end{cases} \tag{9.16}$$

and

$$\bar{T} = \begin{cases} \dfrac{1 + (n\rho - n - 1)\rho^n}{(\mu - \lambda)(1 - \rho^{n+1})}, & \rho \neq 1, \\[2ex] \dfrac{n(1 - \rho^{n+1})}{2\lambda(1 - \rho^n)}, & \rho = 1. \end{cases} \tag{9.17}$$

Table 9.5 summarizes the results for M/M/1/n queues.

9.6   MULTIPLE SERVERS

It is undesirable to refuse service to potential customers or to block service requests on the Internet. In cases such as these, it is better to allocate additional resources to the problem than to simply deny service. In other words, we should add a multiple server facility to our system. This mechanism maintains system integrity and reduces blocking. Thus, there is a trade-off that by incurring an added expense, throughput and quality will increase.

Let us first suppose that there are c servers and no queue. This allows the system to have a maximum of c customers, all of which are being served concurrently. If potential

TABLE 9.5 Summary of M/M/1/n Queues

Description		Symbol	Formula
Probability of:	empty system	π_0	$\dfrac{1-\rho}{1-\rho^{n+1}}$
	k customers in system	π_k	$\dfrac{(1-\rho)\rho^k}{1-\rho^{n+1}}$
	waiting for service		$\dfrac{\rho(1-\rho^n)}{1-\rho}$
	blocking	$A(n,\rho)$	$\dfrac{\rho^n(1-\rho)}{1-\rho^{n+1}}$
Mean number of:	customers in system	$\bar{N}$	$\dfrac{\rho}{1-\rho}\dfrac{1+(n\rho-n-1)\rho^n}{1-\rho^{n+1}}$
	customers in queue	$\bar{L}$	$\dfrac{\rho[(n-1)\rho^{n+1}-n\rho^n+\rho]}{(1-\rho)(1-\rho^{n+1})}$
	customers being served	ρ_a	$\rho\dfrac{1-\rho^n}{1-\rho^{n+1}}$
Mean time:	in system	$\bar{T}$	$\dfrac{1}{\mu-\lambda}\dfrac{1+(n\rho-n-1)\rho^n}{1-\rho^{n+1}}$
	in queue	$\bar{Q}$	$\dfrac{(n+1)\rho^{n+1}-n\rho^n+\rho}{(\mu-\lambda)(1-\rho^{n+1})}$
	being served	τ	$\dfrac{1}{\mu}$
Apparent arrival rate		λ_a	$\dfrac{\lambda(1-\rho^n)}{1-\rho^{n+1}}$

customers arrive while all servers are busy, they are blocked. That is, we consider the M/M/c/c queuing system.

As before, customers enter at a mean rate λ and are serviced in a mean time $\tau = 1/\mu$. However, this time the *customer exit rate is proportional to the number of customers being serviced*. This means that if one customer is being served (and $c-1$ servers are performing no service at all), the departure rate is μ. However, if there are two customers being serviced (and $c-2$ servers doing nothing), the departure rate is 2μ. In general, if k customers are being served, the departure rate is $k\mu$. Therefore, the steady-state transition rate diagram is as shown in Figure 9.16.

The balance equations are as follows:

$$
\begin{aligned}
k=0: &\quad \lambda p_0 = \mu p_1, \\
0 < k < c: &\quad \lambda p_{k-1} + (k+1)\mu p_{k+1} = (\lambda + k\mu)p_k, \\
k=c: &\quad \lambda p_{c-1} = c\mu p_c.
\end{aligned}
\tag{9.18}
$$

The procedure to solve Equation (9.18) is similar to that of the M/M/1 and M/M/1/n systems. We start with the $k=0$ boundary equation, then move to successive nodes to find

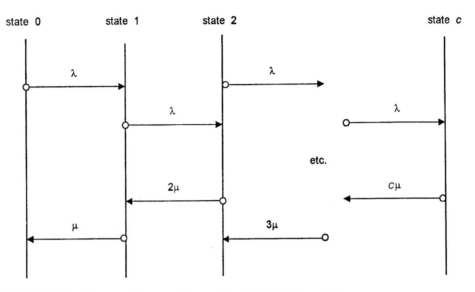

FIGURE 9.16   State transition rate diagram for the M/M/c/c multiple-server queue.

probabilities π_k in terms of π_0:

$$k = 0: \quad \pi_1 = \frac{\lambda}{\mu}\pi_0$$

$$= \rho\pi_0;$$

$$k = 1: \quad \lambda\pi_0 + 2\mu p_2 = (\lambda + \mu)\pi_1$$

$$= (\lambda + \mu)\frac{\lambda}{\mu}\pi_0$$

$$= \frac{\lambda^2}{\mu}\pi_0 + \lambda\pi_0,$$

$$\pi_2 = \frac{1}{2}\frac{\lambda^2}{\mu^2}\pi_0$$

$$= \tfrac{1}{2}\rho^2\pi_0;$$

$$k = 2: \quad \lambda\pi_1 + 3\mu\pi_3 = (\lambda + 2\mu)\pi_2$$

$$= (\lambda + 2\mu)\frac{1}{2}\frac{\lambda}{\mu}\pi_1$$

$$= \frac{1}{2}\frac{\lambda^2}{\mu}\pi_1 + \lambda\pi,$$

$$\pi_3 = \frac{1}{6}\frac{\lambda^2}{\mu^2}\pi_1$$

$$= \tfrac{1}{6}\rho^3\pi_0.$$

Continuing in this manner, in general

$$\pi_k = \frac{1}{k!}\rho^k \pi_0.$$

π_0 can be evaluated by normalizing the sum:

$$\sum_{k=0}^{c} \frac{1}{k!}\rho^k \pi_0 = 1.$$

Although this sum does not result in a closed form, solving for π_0 gives

$$\pi_0 = \frac{1}{\displaystyle\sum_{k=0}^{c} \frac{1}{k!}\rho^k}.$$

Substituting, we find the explicit formula for π_k to be

$$\pi_k = \frac{\rho^k}{k! \displaystyle\sum_{i=0}^{c} \frac{\rho^i}{i!}}. \tag{9.19}$$

If the system is filled with its maximum of c customers, blocking occurs. Of course there is still no waiting, since there are also c servers. But, customers attempting entry at this time will be rejected. Blocking occurs when $k = c$, which leads to the so-called *Erlang B-formula* for the probability of blocking an M/M/c/c queueing system:

$$B(c, \rho) = \frac{\rho^c}{c! \displaystyle\sum_{i=0}^{c} \frac{\rho^i}{i!}}. \tag{9.20}$$

From this, the apparent arrival rate is $\lambda_a = \lambda[1 - B(c, \rho)]$. Since no queue is allowed, $\bar{L} = 0$ and $\bar{Q} = 0$. Therefore, the expected time a customer is in the system is only for service. From this, combined with the fact that the service time for each event is still $\tau = 1/\mu$, it follows that $\bar{T} = 1/\mu$ too. However, in applying Little's formula to find the expected number of customers in the server, the arrival rate must be that seen by the queue: λ_a. Of course, the expected number in the system is the same as that in the server, leading to

$$\bar{N} = \lambda_a \bar{T}$$

$$= \frac{\lambda}{\mu}[1 - B(c, \rho)]$$

$$= \rho \left(1 - \frac{\rho^c}{c! \displaystyle\sum_{k=0}^{c} \frac{\rho^k}{k!}} \right).$$

EXAMPLE 9.7

Suppose a small office has a traditional telephone system with two lines. If one line is busy, the next incoming call "bounces" to the second line. If both lines are busy, any additional calls ring busy and are lost. Calls arrive at mean rate of 10 calls per hour and the average length of a phone call is 4 minutes. Assuming Markovian arrivals and call duration, describe the system behavior. In particular, estimate the number of lost calls over a 9-hour business day's time.

Solution
Since there are only two phones, $c = 2$. Also, calls arrive at $\lambda = \frac{1}{6}$ calls per minute and $\mu = \frac{1}{4}$ calls can be serviced each minute. The steady-state transition rate diagram is shown in Figure 9.17. Notice that when both phones are busy, the service rate is twice that of when there is only one phone busy.

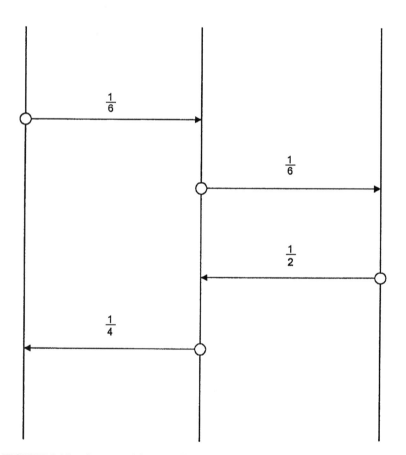

FIGURE 9.17 State transition rate diagram for the M/M/2/2 queue, Example 9.7.

The balance and normalizing equations for each node are

$$\begin{aligned}
\text{node 0:} \qquad & \tfrac{1}{6}\pi_0 = \tfrac{1}{4}\pi_1, \\
\text{node 1:} \qquad & \tfrac{1}{6}\pi_0 + \tfrac{1}{2}\pi_2 = \left(\tfrac{1}{4} + \tfrac{1}{6}\right)\pi_1, \\
\text{node 2:} \qquad & \tfrac{1}{6}\pi_0 = \tfrac{1}{2}\pi_2, \\
\text{normalizing:} \qquad & \pi_0 + \pi_1 + \pi_2 = 1.
\end{aligned}$$ (9.21)

As usual, the nodal equations are redundant in that, of the $n + 1 = 3$ equations, there are only $n = 2$ equations that are independent. Combining these with the normalizing equation presents the unique solution

$$\boldsymbol{\pi} = [\tfrac{9}{17}, \tfrac{6}{17}, \tfrac{2}{17}]$$

The probability of blocking is that of being in the highest state ($n = 2$), so $B(2, \tfrac{2}{3}) = \tfrac{2}{17}$. Also, the expected number in the system is

$$\begin{aligned}
\bar{N} &= \sum_{k=0}^{2} k\pi_k \\
&= (0)(\tfrac{9}{17}) + (1)(\tfrac{6}{17}) + (2)(\tfrac{2}{17}) \\
&= \tfrac{10}{17} \approx 0.59 \text{ calls.}
\end{aligned}$$

The apparent arrival rate is

$$\begin{aligned}
\lambda_a &= \lambda(1 - \pi_2) \\
&= \tfrac{1}{6}(1 - \tfrac{2}{17}) \\
&= \tfrac{5}{34} \approx 0.15 \text{ calls per minute.}
\end{aligned}$$

By Little's formula,

$$\begin{aligned}
\bar{T} &= \frac{1}{\lambda_a}\bar{N} \\
&= (\tfrac{34}{5})(\tfrac{10}{17}) = 4 \text{ minutes per call.}
\end{aligned}$$

In 9 hours, there are $(9)(10) = 90$ calls. Of these, $\tfrac{2}{17}$ will be blocked. Therefore, $\tfrac{2}{17}(90) \approx 11$ calls will be lost. ○

We note that the general equations are convenient, but actual problems can be handled by applying first principles to an accurate state diagram just as easily. Table 9.6 summarizes the results for M/M/c/c queues.

9.7 M/M/c QUEUES

The M/M/c/c queuing system discussed in Section 9.6 dedicates extra servers but does not allow a queue to form. It is much like an old-fashioned telephone answering system where there are c operators to answer calls, but no calls are put on hold and there is

TABLE 9.6 Summary of M/M/c/c Queues

Description		Symbol	Formula
Probability of:	empty system	π_0	$\left(\displaystyle\sum_{i=0}^{c}\frac{\rho^i}{i!}\right)^{-1}$
	k customers in system	π_k	$\dfrac{\rho^k}{k!\displaystyle\sum_{i=0}^{c}\dfrac{\rho^i}{i!}}$
	waiting for service		0
	blocking	$B(c,\rho)$	$\dfrac{\rho^c}{c\displaystyle\sum_{i=0}^{c}\dfrac{\rho_i}{i!}}$
Mean number of:	customers in system	$\bar{N}$	$\rho\left(1-\dfrac{\rho^c}{c!\displaystyle\sum_{i=0}^{c}\dfrac{\rho^i}{i!}}\right)$
	customers in queue	$\bar{L}$	0
	customers being served		$\rho\left(\dfrac{1-\rho^c}{c!\displaystyle\sum_{i=0}^{c}\dfrac{\rho^i}{i!}}\right)$
Mean time:	in system	$\bar{T}$	$\dfrac{1}{\mu}$
	in queue	$\bar{Q}$	0
	being served		$\dfrac{1}{\mu}$
Apparent arrival rate		λ_a	$\lambda\left(1-\dfrac{\rho^c}{c!\displaystyle\sum_{i=0}^{c}\dfrac{\rho^i}{i!}}\right)$

no facility to assemble a queue for incoming phone calls. Callers who call when all lines are busy are simply given a busy signal and are not accommodated by the system; they can either try again or give up. A more modern telephone answering system where callers are automatically queued up and allowed to wait for service can be better modeled with a M/M/c queuing system, which has c operators, but an unlimited-length queue. In this case, if all operators are busy, a queue is formed in which the calls are answered in the order they are received. It follows that blocking will not occur, but customers might have an unexpectedly long wait for service.

The state diagram for an M/M/c queue is shown in Figure 9.18. It will be noted that there are two cases. If less than c customers are in the system, all are served and the exit

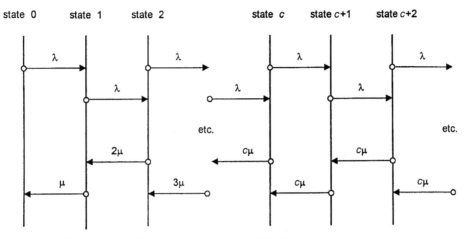

FIGURE 9.18   State transition rate diagram for the M/M/c queue.

rate is $k\mu$, which is proportionate to the number of customers being served. On the other hand, if there are more than c customers, the exit rate is fixed at $c\mu$, since all servers are busy and the remaining customers must wait. The balance equations are as follows:

$$
\begin{aligned}
k = 0: &\quad \lambda\pi_0 = \mu\pi_1, \\
0 < k < c: &\quad \lambda\pi_{k-1} + (k+1)\mu\pi_{k+1} = (\lambda + k\mu)\pi_k, \\
k = c: &\quad \lambda\pi_{c-1} + c\mu\pi_{c+1} = (\lambda + c\mu)\pi_c, \\
k > c: &\quad \lambda\pi_{k-1} + c\mu\pi_{k+1} = (\lambda + c\mu)\pi_k.
\end{aligned}
\tag{9.22}
$$

Proceeding in a manner similar to solving the balance equations above, it can be shown that for an M/M/c queuing system,

$$
\pi_k =
\begin{cases}
\dfrac{\rho^k}{k!}\pi_0, & k = 0, 1, \ldots, c, \\[2ex]
\dfrac{\rho^k}{c^{k-c}c!}\pi_0, & k = c+1, c+2, \ldots,
\end{cases}
\tag{9.23}
$$

where

$$
\pi_0 = \left[\sum_{k=0}^{c-1} \frac{\rho^k}{k!} + \frac{\rho^c}{(1 - \rho/c)c!} \right]^{-1}.
\tag{9.24}
$$

While there can be no blocking for this system, waiting in the queue is a real possibility. It follows that the apparent arrival rate is also the actual arrival rate. The likelihood of waiting in the queue is measured by the *Erlang C-function*. Specifically, the

Erlang C-function gives the probability that all c servers are busy and thus a queue is formed:

$$C(c, \rho) = \sum_{k=c}^{\infty} \pi_k$$

$$= \frac{\rho^c}{(1 - \rho/c)c!} \pi_0, \tag{9.25}$$

where π_0 is defined in Equation (9.24). This permits immediate calculation of the key expected values:

$$\bar{N} = \frac{\rho}{c - \rho} C(c, \rho) + \rho,$$

$$\bar{T} = \frac{\lambda \rho}{c - \rho} C(c, \rho) + \tau,$$

$$\bar{Q} = \frac{\lambda \rho}{c - \rho} C(c, \rho), \tag{9.26}$$

$$\bar{L} = \frac{\rho}{c - \rho} C(c, \rho).$$

Of course, these, and the formulas given earlier, are all fine formulas, and it is tempting to apply them to specific problems as the need arises. However, this should be avoided, since there is no end to variations, and specific problems can always be attacked from first principles. For instance, consider the M/M/2/3 example below.

EXAMPLE 9.8

Reconsider the car wash problem posed in Example 9.6, but this time suppose there are two services bays and only room for a single waiting vehicle. Describe this system and its queuing behavior.

Solution
Now the car wash has two servers and can have up to three customers on the premises at any given time: two in the service bays and one waiting in the parking area. Thus, there are four possible states, each representing the number of cars on site. This is an M/M/2/3 system with $\lambda = 4$ and $\mu = 6$ cars per hour. The steady-state transition diagram is shown in Figure 9.19. The balance and normalizing equations yield

$$
\begin{aligned}
k = 0: &\qquad 4\pi_0 = 6\pi_1, \\
k = 1: &\qquad 4\pi_0 + 12\pi_2 = (6 + 4)\pi_1, \\
k = 2: &\qquad 4\pi_1 + 12\pi_3 = (12 + 4)\pi_2, \\
k = 3: &\qquad 4\pi_2 = 12\pi_3, \\
\text{normalizing:} &\qquad \pi_0 + \pi_1 + \pi_2 + \pi_3 = 1.
\end{aligned}
$$

As before, there is a unique solution determined by these five equations with four

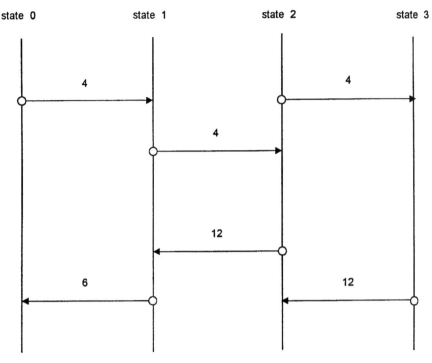

FIGURE 9.19   State transition rate diagram for the M/M/2/3 queue of Example 9.8.

unknowns. Elementary linear algebra leads to the probability vector

$$\boldsymbol{\pi} = [\tfrac{27}{53}, \tfrac{18}{53}, \tfrac{6}{53}, \tfrac{2}{53}].$$

In this system, both blocking and waiting (in the queue) for service can occur. Even so, once the probability state vector is known, all else follows. The probabilities for each come directly from a proper reading of the probability vector. For instance, here are several examples:

- The probability of not getting instant service is $\pi_2 + \pi_3 = \tfrac{6}{53} + \tfrac{2}{53} = \tfrac{8}{53}$.
- The probability of being blocked is $\pi_3 = \tfrac{2}{53}$. Note the difference between this and (1).
- The probability of being in a wait state is $\pi_3 = \tfrac{6}{53}$.
- If at least one customer is already in the system, the probability of waiting is the total probability. Let W be the event that waiting occurs. Then

$$\Pr[W] = \sum_{k=0}^{3} \Pr[W|T(t) = 0] \Pr[T(t) = k]$$
$$= 0 + \tfrac{18}{53} + \tfrac{8}{53} + \tfrac{2}{53} = \tfrac{28}{53}.$$

- The apparent arrival rate is $\lambda_a = (4)(1 - \tfrac{8}{53}) \approx 3.396$ cars per hour.

- The mean number of cars in the system is

$$\bar{N} = (0)(\tfrac{27}{53}) + (1)(\tfrac{18}{53}) + (2)(\tfrac{6}{53}) + (3)(\tfrac{2}{53}) \approx 0.68 \text{ cars.}$$

- Using the apparent arrival rate along with Little's formula, the average time in the system is

$$\bar{T} = \tfrac{36}{53}/4(1 - \tfrac{8}{53}) \approx 0.194 \text{ hours} \approx 11.67 \text{ minutes.}$$

- The mean service time is read from the problem statement: $\tau = 10$ minutes.
- The mean number of cars being serviced is $4(1 - \tfrac{8}{53})/10 \approx 0.334$ cars.
- The mean number of cars in the queue is $\bar{L} = 0.680 - 0.334 \approx 0.346$ cars.
- The mean time for cars to remain in the queue is $\bar{Q} = 11.67 - 10 = 1.67$ minutes. ○

In summary, it is evident that there are many queuing theory formulas that, even though tedious, can be derived. A small sampling has been given in this chapter. In

TABLE 9.7   Summary of M/M/c Queues

Description	Symbol	Formula
Probability of: empty system	π_0	$\left[\dfrac{\rho^c}{(1 - \rho/c)c!} + \displaystyle\sum_{i=0}^{c-1} \dfrac{\rho^k}{i!} \right]^{-1}$
k customers in system	π_k	$\dfrac{\rho^k}{k!}\pi_0, k = 0, 1, \ldots, c$ $\dfrac{\rho^k}{c^{k-c}c!}\pi_0, k = c+1, c+2, \ldots$
waiting for service	$C(c, \rho)$	$\dfrac{\rho^c}{(1 - \rho/c)c!}\pi_0$
blocking		0
Mean number of: customers in system	$\bar{N}$	$\rho + \dfrac{\rho}{c - \rho}C(c, \rho)$
customers in queue	$\bar{L}$	$\dfrac{c}{c - \rho}C(c, \rho)$
customers being served		ρ
Mean time: in system	$\bar{T}$	$\dfrac{1}{\mu}\dfrac{C(c, \rho)}{\mu c - \lambda}$
in queue	$\bar{Q}$	$\dfrac{C(c, \rho)}{\mu c - \lambda}$
being served		$\dfrac{1}{\mu}$
Apparent arrival rate	λ_a	λ

practice, these are rarely used, since special models and circumstances are the order of the day and not the exception. Even so, in all cases the application of first principles is straightforward and usually preferable. First, draw an appropriate steady-state transition diagram. From this diagram, write the balance and normalizing equations. Solving these equations produces the steady-state state vector. This leads to a straightforward computation of the "interesting" quantities described throughout this chapter. Little's formula can be a great time saver in computing mean values. A summary of the $M/M/c$ queuing system is presented for convenience in Table 9.7.

BIBLIOGRAPHY

Banks, J., J. S. Carson, and B. L. Nelson, *Discrete-Event System Simulation*, 2nd edn. Prentice-Hall, 1996.

Buzacott, J. A. and J. G. Shanthikumar, *Stochastic Models of Manufacturing Systems*. Prentice-Hall, 1993.

Freeman, H., *Discrete Time Systems*. Polytechnic Press, 1965.

Hall, R. W., *Queuing Methods: For Services and Manufacturing*. Prentice-Hall, 1991.

Hillier, F. and G. Lieberman, *Introduction to Operations Research Guide*, 6th edn. McGraw-Hill, 1995.

Kleinrock, L., *Queuing System*. Wiley, 1976.

Knuth, D. E., *The Art of Computer Programming*, Vol. 1: *Fundamental Algorithms*, 3rd edn. Addison-Wesley, 1997.

Law, A. and D. Kelton, *Simulation, Modeling and Analysis*. McGraw-Hill, 1991.

MIL 3, Inc., *Modeling and Simulating Computer Networks: A Hands on Approach Using OPNET*. Prentice-Hall, 1999.

Nelson, B., *Stochastic Modeling, Analysis and Simulation*. McGraw-Hill, 1995.

Newell, G. F., *Application of Queuing Theory*. Chapman and Hall, 1982.

Roberts, N., D. Andersen, R. Deal, M. Garet, and W. Shaffer, *Introduction to Computer Simulation*. Addison-Wesley, 1983.

Shoup, T. E., *Applied Numerical Methods for the Micro-Computer*. Prentice-Hall, 1984.

White, J. A., J. W. Schmidt, and G. K. Bennett, *Analysis of Queuing Systems*. Academic Press, 1975.

Wolff, R. W., *Stochastic Modeling and the Theory of Queues*. Prentice-Hall, 1989.

EXERCISES

9.1 Use the simulation of the $M/M/1$ queue given in Listing 9.1 to produce the six summary statistics defined in this chapter. Compare against the theoretical results.

9.2 Implement a simulation of the $M/M/2$ queue. Verify your results against those expected theoretically.

(a) Use $\lambda = 5$ and $\mu = 0.5$.

(b) Use $\lambda = 5$ and $\mu = 1.5$.

(c) Use $\lambda = 5$ and $\mu = 2.5$.

9.3 Consider the following times at which arrivals and departures occur in a system (assume that measurements are taken relative to 2:00):

$$a = [2{:}02, 2{:}04, 2{:}05, 2{:}07, 2{:}10, 2{:}12, 2{:}12, 2{:}12, 2{:}14, 2{:}17],$$
$$d = [2{:}05, 2{:}06, 2{:}09, 2{:}10, 2{:}15, 2{:}18, 2{:}19, 2{:}22, 2{:}23, 2{:}25].$$

(a) Graph $A(t)$ and $D(t)$ versus time (as in Figure 9.13).

(b) Compute how much time T_k each customer k is in the system. Use this to compute the average $\bar{T} = E[T_k]$.

(c) Compute the average number of customers in the system, $\bar{N} = \langle T(t) \rangle$.

(d) Compute the mean arrival rate λ.

9.4 Arrival times of print jobs in a resource sharing computer system are specified as a Poisson process at the rate of 80 print jobs per hour.

(a) Write an expression for the probability mass function, $\Pr[T(t) = k]$, where k is the system state.

(b) Find the probability density function $f_T(t)$ describing the distribution of the inter-event times T_k.

9.5 A mail order company's phone system is equipped with an automatic answering module. Customer's calls arrive as a Poisson process at the rate of 20 calls per hour, whereas the service time is exponentially distributed with a mean of 2.5 minutes. Determine the probability of a customer taking more than 4 minutes of service time.

9.6 Three queues, each with mean service time μ_i, are arranged as shown in the following figure:

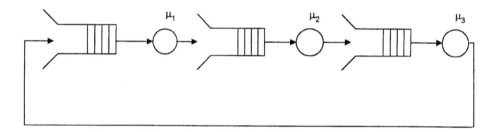

(a) Suppose there is exactly one customer in the loop. Find the mean time $E[T]$ it takes the customer to cycle around the loop. Deduce the mean arrival rate λ_i at each of the queues. Verify Little's formula for this case.

(b) Now suppose there are v customers in the queue. How are the mean arrival rate and the mean cycle time related?

9.7 Consider a queuing system that is empty at time $t = 0$ and where the arrival times of the first 6 customers are $a = [1,4,5,7,8,11]$. The respective service times are $d = [4,3,2,5,5,2,3]$. Sketch a graph of $T(t)$ versus t, and verify Little's formula for each of the following service disciplines:

(a) First-come/first-served

(b) Last-come/first-served

(c) Shortest job first (assuming the service times are known in advance).

9.8 Show that the pth percentile of the waiting time for an M/M/1 queuing system is given by

$$\frac{1}{\mu(1-\rho)} \ln\left(\frac{\rho}{1-p}\right).$$

9.9 Consider an M/M/1 queuing system.

(a) Find a closed-form formula for $\Pr[T(t) \geq k]$

(b) Find the maximum allowable arrival rate in a system with service rate μ if it is required that $\Pr[T(t) \geq 5] = 0.01$.

9.10 Simulate the behavior of a general single-server M/M/1 queuing system G/G/1. That is, write a computer program that exhibits the behavior of a single-server system with arbitrary arrival and service time distributions. Analyze the output statistics for a set of input and server distributions.

   (a) Write programs for the classical single-server queuing systems. The input should be handled entirely by subroutines that are independent of the main program modeling the system. There will need to be two of these: one for system arrivals and one for service times. Specifically, subroutines will be needed for:

   (i) uniform inter-arrival (service) times;
   (ii) exponential inter-arrival (service) times;
   (iii) constant service times.

   (b) The output should include the "interesting" summary measures described in Table 9.3. To begin, the program should create a log of events. That is, the program should make a list of times when customers arrive, when customers begin to be served and when customers exit the system. From these, the program should produce the following:

   (i) $A(t)$ = number of arrivals as a function of time;
   (ii) $D(t)$ = number of departures as a function of time;
   (iii) $T(t)$ = number of customers in the system as a function of time;
   (iv) T_k = time the kth customer remains in the system.

   (c) Create graphs of $A(t)$ and $D(t)$ for each input. Compare the following regimens:

   (i) Uniform inter-arrival time on [0,6] minutes; single server with constant service time 2 minutes.
   (ii) Exponential inter-arrival time with mean 3 minutes; single server with exponential service time of mean 2 minutes.

   Create histograms of $T(t)$ and T_k for each of the regimens defined above. Compare these against the theoretical results using a chi-square test.

9.11 Consider an old-fashioned telephone system with four lines that does not allow queue formation. If all lines are busy, an incoming call is lost. Calls arrive as a Poisson process at the average rate of 20 calls per hour. The service time is exponentially distributes, with a mean of 2.5 minutes per call.

   (a) Compute the probabilities of 0, 1, 2, 3, and 4 calls being handled simultaneously by the system.
   (b) What is the average number of calls in the system?
   (c) Find the probability mass function of the system state, $\Pr[T(t) = k]$.
   (d) Find the probability density function describing the service times, $f_T(t)$.

9.12 A communication network uses two transponder channels to transmit messages. Each channel has the capacity to hold one message in the queue at the same time another call is being transmitted. Messages arrive for transmission as a Poisson process at the rate of 12 messages per hour. Transmission time is exponentially distributed, with a mean of 3.33 minutes per message,

   (a) Compute the probability that there is no message being held.
   (b) Find the probability that a call is waiting for transmission.

9.13 A certain company has two fax machines. It has been found that the time to service each incoming message is exponentially distributed, with a mean of 2 minutes. Assume that requests for service form a Poisson process with rate of 1 request every 2 minutes.

   (a) Find the probability that an incoming fax must wait to be printed.
   (b) If another machine is added, how much improvement in waiting time can be anticipated, assuming the same traffic characteristics?

9.14 Packets of messages are received by two computers connected to a local area network as a Poisson process at the rate of 1 message a minute. The processing time of these messages is exponentially distributed, with a mean of 1 minute per message.

   (a) Compute the probability of having both computers busy simultaneously.

(b) Suppose the two computers are replaced by four "lightweights" that can only process a message every 2 minutes on average. Compute the probability of a message waiting to be processed.

(c) Compare the results of parts (a) and (b).

9.15 Consider an M/M/2 queuing system in which the mean arrival rate is μ and the mean service rate is μ for both servers. Further, let the utilization be $\rho = \lambda/\mu = \lambda$. Using first principles:

(a) Show that Erlang's C-function is

$$C(2, \rho) = \frac{\rho^2}{2 + \rho}.$$

(b) Show that the mean waiting time is

$$E[Q(t)] = \frac{\rho^2}{\mu(4 - \rho^2)}.$$

(c) Show that the mean time a customer is in the system is

$$E[T(t)] = \frac{4}{\mu(4 - \rho^2)}.$$

(d) Show that the probability that the queue is empty is

$$\Pr[T(t) = 0] = \frac{2 - \rho}{2 + \rho}.$$

9.16 Show that the probability of an empty M/M/c queuing system is

$$\pi_0 = \left[\frac{\rho^k}{(c - \rho)(c - 1)!} + \sum_{k=0}^{c-1} \rho^k \right]^{-1}.$$

9.17 Show that the Erlang B-function for an M/M/c queuing system satisfies the following recursive equation:

$$B(c, \rho) = \frac{\rho B(c - 1, \rho)}{c + \rho B(c - 1, \rho)}.$$

9.18 Consider an M/M/1/n system. Verify that the expected number in the system $\bar{N} = E[T(t)]$ and the expected time for service $\bar{T} = \langle T_k \rangle$ are given by

$$\bar{N} = \begin{cases} \dfrac{\rho[1 + (n\rho - n - 1)\rho^n]}{(1 - \rho)(1 - \rho^{n+1})}, & \rho \neq 1, \\[2ex] \frac{1}{2}n, & \rho = 1, \end{cases}$$

$$\bar{T} = \begin{cases} \dfrac{\rho[1 + (n\rho - n - 1)\rho^n]}{\lambda(1 - \rho)(1 - \rho^{n+1})}, & \rho \neq 1, \\[2ex] \dfrac{n(1 - \rho^{n+1})}{2\lambda(1 - \rho^n)}, & \rho = 1. \end{cases}$$

9.19 Consider an M/M/c queuing system in which $c > \rho$. Show that the relationship between the Erlang-B and Erlang C-function is

$$C(c, \rho) = \frac{cB(c, \rho)}{c - \rho B(c, \rho)}.$$

9.20 Consider the basic M/M/1 queuing system. Derive formulas for the variance of each of the following distributions:

(a) The probability mass function $\Pr[T(t) = k]$, where k is the system state.

(b) The probability density function $f_T(t)$ of inter-event times T_k.

9.21 Simulate the behavior of a two-server queuing system. That is, write a computer program that exhibits the behavior and analyze the output statistics for a set of input and server distributions.

(a) Write programs for a two-server queuing system, including subroutines that are independent of the main program modeling the system. Specifically, subroutines will be needed for:

(i) exponential inter-arrival and service times with mean μ;

(ii) Gaussian service times with mean τ and standard deviation σ.

(b) The program should create a log of times when customers arrive, when customers begin to be served and when customers exit the system. From these, the program should produce the following:

(i) $A(t) =$ number of arrivals as a function of time;

(ii) $D(t) =$ number of departures as a function of time;

(iii) $N(t) =$ number of customers in the system as a function of time;

(iv) $T_k =$ time the kth customer remains in the system.

(c) Create graphs of $A(t)$ and $D(t)$ for each input. Compare the following regimens:

(i) Exponential inter-arrival time with mean 3 minutes (single server with exponential service time of mean 2 minutes.

(ii) Gaussian inter-arrival times with mean 3 and variance 9 minutes; Gaussian service times with mean 2 and variance 4 minutes.

Create histograms of $T(t)$ and T_k for each of the regimens defined above. Compare these against the theoretical results

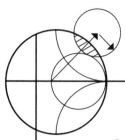

Chapter 10

System Optimization

10.1 SYSTEM IDENTIFICATION

Mathematical modeling usually has two facets. First, the mathematical form must be deduced, and, second, the modeling parameters and constants must be determined. Only after these two things have been determined is the model complete and it is possible for a simulation to be performed. For instance, consider the following modeling examples.

1. **Planetary Motion** Newton's law of gravitation states that the gravitational force of attraction between two masses m_1 and m_2 is given by

$$F = G\frac{m_1 m_2}{r^2},$$

   where r is the distance between the two masses and G is a universal proportionality constant. Assuming this to be a valid model form, the constant G must be determined experimentally, once and for all. This law governs planetary motion throughout the universe.

2. **Population dynamics** The Lotka–Volterra population model describes the population dynamics between a predator x and prey y as follows:

$$\dot{x} = ax - bxy,$$
$$\dot{y} = -cy + dxy,$$

   for positive constant a, b, c, and d, which are system-specific. Unlike the universal gravitational constant G in model 1, a, b, c, and d differ from population to population and therefore from model to model also. While they too must be determined experimentally, an even more interesting question is how the qualitative system behavior changes over the various parametric ranges. That is, are there necessary relationships between a, b, c, and d for stability, periodic, and chaotic behaviors?

3. **Control systems** Although the system *plant* is usually well understood, creating feedback for control purposes also introduces system constants that are to some extent arbitrary. For instance, a system with known plant transfer function $G(D)$

and proportional control has a *closed-loop* transfer function

$$\frac{Z(\boldsymbol{D})}{X(\boldsymbol{D})} = \frac{G(\boldsymbol{D})}{1 + CG(\boldsymbol{D})},$$

where C is a feedback system parameter to be determined subject to design specifications and $\boldsymbol{D}$ is the big-$\boldsymbol{D}$ operator described in Section 6.6. Often these specifications are in the form of optimization principles such as to minimize the step response overshoot or rise time.

4. **Queuing systems**   Queuing models are defined algorithmically rather than by an explicit formula. Even so, system parameters such as the mean service time, arrival rates, and number of servers can be chosen so that the predefined design objective are met. Typical objectives might be to minimize costs or maximize profits. Note the subtle difference between this and model 3. In model 3, the parameter C could only be chosen to best fill a physical need. However, in a queuing system, the need, say, for no customer blocking and instant service could in principle be met. It is just a matter of introducing more resources (money) into the mix. Thus, it is more a matter of a trade-off: Just how much is good service worth?

In a sense, the above four examples represent a continuum of models. Model 1 requires the evaluation of a single universal constant. There is no reason to explore alternatives, since G is physically mandated. On the other hand, model 2 encourages exploration. The parameters a, b, c, and d can range considerably for different species, and it is significantly interesting to compare relative behaviors.

While models 1 and 2 are "scientific" in that they describe nature, models 3 and 4 are "engineering", since they introduce variability and design into the system. Model 3 does this by *tuning*. Tuning can be thought of as continuously varying a parameter to obtain optimal performance. On the other hand, model 4 is more *business*-oriented, since it seems to encourage throwing money (resources) at the problem. While model 3 still has scientific constraints, model 4 can achieve any desired performance if necessary. The only objective is to somehow balance system performance against a resource objective.

- model 1: universal constants;
- model 2: system specific constants;
- model 3: constrained constants;
- model 4: unconstrained constants.

From an engineering perspective, the scientific part of a system (called the *plant*) is known a priori. In other words, all constants have already been determined, and it only remains to assign values to the tuning parameters such as the feedback gain C of model 3. In general, these tuning parameters only work over a limited range. For instance, C will likely be positive due to physical realization constraints and limited from above, $C \leqslant C_{max}$, due to a stability requirement. Thus, C is bounded: $0 \leqslant C \leqslant C_{max}$ on a so-called *constraint interval*.

Since system parameters can be chosen arbitrarily on their constraint intervals, it is natural that they should be chosen so as best to achieve an objective. For instance, in the case of model 3, one might try to optimize a controller's tracking facility by minimizing the feedback error for a given system input. Probably the most popular objective is the *mean square error* (MSE) objective function. Although there are others, the MSE objective is met by minimizing the objective function $f(C)$ defined as follows:

$$f(C) = \int_{t_{min}}^{t_{max}} [z_{mod}(t, C) - z_{des}(t)]^2 \, dt, \qquad (10.1)$$

for input $x(t)$ defined on the time horizon $t_{min} \leqslant t \leqslant t_{max}$. $z_{mod}(t, C)$ is the modeled response to an input $x(t)$, and $z_{des}(t)$ is the desired response to an input $x(t)$. By minimizing $f(C)$ with respect to the constant C, the model is forced to mirror the desired objective, since $f(C) \geqslant 0$, and in the ideal case where $f(C) = 0$, $z_{mod} = z_{des}$ as specified. Such an optimization problem is said to be *well posed*, since

  (i) the parameters are bounded: $0 \leqslant C \leqslant C_{max}$;
  (ii) the objective function is uniquely defined: $f(C) = \int_{t_{min}}^{t_{max}} [z_{mod}(t, C) - z_{des}(t)]^2 \, dt$;

  (iii) the input $x(t)$ is well defined over the time horizon: $0 \leqslant t \leqslant t_{max}$.

From an engineering perspective, there are two major reasons to model a system. First, modeling is useful to demonstrate the feasibility of a proposed design. The alternative – that of building a prototype – usually requires far more effort, and is expensive. By formally specifying the model, a simulation can verify that the design is well defined, consistent, and robust. In principle, the prototype simply implements these specifications. Modeling and simulating the system before creating a prototype should make the prototype a mere validation step that verifies the proposed implementation.

The other reason that engineers model systems is to tune them so as to attain optimum performance. This is possible because proposed models will inevitably introduce a number of arbitrary constants in the system model. To determine these constants so that system performance is optimized in some predefined sense is the nature of the *optimization problem*. System optimization is generally a non-trivial problem. Unlike the contrived problems of elementary calculus, where explicit formulas exist and all functions are differentiable, real systems usually require numerical approaches for their resolution.

EXAMPLE 10.1

Consider the system defined in Figure 10.1, with input $x(t) = 1$ on the interval $0 \leqslant t \leqslant 1$. Owing to physical constraints, the constant C is confined to the interval $1 \leqslant C \leqslant 2$. Using the *mean square error* (MSE) approach defined by Equation (10.1), find the optimal choice for C in each of the following situations.

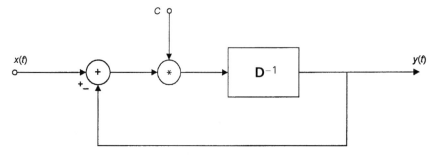

FIGURE 10.1   Block diagram for Example 10.1.

(a) Determine C so that the output differs from $z_{des}(t) = 1$ as little as possible.
(b) Determine C so that the output differs from $z_{des}(t) = \frac{1}{2}$ as little as possible.

Solution
This problem is a *constrained optimization problem*, since the unknown parameter C must be found subject to both a constraint $(1 \leqslant C \leqslant 2)$ and a *minimization* (or *maximization*) principle.

(a) It follows from Figure 10.1 that $y(t) = \int C[x(t) - y(t)] \, dt$. Reformulating this as an equivalent initial-value problem and noting that $x(t) = 1$,

$$\dot{y} + Cy = C,$$
$$y(0) = 0. \tag{10.2}$$

Equation (10.2) is a first-order linear differential equation, and its solution is straightforward: $y(t) = 1 - e^{-Ct}$. We wish to force $y(t)$ close to unity. Using the MSE criteria of Equation (10.1) with $z_{des}(t) = 1$ and $z_{mod}(t, C) = 1 - e^{-Ct}$,

$$f(C) = \int_0^1 [1 - y(t)]^2 \, dt$$

$$= \int_0^1 e^{-2Ct} \, dt$$

$$= \frac{1 - e^{-2C}}{2C}. \tag{10.3}$$

The graph of $f(C)$ is shown in Figure 10.2. Clearly, this is a monotonically decreasing function over the interval $[1,2]$, so the minimum point lies at the right-

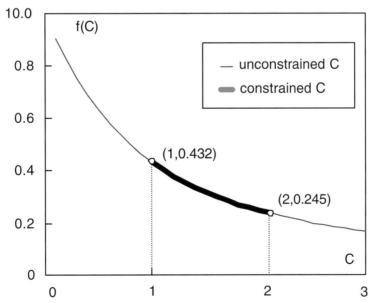

FIGURE 10.2 Constrained objective function $f(C)$ tracking $z_{des} = 1$ on $[1,2]$, Example 10.1.

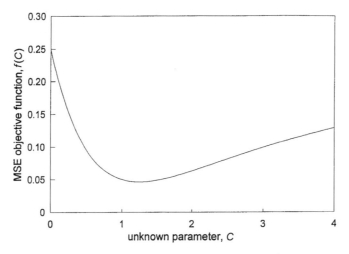

FIGURE 10.3   Constrained objective function $f(C)$ tracking $z_{des} = \frac{1}{2}$ on [1,2], Example 10.1.

hand side of the constraint interval [1,2] at $C = 2$. The corresponding objective functional value is $f(2) = \frac{1}{4}(1 - e^{-4}) \approx 0.245$.

(b) In this case, the output $y(t)$ is unchanged, but the objective function is different. Since we must now attempt to track $z_{des}(t) = \frac{1}{2}$,

$$f(C) = \int_0^1 \left[\tfrac{1}{2} - y(t)\right]^2 dt$$

$$= \int_0^1 \left(\tfrac{1}{2} - e^{-Ct}\right)^2 dt$$

$$= \frac{1}{2C}(1 - e^{-2C}) + \frac{1}{C}(1 - e^{-C}) + \tfrac{1}{4}. \qquad (10.4)$$

In principle, a local minimum can be found by differentiating Equation (10.4), setting it equal to zero, and solving for C. However, even though $f(C)$ can be differentiated, an exact solution for C_{min} cannot be found. Even so, a quick check of the graph in Figure 10.3 reveals a minimum on [1,2] at about $C_{min} = 1.125$. We need a systematic procedure to locate such minima.   ○

It is not always possible to find an explicit solution for the system output $y(t)$, especially if the input isn't cooperative. Thus, a numerical approach is the normal means of attack. We begin by extending the definition of Equation (10.1) to

$$f(t, C) = \int_{t_{min}}^t [z_{des}(t) - z_{mod}(t, C)]^2 dt.$$

Notice that $f(t_{min}) = 0$ and $f(t_{max})$ is the MSE objective to be minimized. Differentiating $f(t, C)$ with respect to t while holding C constant yields the following initial-value problem:

$$\frac{\partial}{\partial t} f(t, C) = [z_{des}(t) - z_{mod}(t, C)]^2,$$

$$f(t_{min}) = 0,$$

(10.5)

where $f(t_{max}, C)$ is to be minimized with respect to C, which is constrained on the interval $[C_{min}, C_{max}]$. Equation (10.5), combined with the model output written as an initial-value problem, defines the system numerically.

Assuming $m + 1$ uniformly distributed evaluation points $C_{min} = C_0$, C_1, $C_2, \ldots, C_m = C_{max}$ for the objective function, the interval size is $\delta = (C_{max} - C_{min})/m$. Also assuming n intervals of length h on time horizon $[t_{min}, t_{max}]$, $h = (t_{max} - t_{min})/n$. Using Euler, the difference equation for Equation (10.5) is

$$t = hk + t_0,$$

$$f(k + 1) = f(k) + h[z_{des}(t) - y(k)]^2,$$

(10.6)

$$f(0) = 0.$$

Listing 10.1 gives a generic algorithm for calculating the objective $f(C)$ at points C_0, C_1, $C_2, \ldots, C_m$. Notice that f is updated before output $y(t)$, since f requires the current value of y prior to its being updated. The constants are set for the parameters of Example 10.1(b).

```
-input signals
        x(t)=1
        z_des(t)=1/2
-simulation
        C_min=1
        C_max=2
        t_min=0
        t_max=1
        m=100
        n=100
        δ=(C_max-C_min)/m
        h=(t_max-t_min)/n
        for i=0 to m
                C=C_min+δ i
                f=0
                y=0
                for k=1 to n
                        t=hk
                        f=f+h[z_des(t)-y]²
                        y=y+hC[x(t)-y]
                next k
                print C, f
        next i
```

LISTING 10.1 Simulation of Example 10.1 using the Euler method.

EXAMPLE 10.2

Again consider the system defined in Example 10.1. However, this time do not assume that an explicit formula for output $y(t)$ is available.

(a) Find C on $[1,2]$ so that $y(t)$ is forced as close as possible to $z_{des}(t) = \frac{1}{2}$ for time t on $[0,1]$. Assume an input of $x(t) = 1$. In other words, repeat Example 10.1(b), but this time do solve it numerically.

(b) Find C on $[0,4]$ so that $y(t)$ is forced as close as possible to $z_{des}(t) = 4$ for time t on $[0,1]$. Assume an input $x(t) = 5(1 + \sin t)$.

(c) Find C on $[0,4]$ so that $y(t)$ is forced as close as possible to $z_{des}(t) = 4$ for time t on $[0,1]$. Assume an input $x(t) = 5(1 + \sin t) + 2\,\text{RND} - 1$. This is a repeat of part (b), but this time with a white noise input.

Solution
A numerical solution requires a nested loop. The outer loop calculates C while the inner loop integrates over time for the specific C-value. Only the values of C and $f(C) = f(t_{max}, C)$ are required, so there is no print inside the inner loop.

(a) The key equations for part (a) are the objective update Equation (10.6) and Euler's update equation $y^{(new)} = y + hC(1 - y)$ from Equation (10.2). These are incorporated in Listing 10.1. Both f and y are initially $f(0) = 0$ and $y(0) = 0$. The results of this simulation are shown in Figure 10.4, where numerical results are given for both the $n = 10$ $(h = 0.1)$ and $n = 100$ $(h = 0.01)$ approximations.

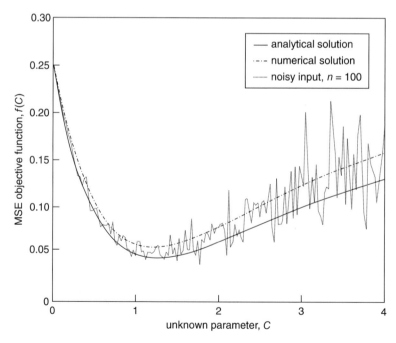

FIGURE 10.4   Numerical calculation of objective function tracking $z_{des} = \frac{1}{2}$, Example 10.2(a).

The accuracy for $n = 100$ is practically identical to that of the ideal solution, but the minimum point C_{min} is practically the same in both instances: $C_{min} = 1.225$.

In a practical situation, the input will more than likely be noisy. Figure 10.4 also shows the result of an ideal input $x_1(t) = 1$ contaminated with additive white noise with unit variance: $x(k) = 1 + \sqrt{3}(2\,RND - 1)$. In this case, there is no choice but to perform the simulation numerically. Note that even though the input is stationary over the entire time interval [0,1], the objective function appears more "noisy" for larger C values. Since C seems to magnify the noise, it seems as though there is yet another reason to maintain a small C.

(b) If the input is $x(t) = 5(1 + \sin t)$, the output can still be found by elementary means. However, it should come as no surprise that an analytical formula for the objective is not a particularly good idea, since Equation (10.6) is rather involved, even though it is still linear. A numerical simulation is produced using the algorithm in Listing 10.1 by letting $x_1(t) = 5(1 + \sin t)$ and $z_{des}(t) = 4$. From the results sketched in Figure 10.5, there is a relative minimum for C at approximately $C_{min} = 2.65$.

(c) The addition of noise to an ideal input signal complicates things. In order to find a minimum, it is first necessary to get past all the high-frequency noise. If we do not take care to eliminate the noise, the absolute minimum might be due to the noise rather than the underlying ideal signal. Also, any algorithm that we might create to find that minimum would likely get trapped in a "well" that is an artifact.

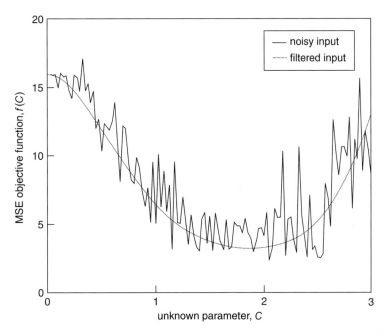

FIGURE 10.5   Numerical calculation of objective function tracking $z_{des} = 4$, Example 10.2(b).

Implementation of the noisy input $x(t) = 5(1 + \sin t + 2\,\mathrm{RND} - 1)$ is handled just like the ideal input of part (b), but the results are quite different. Figure 10.5 illustrates the problem. Given the noisy objective function, it is not obvious what the underlying function even is visually, let alone what sort of algorithm would mechanically produce a useful result.

The answer to this dilemma is to once again apply a low-pass filter (LPF). Recall that one of the basic LPFs is defined by the equation

$$\tau\dot{z} + z = f(C), \tag{10.7}$$

where $f(C)$ is the objective function that is input to the filter, z is the filter output, and τ is an LPF parameter yet to be chosen. Equation (10.7) can be handled digitally by the update formula $z^{(\mathrm{new})} = z + h(f - z)/\tau$. This update should be imbedded in Listing 10.1 just before the objective f is updated, since z requires the "current" f and not the new f. Since z is a linear filter, any initial condition will work in principle. This is because after the transients die out, all that remains is the ideal signal and the filtered noise, which is stationary. However, it is best to note that since z approximates f, the transients can effectively be eliminated all together by simply initializing z to the same value as f. It is apparent from Figure 10.5 that $z(0) = 16$ would be an excellent choice, since there are virtually no transients.

The only real question in implementing the LPF of Equation (10.7) is what the parameter τ should be. There are good theoretical reasons to consider a family of exponentially related filters. For instance, Figure 10.6 gives the filtered results for $\tau = 2, 20, 100$, and 200. This results from observing $\tau = 2$ to be far too noisy and $\tau = 200$ to be about right, then backing off to $\tau = 100$ to find the best shape-preserving curve that is at the same time smooth enough to provide a unique minimum point. The results of this filter should be compared against the graph of part (b), where it is evident that this is an excellent choice for parameter τ. From this curve, the minimum $C_{\min}$ can be estimated visually or an algorithm (to be discussed shortly) can be implemented by which to accomplish the task automatically.   ○

Example 10.2 illustrates several lessons. First, numerical optimization techniques are not only inevitable, but desirable. Realistic simulations with additive noise cannot be handled any other way. At the same time, it is evident that inaccuracies are introduced in the integration. Even so, the biggest problem is not so much the inaccuracy, which can always be improved on by either using another integration technique or reducing the step size, but the fact that by establishing too coarse a mesh, "holes" can be overlooked in the objecting function. It should also be clear that we need to automate the entire optimization process.

Even though noise is probably inevitable, the quadratic form of the MSE objective function, Equation (10.1), means that there is most likely an underlying concavity for the ideal signal. Thus, if the noise can be dealt with by using a filter, the objective function will usually be locally convex. This leads to the following workable strategy for attacking constrained optimization problems:

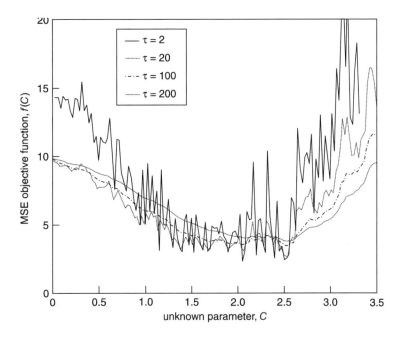

FIGURE 10.6   Effect of low-pass filters on noisy objective functions tracking $z_{des} = 4$, Example 10.2(c).

1. Search the entire constraint interval using a uniformly distributed course mesh of evaluation points. Find the point C_{min} where $f(C)$ is minimized.
2. Using the local interval around C_{min}, perform a sequential search that narrows in on the "bottom" of the "well" to find a better approximation for C_{min}.

Rather than just visually inspecting the graph of coarsely selected evaluation points, we can narrow in on a valley and locate the minimum point sequentially. For the most part, we assume that derivatives are not available and that all data is suspect with high-frequency noise contamination. Here we consider a set of very powerful search algorithms by which to optimize objective functions, and in turn identify optimal system parameters.

10.2   NON-DERIVATIVE METHODS

In general, one knows little about the objective function except that it is non-negative and well defined over the constraint interval. In the case of system identification problems where the MSE criteria require the output to be re-computed for each evaluation point, it can also be costly (in the sense of computer time) to evaluate. Even though values can be determined, an explicit formula may not exist – and if it does, it will likely be non-deterministic. Also, either the function may not be differentiable or, if it is differentiable, derivative information may not exist. In these cases, it is not possible to do more than estimate this derivative numerically. However, recalling earlier work, this is a process that is dubious at best.

Even so, it is often the case that a problem's context will be helpful. In the case of Example 10.1, it is intuitive that the objective f, which is the MSE deviation of the output from the input, will be *convex* in that its concavity will not change. Further, since the objective is obtained by the squaring function (as opposed to the absolute value), it is *differentiable* if the input x is differentiable. Therefore, the function will ideally appear parabolic. In this sense, f is well behaved.

The convexity property referred to above will be found to be extremely helpful. However, in reality, all that is necessary is for the function to be *locally convex*. That is, in the neighborhood of the optimal value, the objective is convex, and thus approximately parabolic.

EXAMPLE 10.3

Consider the objective function $f(C) = e^{-C} \cos C$ defined on the interval $0 \leqslant C \leqslant 15$. It is clear that this function is infinitely differentiable for all real C, so its concavity is well defined on [0,15]. In particular, the concavity of such an analytic function is completely characterized by its zeroth, first, and second derivatives. This is shown in Table 10.1. Characterize the convexity of $f(C)$.

Solution
The derivatives of $f(C)$ are straightforward. Noting that the "zeroth" derivative of a function is the function itself,

$$f(C) = e^{-C} \cos C,$$
$$f'(C) = -e^{-C}(\cos C + \sin C),$$
$$f''(C) = 2e^{-C} \sin C.$$

Since $e^{-C} > 0$ for all real C, the objective $f(C)$ is completely characterized by the positiveness or negativeness of $\sin C$. Crossing points on the constraint interval [0,15] are at $C_{\text{crossing}} = \frac{1}{2}\pi, \frac{3}{2}\pi, \frac{5}{2}\pi$, and $\frac{7}{2}\pi$. In between each subinterval, the function goes from being positive on $[0, \frac{1}{2}\pi]$, negative on $[\frac{1}{2}\pi, \frac{3}{2}\pi]$, and alternating thereafter. This is shown graphically in Figure 10.7.

Similarly, the critical points, which are candidates for local maxima and minima, are where the derivative is zero. Again ignoring the exponential term, the critical points are at $C_{\text{critical}} = \frac{3}{4}\pi, \frac{7}{4}\pi, \frac{11}{4}\pi$, and $\frac{15}{4}\pi$. On the subintervals, the

TABLE 10.1   Concavity Characteristics of the Objective Function $f(C)$
as Defined by Derivative Properties

	Positive	Zero	Negative
$f(C)$	Positive	Crossing point	Negative
$f'(C)$	Increasing	Critical point	Decreasing
$f''(C)$	Concave up	Inflection point	Concave down

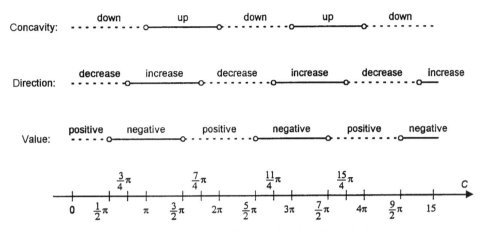

FIGURE 10.7   Concavity characteristics of objective function, Example 10.3.

function is either increasing or decreasing, as shown in Figure 10.7. The concavity is given where the second derivative is zero. As above, this is at $C_{inflection} = 0$, π, 2π, 3π, and 4π. As indicated in Figure 10.7, $f(C)$ is either concave up or concave down on alternate intervals between the inflection points.

In any study of sequential algorithms, it is important to consider efficiency. This is especially true in system identification procedures, since every evaluation of the objective function requires performing the entire simulation. So, to measure the performance of a sequential algorithm, a favorable ratio of progress to number of evaluations is important.

Consider a convex function $f(C)$ with a global minimum at $C = C_{min}$. In establishing this minimum, the function has to be evaluated at a set of evaluation points $C_0, C_1, C_2, \ldots,$ C_m and objective values $f(C_0), f(C_1), f(C_2), \ldots, f(C_m)$, of which the smallest f-value is at C_K. Then we are assured that the global minimum C_{min} is on the interval $C_{K-1} < C_{min} < C_{K+1}$. This interval is called the *interval of uncertainty* and the difference $C_{K+1} - C_{K-1}$ is the *length of uncertainty*. Since this requires m evaluations of the objective function $f(C)$, which can generally be expensive in computer time, it is useful to define an *effective ratio*:

$$R_m = \frac{\text{length of uncertainty after } m \text{ evaluations}}{\text{length of uncertainty after zero evaluations}}, \tag{10.8}$$

which measures the rate at which the length of uncertainty decreases compared with the number of evaluations required. Clearly, small effective ratios are preferred.

To illustrate the application of Equation (10.8), consider the type of search used in Examples 10.1 and 10.2. Such a search for the minimum point is called *uniform*, since it involves simply spreading a uniformly distributed mesh over the entire constraint interval, evaluating the objective at each point and choosing the minimum. Intuitively, we imagine that such an idea is probably not efficient, but we can still compute its effective ratio. In particular, the $m + 1$ mesh points are chosen as

$$C_i = a + \frac{i}{m}(b - a) \quad \text{for} \quad i = 0, 1, \dots, m,$$

giving

$$C_0 = a,$$
$$C_1 = a + \frac{1}{m}(b - a),$$
$$C_2 = a + \frac{2}{m}(b - a),$$
$$\vdots$$
$$C_{m-1} = a + \frac{m - 1}{m}(b - a),$$
$$C_m = b.$$

The minimum is chosen by evaluating the function at each point and finding the smallest value:

$$\hat{f} = \min_i f(C_i),$$
$$\hat{C} = C_i.$$

$\hat{C}$ is an estimate of the actual minimum $C_{\min}$, which is apparently somewhere in the interval defined by adjacent mesh points on either side: $C_{i-1} \leqslant C_{\min} \leqslant C_{i+1}$. Accordingly, the length of uncertainty is

$$L_i = C_{i+1} - C_{i-1}$$
$$= \left[a + \frac{i + 1}{m}(b - a) \right] - \left[a + \frac{i - 1}{m}(b - a) \right]$$
$$= \frac{2(b - a)}{m}.$$

Since this calculation requires $m + 1$ evaluations, the effective ratio is

$$R_{m+1} = \frac{L_i}{b - a} = \frac{2}{m}.$$

Equivalently, on replacing m by $m - 1$,

$$R_m(\textit{uniform}) = \frac{2}{m - 1}. \tag{10.9}$$

10.3   SEQUENTIAL SEARCHES

Uniform searches are good in situations where the objective function's convexity is unknown over the constraint interval. By using a coarse mesh over the interval of uncertainty, a crude estimate can be made of the minimum point, along with an interval of uncertainty over which the objective is more likely to be convex. The reason that convexity is so important can be summarized by the following convexity theorem:

> **Convexity Theorem**   Let the objective function f be convex (without loss of generality, concave up) on the constraint interval [a,b]. Further, if λ and μ are two evaluation points (that is, $f(\lambda)$ and $f(\mu)$ are evaluations) such that $\lambda < \mu$, then exactly one of the following conclusions holds:
>
> (i) If $f(\lambda) < f(\mu)$ then the minimum exists on (a, μ).
> (ii) If $f(\lambda) > f(\mu)$ then the minimum exists on (λ, b).
> (iii) If $f(\lambda) = f(\mu)$ then the minimum exists on (λ, μ).

The proof of this theorem is straightforward and easily shown graphically. However, a formal demonstration will not be given here.

Consider a typical constrained optimization problem of the form

$$\text{minimize} \quad f(C)$$
$$\text{subject to} \quad a \leqslant C \leqslant b.$$

If it is known that $f(C)$ is convex on $[a, b]$, then one can apply the convexity theorem iteratively with a so-called *sequential line search* to obtain a value $\hat{C}$ that is arbitrarily close to the actual minimum. This is done by starting with the initial *interval of uncertainty* $[a_0, b_0] = [a, b]$ and length of uncertainty $b_0 - a_0$. Using the convexity theorem, it is possible to generate a sequence of new uncertainty intervals:

$$[a_{i+1}, \ b_{i+1}] = \begin{cases} [a_i, \ \mu], & f(\lambda) < f(\mu), \\ [\lambda, \ b_i], & f(\lambda) < f(\mu), \\ [\lambda, \ \mu], & f(\lambda) = f(\mu). \end{cases} \qquad (10.10)$$

The new length of uncertainty $b_i - a_i$ is no larger than the previous length. This process continues until the length of uncertainty is arbitrarily small: $b_i - a_i < L$, for some predefined tolerance L. In general, the intervals defined by Equation (10.10) will be nested, that is, $[a_{i+1}, b_{i+1}] \subset [a_i, b_i]$, and with monotonically decreasing uncertainty lengths $b_{i+1} - a_{i+1} < b_i - a_i$.

As a first example of the sequential technique, suppose the evaluation points are equally spaced in the interval $[a_i, b_i]$. This is shown in Figure 10.8, where $[a_i, b_i]$ is divided into thirds. It follows that the evaluation points are computed by

$$\lambda_i = \tfrac{1}{3}(2a_i + b_i),$$
$$\mu_i = \tfrac{1}{3}(a_i + 2b_i), \qquad (10.11)$$

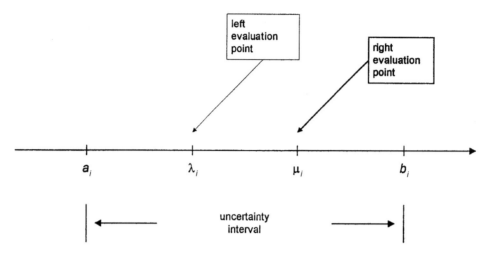

FIGURE 10.8   Evaluation points of the *thirds* sequential search algorithm.

and, by the convexity theorem,

$$\text{if } f(\lambda_i) < f(\mu_i), \text{ then } a_{i+1} = a_i, \ b_{i+1} = \mu_i;$$
$$\text{if } f(\lambda_i) > f(\mu_i), \text{ then } a_{i+1} = \lambda_i, \ b_{i+1} = b_i;$$
$$\text{if } f(\lambda_i) = f(\mu_i), \text{ then } a_{i+1} = \lambda_i, \ b_{i+1} = \mu_i.$$

For a given initial interval $[a, b]$ and length of uncertainty tolerance L, Listing 10.2 presents the *thirds* sequential search. Notice that there is no need to maintain arrays for either the sequence of evaluation points or endpoints.

EXAMPLE 10.4

Consider the objective function $f(C) = e^{-C} \cos C$ defined on the uncertainty interval $0 \leqslant C \leqslant \pi$. We know from Example 10.3 that the objective is convex on the given initial interval, so the convexity theorem guarantees a global minimum somewhere on the interval $[0, \pi]$. Find $C_{\min}$ using the *thirds* algorithm.

Solution
Applying the algorithm in Listing 10.2 gives the data listed in Table 10.2. It should be noted that the intervals of uncertainty are indeed nested and that the lengths of uncertainty also decrease monotonically. As expected, the sequence converges to the actual minimum at $C_{\min} = \frac{3}{4}\pi\mu \approx 2.356$.   ○

Listing 10.2 shows that there are two objective function computations required for each iteration. Also, each iteration yields an interval of uncertainty that is two-thirds the

```
     -given formulas and data
          f(C)="objective function"
          input L, a, b
     -optimization procedure
          λ=⅓(2a+b)
          µ=⅓(a+2b)
          f_left=f(λ)
          f_right=f(µ)
          while b-a>L
               if f_left<f_right then b=µ
               if f_left>f_right then a=λ
               if f_left=f_right then a=λ : b=µ
               λ=⅓(2a+b)
               µ=⅓(a+2b)
               f_left=f(λ)
               f_right=f(µ)
          end while
          print ½(a+b)
```

LISTING 10.2 The *thirds* algorithm for a sequential line search.

previous interval length. (Recall the name of this technique: *thirds*.) Thus, there are $2m$ computations required to achieve the mth interval of uncertainty. This yields

$$R_{2m} = \frac{\left(\frac{2}{3}\right)^m (b-a)}{b-a} = \left(\frac{2}{3}\right)^m.$$

Replacing $2m$ by m, an equivalent expression for the effective ratio is

$$R_m(thirds) = \left(\tfrac{2}{3}\right)^{m/2} \tag{10.12}$$

Since the effective ratio for *thirds* is exponential and the effective ratio for *uniform* is hyperbolic (recall Equation (10.9)), it is evident that, from an efficiency viewpoint, *thirds*

TABLE 10.2 Results of the *thirds* Sequential Line Search of Example 10.3

i	a_i	b_i	λ_i	μ_i	$b_i - a_i$
0	0.0000	3.1416	1.0472	2.0944	3.1416
1	1.0472	3.1416	1.7453	2.4435	2.0944
2	1.7453	3.1416	2.2107	2.6762	1.3963
3	1.7453	2.6762	2.0556	2.3659	0.9308
4	2.0556	2.6762	2.2625	2.4693	0.6206
5	2.0556	2.4693	2.1935	2.3314	0.4137
6	2.1935	2.4693	2.2854	2.3774	0.2758
7	2.2854	2.4693	2.3467	2.4080	0.1839
8	2.2854	2.4080	2.3263	2.3672	0.1226
9	2.3263	2.4080	2.3535	2.3808	0.0817

$$C(\text{approximate}) = \tfrac{1}{2}(a+b) = 2.3672 \qquad C(\text{actual}) = \tfrac{3}{4}\pi \approx 2.3562$$

is a superior algorithm. This is not to say that the uniform search isn't a good algorithm. If convexity information is unknown, simply applying a sequential search can yield a *local* minimum, but there is no reason to believe that this will also be the *global* minimum. In order to achieve the global minimum, one usually applies the uniform search algorithm first to obtain convexity information. This allows the subsequent sequential search to focus on the appropriate convex region and finish the task most efficiently within this subinterval.

10.4 GOLDEN RATIO SEARCH

The *thirds* algorithm's partitioning of the uncertainty interval into thirds was somewhat arbitrary. It is possible to be more systematic in this choice, thus improving efficiency. This can be accomplished by using one of the previous evaluation points in the new iteration. It turns out that this technique, called the *golden ratio* line search, is most efficient, yet is easy to apply. Suppose the evaluation points are defined in terms of the uncertainty endpoints as follows:

$$\lambda_i = \alpha a_i + (1 - \alpha)b_i,$$
$$\mu_i = (1 - \alpha)a_i + \alpha b_i, \tag{10.13}$$

for a constant α yet to be determined. Evidently if $\alpha = 1$, each evaluation point coincides with the interval endpoints, and if $\alpha = \frac{1}{2}$, the evaluation points are identical. For values of α in the range $\frac{1}{2} < \alpha < 1$, the evaluation points are somewhere in between the two endpoints a_i and b_i. We wish to find the value of α that results in a need for no new evaluation at the next iteration. The two possible configurations are shown in Figure 10.9, where

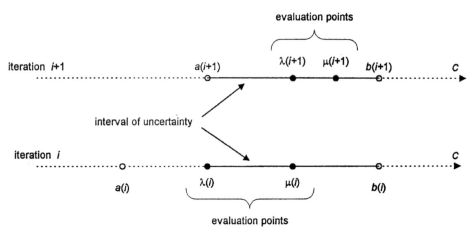

FIGURE 10.9 Geometry of re-evaluation points for the *golden ratio* sequential line search assuming $f(\lambda) < f(\mu)$.

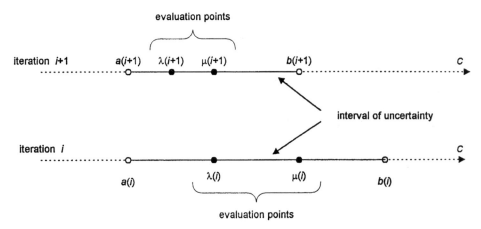

FIGURE 10.10   Geometry of re-evaluation points for the *golden ratio* sequential line search assuming $f(\lambda) \geqslant f(\mu)$.

$f(\lambda) < f(\mu)$, and Figure 10.10, where $f(\lambda) \geqslant f(\mu)$. Notice that the third case of the convexity theorem has been included in Figure 10.10.

Let us assume that $f(\lambda_i) < f(\mu_i)$. As seen in Figure 10.9, $\mu_{i+1} = \lambda_i$. Using Equations (10.13),

$$(1 - \alpha)a_{i+1} + \alpha b_{i+1} = \alpha a_i + (1 - \alpha)b_i.$$

Again referring to Figure 10.9, $a_{i+1} = a_i$ and $b_{i+1} = \mu_i$. Therefore,

$$(1 - \alpha)a_i + \alpha[(1 - \alpha)a_i + \alpha b_i] = \alpha a_i + (1 - \alpha)b_i,$$

from which it follows that $\alpha^2 + \alpha = 1$. The positive solution of this quadratic equation is $\alpha = \frac{1}{2}(-1 + \sqrt{5}) \approx 0.618$, which is the famous golden ratio from Greek antiquity! The Athenians spent considerable effort on esthetics, and one of their pursuits was to find what they considered to be the ideal length-to-width ratio for their temples. They felt that the golden ratio of $\alpha \approx 0.618$ was that ideal. In any case, using this value for α, Equations (10.13) may be rewritten as approximately

$$\lambda_i = (0.618)a_i + (0.382)b_i,$$
$$\mu_i = (0.382)a_i + (0.618)b_i. \tag{10.14}$$

Further, it can be shown that the same value of α occurs under the assumption $f(\lambda_i) \geqslant f(\mu_i)$. This is done as shown above, but referencing Figure 10.10 in place of the Figure 10.9 geometry.

```
          -given formulas and data
                f(C)="objective function"
                input L, a, b
          -optimization procedure
                λ=(0.618)a+(0.382)b
                μ=(0.382)a+(0.618)b
                f_left=f(λ)
                f_right=f(μ)
                while b-a>L
                        if f_left<f_right then
                                b=μ
                                μ=λ
                                λ=(0.618)a+(0.382)b
                                f_right=f_left
                                f_left=f(λ)
                        else
                                a=λ
                                λ=μ
                                μ=(0.382)a+(0.618)b
                                f_left=f_right
                                f_right=f(μ)
                        end if
                end while
                print ½(a+b)
```

LISTING 10.3   The *golden ratio* algorithm for a sequential line search.

By using Equations (10.14) as the evaluation points, it is possible to recompute just once per iteration rather than twice as is done using the *thirds* algorithm. This results in a significantly increased net efficiency.

The *golden ratio* algorithm is given in Listing 10.3. Data names f_{left} and f_{right} are used to swap values of evaluation points λ and μ. By using this device, it will be seen that aside from the two evaluations in the initialization section, there is but one evaluation per iteration. It can further be shown that the length of uncertainty L_i after i iterations is $L_i = \alpha^i(b_0 - a_0)$. Therefore, after m computations ($i + 1$ iterations),

$$R_m(golden\ ratio) = \left(\frac{-1 + \sqrt{5}}{2}\right)^{m-1} \qquad (10.15)$$

EXAMPLE 10.5

Again consider the objective function $f(C) = e^{-C}\cos C$ on the uncertainty interval $[0, \pi]$, but this time use the *golden ratio* algorithm. In other words, repeat Example 10.4. The results of a golden ratio line search are shown in Table 10.3. The sequence once again converges to the known optimal point $C(\text{actual}) = \frac{3}{4}\pi \approx 2.3562$. Notice how there are not only half the computations but fewer iterations as well. The *golden ratio* algorithm is very efficient.

TABLE 10.3   Results of the *golden ratio* Sequential Line Search of Example 10.5

i	a_i	λ_i	μ_i	b_i	$b_i - a_i$
0	0.0000	1.2001	1.9415	3.1416	3.1416
1	1.2001	1.9415	2.3999	3.1416	1.9415
2	1.9415	2.3999	2.6832	3.1416	1.2001
3	1.9415	2.2248	2.3999	2.6832	0.7417
4	2.2248	2.3999	2.5081	2.6832	0.4583
5	2.2248	2.3330	2.3999	2.5081	0.2833
6	2.2248	2.2917	2.3330	2.3999	0.1751
7	2.2917	2.3330	2.3586	2.3999	0.1082
8	2.3330	2.3586	2.3744	2.3999	0.6692

$$C(\text{approximate}) = \tfrac{1}{2}(a+b) = 2.3665 \qquad C(\text{actual}) = \tfrac{3}{4}\pi \approx 2.3562$$

TABLE 10.4   Comparative Effective Ratios for Search Algorithms

Iteration m	Effective ratio		
	uniform	*thirds*	*golden ratio*
2	2.000	0.666	0.618
3	1.000	0.543	0.382
4	0.667	0.443	0.236
5	0.500	0.362	0.146
6	0.400	0.295	0.090

EXAMPLE 10.6

It is useful to compare the effective ratios for the three line searches discussed in this chapter. Recalling Equations (10.9), (10.12), and (10.15),

$$R_m(\text{uniform}) = \frac{2}{m-1},$$
$$R_m(\text{thirds}) = \left(\tfrac{2}{3}\right)^{m/2} \approx (0.816)^m,$$
$$R_m(\text{golden ratio}) = \alpha^{m-1} \approx (0.618)^{m-1}.$$

These ratios are tabulated in Table 10.4 for $m = 2, 3, 4, 5, 6$ computations. Since the effective ratio is best kept small, it is evident that the *golden ratio* algorithm is, in general, the superior method.

10.5   ALPHA/BETA TRACKERS

As a specific modeling example, let us consider a tracking system, the so-called α/β tracker. This device acts like a filter in that a discrete output $u(k)$ tends to follow or *track* an ideal continuous input $x_1(t)$ that is subject to an unwanted *disturbance* or *noise* $x_2(t)$. These

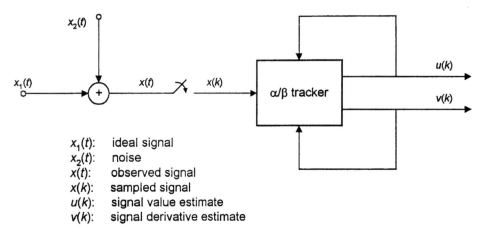

$x_1(t)$:   ideal signal
$x_2(t)$:   noise
$x(t)$:   observed signal
$x(k)$:   sampled signal
$u(k)$:   signal value estimate
$v(k)$:   signal derivative estimate

FIGURE 10.11   Alpha/beta tracker.

signals are summed, then sampled at a sampling frequency $f = 1/\delta$, as shown in the block diagram of Figure 10.11. This produces an observed digital signal $x(k)$. The signal $x(k)$ then passes through the α/β tracker filter, which produces an output $u(k)$, which is an estimate of the original signal $x(k)$. As an added benefit, the α/β tracker estimates the derivative of $x(k)$ as well with a signal $v(k)$.

As indicated in Figure 10.11, the internals of the α/β tracker use the sampled observed signal $x(k)$, and feedback estimates $u(k)$ and $v(k)$. First, a preliminary estimate of the future is found: $u(k) \approx u(k-1) + \delta v(k-1)$, where δ is the sampling interval. This estimate is refined by taking a weighted average of this estimate with the observed signal:

$$u(k) = \alpha x(k) + (1-\alpha)[u(k-1) + \delta v(k-1)],$$

where α is a constant yet to be determined. Next, an estimate for the derivative is given by

$$v(k) = v(k-1) + \frac{\beta}{\delta}[x(k) - u(k)],$$

where β is another weighting constant. In summary, the α/β tracker algorithm is as follows:

$$
\begin{aligned}
t(k) &= \delta k + t_0, \\
x(k) &= x_1(k) + x_2(k), \\
u(k) &= \alpha x(k) + (1-\alpha)[u(k-1) + \delta v(k-1)], \\
v(k) &= v(k-1) + \frac{\beta}{\delta}[x(k) - u(k)].
\end{aligned}
\tag{10.16}
$$

It remains to compute α and β. Once these design constants are known, the algorithm is complete.

EXAMPLE 10.7

Suppose an ideal signal $x_1(t) = \sin t + \sin 2t$ is combined with uniformly distributed white noise with unit variance to form an observed signal. This observed signal is sampled at a sampling frequency $f = 1/\delta = 10$ samples per unit time. Compute and compare the estimates $u(k)$ and $v(k)$ against the ideal functional value and the derivative of the signal.

Solution
From the specifications, the noise is produced by $x_2(t) = \sqrt{3}(2\,\text{RND} - 1)$. This, combined with the ideal signal $x_1(t)$, forms the observed signal

$$x(t) = \sin t + \sin 2t + \sqrt{3}(2\,\text{RND} - 1).$$

This is incorporated in Listing 10.4, which is a program segment implementing the α/β tracker using $\alpha = 0.5$ and $\beta = 0.2$ for 10 time units. Since no initial conditions were specified, $u(0)$ and $v(0)$ are set to zero, since this will affect only the transient phase anyway.

The results comparing the estimated signal $u(k)$ with the ideal $x_1(t)$ are given in Figure 10.12. It will be noted that u has a smoothing effect on the noise, and yet serves as an estimate of x_1. This is in contrast to the derivative estimator $v(k)$, which is shown along with $\dot{x}_1(t) = \cos t + 2\cos 2t$ in Figure 10.13. Clearly $v(k)$ follows $\dot{x}_1(t)$, but still retains high-frequency components, and there appears to be a small phase shift to the right. Since the α/β tracker acts as a filter, it should be possible to reduce these components by a more judicious choice of the parameters α and β. ◯

The challenge is to determine the best values for α and β, which are always problem-specific. In order to do this, one solution is to *train* the system to learn the best values for

```
-given formulas and data
      x(t)=sin(t)+sin(2t)+√3(2*RND-1)
      δ=0.1
      α=0.5
      β=0.2
-optimization procedure
      n=10/δ
      t=0
      u=0
      v=0
      print t,x(0),u,v
      for k=1 to n
            t=δk
            x=x(t)
            u=αx+(1-α)(u+δv)
            v=v+β(x-u)/δ
            print t, x, u, v
      next k
```

LISTING 10.4 Implementation of the α/β tracker for $\alpha = 0.5$ and $\beta = 0.2$, Example 10.6.

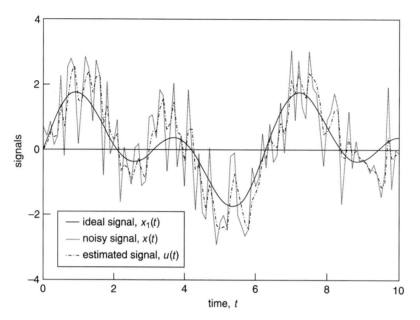

FIGURE 10.12   Input signals to the α/β tracker, Example 10.6.

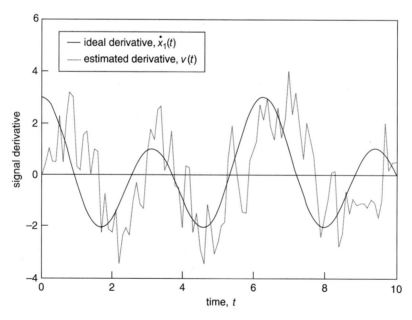

FIGURE 10.13   Response to input signal of Figure 10.12 for the α/β tracker with $\alpha = 0.5$ and $\beta = 0.2$, Example 10.6.

each parameter. By taking a typical archetype signal over a suitable time horizon, along with typical noise, it is possible to find α and β so that the resulting $u(k)$ follows the ideal input $x_1(t)$ more closely than any other choice of α and β. This is usually done in the MSE sense.

In order to determine the optimal filter, we will apply the MSE objective function (Equation (10.1)) to the ideal signal value and its derivative. The ideal signal will be used to train the α/β tracker by either of the following objective functions:

$$\text{signal value alone:} \quad f(\alpha, \beta) = \int_{t_{\min}}^{t_{\max}} [x_1(t) - u(t)]^2 \, dt,$$

$$\text{signal value and derivative:} \quad f(\alpha, \beta) = \int_{t_{\min}}^{t_{\max}} \left\{ [x_1(t) - u(t)]^2 + [\dot{x}_1(t) - v(t)]^2 \right\} dt.$$

In an even more general setting, the two terms of the second objective function can be weighted relative to each other so that any desired balance can be struck. It is not the actual weight of each term that is important, but the relative weight. Thus, for suitable weight W, where $0 \leqslant W \leqslant 1$, the most general objective function is

$$f(\alpha, \ \beta) = \int_{t_{\min}}^{t_{\max}} \left\{ W[x_1(t) - u(t)]^2 + (1 - W)[\dot{x}_1(t) - v(t)]^2 \right\} dt. \tag{10.17}$$

If $W = 1$, Equation (10.17) corresponds to the objective for the signal values alone. If $W = 0.5$, both the signal value and its derivative are equally weighted, so Equation (10.17) reduces to the second case.

There is one special case of the α/β tracker that is especially important. This is the so-called *critical damping hypothesis*. The details are omitted here, but suffice it to say that this case requires that

$$\alpha = 2\sqrt{\beta} - \beta. \tag{10.18}$$

Under the critical damping hypothesis, the net error is a function of only one variable rather than two, since once β has been determined, α follows immediately.

EXAMPLE 10.8

Again consider the situation in which an ideal input signal $x_1(t) = \sin t + \sin 2t$ is contaminated by uniformly distributed white noise over the time horizon [0,10]. We wish to study the effects of the white noise variance σ_w^2 and the two major objective functions outlined above, assuming the critical damping hypothesis.

Solution
Listing 10.5 performs a uniform search by graphing the objective function given by Equation (10.17) with $W = 1$. This minimizes the deviation of the signal value against the ideal signal, $x_1(t)$. By varying β from 0 to 3.5 in steps of 0.01 and computing α within the loop, it is possible to construct a graph of the objective function $f(\alpha, \beta)$ as a function of β. The results of this procedure are shown in Figure 10.14 for white noise variances of $\sigma_w^2 = 0.4, 0.6, 0.8,$ and 1.0. It is clear from the graph that optimal values exist, and vary as a function of β. Assuming

```
-given formulas and data
      x₁(t) = sin(t) + sin(2t)
      ẋ₁(t) = cos(t) + 2cos(2t)
      tmax = 10
      δ = 0.1
-optimization procedure
      x₂(t) = √3(2*RND − 1)
      n = tmax/δ
      for β = 0 to 3 step 0.01
            α = 2√β − β
            t = 0
            Calculate x = x₁(t) + x₂(t)σw
            u = 0
            v = 0
            f = 0
            for k = 1 to n
                  t = δk
                  Calculate x = x₁(t) + x₂(t)σw
                  u = αx + (1 − α)(u + δv)
                  v = v + β(x − u)/δ
                  f=f+δ[x₁(t) − u]²
            next k
            print α, β, f
      next β
```

LISTING 10.5 Program segment to train the α/β tracker according to the specifications of Example 10.8.

convexity, which seems reasonable from the graph, the intervals of uncertainty show an underlying convexity, even though the function is obviously somewhat noisy. With a proper filter, this can be eliminated and a sequential search such as the *golden ratio* algorithm implemented.

If an objective function incorporating both the signal value and the signal derivative ($W = \frac{1}{2}$) is used, the graph changes. Figure 10.15 shows a graph with a similar shape but some significant differences too. In the signal-value-only objective function, the acceptable white noise variance is higher. Also, the optimal β is generally larger for the signal-value-only objective than it is for the signal value and derivative combined objective. In any case, there is an apparent (and for some variances) obvious optimum β-value for the α/β tracker.

○

Clearly, it is desirable to automate this process further. Finding β, and in turn α, by means of a sequential search would be faster and more reliable. However, to do this requires that the objective function $f(\alpha, \beta)$ first be filtered. If this is not done, a sequential search such as the *golden ratio* algorithm would become stuck in a local valley created by the noisy function, and the global minimum would never be found. There are good theoretical reasons to implement a moving-average (MA) filter in this situation. For

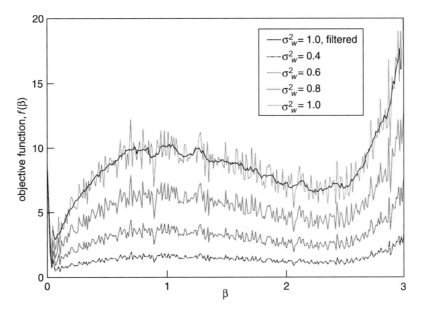

FIGURE 10.14   Objective function f using Equation (10.17) with $W = 1$.

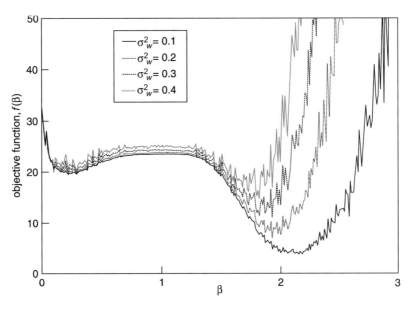

FIGURE 10.15   Objective function f using Equation (10.17) with $W = \frac{1}{2}$.

instance, since the objective function does not present a causal system, the typical filter choice is a balanced odd filter such as the following MA($2m + 1$) filter:

$$z(k) = \frac{1}{2m + 1} \sum_{i=-m}^{m} x(k + i).$$

For instance, Figure 10.14 shows the results of implementing a ninth-order ($m = 4$) MA filter for the case of $\sigma_w^2 = 1$. Clearly, this is much smoother and manageable. This will tend to smooth out the objective's graph and make the minimum easily accessible. From the filtered $z(k)$, automated procedures are fast, true, and straightforward.

10.6   MULTIDIMENSIONAL OPTIMIZATION

Line searches are methods for optimizing an objective function $f(C)$ of a single unknown parameter. Geometrically, this is analogous to searching along a single dimension for an optional point. However, in practice, higher dimensions are necessary, since there is often more than one parameter to determine. While this is easy to visualize for two dimensions, there are many cases where the dimension exceeds three and one must work formally. Here we consider the case of two dimensions, but extensions should be obvious.

Let us begin by considering the two-dimensional contour of an objective function f as shown in Figure 10.16. Each contour is a path where f has the same constant value, so that the view shown resembles a topographical map of a valley. The goal is to find the point $\mathbf{C}_{min}$, where $f_{min} = f(\mathbf{C}_{min})$ is the point of convergence of the innermost contour line. If, for some reason, f_{max} were needed instead of f_{min}, we would apply the same techniques to $-f$ and proceed in the same manner. As with one-dimensional optimization methods, we consider a sequence of points, which will always converge to a local minimum of a well-behaved objective function. Of course we want the global minimum, but for now let's settle for the local one.

In Figure 10.16, notice that after an initial trial at point $\mathbf{C}^{(0)}$, the points $\mathbf{C}^{(1)}$, $\mathbf{C}^{(2)}$, $\mathbf{C}^{(3)}$, ... progress sequentially to point $\mathbf{C}^{(n)}$, where $f(\mathbf{C}^{(n)}) \approx f_{min}$. At each point along the way, two steps are performed: choose a direction of travel, and perform a line search in that direction to the minimum. By repeating these steps over and over, the search will eventually find its way to the bottom of a valley. Thus, the basic algorithm takes the following form:

```
choose an initial test point C(0)
do until no more progress is to be made
      choose a direction to travel
      travel along a line in that direction until a
        minimum is achieved
      update the point C(i)
   repeat
```

Choosing a direction can be done in many different ways. For instance, if we have derivative information, the gradient can be computed to give the direction of steepest descent. This method, called the *method of steepest descent*, is classical but of little value

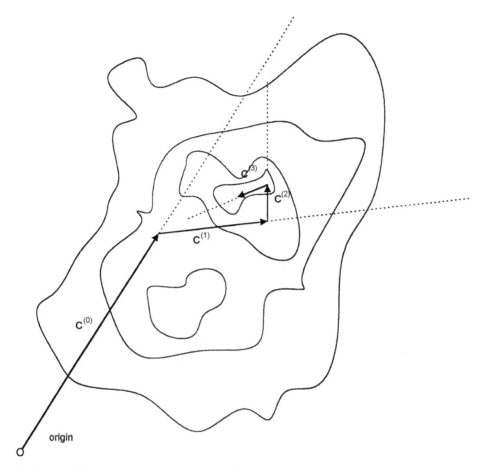

FIGURE 10.16   The two-dimensional sequential search path.

in practice, since we rarely have derivative information is system identification problems. Thus, we restrict ourselves to methods that only require information regarding the *value* of the objective f. For instance, there is nothing wrong with simply choosing random directions or simply alternating between north–south lines and east–west lines at each step of the sequence.

The second step, that of traveling along the line in the chosen direction to the minimum point, can be handled by any of the line search methods outlined earlier in this chapter. Since these steps repeat over and over, the only real question remaining is just when to terminate the search sequence. There is no easy answer for all cases, but the so-called *Cauchy criterion* is most often employed. In this case, once a point $\mathbf{C}^{(i)}$ and its successor $\mathbf{C}^{(i+1)}$ are close enough, the sequence terminates and victory is declared. Specifically, if $\|\mathbf{C}^{(i)} - \mathbf{C}^{(i+1)}\|$, where $\|\cdot\|$ is the norm of $\mathbf{C}^{(i)} - \mathbf{C}^{(i+1)}$, is smaller than some predefined small number, the procedure terminates.

It is convenient to represent both vectors and points as ordered n-tuples. For instance in two dimensions, the point $\mathbf{C} = (c_1, c_2)$ can be written as either an argument or a position

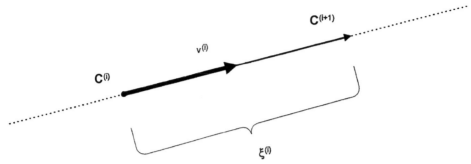

FIGURE 10.17　A multidimensional line search proceeds in direction $\mathbf{v}^{(i)}$ a distance $\xi^{(i)}$.

vector of a function f of two variables: $f(C_1, C_2) = f(\mathbf{C})$. This makes it natural to visualize f by its contour lines over the constraint domain. Since contour lines are, by definition, lines where f is constant (i.e., where the height is level), these must be non-intersecting closed curves, as shown in Figure 10.16.

Each step of the sequence of points is specified by a direction and a distance. Going from $\mathbf{C}^{(i)}$ to $\mathbf{C}^{(i+1)}$ is shown geometrically in Figure 10.17. Mathematically, these updates are written as vectors, and points are updated by

$$\mathbf{C}^{(i+1)} = \mathbf{C}^{(i)} + \xi^{(i)}\mathbf{v}^{(i)}, \tag{10.19}$$

where $\mathbf{C}^{(i)}$ is the ith point in the sequence, $\mathbf{v}^{(i)}$ is a *vector* in the chosen direction of travel, and $\xi^{(i)}$ is a scalar denoting the distance traveled in the direction $\mathbf{v}^{(i)}$. If $\mathbf{v}^{(i)}$ is a *unit vector*, then $\xi^{(i)}$ is the actual distance traveled. Since this equation looks rather unwieldy, if the specific point is understood, we will write this as $\mathbf{C}^{(new)} = \mathbf{C} + \xi\mathbf{v}$.

10.7　NON-DERIVATIVE METHODS FOR MULTIDIMENSIONAL OPTIMIZATION

Non-derivative methods determine the new direction at each step by cycling through a protocol of directions, independent of function f. The simplest non-derivative method is called *cyclic coordinates*. In this method, directions are chosen parallel to the coordinate axis; first C_1, then C_2, then C_1 again, and so forth, as illustrated in Figure 10.18. (The extension to higher-degree systems should be obvious.) This is best illustrated by an example.

EXAMPLE 10.9

Consider the unconstrained objective $f(\mathbf{C}) = 3C_1^2 + 2C_1C_2 + 3C_2^2 - 16C_1 - 8C_2$. Since $f(\mathbf{C})$ is quadratic and its discriminant $B^2 - 4AC = 2^2 - (4)(3)(3) = -20 < 0$ is negative, it should be clear that it is shaped like a

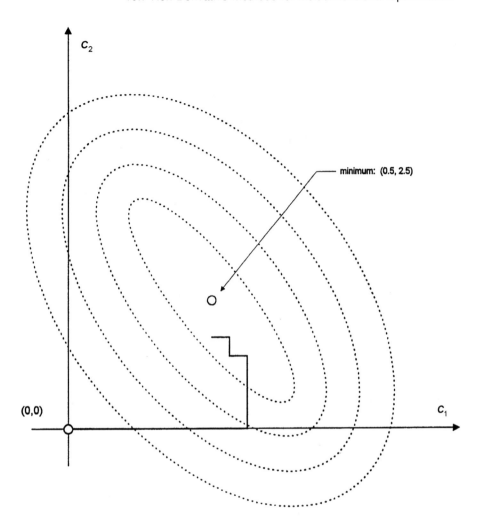

FIGURE 10.18   The path taken in a *cyclic coordinates* line search with diagonally oriented objective contours

bowl with elliptical cross-sections. Accordingly, there should be one unique global minimum.

The exact minimum can be found by setting the two partial derivatives equal to zero, and solving for C_1 and C_2:

$$\frac{\partial f}{\partial C_1} = 6C_1 + 2C_2 - 16 = 0,$$

$$\frac{\partial f}{\partial C_2} = 2C_1 + 6C_2 - 8 = 0. \tag{10.20}$$

Solving Equations (10.20) simultaneously gives $\mathbf{C} = \left(\frac{5}{2}, \frac{1}{2}\right)$. With elementary calculus, this is a trivial problem. However, assuming that we have no access to derivative information, solve the problem numerically using the *cyclic coordinates* algorithm with initial trial point at $\mathbf{C}^{(0)} = (0,0)$.

Solution

The projection of objective f onto the $C_1 - C_2$ plane (commonly referred to as a *contour plot*) is shown in Figure 10.18. It will be noted that all directions traveled in the cyclic coordinate search are parallel to the C_1 and C_2 axes. The first direction is in the direction parallel to the C_1 axis, so the initial direction vector is $\mathbf{v}^{(0)} = (1,0)$. The update to the next point $\mathbf{C}^{(1)}$ is

$$\begin{aligned}
\mathbf{C}^{(1)} &= \mathbf{C}^{(0)} + \xi \mathbf{v}^{(0)} \\
&= (0,0) + \xi(1,0) \\
&= (\xi, 0),
\end{aligned}$$

where $\xi = \xi^{(1)}$ for simplicity of notation. It follows that the objective function at the new points is $f(\mathbf{C}^{(1)}) = f(\xi, 0) = 3\xi^2 - 16\xi$. Instead of using a line search, such as the *golden ratio* algorithm, we can find the minimum by setting the derivative of f equal to zero. Since f is now a function only of the variable ξ,

$$\frac{d}{d\xi} f(\mathbf{C}^{(1)}) = 6\xi - 16 = 0,$$

$$\xi = \tfrac{8}{3}.$$

Therefore, $\mathbf{C}^{(1)} = \left(\tfrac{8}{3}, 0\right)$.

The question now is whether or not the process terminates. The Cauchy criterion compares $\mathbf{C}^{(1)}$ against $\mathbf{C}^{(0)}$ by simply considering the Euclidean two-norm, which is the same as the Euclidean distance between the two points. In this case,

$$\|\mathbf{C}^{(1)} - \mathbf{C}^{(0)}\| = \left\|\left(\tfrac{8}{3}, 0\right) - (0,0)\right\| = \tfrac{8}{3}.$$

Since the norm is relatively large, we conclude that a significant change has been made in the distance traveled, and thus we continue on to the next iteration. However, this time we go in the C_2 direction: $\mathbf{v}^{(1)} = (0,1)$. Proceeding as in the first iteration,

$$\begin{aligned}
\mathbf{C}^{(2)} &= \mathbf{C}^{(1)} + \xi \mathbf{v}^{(1)} \\
&= \left(\tfrac{8}{3}, 0\right) + \xi(0,1) \\
&= \left(\tfrac{8}{3}, \xi\right), \\
f(\mathbf{C}^{(2)}) &= f\left(\tfrac{8}{3}, \xi\right) \\
&= 3\xi^2 - \tfrac{8}{3}\xi - \tfrac{64}{3}, \\
\frac{d}{d\xi} f(\mathbf{C}^{(1)} + \xi \mathbf{v}^{(1)}) &= 6\xi - \tfrac{8}{3} = 0.
\end{aligned}$$

TABLE 10.5   Sequence of Iterations for Example 10.8 Using the
                  cyclic coordinates Algorithm.

i	$\mathbf{C}^{(i)}$	$\mathbf{v}^{(i)}$	ξ	$f(\mathbf{C}^{(i)})$
0	$(0.00, 0.00)$	$(1, 0)$	2.67	0.000
1	$(2.67, 0.00)$	$(0, 1)$	0.44	-21.333
2	$(2.67, 0.44)$	$(1, 0)$	-0.15	-21.923
3	$(2.52, 0.44)$	$(0, 1)$	0.05	-21.990
4	$(2.52, 0.49)$	$(1, 0)$	-0.02	-22.000
5	$(2.50, 0.49)$	$(0, 1)$	0.01	-22.000
6	$(2.50, 0.50)$	$(1, 0)$	0.00	-22.000

It follows from this that $\xi = \frac{4}{9}$ and $\mathbf{C}^{(2)} = \left(\frac{8}{3}, \frac{4}{9}\right)$. The norm defining the stopping
criterion is now

$$\|\mathbf{C}^{(2)} - \mathbf{C}^{(1)}\| = \|\left(\tfrac{8}{3}, \tfrac{4}{9}\right) - \left(\tfrac{8}{3}, 0\right)\| = \tfrac{4}{9},$$

but this is still not "small".

Therefore, we continue – but we cycle back to the vector $\mathbf{v}^{(2)} = (1, 0)$. This
gives

$$f(\mathbf{C}^{(2)} + \xi \mathbf{v}^{(2)}) = 3\xi^2 + \tfrac{8}{9}\xi + \tfrac{1264}{81}.$$

This is minimized when $\xi = -\frac{4}{27} \approx -0.1481$, which implies that

$$\mathbf{C}^{(3)} = \mathbf{C}^{(2)} - \tfrac{4}{27}\mathbf{v}^{(2)} \approx (2.5115, \ 0.4444).$$

It follows that the norm is $\|\mathbf{C}^{(3)} - \mathbf{C}^{(2)}\| \approx 0.1481$, which we might consider to
be sufficiently small enough a change in position, and therefore terminate the
process. Table 10.5 lists the sequence of iterates using two-decimal-place
accuracy. It will be noted that even though approximation liberties were taken
at intermediate steps, convergence to the correct solution (2.5,0.5) is still
accomplished in a timely manner. Also, we observe that in the case of cyclic
coordinates, the norm is simply $|\xi|$.   ○

It is possible to automate the entire search procedure outlined thus far. Since the *cyclic
coordinates* algorithm ultimately requires a line search, it is a good idea to implement one
of the sequential searches described earlier in this chapter in the form of a subroutine.
Function evaluation, which we have already discovered is often much more than a simple
formula, will likely require a subroutine as well. As a rough idea, Listing 10.6 illustrates
the approach. After an initial trial point $\mathbf{C}$ is given and a direction decision $\mathbf{v}$ is made, the
minimum point is derived and a modified $\mathbf{C}$ is computed. This new point is compared
against the old point, called $\mathbf{C}^*$ in the program, and the iteration continues until the norm
$\|\mathbf{C} - \mathbf{C}^*\| < \epsilon$, where ϵ is the Cauchy tolerance.

However, let us be more precise. A main program called *cyclic* requires the initial
point $\mathbf{C} = (C_1, C_2, \ldots, C_n)$, the Cauchy tolerance ϵ (by which we can test the norm), and L

```
input ε, L
input C
C* =some point distant from C
print C
do while ‖C − C*‖ ⩾ ε
        C* = C
        v = (1, 0)
        find ξ so that f(C + ξv) is minimal
        C = C + ξv
        print C
        v = (0, 1)
        find ξ so that f(C + ξv) is minimal
        C = C + ξv
        print C
end do
```

LISTING 10.6 The *cyclic coordinates* algorithm.

(the length of uncertainty tolerated during a line search). In order that a line search be initiated, it is also necessary to know just where to place the initial interval of uncertainty. This is done using a fixed distance d from **C**: $a = C_i - d$ and $b = C_i + d$. The program *cyclic*, in turn, needs to call a function evaluation subroutine. The output from *cyclic* is simply the final (minimum) point found by the algorithm.

The program *cyclic* must call a line search routine, for example, *golden* (after all, this is the best of the sequential approaches), to perform the actual line search. It is necessary that the point **C**, tolerance L, and distance d be passed from *cyclic* to *golden*. Also, there is a need for a flag η that indicates which coordinate is to be subjected to the line search algorithm. Knowledge of η is equivalent to specifying **v**, since *cyclic* retains the same cycle in each step of the procedure. At the end, *golden* will pass the best point **C** back to *cyclic* as well. In order for *cyclic* to produce a result, it is necessary to evaluate the objective function $f(\mathbf{C})$. This is done by simply passing the point **C** from *golden* to the routine *function*, which in turn returns the value of f. All these communication requirements are illustrated in Figure 10.19.

As a specific case, consider a two-dimensional optimization problem for a function $f = f(C_1, C_2)$. A working program segment for *cyclic* is shown in Listing 10.7, where it will be noticed that the external communication is shown as input and print statements. Communication to *golden* is in the form of subroutine calls as

$$golden(C_1, \ C_2, \ \eta, \ d, \ L)$$

where the arguments act as both input and output. Also, as is traditional in implementing the *cyclic coordinates* algorithm, the iteration cycle is considered to include both updates in both directions before checking the Cauchy condition.

A subroutine built around *cyclic* in Listing 10.7 is shown as *golden* in Listing 10.8. Notice that the indicator η is such that $\eta = 1$ indicates the direction $\mathbf{v} = (1, 0)$ and $\eta = 2$ indicates $\mathbf{v} = (0,1)$. By astute use of this device, it is possible to avoid explicit use of **v** in the computation of the function f. The function $f(C_1, C_2)$ is evaluated in turn by a separate

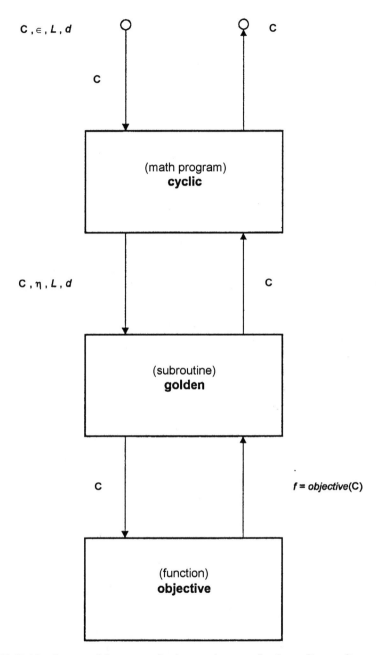

FIGURE 10.19 Inter-module communication requirements for the *cyclic coordinates* algorithms.

```
            input ε, L, d
            input C₁, C₂
            C₁* = C₁ + 2ε,  C₂* = C₂ + 2ε
            print C₁, C₂
            i = 0
            while |(C₁ − C₁*)² + (C₂ − C₂*)²| ⩾ ε
                    i = i + 1
                    C₁* = C₁
                    C₂* = C₂
                    call golden(C₁, C₂, 1, d, L)
                    print i, C₁, C₂
                    call golden(C₁, C₂, 2, d, L)
                    print C₁, C₂
            end while
```

LISTING 10.7 Main program *cyclic* to perform two-dimensional optimization.

```
    subroutine golden(C₁, C₂, η, d, L)
        if η = 1 then a = C₁ − d,  b = C₁ + d
        if η = 2 then a = C₂ − d,  b = C₂ + d
        λ = (0.618)a + (0.382)b
        μ = (0.382)a + (0.618)b
        if η = 1 then f_left = objective(λ, C₂): f_right = objective(μ, C₂):
        if η = 2 then f_left = objective(C₁, λ): f_right = objective(C₁, μ):
        end if
        while b−a > L
                if f_left < f_right then
                            b = μ
                            μ = λ
                            λ = (0.618)a + (0.382)b
                            f_right = f_left
                            if η = 1 then f_left = objective(λ, C₂):
                                          f_right = objective(μ, C₂)
                            if η = 2 then f_left = objective(C₁, λ):
                                          f_right = objective(C₁, μ)
                else
                            a = λ
                            λ = μ
                            μ = (0.382)a + (0.618)b
                            f_left = f_right
                            if η = 1 then f_left = objective(λ, C₂):
                                          f_right = objective(μ, C₂)
                            if η = 2 then f_left = objective(C₁, λ):
                                          f_right = objective(C₁, μ)
                end if
        end while
        if η = 1 then C₁ = ½(a + b)
        if η = 2 then C₂ = ½(a + b)
    return
```

LISTING 10.8 The *golden* subroutine.

function called *objective*. This method is easily expandable to other non-derivative methods, as illustrated below.

Regardless of the number of parameters to be determined, we can use the MSE criterion to force behavior into a system. For instance, consider a second-order mechanical system in which position and velocity are the only state variables. If it is desired to force the position $y(t)$ to track a variable $y_{des}(t)$, an appropriate objective function is

$$f(C_1,\ C_2) = \int_{t_0}^{t_{max}} [y_{des}(t) - y(C_1,\ C_2,\ t)]^2 \, dt. \tag{10.21}$$

On the other hand, if both the position and velocity are to be tracked, we use the objective function

$$f(C_1,\ C_2) = \int_{t_0}^{t_{max}} \left\{[y_{des}(t) - y(C_1,\ C_2,\ t)]^2 + [v_{des}(t) - v(C_1,\ C_2,\ t)]^2\right\} \, dt. \tag{10.22}$$

By using objective functions such as those of Equations (10.21) and (10.22), we can try to impose our will on any system. Of course, the more variables there are to track and the fewer degrees of freedom there are in our system, the poorer will be the tracking. It is also possible to weight the objective function so that some criteria are more important that others. For instance, if the first term in Equation (10.22) were multiplied by 10, there would be a 10-fold importance in the resulting parameter estimation. Of course, care should be taken not to try to force an unnatural situation on the system model. If an optimal solution doesn't fall into place, the chances are that the underlying model is suspect and there is really a modeling problem.

EXAMPLE 10.10

Consider the servo system whose block diagram is given in Figure 10.20. There are two tuning parameters C_1 and C_2 that weight the position and velocity components of the feedback loops. We wish to determine C_1 and C_2 so that the

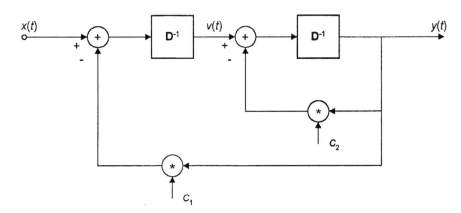

FIGURE 10.20   Servo system with position and velocity feedback loops, Example 10.10.

steady-state response to the step input $x(t) = 10$ for $t > 10$ tracks the position $y_{ss} = 3$ and $v_{ss} = 6$.

(a) Since this is a linear system, a theoretical solution is possible. Determine C_1 and C_2 from theoretical considerations.

(b) Since two tracking criteria are given, use Equation (10.22) to determine C_1 and C_2.

(c) Implement the *cyclic coordinates* algorithm to find C_1 and C_2.

Solution

Direct interpretation of the block diagram gives

$$y(t) = \int_0^t [v(t) - C_2 y(t)] \, dt, \qquad v(t) = \int_0^t [x(t) - C_1 y(t)] \, dt.$$

Differentiation results in the following differential equations:

$$\dot{y}(t) = v(t) - C_2 y(t),$$
$$\dot{v}(t) = x(t) - C_1 y(t). \tag{10.23}$$

(a) Equations (10.23) describe a linear system. As such, the output will tend to follow the input, so the choice to force the output to approach constant values was astute, since the input is also constant. At steady state, the derivatives go to zero, so $x_{ss} - C_1 y_{ss} = 0$ and $v_{ss} - C_2 y_{ss} = 0$. Since the steady-state output is specified in this problem, $C_1 = x_{ss}/y_{ss}$ and $C_2 = v_{ss}/y_{ss}$. Substituting $x_{ss} = 10$, $y_{ss} = 3$, and $v_{ss} = 6$, it follows that $C_1 = \frac{10}{3} \approx 3.33$ and $C_2 = 2$, and the system becomes

$$\dot{y}(t) = v(t) - 2y(t),$$
$$\dot{v}(t) = 10[1 - \tfrac{1}{3} y(t)]. \tag{10.24}$$

Equations (10.24) can be solved either analytically or numerically. Regardless of the method, the solution is straightforward and the graph is sketched in Figure 10.21 over the time interval [0,5]. Clearly, theory is confirmed by the system behavior.

(b) Since there are two criteria imposed on the system – one for position and one for velocity – and the time horizon is [0,10], the objective function used is

$$f(C_1, \; C_2) = \int_0^{10} [(3 - y)^2 + (6 - v)^2] \, dt.$$

A program segment producing functional values for this objective over the region $2 \leqslant C_1 \leqslant 5$, $0 \leqslant C_2 \leqslant 3$ with a mesh of resolution $\delta C_1 = 0.2$ and $\delta C_2 = 0.5$ is given in Listing 10.9, and the corresponding output is given in Table 10.6. It is clear that the minimum value of the objective function is in the neighborhood of $C_1 = 3.4$, $C_2 = 2.0$, but just exactly where is unknown. It seems fairly reasonable from the table that this point is in the region of uncertainty

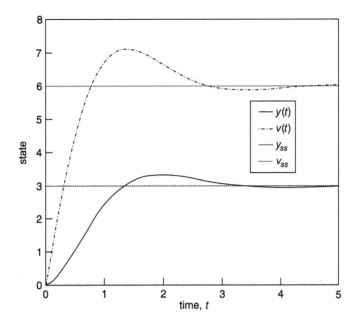

FIGURE 10.21   Solution to the servo system forcing the steady-state requirement, Example 10.10(a).

```
t_max = 10
n = 100
h = t_max/n
for C₁ = 2 to 5 step .2
        for C₂ = 0 to 3 step .5
                t = 0
                v = 0
                y = 0
                f = 0
                for k = 1 to n
                        t = hk
                        f = f + h[(3 − y)² + (6 − v)²]
                        v = v + h(10 − C₁y)
                        y = y + h(v − C₂y)
                next k
                print f,
        next C₂
        print
next C₁
```

LISTING 10.9   Creating a uniform grid over the constraint region $[2,5] \times [0,3]$, Example 10.10(b).

TABLE 10.6 Functional Values over a Uniform Grid [2,5] × [0,3], Example 10.10(b).

C_1 \ C_2	0.0	0.5	1.0	1.5	2.0	2.5	3.0
2.0	704	199	73	81	193	389	655
2.2	625	192	64	50	124	273	485
2.4	572	190	63	33	80	192	360
2.6	544	192	66	25	50	134	269
2.8	538	197	73	23	32	93	200
3.0	544	204	81	25	22	65	150
3.2	551	211	90	30	17	45	112
3.4	550	219	99	36	15	32	84
3.6	539	226	109	43	17	24	63
3.8	522	232	118	52	20	20	48
4.0	503	238	128	60	25	18	37
4.2	485	243	137	69	30	18	30
4.4	469	248	145	78	37	20	26
4.6	455	253	154	86	44	23	24
4.8	446	258	162	95	51	27	23
5.0	442	263	170	103	58	32	24

$3.2 \leqslant C_1 \leqslant 3.6$, $1.5 \leqslant C_2 \leqslant 2.5$. This is a good region in which to proceed with the *cyclic coordinates* algorithm.

(c) The *cyclic coordinates* algorithm is applied to the system defined by Equations (10.23) in Listings 10.7, 10.8, and 10.10. In the main program *cyclic* (Listing 10.7), the initial point is arbitrarily chosen at $C_1 = 6$, $C_2 = 5$. The comparison point is also arbitrary at $C_1{}^* = 100$, $C_2{}^* = 100$, since this is sufficiently distant from (6,5) to force the Cauchy test to fail on the initial iteration and thus allow the algorithm to proceed. The parameters L, d, and ϵ are

```
function objective (C₁, C₂)
        t_max = 30
        n = 100
        h = t_max/n
        t = 0
        y = 0
        v = 0
        f = 0
        for k = 1 to n
                t = hk
                f = f + h[(3 − y)² + (6 − v)²]
                y = y + h(v − C₂y)
                v = v + h(10 − C₁y)
        next k
        objective = f
    return
```

LISTING 10.10 The function *objective* for Example 10.10.

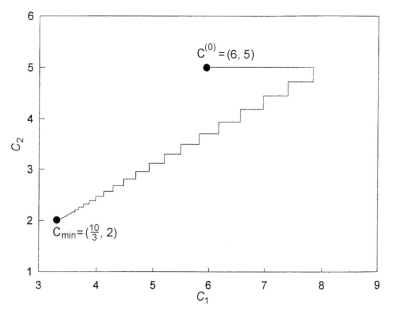

FIGURE 10.22   Search point sequence using the *cyclic coordinates* algorithm, Example 10.10(c).

all chosen as shown. The subroutine *golden* (Listing 10.8) requires no special parameter values except those sent from *cyclic*. Notice that d, which is half the initial length of uncertainty, was chosen large enough to more than encompass the 3 units shown in Table 10.6. The function *objective* performs the actual simulation, and is listed in Listing 10.10.

The output from *cyclic* forms a sequence of points C_1, C_2. These points are graphed in Figure 10.22. This sequence begins at the initial point, and progresses to the known optimal point $C_1 = 3.33$, $C_2 = 2$. Notice how the Cauchy condition doesn't always work well. Unless ϵ is chosen extremely small, the process will terminate far too soon, since there is no general relationship between $\|\mathbf{C}^{(i)} - \mathbf{C}^{(i+1)}\|$ and $\|\mathbf{C}^{(i)} - \mathbf{C}_{\min}\|$. In other words, just because we haven't traveled far doesn't mean that we have a short distance yet to go!   ○

Example 10.10 gives convincing proof that system identification using techniques such as *cyclic coordinates* work. Indeed, it is easy to get carried away and blindly apply this approach without investigating theoretical, numerical, and statistical aspects beforehand, as emphasized throughout this text. Be warned that these preliminaries are almost always necessary and that they should be part of any simulation–modeling methodology.

EXAMPLE 10.11

Repeat Example 10.10 for the case where tracking is desired for $y_{ss} = 6$ and $v_{ss} = 3$. Again assume that input to be $x(t) = 10$ and the time horizon to be $[0,20]$.

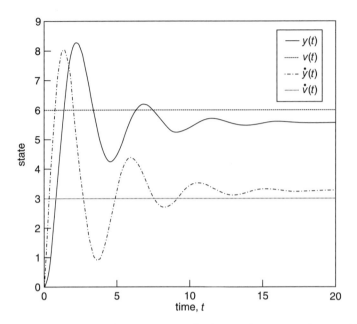

FIGURE 10.23   Solution to the servo system forcing steady-state requirement, Example 10.11.

Solution

Using the criteria for steady-state linear systems, $C_1 = x_{ss}/y_{ss} = \frac{10}{6} \approx 1.67$ and $C_2 = v_{ss}/y_{ss} = \frac{3}{6} = 0.5$. Unfortunately, the sequence does not appear to converge to these values. The reason for this can be seen from observing the response shown in Figure 10.23, where the error is evident. In this case, the system response has a significant amount of overshoot and there is considerable error accumulated before the steady-state phase actually begins. This is true even though the time horizon in this example is [0,2] instead of [0,10] as it was in Example 10.10.

There are two ways in which this problem can be remedied. First, as was done in this example, we can extend the time horizon. An even better approach is to trim the objective function limits to the range [10,20] so that the error is not measured so long over the transient phase. Application of the *cyclic coordinates* technique over the horizon [10,20] shows how effective this is. The *cyclic coordinates* sequence shown in Figure 10.24 indicates that convergence seems swift and true.   ○

The *cyclic coordinates* protocol used for determining search directions is elementary. Even so, Examples 10.10 and 10.11 show powerful results. However, if the contour lines of the objective function are oriented diagonally, it will converge rather slowly, as illustrated in Figure 10.22. This suggests that choosing a diagonal direction might also be useful. Diagonal directions is precisely the idea behind the method of Hook and Jeeves. For $d = 2$ dimensions, *cyclic coordinates* has but two direction vectors, while *Hook and Jeeves* has three:

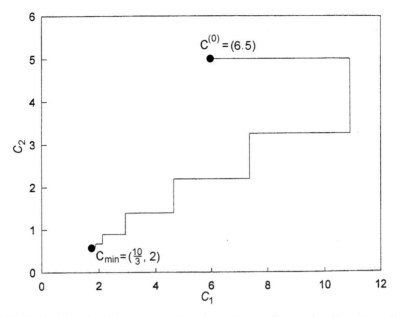

FIGURE 10.24   Search point sequence using the *cyclic coordinates* algorithm, Example 10.11.

cyclic coordinates	Hook and Jeeves
$(1,0)$	$(1,0)$
$(0,1)$	$(0,1)$
	$(1,1)$

In the $d = 3$ case, there are even more:

cyclic coordinates	Hook and Jeeves
$(1,0,0)$	$(1,0,0)$
$(0,1,0)$	$(0,1,0)$
$(0,0,1)$	$(0,0,1)$
	$(1,1,0)$
	$(1,0,1)$
	$(0,1,1)$
	$(1,1,1)$

and in the general d-dimensional case, there are d cyclic coordinate vectors and $2^d - 1$ vectors for *Hook and Jeeves*. Both are implemented in a straightforward manner. This is illustrated in the following example.

TABLE 10.7   Sequence of Iterations for Example 10.12 Using the *Hook and Jeeves* Algorithm

i	$\mathbf{C}^{(i)}$	$\mathbf{v}^{(i)}$	ξ	$f(\mathbf{C}^{(i)})$
0	$(0.00, 0.00)$	$(1, 0)$	1.05	0.000
1	$(2.67, 0.00)$	$(0, 1)$	0.44	-21.333
2	$(2.67, 0.44)$	$(1, 1)$	-0.15	-21.923
3	$(2.62, 0.39)$	$(1, 0)$	0.05	-21.990
4	$(2.54, 0.39)$	$(0, 1)$	-0.02	-22.000
5	$(2.54, 0.49)$	$(1, 1)$	0.01	-22.000
6	$(2.53, 0.48)$	$(1, 0)$	0.00	-22.000
7	$(2.51, 0.48)$	$(0, 1)$	0.00	-22.000
8	$(2.51, 0.50)$	$(1, 1)$	0.00	-22.000
9	$(2.51, 0.50)$	$(1, 0)$	0.00	-22.000
10	$(2.50, 0.50)$	$(0, 1)$	0.00	-22.000

EXAMPLE 10.12

Apply the method of Hook and Jeeves to the objective function defined in Example 10.9, $f(\mathbf{C}) = 3C_1^2 + 2C_1C_2 + 3C_2^2 - 16C_1 - 8C_2$.

Solution
Since *Hook and Jeeves* has three directions, there are now three direction vector indicators: $\eta = 1$, $\eta = 2$, and $\eta = 3$. These, along with the consequences of each direction vector, need to be incorporated in both the main program *cyclic* and *golden*. The function objective reduces to a single line since the explicit formula for $f(C)$ is given.

The resulting sequence of points is given in Table 10.7 and graphed in Figure 10.25. As should be expected, the method of Hook and Jeeves gives results better than those of cyclic coordinates since the optimal point is approached from a variety of directions.   ○

10.8   MODELING AND SIMULATION METHODOLOGY

System optimization is inherently part of engineering. While the scientific components of a system might be well understood and their models uniquely defined, the interactions, communication links, and environments must all be designed. This flexibility allows a system designer the opportunity to optimize the system performance. However, this too can be deceiving. Training and tuning a system can only be done for a single nominal input set. After this input has been used to determine the constants, the system is fixed. Thus begins the simulation phase. Even though everything might work as advertised for the nominal input, how does it work for slightly different input? For that matter, how does it work for significantly different input? Unless the system is robust, realistic

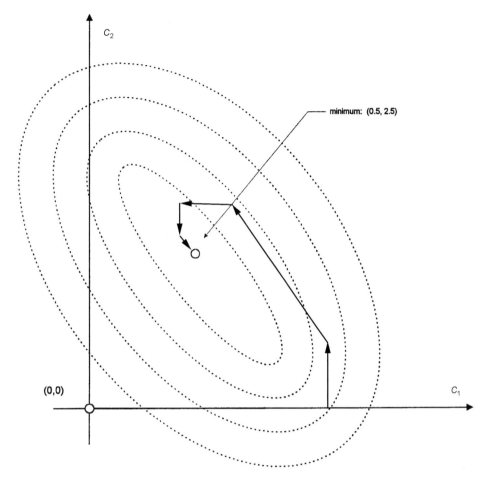

FIGURE 10.25 The path taken in a *Hook and Jeeves* line search with diagonally oriented contours.

performance will be vastly different from the nominal, and will likely be unsatisfactory. It follows that two of the major features sought after in performing simulations are parameter and input sensitivity.

Parameter sensitivity is a property illustrated for the objective $f(c)$ sketched in Figure 10.26. In a global minimum sense, C_2 is the optimal point of choice for a proposed tuning parameter C. However, it is clear that a small deviation in C due to either a degraded system or deterioration of a component over time can lead to very non-optimal performance. Even though C_1 provides slightly inferior results, perturbing C slightly from C_1 does not significantly degrade performance at all.

Input sensitivity, as the name implies, measures the performance variability due to non-nominal input. This has two aspects. First, the system design and parameter

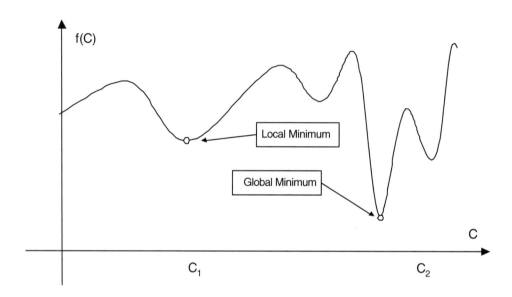

FIGURE 10.26   An arbitrary objective function in parameter space.

determination should be done using stochastic inputs. Recalling Example 10.8, it will be noted that, even assuming additive white noise, the objective function resulted in reasonable statistics. (See Figures 10.14 and 10.15.) A reasonable system will demonstrate this kind of behavior, where a stationary input results in a stationary output and a stationary objective function.

Figures 10.14 and 10.15 also demonstrate how the α/β tracker system responds to variability in the input statistical parameters. For example, even though a system might be designed using a step function, it should also be tested (using simulation) using ramps, sinusoids, and exponential inputs, along with stochastic variants as well.

Even though we have stayed clear of advocating a modeling or simulation methodology, good results obviously require a systematic approach. Simulation and modeling methods are closely akin to the scientific method, where experiment leads to insight, which leads to a hypothesis, and on to validation experiments. Validation experiments are sometimes referred to as *interpolation experiments*, since they do not extend the domain of the original hypothesis. After adjusting the detailed model to accommodate these validation experiments, an attempt is made to extend the domain into new territory. This is done by making predictions, which are hopefully verified by *extrapolation experiments*. As more of these experiments are successful, this cycle is repeated until at long last victory is declared and the hypothesis is declared an accepted theory.

Modeling and simulation works in an analogous manner. However, rather than create hypotheses, we create models; rather than perform physical experiments, we perform simulations. See Figure 10.27. In the end, models are used either for predictions or for concept verification. Regardless, this requires a number of simulation experiments by

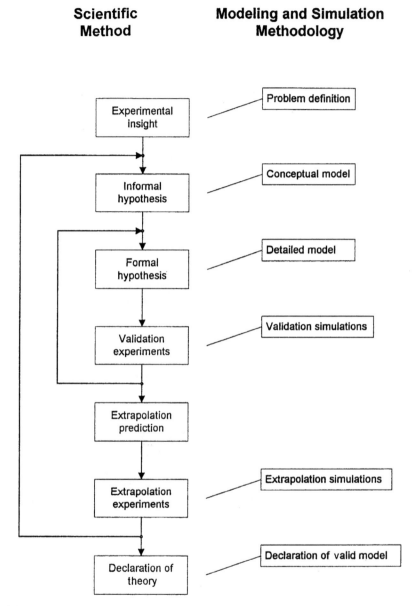

FIGURE 10.27   Correlation between the scientific method and generic modeling and simulation methodology.

which conclusions can be inferred. Not surprisingly, these experiments should not be performed arbitrarily. Rather, they need systematic planning so that statistically valid conclusions may be reached. Therefore, disciplines such as experimental design and statistical inference are the order of the day. Such issues must always be considered by experimenters and designers.

—————— **BIBLIOGRAPHY**

Bazaraa, M. S. and C. M. Shetty, *Nonlinear Programming: Theory and Algorithms*. Wiley, 1979.
Beveridge, G. S. G. and R. S. Schechter, *Optimization: Theory and Practice*. McGraw-Hill, 1970.
Converse, A. O., *Optimization*. Holt, Rinehart, and Winston, 1970.
Cooper, L. and D. Steinberg, *Introduction to Methods of Optimization*. W. B. Saunders, 1970.
Jacoby, S. L. S., J. S. Kowalik and J. T. Pizzo, *Iterative Methods for Nonlinear Optimization Problems*. Prentice-Hall, 1972.
Juang, J.-N., *Applied System Identification*. Prentice-Hall, 1994.
Kuester, J. L. and J. H. Mize, *Optimization Techniques with Fortran*. McGraw-Hill, 1973.
Ljung, L., *System Identification – Theory for the User*, 2nd edn. Prentice-Hall, 1999.

—————— **EXERCISES**

10.1 Write a computer program to perform a *uniform* search in order to find a global minimum for each of the following objective functions over the indicated interval:
(a) $f(C) = e^{-C/2} \sin C$, $0 \leqslant C \leqslant 10$;
(b) $f(C) = \dfrac{10}{C} \sin C$, $1 \leqslant C \leqslant 12$.

Execute the program using a mesh of 10 evaluation points; 20 evaluation points; 100 evaluation points. Compare your results.

10.2 Write a computer program to perform the *thirds* sequential search technique, thus finding a local minimum for each of the following objective functions:
(a) $f(C) = e^{-C/2} \sin C$, $0 \leqslant C \leqslant 10$;
(b) $f(C) = 3 + \sin C + 2 \cos 2C$, $0 \leqslant C \leqslant 8$.
Print the sequence of uncertainty intervals using the endpoint definitions as the initial search interval.

10.3 Repeat Exercise 10.2 using the *golden ratio* sequential search algorithm.

10.4 Create a computer program to minimize an objective function with two phases:
- Phase 1: Performs a uniform search over $[a, b]$ in which a locally convex region in which the global minimum resides is (hopefully) discovered.
- Phase 2: Performs a sequential search over the reduced interval of uncertainty uncovered in phase 1 to establish the global minimum.

Apply your results to the following objective functions:
(a) $f(C) = \dfrac{10}{C} \sin C$ over $[1,10]$.
(b) $f(C) = Ce^{-C/2} \sin C$ over $[0,4]$.

10.5 Create a sequential search algorithm called *dichotomous* that splits the interval of uncertainty into roughly half rather than thirds as does the *thirds* algorithm. Specifically, make the evaluation points $\lambda = \frac{1}{2}(a+b) - \epsilon$ and $\mu = \frac{1}{2}(a+b) + \epsilon$, where ϵ is a small parameter, probably just greater than machine zero. Note that ϵ cannot be zero, since this would make both interior evaluation points λ and μ the same. Accordingly, the convexity theorem would no longer apply.
(a) State and implement the algorithm.
(b) Show that the effective ratio is approximately

$$R_m(dichotomous) \approx \left(\tfrac{1}{2}\right)^{m/2}.$$

(c) Test your results on both functions of Exercise 10.2.

10.6 Automate the α/β tracker using the *golden ratio* search technique to find the optimal values for α and β in the least squares sense. Assume an archetypical signal over a fixed time interval for training purposes.

    (i) Assume a given $\alpha = 0.5$.
    (ii) Assume a fixed $\beta = 0.2$.
    (iii) Assume the critical damping hypothesis.

Apply your results to each of the following scenarios:

(a) The system and signal defined in Example 10.7.

(b) An ideal input signal of $x_1(t) = 3 \sin t - 2 \cos 5t$ and an additive noise signal of $x_2(t) = \frac{1}{10} \sin 10t + \frac{1}{5} \sin 20t$ over the time horizon $[0,12]$.

(c) An input signal of $x_1(t) = 3 \sin t - 2 \cos 5t$, $0 \leqslant t \leqslant 12$, and an additive noise signal that is uniformly distributed on $[0.0, 0.1]$.

(d) An input signal of $x_1(t) = 3 \sin t - 2 \cos 5t$, $0 \leqslant t \leqslant 12$, and an additive noise signal that is Gaussian with mean 0 and variance 0.2.

10.7 Implement the *thirds* algorithm as a subroutine with

input	$a, b, L,$
output	$t_{optimal},$
calling function	$eval(C),$

where the initial interval of uncertainty is $[a, b]$ and the tolerated length of uncertainty is L. Verify your results using the following objective functions:

(a) $f(C) = e^{-C/2} \sin C$, $0 \leqslant C \leqslant 10$;

(b) $f(C) = 3 + \sin C + 2 \cos 2C$, $0 \leqslant C \leqslant 8$.

10.8 Let $[a, b]$ be the endpoints and $[\lambda, \mu]$ the interior evaluation points in the *golden ratio* algorithm. Show that if $f(\lambda) > f(\mu)$, then $\alpha^2 + \alpha = 1$.

10.9 Using numerical differentiation (see Chapter 3), create an algorithm to "mark" each evaluation point of the uniform search as "increasing" $(+)$ or "decreasing" $(-)$. Verify your results using the objective function $f(C) = e^{-C} \cos C$.

10.10 Using numerical differentiation (see Chapter 3), create an algorithm to "mark" each evaluation point of the uniform search as "concave up" $(+)$ or "concave down" $(-)$. Verify your results using the objective function $f(C) = e^{-C} \cos C$.

10.11 Using the results of Exercises 10.9 and 10.10, recreate Figure 10.7:

(a) using $f(C) = e^{-C} \cos C$ on $[0,15]$;

(b) using $f(C) = C(5 - C) \sin C$ on $[0, 2\pi]$.

10.12 Consider a system with a nonlinear "mystery box" and an α/β tracking unit:

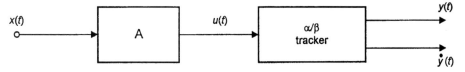

Your task is to design the α/β filter so that $y(t)$ tracks $x(t)$ over a given time interval. In particular, you should determine optimal values for α and β for an input signal

$$x(t) = \cos 1500t + 3 \sin 2000t + 2 \cos 5000t,$$

assuming a sample rate of 10 000 samples per second over the time horizon $0 \leqslant t \leqslant 0.01$ seconds. Box A is the nonliner algebraic transformation $u = x/(1 + x^2)$. Do this by minimizing the following objective function:

$$E(\alpha, \beta) = \sum_{k=0}^{100} (y_k - x_k)^2 + (\dot{y}_k - \dot{x}_k)^2.$$

(a) Without assuming any relationship between α and β, graph $E(\alpha, \beta)$ for $\beta = 0.1, 0.3, 0.5, 0.7,$ and 0.9 as a family of curves of α versus $E(\alpha, \beta)$.
(b) Now assume the *critical damping hypothesis* for the tracker. Optimize using the *golden ratio* line search technique. Clearly show the search sequence.
(c) Repeat part (b) for several different initial conditions.

10.13 Repeat Exercise 10.8 for a pulse input $x(t)$ as defined in the following figure over the time interval [0,3]:

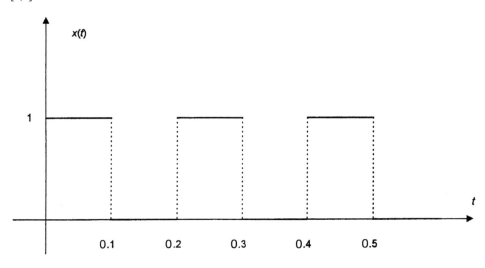

10.14 Implement the two-dimensional *cyclic coordinates* multidimensional search algorithm in a program that calls the *golden ratio* line search, which in turn calls the *function evaluation* routine. Apply this to the objective function $f(C_1, C_2) = 9C_1^2 + 7C_2^2 - 5C_1 - 3C_2 + C_1C_2$. Create a list of successive iterates:
(a) beginning at $C^{(0)} = (0,0)$;
(b) beginning at $C^{(0)} = (1,-1)$.

10.15 Implement the three-dimensional *cyclic coordinates* algorithm in a program that calls the *golden ratio* line search, which in turn calls the *function evaluation* routine. Apply this to the objective function

$$f(C_1, C_2, C_3) = 16C_1^2 + 9C_2^2 - 4C_3^2 - 7C_1C_2 + 3C_2C_3 - 5C_1C_3 + 8C_1C_2C_3.$$

Create a list of successive iterates:
(a) beginning at $C^{(0)} = (0,0,0)$;
(b) beginning at $C^{(0)} = (1,-1,2)$.

10.16 Implement the two-dimensional version of the *Hook and Jeeves* multidimensional search algorithm in a program that calls the *golden ratio* line search, which in turn calls the *function evaluation* routine. Apply this to the objective function

$$f(C_1, C_2) = 9C_1^2 + 7C_2^2 - 5C_1 - 3C_2 + C_1C_2.$$

Create a list of successive iterates:
(a) beginning at $C^{(0)} = (0, 0)$;
(b) beginning at $C^{(0)} = (1, -1)$.

10.17 Implement the three-dimensional *Hook and Jeeves* search algorithm in a program that calls the *golden ratio* line search, which in turn calls the *function evaluation* routine. Apply this to the objective

$$f(C_1,\ C_2,\ C_3) = 16C_1^2 + 9C_2^2 - 4C_3^2 - 7C_1C_2 + 3C_2C_3 - 5C_1C_3 + 8C_1C_2C_3.$$

Create a list of successive iterates:
(a) beginning at $C^{(0)} = (0,0,0)$;
(b) beginning at $C^{(0)} = (1,-1, 2)$.

10.18 Consider the objective function $f(C_1, C_2) = e^{-(C_1^2 + C_2^2)} \sin(C_1^2 + C_2^2)$.
(a) Using calculus methods, find the families of local minima.
(b) Apply the *cyclic coordinates* algorithm to find a local minimum point. Explain your results.
(c) Apply the *Hook and Jeeves* algorithm to find a local minimum point. Explain your results as compared with part (b).

10.19 Consider the following objective function:

$$f(C_1,\ C_2,\ C_3) = 2C_1^2 + 8C_2^2 + 10C_3^2 + 2C_1C_2 + C_1C_3 - 6C_2C_3 - 11C_1 + 4C_2 - 25C_3 + 110.$$

(a) Implement the three-dimensional *cyclic coordinates* algorithm as a program that calls the *golden ratio* line search subroutine, which in turn calls the *function evaluation* subroutine. The input should be the initial point and an acceptable error tolerance of the norm.
(b) Apply your program of part (a) to print a list of successive iterates, beginning at $C^{(0)} = (5,-4,-13)$.
(c) Implement the three-dimensional version of the *Hook and Jeeves* algorithm as a computer program that calls the *golden ratio* line search subroutine, which in turn calls the *function evaluation* subroutine. The input should be the initial point and an acceptable error tolerance of the norm.
(d) Apply your program of part (c) to print a list of successive iterates, beginning at $C^{(0)} = (5,-4,-13)$.
(e) Compare the efficiency and accuracy of these two approaches.

10.20 Reconsider the α/β tracker of Exercise 10.12. But, this time, do not assume the critical damping hypothesis; that is, both α and β are independent variables. Using either the *cyclic coordinates* or *Hook and Jeeves* algorithm, find the optimal value for each of the two tuning parameters.

10.21 Repeat Exercise 10.20, but this time use the pulse-train input of Exercise 10.13.

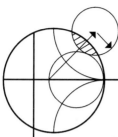

Appendices

Appendix A   STOCHASTIC METHODOLOGY

Both natural and engineered systems are more often than not, non-deterministic. Such systems have no precisely predictable states, and are said to be *stochastic*. In spite of this, even a stochastic system is far from being totally random – it has a distinct structure. Just as with deterministic systems, it is necessary to apply a scientific method to discover and demonstrate knowledge of this underlying structure. Toward this end, recall the classical scientific method as espoused by Sir Francis Bacon in the sixteenth century:

1. Obtain experimental evidence
2. Form a hypothesis characterizing the observed phenomena.
3. Test the hypothesis against further experimental evidence.
4. If the test fails, go to step 2.
5. Obtain Nobel Prize.

It will be noticed that Sir Francis was really a proponent of algorithmic design. As in all algorithmic processes, his program has an initialization step (1) performed exactly once, a number of procedural steps performed several times (2, 3, and 4), and a halting step (5).

This procedure is just as valid for stochastic models as for deterministic ones – perhaps more so. However, in order to apply it, one must first learn to measure stochastic data, formulate stochastic hypotheses, and generate and test stochastic results. This methodology, a stochastic way of thinking, is the subject of this appendix.

Determining the Density Function

To model a random variable, one must first know how the variable is distributed. This requires that a metric be established that can be validated experimentally. If an underlying principle is known a priori, it is necessary to confirm the hypothesis. If one knows nothing regarding the phenomena, data must be obtained, from which to infer a hypothesis. This hypothesis can then be tested from experimental data. The usual means for representing a distribution empirically is the *histogram*. A histogram is essentially a graphical approximation to the density function's graph. Since density functions tend to have recognizable shapes, the histogram can provide a valuable clue to the distributions that may best model data.

To make a histogram, one must first specify two (usually *artfully* determined) constants: ξ and δ. The range of random variables is partitioned into m intervals I_k of equal length called *class intervals* from ξ to $\xi + m\delta$ as follows:

$$
\begin{aligned}
I_1: & \quad \xi \leqslant x < \xi + \delta, \\
I_2: & \quad \xi + \delta \leqslant x < xi + 2\delta, \\
& \quad \vdots \\
I_m: & \quad \xi + (m-1)\delta \leqslant x < \xi + m\delta.
\end{aligned}
$$

Each datum $x_1, x_2, \ldots, x_n$ is assigned to the appropriate interval and a frequency distribution table is formed. The graph of the frequency ϕ_k versus I_k is called a *histogram*. The frequency ϕ_k is then normalized with respect to the total frequency: $n = \sum_{i=1}^{m} \phi_i$. Thus, each interval I_k is associated with the *proportion* p_k of data in that class. The graph

TABLE A.1 Frequency Distribution Table for Continuous Data

Interval number k	Interval I_k	Frequency ϕ_k	Empirical proportion p_k	Empirical density f_k
1	$\xi \leqslant x < \xi + \delta$	ϕ_1	$p_1 = \phi_1/n$	$f_1 = p_1/\delta$
2	$\xi + \delta \leqslant x < \xi 2\delta$	ϕ_2	$p_2 = \phi_2/n$	$f_2 = p_2/\delta$
3	$\xi + 2\delta \leqslant x < \xi + 3\delta$	ϕ_3	$p_3 = \phi_3/n$	$f_3 = p_3/\delta$
$\vdots$		$\vdots$	$\vdots$	$\vdots$
m	$\xi + (m-1)\delta \leqslant x < \xi + m\delta$	ϕ_m	$p_m = \phi_m/n$	$f_m = p_m/\delta$
		$n = \sum \phi_i$	$1 = \sum p_i$	$1/\delta = \sum f_i$

of p_k versus k is called the *empirical probability function*. If the frequency is normalized with respect to the total area, the resulting graph of f_k versus I_k is the *empirical density function*. It is easy to see that both p_k and f_k are related by interval size, since, in general, for the kth interval,

$$I_k: \quad \xi + (k-1)\delta \leqslant x < \xi + k\delta, \tag{A.1}$$

$$p_k = \delta f_k = \frac{\phi_k}{n} = \frac{\phi_k}{\sum\limits_{i=1}^{m} \phi_k}, \tag{A.2}$$

$$f_k = \frac{p_k}{\delta} = \frac{\phi_k}{\delta \sum\limits_{i=1}^{m} \phi_k}. \tag{A.3}$$

The general frequency distribution table described above is shown as Table A.1. The histogram is simply the bar graph of the frequency ϕ_k over the interval I_k for the data. This is shown in Figure A.1, as are graphs for the proportion p_k and density f_k. It should be noted that the area under the histogram is not unity. However, the heights of the proportion graph and the areas of the density graph sum to one. If the data is discrete and each class corresponds to a single point, the proportion is an *empirical mass*.

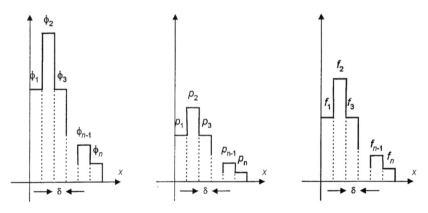

FIGURE A.1 Relationship between the histogram, proportion, and empirical density graphs.

EXAMPLE A.1

Consider the following sorted data: {0.1, 0.3, 0.3, 0.5, 0.6, 0.6, 0.7, 0.8, 0.8, 0.8, 0.9, 1.2, 1.3, 1.4, 1.7, 1.9, 2.2, 2.6, 2.9, 3.4}. This data might be the recorded length of long-distance telephone calls (measured in minutes) at a residence over a one-month period. It is desired to plot the histogram for this data and infer an appropriate density function.

Solution
The unbiased estimates for the population mean and standard deviation are calculated to be

$$\hat{\mu} = \frac{1}{n} \sum_{i=1}^{n} x_i = 1.25, \tag{A.4}$$

$$\hat{\sigma} = \sqrt{\frac{1}{n-1} \sum_{i=1}^{n} (x_i - \hat{\mu})^2}$$

$$= \sqrt{\frac{1}{n-1} \left(\sum_{i=1}^{n} x_i^2 - n\hat{\mu}^2 \right)} = 0.9259. \tag{A.5}$$

Using $\xi = 0$ and $\delta = 0.5$ along with $m = 7$ intervals, the frequency distribution table is given in Table A.2. The corresponding empirical probability and density functions are plotted in Figure A.2. Since the graph resembles a Gamma distribution, this may be assumed to be a fair picture of the underlying (actual) distribution. Assuming this to be the case, it remains to estimate the parameters that characterize the Gamma distribution: α and β.

One method to estimate the parameters of the Gamma distribution is to simply note (using Appendix C) that its mean and variance are given by $\mu = \alpha\beta$ and $\sigma^2 = \alpha\beta^2$, respectively. Solving for α and β in terms of μ and σ, then using the unbiased estimators found by Equations (A.4) and (A.5), the estimators for α and β are found to be $\alpha = \hat{\mu}^2/\hat{\sigma}^2 \approx 1.8224$ and $\beta = \hat{\sigma}^2/\hat{\mu} \approx 0.6859$.

TABLE A.2   Frequency Distribution Table, Example A.1

Number	Interval	ϕ_k	p_k	f_k
0	$0.0 \leqslant x < 0.5$	3	0.15	0.30
1	$0.5 \leqslant x < 1.0$	8	0.40	0.80
2	$1.0 \leqslant x < 1.5$	3	0.15	0.30
3	$1.5 \leqslant x < 2.0$	2	0.10	0.20
4	$2.0 \leqslant x < 2.5$	1	0.05	0.10
5	$2.5 \leqslant x < 3.0$	2	0.10	0.20
6	$3.0 \leqslant x < 3.5$	1	0.05	0.10
		20	1.00	2.00

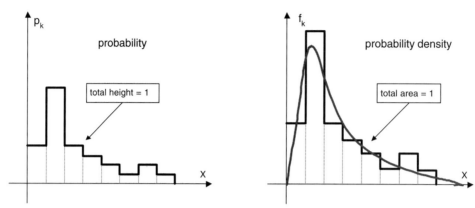

FIGURE A.2   Empirical probability and density graphs for Example A.1.

Again referring to appendix C, the density function estimator is

$$\hat{f}(x) = \frac{\beta^{-\alpha}}{\Gamma(\alpha)} x^{\alpha-1} e^{-x/\beta}$$

$$= \frac{(0.6859)^{-1.8224}}{\Gamma(1.8224)} x^{0.8224} e^{-x/0.6859}$$

$$= 2.122 \frac{x^{0.8224}}{e^{1.4579x}}, \qquad x \geqslant 0. \tag{A.6}$$

This function is graphed and compared against the empirical density function in Figure A.2. It should be noticed that the artistic comparison (smoothness/sharpness eyeball test) agrees fairly well with the inferred density function, especially considering the small sample size. ○

If the random variable is discrete it is possible to treat the proportion graph as a probability mass function, rather than as a density function. In this case, one does not consider class intervals, but rather $m + 1$ *class marks* M_k:

$$\begin{aligned} M_0: &\quad \xi, \\ M_1: &\quad \xi + \delta, \\ &\quad \vdots \\ M_m: &\quad \xi + m\delta. \end{aligned}$$

The histogram is then usually graphed as a line graph, rather than a bar graph, and corresponds directly to the mass function. There is no analogy with the empirical density function.

EXAMPLE A.2

Suppose that a study of the number of phone calls received at a certain site is to be made. To do this, the number of phone calls received over a specified 30-minute period is recorded for 25 days. In this study, it is important to not only

TABLE A.3   Frequency Distribution Table, Example A.2

Number k	Score x_k	ϕ_k	p_k
0	0	2	0.08
1	1	4	0.16
2	2	8	0.32
3	3	5	0.20
4	4	3	0.12
5	5	2	0.08
6	6	1	0.04
		25	1

estimate the average number of calls, but the distribution of such calls as well. For instance, there may be a large number of days where few calls are received, but many days where the line is especially busy. We want to analyze the difference.

The results of an empirical study are recorded and summarized in Table A.3. As can be seen, this is a frequency distribution table, using $\xi = 0$, $\delta = 1$, and $m = 7$ classes. Analyze this data statistically and infer an appropriate mass function.

Solution

Since, by the summary description, the variates must be integral, the distribution must be discrete. We therefore consider the mass function to be analogous to the empirical probability, which is shown in Figure A.3. The *unbiased* mean and variance estimators are computed as follows:

$$\hat{\mu} = \frac{1}{n}\sum_{i=1}^{n} \phi_i x_i = 2.52,$$

$$\hat{\sigma}^2 = \frac{1}{n-1}\sum_{i=1}^{n} \phi_i(x_i - \hat{\mu})^2$$

$$= \frac{1}{n-1}\sum_{i=1}^{n} n\phi_i(x_i - \hat{\mu})^2 \approx 2.343.$$

In order to determine the actual distribution of this random variable, a slightly artistic approach is called for. For instance, by noting that the distribution has range [0,6] and is slightly skewed to the right and that there are no "tails", the binomial distribution as described in Appendix B is a realistic possibility. Conveniently, formulas for the mean and variance of the binomial distribution are also given in Appendix B. They are

$$\mu = np,$$
$$\sigma^2 = np(1-p) \tag{A.7}$$

Solving Equations (A.7) for n and p and, then substituting the known mean and standard deviation estimators gives values of $p = 1 - \hat{\sigma}^2/\hat{\mu} \approx 0.07$ and

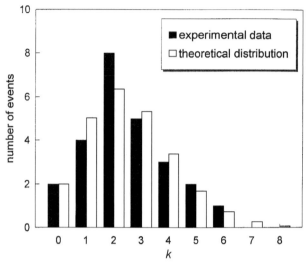

FIGURE A.3   Histogram and Poisson fit for Example A.2.

$n = \hat{\mu}^2/(\hat{\mu} - \hat{\sigma}^2) \approx 2.548$, from which the estimated mass function can in principle be computed. The value of $\hat{p} \approx 0.07$ makes good sense, since p is a probability. However, there is no integral value for $\hat{n}$ that is obvious (which should it be: $n = 2$ or $n = 3$?). Also, it should be noted that the binomial distribution has but a finite range between 0 and n (in this computation, [0.0,2.548]). From the data, this is certainly not the case.

On the other hand, it can be argued that the true distribution should have a right tail, since theoretically one can receive an unlimited number of calls over a 30-minute period. Also, the mean and variance are nearly equal. Thus, a Poisson distribution may be a better bet! Since the Poisson parameter μ equals the mean, an estimated mass function is

$$p(x) = e^{-\hat{\mu}}\frac{\hat{\mu}^x}{x!}$$
$$= (0.08)\frac{(2.52)^x}{x!}, \qquad x = 0, 1, 2, \ldots.$$

However, since the mean and variance are equal for Poisson distributions, couldn't we use $\mu = \sigma^2$ too? The reason that we do this is reflected in Equation (B.13) of Appendix B. The so-called maximum likelihood estimator is $\hat{\lambda} = n^{-1}\sum_{i=1}^{n} x_i$, which is the mean and not the variance. We verify this result and the fact that $\mu = 2.52$ is a better estimator in the next section.

Also, by recalling that the original data in Table A.3 were the result of incoming telephone calls, the Poisson hypothesis of Chapter 8 is reasonable. Therefore, it should be argued that we will use the Poisson if it is good enough. This, too, will be considered in a later section. The graph of the Poisson estimator is sketched along with the empirical mass function in Figure A.3. Clearly, the results are reasonable.   ○

Estimation of Parameters

Once a particular distribution has been chosen, it is necessary to determine the parameters of that distribution. There are a number of ways to do this, each of which is superior to the overly simple techniques illustrated in Examples A.1 and A.2. Here we present what is probably the most popular approach, that of *maximum likelihood estimators* (MLEs). MLEs have many powerful properties, including the following:

1. They make intuitive sense.
2. They are most important in using the popular *chi-square test*.
3. They are strongly consistent. That is, an MLE asymptotically approximates the actual parameter value as the number of trials n becomes large: $\lim_{n \to \infty} \hat{\theta} = \theta$ for estimator $\hat{\theta}$ and true parameter θ.

Therefore, we consider only MLEs in this presentation.

Consider a data set $x_1, x_2, \ldots, x_n$ for which a probability mass function has been chosen. We consider this data set to be derived from n independent and identically distributed (i.i.d.) distributions. Suppose further that the mass function has m undetermined parameters $\theta_1, \theta_2, \ldots, \theta_m$. We write the mass function as

$$p(x) = p(x; \theta_1, \theta_2, \ldots, \theta_m).$$

Since the data results from n independent trials of an experiment, the joint probability mass function is

$$
\begin{aligned}
L(\theta_1, \ldots, \theta_m) &= \Pr[X_1 = x_1, X_2 = x_2, \ldots, X_n = x_n] \\
&= \Pr[X = x_1]\, \Pr[X = x_2] \ldots \Pr[X = x_n] \\
&= p(x_1; \theta_1, \ldots, \theta_m) p(x_2; \theta_1, \ldots, \theta_m) \ldots p(x_n; \theta_1, \ldots, \theta_m) \\
&= \prod_{i=1}^{n} p(x_i; \theta_1, \ldots, \theta_m).
\end{aligned}
\tag{A.8}
$$

The function L is called the *likelihood function*, and is a function of data $x_1, x_2, \ldots, x_n$ and parameters $\theta_1, \theta_2, \ldots, \theta_m$. It represents the probability that the data x_i are realized under the proposed mass function with parameters θ_i. Thus, if this mass function is "good", a large value of L results. Similarly, a small L means that the parameters of the mass function were chosen poorly. Thus, we wish to maximize this with respect to each of the parameters $\theta_1, \theta_2, \ldots, \theta_m$.

If the random variable is continuous, rather than discrete, it can be shown that the joint density function $f(x_i; \theta_1, \ldots, \theta_m)$ has likelihood function

$$L(\theta_1, \ldots, \theta_m) = \prod_{i=1}^{n} f(x_i; \theta_1, \ldots, \theta_m).
\tag{A.9}$$

Regardless of whether the variable is discrete or continuous, it is necessary to maximize L. Often this can be done using traditional calculus methods, as the following example illustrates.

EXAMPLE A.3

Find the maximum likelihood estimator for the parameter μ in the Poisson distribution:

$$p(x) = e^{-\mu}\frac{\mu^x}{x!}, \qquad x = 0, 1, 2, \ldots.$$

Solution

Assume n data points, $x_1, x_2, \ldots, x_n$. Since the Poisson is a discrete distribution and there is only one unknown parameter μ, $m = 1$, and the likelihood function is defined by Equation (A.8) as

$$L(\mu) = \left(e^{-\mu}\frac{\mu^{x_1}}{x_1!}\right)\left(e^{-\mu}\frac{\mu^{x_2}}{x_2!}\right)\cdots\left(e^{-\mu}\frac{\mu^{x_n}}{x_n!}\right)$$

$$= e^{-\mu n}\frac{\mu^{x_1+x_2+\ldots+x_n}}{(x_1!)(x_2!)\ldots(x_n!)}$$

$$= e^{-\mu n}\frac{\mu^{\sum_{i=1}^{n}x_i}}{\prod_{i=1}^{n}x_i!},$$

where it should be noted that L is a function only of μ. It is best to take the logarithm of both sides before differentiating:

$$\ln L(\mu) = -\mu n + \left(\sum_{i=1}^{n}x_i\right)\ln\mu - \ln\left(\prod_{i=1}^{n}x_i!\right),$$

$$\frac{1}{L}\frac{d}{d\mu}L(\mu) = -n + \frac{1}{\mu}\sum_{i=1}^{n}x_i.$$

The maximum occurs at $dL/d\mu = 0$. It follows that

$$-n + \frac{1}{\mu}\sum_{i=1}^{n}x_i = 0$$

and

$$\hat{\mu} = \frac{1}{n}\sum_{i=1}^{n}x_i. \tag{A.10}$$

The second derivative confirms this to be a maximum. Therefore, by Equation (A.10), the best estimate for μ turns out to be the average of the data points. This discovery justifies our parameter choice in Example A.2 of using the computed mean estimate as opposed to using the variance. ○

Unfortunately, it is not always possible to use simple calculus to find the MLE. Even the simple uniform distribution requires other means. Sometimes, when calculus is appropriate, the algebra can be non-trivial. Even so, the MLE is accessible and a good way by which to obtain parameter values. All of the distributions in Appendices B and C have their MLEs included for convenience.

EXAMPLE A.4

Find the maximum likelihood estimator for the parameters α and β in the Gamma distribution:

$$f(x) = \begin{cases} \dfrac{\beta^{-\alpha} x^{\alpha-1} e^{-x/\beta}}{\Gamma(\alpha)}, & x \geqslant 0, \\ 0, & x < 0. \end{cases}$$

Since this is a continuous distribution, use the MLE defined by Equation (A.9) instead of Equation (A.8) as was done for the discrete case in Example A.3.

Solution
There are two parameters: α and β. Therefore, the likelihood is a function of two variables, $L(\alpha, \beta)$. Assuming data points $x_1, x_2, \ldots, x_n$, the likelihood function is given by

$$
\begin{aligned}
L(\alpha, \beta) &= \left(\frac{\beta^{-\alpha} x_1^{\alpha-1} e^{-x_1/\beta}}{\Gamma(\alpha)} \right) \left(\frac{\beta^{-\alpha} x_2^{\alpha-1} e^{-x_2/\beta}}{\Gamma(\alpha)} \right) \cdots \left(\frac{\beta^{-\alpha} x_n^{\alpha-1} e^{-x_n/\beta}}{\Gamma(\alpha)} \right) \\
&= \beta^{-n\alpha} \frac{(x_1 x_2 \cdots x_n)^{\alpha-1} e^{-(x_1 + x_2 + \ldots + x_n)/\beta}}{\Gamma^n(\alpha)} \\
&= \frac{\beta^{-n\alpha} \left(\prod\limits_{i=1}^{n} x_i \right)^{\alpha-1} \exp\left(-\dfrac{1}{\beta} \sum\limits_{i=1}^{n} k_i \right)}{\Gamma^n(\alpha)}.
\end{aligned}
$$

Again, it is easiest to first take the logarithm of each side:

$$\ln L(\alpha, \beta) = -n\alpha \ln \beta + (\alpha - 1) \ln\left(\prod_{i=1}^{n} x_i \right) - \frac{1}{\beta} \sum_{i=1}^{n} x_i - n \ln \Gamma(\alpha).$$

However, this time it is required to set each partial derivative to zero. First, with respect to β,

$$\frac{1}{\Gamma} \frac{\partial L}{\partial \beta} = -\frac{n\alpha}{\beta} + \frac{1}{\beta^2} \sum_{i=1}^{n} x_i = 0, \tag{A.11}$$

$$\alpha\beta = \frac{1}{n} \sum_{i=1}^{n} x_i.$$

Next, with respect to α,

$$\frac{1}{L} \frac{\partial L}{\partial \alpha} = -n \ln \beta + \ln\left(\prod_{i=1}^{n} x_i \right) - n \frac{\Gamma'(\alpha)}{\Gamma(\alpha)} = 0,$$

$$
\begin{aligned}
\ln \beta &= \frac{1}{n} \ln\left(\prod_{i=1}^{n} x_i \right) - \frac{\Gamma'(\alpha)}{\Gamma(\alpha)} \\
&= \frac{1}{n} \sum_{i=1}^{n} \ln x_i - \Psi(\alpha),
\end{aligned}
\tag{A.12}
$$

where $\Psi(\alpha) = \Gamma'(\alpha)/\Gamma(\alpha)$ is the so-called *digamma function*.
Combining Equations (A.11) and (A.12) to eliminate β,

$$\Psi(\alpha) - \ln \alpha = \ln\left(\frac{1}{n} \sum_{i=1}^{n} x_i \right) - \frac{1}{n} \sum_{i=1}^{n} \ln x_i. \tag{A.13}$$

Unfortunately, no explicit solution of Equation (A.13) for α is known to exist. However, a very accurate approximation can be shown to be

$$\hat{\alpha} \approx 0.153495 + \frac{0.5}{\ln\left(\frac{1}{n}\sum\limits_{i=1}^{n} x_i\right) - \frac{1}{n}\sum\limits_{i=1}^{n} \ln x_i}. \tag{A.14}$$

Once α is known, β is determined by

$$\hat{\beta} \approx \frac{1}{n\hat{\alpha}}\sum_{i=1}^{n} x_i. \tag{A.15}$$

○

The MLE need not be an unbiased estimator. That is, if $\hat{\theta}$ is the estimator and θ is the (true) parameter value, $E[\hat{\theta}] = \theta$ does not always hold. However, the MLE is *strongly consistent* in the sense that $\lim_{n\to\infty} \hat{\theta} = \theta$. While each of these properties is most desirable, there are no known estimators that possess both of them. Therefore, in practice, we simply go with MLEs and call them sufficient.

Goodness of Fit

Even though a parameter is best-fitting in some sense (e.g., maximum likelihood or least squares), the theoretical density function may not fit the empirical density function closely overall. In order to measure the quality of fit, the relative deviation between the two probability measures is compared and totaled. This total, called χ^2 (chi-square), is then compared against a standard, and the decision to accept or reject the fit is made.

First consider the case of a continuous random variable. Suppose the histogram has defining partition $\xi, \xi + \delta, \xi + 2\delta, \ldots, \xi + m\delta$. Further suppose there is frequency ϕ_k in the interval $I_k : x + (k-1)\delta \leqslant x < \xi + k\delta$. Then, under the hypothesis that $\hat{f}(x)$ is a reasonable estimate of the true density function, the probability of occurrence on I_k is

$$\Pr[\xi + (k-1)\delta \leqslant x < \xi + k\delta] = \int_{\xi+(k-1)\delta}^{\xi+k\delta} \hat{f}(x)\, dx.$$

It follows that the expected frequency over the interval I_k under the density function $\hat{f}(x)$ is

$$e_k = n\int_{\xi+(k-1)\delta}^{\xi+k\delta} \hat{f}(x)\, dx.$$

To be precise, the χ^2 statistic measures the relative difference between the expected frequency e_k and the actual frequency ϕ_k. This difference is squared so as to make χ^2 positive and summed over each of the classes. Formally,

$$\chi^2 = \sum_{k=1}^{m} \frac{(\phi_k - e_k)^2}{e_k}. \tag{A.16}$$

Certainly, an ideal χ^2 statistic would be one where the observed and relative frequencies are the same, in which case χ^2 is zero. Therefore, it is desirable that χ^2 should be as small as possible. The question naturally arises as to how small χ^2 should be in order to accept $f(x)$ as the "true" density function? The answer to this question lies in using the distribution function for χ^2, which is tabulated in Appendix F.

From Appendix C, the χ^2 statistic with v degrees of freedom is a random variable with density function

$$f(x) = \begin{cases} \dfrac{x^{(v-2)/2}e^{-x/2}}{2^{v/2}\Gamma(\frac{1}{2}v)}, & x \geqslant 0, \\ 0, & x < 0. \end{cases}$$

The term "degrees of freedom" requires some explanation. There are m intervals or classes associated with the calculation of χ^2. However, the additional requirement that the total $\sum_{i=1}^{m} \phi_i = n$ poses an additional constraint. Thus, there are $v = m - 1$ ways in which this can be done. This is the case if there is no estimation of parameters. However, if in the establishment of the density function, l parameters were estimated too (each further constrains the problem), there is a set of $v = m - l - 1$ degrees of freedom.

In practice, we begin by assuming that the estimated density function $f(x)$ fits the data satisfactorily. Under this hypothesis, we calculate χ^2 (using Equation (A.16)) and v. If the calculated value is greater than some critical value (such as $\chi^2_{0.05}$, which is the 5% significance level using Appendix F), we conclude that empirical and estimated density functions differ "significantly". If this critical value is not exceeded, the estimated density function $\hat{f}(x)$ is essentially the same as the true density function $\hat{f}(x)$. This test is referred to as the *chi-square hypothesis test*.

EXAMPLE A.5

Consider the data {2, 2, 2, 4, 4, 7, 10, 12, 13, 14}. Estimate an appropriate continuous uniform density function and test to see if such a hypothesis is reasonable.

Solution
The frequency distribution table and empirical density functions are shown in Table A.4 and Figure A.4 respectively. The partition defined by $\xi = 0$ and $\delta = 2.5$ seems reasonable, but is relatively arbitrary.

The uniform density function is defined by parameters a and b as given in Appendix C. The MLEs for a and b are 2 and 14 respectively. Therefore, the density function is

$$\hat{f}(x) = \begin{cases} \frac{1}{12}, & 2 \leqslant x \leqslant 14, \\ 0, & \text{otherwise.} \end{cases} \tag{A.17}$$

TABLE A.4 Frequency Distribution Table, Example A.5

Number	Interval	ϕ_k	p_k	f_k
0	$0.0 \leqslant x < 2.5$	3	0.3	0.12
1	$2.5 \leqslant x < 5.0$	2	0.2	0.08
2	$5.0 \leqslant x < 7.5$	1	0.1	0.04
3	$7.5 \leqslant x < 10.0$	0	0.0	0.00
4	$10.0 \leqslant x < 12.5$	2	0.2	0.08
5	$12.5 \leqslant x < 15.0$	2	0.1	0.08
		10	1.0	0.40

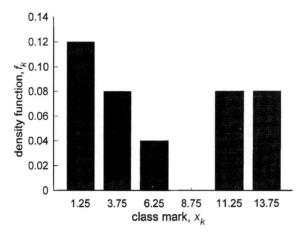

FIGURE A.4   Empirical density function, Example A.5.

Using this estimator, we proceed to calculate the expected frequency for each interval, being careful to integrate only over the interval defined by Equation (A.17):

$$I_1: \quad e_1 = n \int_0^{2.5} \hat{f}(x)\, dx$$

$$= 10 \int_2^{2.5} \tfrac{1}{12}\, dx \approx 0.417;$$

$$I_2: \quad e_2 = n \int_{2.5}^{5.0} \hat{f}(x)\, dx$$

$$= 10 \int_{2.5}^{5.0} \tfrac{1}{12}\, dx \approx 2.083;$$

similarly, it may be shown that $e_3 = e_4 = e_5 \approx 2.083$;

$$I_6: \quad e_6 = n \int_{12.5}^{15.0} \hat{f}(x)\, dx$$

$$= 10 \int_{12.5}^{14.0} \tfrac{1}{12}\, dx = 1.25$$

Comparing these calculated frequencies with the actual ones listed in Table 3.4, we next compute χ^2:

$$\chi^2 = \sum_{k=1}^{6} \frac{(\phi_k - e_k)^2}{e_k}$$

$$= \frac{(3 - 0.417)^2}{0.417} + \frac{(2 - 2.083)^2}{2.083} + \frac{(1 - 2.083)^2}{2.083}$$

$$+ \frac{(0 - 2.083)^2}{2.083} + \frac{(2 - 2.083)^2}{2.083} + \frac{(2 - 1.250)^2}{1.250} \approx 19.124.$$

Since there were two parameters estimated in the density function, $l = 2$ and $v = 6 - 2 - 1 = 3$. Checking Appendix F, the critical χ^2-values at the 5% and 1% significance levels are $\chi^2_{0.05} = 7.815$ and $\chi^2_{0.01} = 11.341$. In both cases, χ^2 exceeds the critical values. Therefore, the estimated uniform density function defined by Equation (A.17) is rejected.   ○

EXAMPLE A.6

Using the same data as in Example A.5, estimate the density function using an exponential distribution. Test the hypothesis that this is a reasonable estimate.

Solution

The mean of the data is $\mu = 7$. Since the mean is also the MLE for the exponential distribution, the estimated density function is

$$\hat{f}(x) = \begin{cases} \frac{1}{7}e^{-x/7}, & x \geq 0, \\ 0, & x < 0. \end{cases}$$

Notice that for an exponential distribution, one must use an additional interval $x \geq 15$ since the exponential is right-tailed. The expected frequencies are calculated as follows:

$$e_1 = n \int_0^{2.5} \hat{f}(x)\, dx$$

$$= 10 \int_0^{2.5} \frac{1}{7}e^{-x/7} dx$$

$$= -10e^{-2.5/7} \Big|_0^{2.5}$$

$$= 10(1 - e^{-2.5/7}) \approx 3.00,$$

$$e_2 = 10 \int_{2.5}^{5.0} \frac{1}{7}e^{-x/7}\, dx \approx 2.101,$$

$$e_3 = 10 \int_{5.0}^{7.5} \frac{1}{7}e^{-x/7}\, dx \approx 1.470,$$

$$e_4 = 10 \int_{7.5}^{10.0} \frac{1}{7}e^{-x/7}\, dx \approx 1.029,$$

$$e_5 = 10 \int_{10.0}^{12.5} \frac{1}{7}e^{-x/7}\, dx \approx 0.720,$$

$$e_6 = 10 \int_{12.5}^{15.0} \frac{1}{7}e^{-x/7}\, dx \approx 0.506,$$

$$e_7 = 10 \int_{15.0}^{\infty} \frac{1}{7}e^{-x/7}\, dx \approx 1.172.$$

Again calculating χ^2,

$$\chi^2 = \sum_{k=1}^{7} \frac{(\phi_k - e_k)^2}{e_k}$$

$$= \frac{(3 - 3.000)^2}{3.000} + \frac{(2 - 2.101)^2}{2.101} + \frac{(1 - 1.470)^2}{1.470} + \frac{(0 - 1.029)^2}{1.029}$$

$$+ \frac{(2 - 0.720)^2}{0.720} + \frac{(2 - 0.506)^2}{0.506)} + \frac{(0 - 1.172)^2}{1.172} \approx 9.043.$$

Since there is but one parameter determined ($l = 1$) and 7 classes ($m = 7$), this time the number of degrees of freedom is $v = 7 - 1 - 1 = 5$. Checking the critical values, $\chi^2_{0.05} = 11.070$ and $\chi^2_{0.01} = 15.086$. Since $\chi^2 < \chi^2_{0.05}$ and $\chi^2 < \chi^2_{0.01}$, we will accept this hypothesis at either the 5% or 1% significance level.   ○

The χ^2 statistic and the corresponding test are very general. In situations where a goodness of fit over a range of values is called for, they are considered the standard. Just as was illustrated here, there is a straightforward methodology called for:

1. From a frequency distribution table, sketch the histogram of the empirical data.
2. Using Appendices B and C, select a candidate distribution.
3. Use the maximum likelihood estimator to compute the parameters of the selected distribution.
4. Find the goodness of fit using the χ^2 statistic.
5. Apply the χ^2 test to decide whether or not the selected distribution is up to the task.

This sequence should be repeated until the distribution is found such that the goodness-of-fit requirements are satisfactory.

BIBLIOGRAPHY

Hoel, P. G., *Introduction to Mathematical Statistics*, 4th edn. Wiley, 1971.
Montgomery, D. C. and G. C. Runger, *Applied Statistics and Probability for Engineers*. Wiley, 1994.
Petruccelli, J. D., B. Nandram, and M. Chen, *Applied Statistics for Engineers and Scientists*. Prentice-Hall, 1999.
Spiegel, M. R., *Probability and Statistics* [Schaum's Outline Series]. McGraw-Hill, 1975.

Appendix B   POPULAR DISCRETE PROBABILITY DISTRIBUTIONS

Discrete Uniform

Parameters: i: left endpoint;
               j: right endpoint.

Mass function:

$$p(x) = \begin{cases} \dfrac{1}{j-i+1}, & x = i, i+1, \ldots, j, \\ 0, & \text{otherwise.} \end{cases} \tag{B.1}$$

Distribution function:

$$F(x) \begin{cases} 0, & x < i, \\ \dfrac{\lfloor x \rfloor - i + 1}{j - i + 1}, & i \leqslant x \leqslant j, \\ 1, & x > j. \end{cases} \tag{B.2}$$

Mean: $\frac{1}{2}(i+j)$.

Mode: Does not uniquely exist.

Variance: $\frac{1}{12}[(j-1+1)^2 - 1]$.

Maximum likelihood estimators:

$$i = \min_{1 \leqslant k \leqslant n} \{x_k\} \tag{B.3}$$

$$\hat{j} = \max_{1 \leqslant k \leqslant n} \{x_k\}. \tag{B.4}$$

Binomial

Parameters: n (a positive integer): the number or trials;
$p(0 < p < 1)$: the probability of a single success.

Mass function:

$$p(x) = \begin{cases} \dbinom{n}{x} p^x (1-p)^{n-x}, & x = 0, 1, \ldots n, \\ 0, & \text{otherwise.} \end{cases} \tag{B.5}$$

Distribution function:†

$$F(x) = \begin{cases} 0, & x < 0, \\ \displaystyle\sum_{k=0}^{\lfloor x \rfloor} \dbinom{n}{k} p^k (1-p)^{n-k}, & i \leqslant x \leqslant j, \\ 1, & x > n. \end{cases} \tag{B.6}$$

Mean: np.

Modes: $\begin{cases} p(n+1) - 1 \text{ and } p(n+1), & \text{if } p(n+1) \text{ is an integer,} \\ \lfloor p(n+1) \rfloor, & \text{otherwise.} \end{cases}$

Variance: $np(1-p)$.

Maximum likelihood estimator: if n is known, the MLE for p is

$$\hat{p} = \frac{1}{n^2} \sum_{k=0}^{n} x_k. \tag{B.7}$$

† The notation $\lfloor x \rfloor$ indicates the integer part of x.

If both n and p are unknown, the problem is somewhat more difficult, and is not treated here.

Geometric

Parameter: p ($0 < p < 1$): the probability of a single success.

Mass function:

$$p(x) = \begin{cases} p(1-p)^{x-1}, & x = 1, 2, \ldots, \\ 0, & \text{otherwise.} \end{cases} \qquad (\text{B.8})$$

Distribution function:

$$F(x) \begin{cases} 0, & x < 1, \\ 1 - (1-p)^{\lfloor x \rfloor}, & x \geqslant 1. \end{cases} \qquad (\text{B.9})$$

Mean: $\dfrac{1}{p}$.

Mode: 1.

Variance: $\dfrac{1-p}{p^2}$.

Maximum likelihood estimator:

$$\hat{p} = \frac{1}{\dfrac{1}{n}\displaystyle\sum_{i=1}^{n} x_i + 1}. \qquad (\text{B.10})$$

Poisson

Parameter: μ (positive real number): mean.

Mass function:

$$p(x) = \begin{cases} \dfrac{e^{-\mu}\mu^x}{x!}, & x = 0, 1, \ldots, \\ 0, & \text{otherwise.} \end{cases} \qquad (\text{B.11})$$

Distribution function:

$$F(x) = \begin{cases} 0, & x < 1, \\ \dfrac{\lfloor x \rfloor - i + 1}{j - i + 1}, & i \leqslant x \leqslant j, \\ 1, & x > j. \end{cases} \qquad (\text{B.12})$$

Mean: μ.

Mode : $\begin{cases} \mu \text{ and } \mu - 1, & \mu \text{ is an integer,} \\ \lfloor \mu \rfloor, & \text{otherwise.} \end{cases}$

Variance: μ.

Maximum likelihood estimator:

$$\hat{\lambda} = \frac{1}{n}\sum_{i=1}^{n} x_i. \tag{B.13}$$

Appendix C POPULAR CONTINUOUS PROBABILITY DISTRIBUTIONS

Continuous Uniform

Parameters: a: left endpoint;
$\quad\quad\quad\quad\quad$ b: right endpoint.

Density function:

$$f(x) = \begin{cases} \dfrac{1}{b-a}, & a \leqslant x \leqslant b, \\ 0, & \text{otherwise.} \end{cases} \tag{C.1}$$

Distribution function:

$$F(x) = \begin{cases} 0, & x < a, \\ \dfrac{x-a}{b-a}, & a \leqslant x \leqslant b, \\ 1, & x > b. \end{cases} \tag{C.2}$$

Mean: $\frac{1}{2}(a+b)$.

Mode: Does not uniquely exist.

Variance: $\frac{1}{12}(b-a)^2$.

Maximum likelihood estimators:

$$\hat{a} = \min_{1 \leqslant k \leqslant n} \{x_k\} \tag{C.3}$$

$$\hat{b} = \max_{1 \leqslant k \leqslant n} \{x_k\}. \tag{C.4}$$

Gamma

Parameters: α ($\alpha > 0$),
$\quad\quad\quad\quad\quad$ β ($\beta > 0$).

Density function:

$$f(x) = \begin{cases} 0, & x < 0, \\ \dfrac{x^{\alpha-1}e^{-x/\beta}}{\beta^\alpha \Gamma(\alpha)}, & x \geqslant 0. \end{cases} \tag{C.5}$$

Distribution function: no general closed form.

Mean: $\alpha\beta$

Mode : $\begin{cases} \beta(\alpha - 1), & \alpha \geqslant 1, \\ 0, & \alpha < 1. \end{cases}$

Variance: $\alpha\beta^2$

Maximum likelihood estimators:

$$\hat{\alpha} = 0.153495 + \frac{0.5}{\ln\left(\frac{1}{n}\sum_{i=1}^{n}x_i\right) - \frac{1}{n}\sum_{n-1}^{n}\ln x_i},\tag{C.6}$$

$$\hat{\beta} = \frac{1}{n\hat{\alpha}}\sum_{i=1}^{n}x_i.\tag{C.7}$$

Exponential

This is a special case of the Gamma distribution with $\alpha = 1$ and $\beta = 1/\lambda$.

Parameter: $\lambda(\lambda > 0)$

Density function:

$$f(x) = \begin{cases} \lambda e^{-\lambda x}, & x \geqslant 0, \\ 0, & \text{otherwise.} \end{cases}\tag{C.8}$$

Distribution function:

$$F(x) = \begin{cases} 0, & x < 0, \\ 1 - e^{-\lambda x}, & x \geqslant 0. \end{cases}\tag{C.9}$$

Mean: $\dfrac{1}{\lambda}$.

Mode: 0.

Variance: $\dfrac{1}{\lambda^2}$.

Maximum likelihood estimator:

$$\hat{\lambda} = \frac{n}{\sum_{i=1}^{n}x_i}.\tag{C.10}$$

Chi-Square

This is a special case of the Gamma distribution with $\alpha = \frac{1}{2}v$, v a positive integer, and $\beta = 2$.

Parameter: v (a positive integer): degrees of freedom.

Density function:

$$f(x) = \begin{cases} 0, & x < 0, \\ \dfrac{x^{v/2-1}e^{-x/2}}{2^{v/2}\Gamma(\frac{1}{2}v)}, & x \geqslant 0. \end{cases}\tag{C.11}$$

Distribution function: no general closed form.

Mean: v.

Mode : $\begin{cases} v - \frac{1}{2}, & v \geqslant 2, \\ 0, & v < 2. \end{cases}$

Variance: $2v$.

Maximum likelihood estimator:

$$\hat{v} = \frac{1}{n} \sum_{i=1}^{n} x_i, \tag{C.12}$$

m-Erlang

This is a special case of the Gamma distribution with $\alpha = m$ a positive integer ( ) and $\beta = 1/\lambda$. m (a positive integer) is the number of IID exponential variates;

Parameters:

$$\lambda(\lambda > 0).$$

Density function:

$$f(x) = \begin{cases} 0, & x < 0, \\ \dfrac{\lambda^m x^{m-1} e^{-\lambda x}}{(m-1)!}, & x \geqslant 0. \end{cases} \tag{C.13}$$

Distribution function:

$$F(x) = \begin{cases} 0, & x < 0, \\ 1 - e^{-\lambda x} \sum_{i=0}^{m-1} \dfrac{(\lambda x)^i}{i!}, & x \geqslant 0. \end{cases} \tag{C.14}$$

Mean: $\dfrac{m}{\lambda}$.

Mode: $\dfrac{m-1}{\lambda}$.

Variance: $\dfrac{m}{\lambda^2}$.

Maximum likelihood estimators:

$$\hat{\lambda} = \frac{mn}{\sum_{i=1}^{n} x_i}. \tag{C.15}$$

Gaussian

Parameters: μ: mean;
σ: standard deviation.

Density function:

$$f(x) = \frac{1}{\sqrt{2\pi}\sigma}\exp\left[-\frac{1}{2}\left(\frac{x-\mu}{\sigma}\right)^2\right]. \tag{C.16}$$

Distribution function: no general closed form.

Mean: μ.

Mode: μ.

Variance: σ^2.

Maximum likelihood estimators:

$$\hat{\mu} = \frac{1}{n}\sum_{i=1}^{n} x_i, \tag{C.17}$$

$$\hat{\sigma} = \sqrt{\frac{1}{n-1}\sum_{i=1}^{n}(x_i - \hat{\mu})^2}. \tag{C.18}$$

Beta[†]

Parameters: $a\ (a > 0)$,
$\qquad\qquad b\ (b > 0)$.

Density function:

$$f(x) = \begin{cases} \dfrac{\Gamma(a+b+2)}{\Gamma(a+1)\Gamma(b+1)}x^a(1-x)^b, & 0 \leqslant x \leqslant 1, \\ 0, & \text{otherwise.} \end{cases} \tag{C.19}$$

Distribution function: no general closed form.

Mean: $\dfrac{a+1}{a+b+2}$.

Mode: $\dfrac{a}{a+b}$.

Variance: $\dfrac{(a+1)(b+1)}{(a+b+2)^2(a+b+3)}$.

Maximum likelihood estimators: solve the following pair of equations simultaneously for $\hat{a}$ and $\hat{b}$:

$$\Psi(\hat{a}+1) - \Psi(\hat{a}+\hat{b}+2) = \left[\ln\left(\prod_{i=1}^{n} x_i\right)\right]^{1/n}, \tag{C.21}$$

$$\Psi(\hat{b}+1) - \Psi(\hat{a}+\hat{b}+2) = \left[\ln\left(\prod_{i=1}^{n}(1-x_i)\right)\right]^{1/n}, \tag{C.22}$$

where Ψ is the digamma function.

[†]This definition of the Beta distribution is somewhat non-standard. Usually, it is defined for $a > -1$ and $b > -1$. However, most practical applications use only the case of positive parameters a and b.

Appendix D   THE GAMMA FUNCTION

x	$\Gamma(x)$	x	$\Gamma(x)$	x	$\Gamma(x)$
1.00	1.00000	1.40	0.88726	1.80	0.93138
1.01	0.99433	1.41	0.88676	1.81	0.93408
1.02	0.98884	1.42	0.88636	1.82	0.93685
1.03	0.98355	1.43	0.88604	1.83	0.93969
1.04	0.97844	1.44	0.88581	1.84	0.94261
1.05	0.97350	1.45	0.88566	1.85	0.94561
1.06	0.96874	1.46	0.88560	1.86	0.94869
1.07	0.96415	1.47	0.88563	1.87	0.95184
1.08	0.95973	1.48	0.88575	1.88	0.95507
1.09	0.95546	1.49	0.88595	1.89	0.95838
1.10	0.95135	1.50	0.88623	1.90	0.96177
1.11	0.94740	1.51	0.88659	1.91	0.96523
1.12	0.94359	1.52	0.88704	1.92	0.96877
1.13	0.93993	1.53	0.88757	1.93	0.97240
1.14	0.93642	1.54	0.88818	1.94	0.97610
1.15	0.93304	1.55	0.88887	1.95	0.97988
1.16	0.92980	1.56	0.88964	1.96	0.98374
1.17	0.92670	1.57	0.89049	1.97	0.98768
1.18	0.92373	1.58	0.89142	1.98	0.99171
1.19	0.92089	1.59	0.89243	1.99	0.99581
1.20	0.91817	1.60	0.89352	2.00	0.1.00000
1.21	0.91558	1.61	0.89468		
1.22	0.91311	1.62	0.89592		
1.23	0.91075	1.63	0.89724		
1.24	0.90852	1.64	0.89864		
1.25	0.90640	1.65	0.90012		
1.26	0.90440	1.66	0.90167	for $x > 2$,	
1.27	0.90250	1.67	0.90330	$\Gamma(x) = (x-1)\Gamma(x-1)$	
1.28	0.90072	1.68	0.90500		
1.29	0.89904	1.69	0.90678		
1.30	0.89747	1.70	0.90864	for $x < 1$,	
1.31	0.89600	1.71	0.91057	$\Gamma(x) = \Gamma(x+1)/x$	
1.32	0.89464	1.72	0.91258		
1.33	0.89338	1.73	0.91467		
1.34	0.89222	1.74	0.91683		
1.35	0.89115	1.75	0.91906		
1.36	0.89017	1.76	0.92137		
1.37	0.88931	1.77	0.92376		
1.38	0.88854	1.78	0.92623		
1.39	0.88785	1.79	0.92877		

Appendix E THE GAUSSIAN DISTRIBUTION FUNCTION

Values of $F(z)=Pr[Z<z]$ for the Normalized Gaussian Distribution

z	0.00	0.01	0.02	0.03	0.04	0.05	0.06	0.07	0.08	0.09
0.0	0.5000	0.5040	0.5080	0.5120	0.5160	0.5199	0.5239	0.5279	0.5319	0.5359
0.1	0.5398	0.5438	0.5478	0.5517	0.5557	0.5596	0.5636	0.5675	0.5714	0.5753
0.2	0.5793	0.5832	0.5871	0.5910	0.5948	0.5987	0.6026	0.6064	0.6103	0.6141
0.3	0.6179	0.6217	0.6255	0.6293	0.6331	0.6368	0.6406	0.6443	0.6480	0.6517
0.4	0.6554	0.6591	0.6628	0.6664	0.6700	0.6736	0.6772	0.6808	0.6844	0.6879
0.5	0.6915	0.6950	0.6985	0.7019	0.7054	0.7088	0.7123	0.7157	0.7190	0.7224
0.6	0.7257	0.7291	0.7324	0.7357	0.7389	0.7422	0.7454	0.7486	0.7517	0.7549
0.7	0.7580	0.7611	0.7642	0.7673	0.7704	0.7734	0.7764	0.7794	0.7823	0.7852
0.8	0.7881	0.7910	0.7939	0.7967	0.7995	0.8023	0.8051	0.8078	0.8106	0.8133
0.9	0.8159	0.8186	0.8212	0.8238	0.8264	0.8289	0.8315	0.8340	0.8365	0.8389
1.0	0.8413	0.8438	0.8461	0.8485	0.8508	0.8531	0.8554	0.8577	0.8599	0.8621
1.1	0.8463	0.8665	0.8686	0.8608	0.8729	0.8749	0.8770	0.8790	0.8810	0.8830
1.2	0.8849	0.8869	0.8888	0.8907	0.8925	0.8944	0.8962	0.8980	0.8997	0.9015
1.3	0.9032	0.9049	0.9066	0.9082	0.9099	0.9115	0.9131	0.9147	0.9162	0.9177
1.4	0.9192	0.9207	0.9222	0.9236	0.9251	0.9265	0.9279	0.9292	0.9306	0.9319
1.5	0.9332	0.9345	0.9357	0.9370	0.9382	0.9394	0.9406	0.9418	0.9429	0.9441
1.6	0.9452	0.9463	0.9474	0.9484	0.9495	0.9505	0.9515	0.9525	0.9535	0.9545
1.7	0.9554	0.9564	0.9573	0.9582	0.9591	0.9599	0.9608	0.9616	0.9625	0.9633
1.8	0.9641	0.9649	0.9656	0.9664	0.9671	0.9678	0.9686	0.9693	0.9699	0.9706
1.9	0.9713	0.9719	0.9726	0.9732	0.9738	0.9744	0.9750	0.9756	0.9761	0.9767
2.0	0.9772	0.9778	0.9783	0.9788	0.9793	0.9798	0.9803	0.9808	0.9812	0.9817
2.1	0.9821	0.9826	0.9830	0.0934	0.9838	0.9842	0.9846	0.9850	0.9854	0.9857
2.2	0.9861	0.9864	0.9868	0.9871	0.9875	0.9878	0.9881	0.9884	0.9887	0.9890
2.3	0.9893	0.9896	0.9898	0.9901	0.9904	0.9906	0.9909	0.9911	0.9913	0.9916
2.4	0.9918	0.9920	0.9922	0.9925	0.9927	0.9929	0.9931	0.9932	0.9934	0.9936
2.5	0.9938	0.9940	0.9941	0.9943	0.9945	0.9946	0.9948	0.9949	0.9951	0.9952
2.6	0.9953	0.9955	0.9956	0.9957	0.9959	0.9960	0.9961	0.9962	0.9963	0.9964
2.7	0.9965	0.9966	0.9967	0.9968	0.9969	0.9970	0.9971	0.9972	0.9973	0.9974
2.8	0.9974	0.9975	0.9976	0.9977	0.9977	0.9978	0.9979	0.9979	0.9980	0.9981
2.9	0.9981	0.9982	0.9982	0.9983	0.9984	0.9984	0.9985	0.9985	0.9986	0.9986
3.0	0.9987	0.9987	0.9987	0.9988	0.9988	0.9989	0.9989	0.9989	0.9990	0.9990
3.1	0.9990	0.9991	0.9991	0.9991	0.9992	0.9992	0.9992	0.9992	0.9993	0.9993
3.2	0.9993	0.9993	0.9994	0.9994	0.9994	0.9994	0.9994	0.9995	0.9995	0.9995
3.3	0.9995	0.9995	0.9996	0.9996	0.9996	0.9996	0.9996	0.9996	0.9996	0.9997
3.4	0.9997	0.9997	0.9997	0.9997	0.9997	0.9997	0.9997	0.9997	0.9998	0.9998
3.5	0.9998	0.9998	0.9998	0.9998	0.9998	0.9998	0.9998	0.9998	0.9998	0.9998
3.6	0.9998	0.9998	0.9999	0.9999	0.9999	0.9999	0.9999	0.9999	0.9999	0.9999

Appendix F THE CHI-SQUARE DISTRIBUTION FUNCTION

Values of x for given $1 - F(x)$ with ν degrees of freedom

Degrees of freedom ν	Complemented distribution, $1 - F(x)$			
	0.10	0.05	0.02	0.01
1	2.706	3.841	5.412	6.635
2	4.605	5.991	7.824	9.210
3	6.251	7.815	9.837	11.341
4	7.779	9.488	11.688	13.277
5	9.236	11.070	13.388	15.086
6	10.645	12.592	15.033	16.812
7	12.017	14.067	16.622	18.475
8	13.362	15.507	18.168	20.090
9	14.684	16.919	19.679	21.666
10	15.987	18.307	21.161	23.209
11	17.275	19.675	22.618	24.725
12	18.549	21.026	24.054	26.217
13	19.812	22.362	25.472	27.688
14	21.064	23.685	26.873	29.141
15	22.307	24.996	28.259	30.578
16	23.542	26.296	29.633	32.000
17	24.769	27.587	30.995	33.409
18	25.989	28.869	32.346	34.805
19	27.204	30.144	33.687	36.191
20	28.412	31.410	35.020	37.566
21	29.615	32.671	36.343	38.932
22	30.813	22.924	37.659	40.289
23	32.007	35.172	38.968	45.638
24	33.196	36.415	40.270	42.980
25	34.382	37.652	41.566	44.314
26	35.563	38.885	42.856	45.642
27	36.741	40.113	44.140	46.963
28	37.916	41.337	45.419	48.278
29	39.087	42.557	46.693	49.588
30	40.256	43.773	47.962	50.892

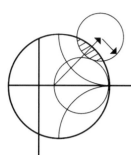

Index

Systems Modeling and Simulation